PRINCIPLES OF
SEQUENCE STRATIGRAPHY

PRINCIPLES OF
SEQUENCE STRATIGRAPHY

OCTAVIAN CATUNEANU
DEPARTMENT OF EARTH AND ATMOSPHERIC SCIENCES
UNIVERSITY OF ALBERTA
EDMONTON, ALBERTA, CANADA

ELSEVIER

Amsterdam – Boston – Heidelberg – London – New York – Oxford
Paris – San Diego – San Francisco – Singapore – Sydney – Tokyo

Elsevier
The Boulevard, Langford Lane, Kidlington, Oxford, OX5 1GB
Linacre House, Jordan Hill, Oxford, OX2 8DP, UK

First edition 2006
Reprinted 2006, 2007 (twice), 2008, 2011 (twice), 2012, 2013 (twice)

British Library Cataloguing in Publication Data
A catalogue record for this book is available from the British Library

Library of Congress Cataloging-in-Publication Data
A catalog record for this book is available from the Library of Congress

ISBN: 978-0-444-51568-1

For information on all Elsevier publications
visit our website at www.elsevierdirect.com

Working together to grow
libraries in developing countries

www.elsevier.com | www.bookaid.org | www.sabre.org

ELSEVIER   BOOK AID International   Sabre Foundation

Preface

Sequence stratigraphy analyzes the sedimentary response to changes in base level, and the depositional trends that emerge from the interplay of accommodation (space available for sediments to fill) and sedimentation. Sequence stratigraphy has tremendous potential to decipher the Earth's geological record of local to global changes, and to improve the predictive aspect of economic exploration and production. For these reasons, sequence stratigraphy is currently one of the most active areas of research in both academic and industrial environments.

'Principles' of sequence stratigraphy are to a large extent independent of the type of depositional environments established within a sedimentary basin (e.g., siliciclastic *vs.* carbonate), and clastic systems are generally used by default to explain and exemplify the concepts. However, the difference in stratigraphic responses to changes in base level between clastic and carbonate systems is discussed in the book, and the departure of the carbonate sequence stratigraphic model from the 'standard' model developed for clastic rocks is examined. The principles of sequence stratigraphy are also independent of scale. The resolution of the sequence stratigraphic work can be adjusted as a function of the scope of observation, from sub-depositional system scales to the scale of entire sedimentary basin fills. Between these end members, processes that operate over different spatial and temporal scales are interrelated. The sequence stratigraphic framework of facies relationships provides a template that allows one to see how smaller-scale processes and depositional elements fit into the bigger picture. As such, sequence stratigraphy is an approach to understanding the 4D development of sedimentary systems, integrating cross-sectional information (stratigraphy) with plan-view data (geomorphology) and insights into the evolution of sedimentation regimes through time (process sedimentology). Any of these 'conventional' disciplines may show a more pronounced affinity to sequence stratigraphy, depending on case study, scale, and scope of observation. The application of the sequence stratigraphic method also relies on the integration of multiple data sets that may be derived from outcrops, core, well logs, and seismic volumes.

Even though widely popular among all groups interested in the analysis of sedimentary systems, sequence stratigraphy is yet a difficult undertaking due to the proliferation of informal jargon and the persistence of conflicting approaches as to how the sequence stratigraphic method should be applied to the rock record. This book examines the relationship between such conflicting approaches from the perspective of a unifying platform, demonstrating that sufficient common ground exists to eliminate terminology barriers and to facilitate communication between all practitioners of sequence stratigraphy. The book is addressed to anyone interested in the analysis of sedimentary systems, from students to geologists, geophysicists, and reservoir engineers.

The available sequence stratigraphic literature has focussed mainly on (1) promoting particular models; (2) criticizing particular models or assumptions; and (3) providing comprehensive syntheses of previous work and ideas. This book builds on the existing literature and, avoiding duplication with other volumes on the same topic, shifts the focus towards making sequence stratigraphy a more user-friendly and flexible method of analysis of the sedimentary rock record. This book is not meant to be critical of some models in favor of others. Instead, it is intended to explain how models relate to each other and how their applicability may vary with the case study. There is, no question, value in all existing models, and one has to bear in mind that their proponents draw their experience from sedimentary basins placed in different tectonic settings. This explains in part the variety of opinions and conflicting ideas. The refinement of the sequence stratigraphic model to account for the variability of tectonic and sedimentary regimes across the entire spectrum of basin types is probably the next major step in the evolution of sequence stratigraphy.

Research support during the completion of this work was provided by the Natural Sciences and Engineering Research Council of Canada (NSERC), and by the University of Alberta. Generous financial support from NSERC, Marathon Oil Company and Real Resources Inc. allowed for the publication of this book in full colour. I wish to thank Tirza van Daalen, the Publishing Editor on behalf of Elsevier, for her constant support ever since we decided to produce this book back in 2003. I am most grateful to Pat Eriksson and Tom van Loon, who critically read the entire manuscript, for undertaking this enormously time-consuming and painstaking task and for their thoughtful and constructive comments. Pat's support over the past decade has been an outstanding measure of friendship and professionalism – many thanks! Tom, who is the Editor of Elsevier's 'Developments in Sedimentology' series, has also offered exceptional editorial guidance I am also in debt to Henry Posamentier, Art Sweet, and Alex MacNeil for reading and giving me feedback on selected chapters of the manuscript.

Fruitful discussions over the years with Andrew Miall, Ashton Embry, Henry Posamentier, Bill Galloway, Dale Leckie, Mike Blum, Guy Plint, Janok Bhattacharya, Keith Shanley, Pat Eriksson, Darrel Long, Nicholas Christie-Blick, Bruce Ainsworth, Martin Gibling, Simon Lang, and many others, allowed me to see the many facets and complexities of sequence stratigraphy, as seen from the perspective of the different 'schools.' Henry Posamentier contributed significantly to the quality of this book, by providing an outstanding collection of numerous seismic images. Additional images or field photographs have been made available by Martin Gibling, Guy Plint, Art Sweet, Murray Gingras, Bruce Hart, Andrew Miall, and the geoscientists of the Activo de Exploracion Litoral of PEMEX. While thanking all these colleagues for their help and generosity, I remain responsible for the views expressed in this book, and for any remaining errors or omissions.

I dedicate this book to Ana, Andrei, Gabriela, and my supportive parents.

Octavian Catuneanu
University of Alberta
Edmonton, 2005

Contents

7. Time Attributes of Stratigraphic Surfaces

8. Hierarchy of Sequences and Sequence Boundaries

9. Discussion and Conclusions

CHAPTER

1

Introduction

SEQUENCE STRATIGRAPHY— AN OVERVIEW

Sequence Stratigraphy in the Context of Interdisciplinary Research

Sequence stratigraphy is the most recent revolutionary paradigm in the field of sedimentary geology. The concepts embodied by this discipline have resulted in a fundamental change in geological thinking and in particular, the methods of facies and stratigraphic analyses. Over the past fifteen years, this approach has been embraced by geoscientists as the preferred style of stratigraphic analysis, which has served to tie together observations from many disciplines. In fact, a key aspect of the sequence stratigraphic approach is to encourage the integration of data sets and research methods. Blending insights from a range of disciplines invariably leads to more robust interpretations and, consequently, scientific progress. Thus, the sequence stratigraphic approach has led to improved understanding of how stratigraphic units, facies tracts, and

depositional elements relate to each other in time and space within sedimentary basins (Fig. 1.1). The applications of sequence stratigraphy range widely, from predictive exploration for petroleum, coal, and placer deposits, to improved understanding of Earth's geological record of local to global changes.

The conventional disciplines of process sedimentology and classical stratigraphy are particularly relevant to sequence stratigraphy (Fig. 1.2). Sequence stratigraphy is commonly regarded as only one other type of stratigraphy, which focuses on changes in depositional trends and their correlation across a basin (Fig. 1.3). While this is in part true, one should not neglect the strong sedimentological component that emphasizes on the facies-forming processes within the confines of individual depositional systems, particularly in response to changes in base level. At this scale, sequence stratigraphy is generally used to resolve and explain issues of facies cyclicity, facies associations and relationships, and reservoir compartmentalization, without necessarily applying this information for larger-scale correlations.

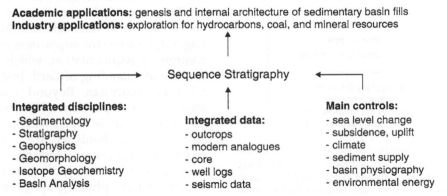

Academic applications: genesis and internal architecture of sedimentary basin fills
Industry applications: exploration for hydrocarbons, coal, and mineral resources

Sequence Stratigraphy

Integrated disciplines:
- Sedimentology
- Stratigraphy
- Geophysics
- Geomorphology
- Isotope Geochemistry
- Basin Analysis

Integrated data:
- outcrops
- modern analogues
- core
- well logs
- seismic data

Main controls:
- sea level change
- subsidence, uplift
- climate
- sediment supply
- basin physiography
- environmental energy

FIGURE 1.1 Sequence stratigraphy in the context of interdisciplinary research—main controls, integrated data sets and subject areas, and applications.

1

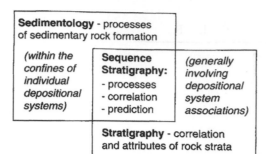

FIGURE 1.2 Sequence stratigraphy and its overlap with the conventional disciplines of sedimentology and stratigraphy (definitions modified from Bates and Jackson, 1987). When applied to a specific depositional system, sequence stratigraphy helps to understand processes of facies formation, facies relationships, and facies cyclicity in response to base-level changes. At larger scales, the lateral correlation of coeval depositional systems becomes a more significant issue, which also brings in a component of facies predictability based on the principle of common causality related to the basin-wide nature of the allogenic controls on sedimentation.

Owing to the 'genetic' nature of the sequence stratigraphic approach, process sedimentology is an important prerequisite that cannot be separated from, and forms an integral part of sequence stratigraphy. The importance of process sedimentology in sequence stratigraphic analysis becomes evident when attempting to identify sequence stratigraphic surfaces in the rock record. As discussed in detail throughout the book, most criteria involved in the interpretation of stratigraphic surfaces revolve around the genetic nature of facies that are in contact across the surface under analysis, which in turn requires a good understanding of depositional processes and environments. The importance of process sedimentology is also evident when it comes to understanding the origin and distribution of the various types of unconformities that may form in nonmarine, coastal, or fully marine

environments, as well as the facies characteristics and variability that may be encountered within the different portions of systems tracts. The stratigraphic component of sequence stratigraphy consists of its applicability to correlations in a time framework, usually beyond the scale of individual depositional systems, in spite of the lateral changes of facies that are common in any sedimentary basin. In addition to its sedimentological and stratigraphic affinities, sequence stratigraphy also brings a new component of facies predictability which is particularly appealing to industry-oriented research (Fig. 1.2).

The conventional types of stratigraphy, such as biostratigraphy, lithostratigraphy, chemostratigraphy, or magnetostratigraphy, involve both data collection and interpretation based on the data, just as does sequence stratigraphy, but no sophisticated interpretation is required in order to do conventional stratigraphic correlations. In contrast, sequence stratigraphic correlations depend on interpretation to develop the correlation model. Therefore, sequence stratigraphy has an important built-in interpretation component which addresses issues such as the reconstruction of the allogenic controls at the time of sedimentation, and predictions of facies architecture in yet unexplored areas. The former issue sparked an intense debate, still ongoing, between the supporters of eustatic *vs.* tectonic controls on sedimentation, which is highly important to the understanding of Earth history and fundamental Earth processes. Beyond sea-level change and tectonism, the spectrum of controls on stratigraphic patterns is actually much wider, including additional subsidence mechanisms (e.g., thermal subsidence, sediment compaction, isostatic, and flexural crustal loading), orbital forcing of climate changes, sediment supply, basin physiography, and environmental energy (Fig. 1.1). The second issue, on the economic

Stratigraphy	Property
Lithostratigraphy	lithology
Biostratigraphy	fossils
Magnetostratigraphy	magnetic polarity
Chemostratigraphy	chemical properties
Chronostratigraphy	absolute ages
Allostratigraphy	discontinuities
Seismic stratigraphy	seismic data
Sequence stratigraphy	depositional trends

Depositional trends refer to aggradation versus erosion, and progradation versus retrogradation. Changes in depositional trends are controlled by the interplay of sedimentation and base-level shifts.

FIGURE 1.3 Types of stratigraphy, defined on the basis of the property they analyze. The interplay of sedimentation and shifting base level at the shoreline generates changes in depositional trends in the rock record, and it is the analysis and/or correlation of these changes that defines the primary objectives of sequence stratigraphy.

aspect of facies predictability, provides the industry community with a powerful new analytical and correlation tool of exploration for natural resources.

In spite of its inherent genetic aspect, one should not regard sequence stratigraphy as the triumph of interpretation over data, or as a method developed in isolation from other geological disciplines. In fact sequence stratigraphy builds on many existing data sources, it requires a good knowledge of sedimentology and facies analysis, and it integrates the broad field of sedimentary geology with geophysics, geomorphology, absolute and relative age-dating techniques, and basin analysis. As with any modeling efforts, the reliability of the sequence stratigraphic model depends on the quality and variety of input data, and so integration of as many data sets as possible is recommended. The most common data sources for a sequence stratigraphic analysis include outcrops, modern analogues, core, well logs, and seismic data (Fig. 1.1).

In addition to the facies analysis of the strata themselves, which is the main focus of conventional sedimentology, sequence stratigraphy also places a strong emphasis on the contacts that separate packages of strata characterized by specific depositional trends. Such contacts represent event-significant bounding surfaces that mark changes in sedimentation regimes, and are important both for regional correlation, as well as for understanding the facies relationships within the confines of specific depositional systems. The study of stratigraphic contacts may not, however, be isolated from the facies analysis of the strata they separate, as the latter often provide the diagnostic criteria for the recognition of bounding surfaces.

Sequence Stratigraphy—A Revolution in Sedimentary Geology

Sequence stratigraphy is the third of a series of major revolutions in sedimentary geology (Miall, 1995). Each revolution resulted in quantum paradigm shift that changed the way geoscientists interpreted sedimentary strata. The first breakthrough was marked by the development of the flow regime concept and the associated process/response facies models in the late 1950s and early 1960s (Harms and Fahnestock, 1965; Simons et al., 1965). This first revolution provided a unified theory to explain, from a hydrodynamic perspective, the genesis of sedimentary structures and their predictable associations within the context of depositional systems. Beginning in the 1960s, the incorporation of plate tectonics and geodynamic concepts into the analysis of sedimentary processes at regional scales, marked the second revolution in sedimentary geology.

Ultimately, these first two conceptual breakthroughs or revolutions led to the development of Basin Analysis in the late 1970s, which provided the scientific framework for the study of the origins and depositional histories of sedimentary basins. Sequence stratigraphy marks the third and most recent revolution in sedimentary geology, starting in the late 1970s with the publication of AAPG Memoir 26 (Payton, 1977), although its roots can be traced much further back in time as explained below. Sequence stratigraphy developed as an interdisciplinary method that blended both autogenic (i.e., from within the system) and allogenic (i.e., from outside the system) processes into a unified model to explain the evolution and stratigraphic architecture of sedimentary basins (Miall, 1995).

The success and popularity of sequence stratigraphy stems from its widespread applicability in both mature and frontier hydrocarbon exploration basins, where data-driven and model-driven predictions of lateral and vertical facies changes can be formulated, respectively. These predictive models have proven to be particularly effective in reducing lithology-prediction risk for hydrocarbon exploration, although there is an increasing demand to employ the sequence stratigraphic method for coal and mineral resources exploration as well.

HISTORICAL DEVELOPMENT OF SEQUENCE STRATIGRAPHY

Early Developments

Sequence stratigraphy is generally regarded as stemming from the seismic stratigraphy of the 1970s. In fact, major studies investigating the relationship between sedimentation, unconformities, and changes in base level, which are directly relevant to sequence stratigraphy, were published prior to the birth of seismic stratigraphy (e.g., Grabau, 1913; Barrell, 1917; Sloss et al., 1949; Wheeler and Murray, 1957; Wheeler, 1958, 1959, 1964; Sloss, 1962, 1963; Curray, 1964; Frazier, 1974). As early as the eighteenth century, Hutton recognized the periodic repetition through time of processes of erosion, sediment transport, and deposition, setting up the foundation for what is known today as the concept of the 'geological cycle.' Hutton's observations may be considered as the first account of stratigraphic cyclicity, where unconformities provide the basic subdivision of the rock record into repetitive successions. The link between unconformities and base-level changes was explicitly emphasized by Barrell (1917), who stated that 'sedimentation controlled by base level will result in divisions of the stratigraphic series separated by breaks.'

The term 'sequence' was introduced by Sloss *et al.* (1949) to designate a stratigraphic unit bounded by subaerial unconformities. Sloss emphasized the importance of such sequence-bounding unconformities, and subsequently subdivided the entire Phanerozoic succession of the interior craton of North America into six major sequences (Sloss, 1963). Sloss also emphasized the importance of tectonism in the generation of sequences and bounding unconformities, an idea which is widely accepted today but was largely overlooked in the early days of seismic stratigraphy. It is noteworthy that the original 'sequence' of Sloss referred to 'unconformity-bounded masses of strata of greater than group or supergroup rank' (Krumbein and Sloss, 1951), which restricted the applicability of the 'sequence' concept only to regional-scale stratigraphic studies. The meaning of a stratigraphic 'sequence' has been subsequently expanded to include any 'relatively conformable succession of genetically related strata' (Mitchum, 1977), irrespective of temporal and spatial scales. In parallel with the development of the 'sequence' concept in a stratigraphic context, sedimentologists in the 1960s and 1970s have redefined the meaning of the term 'sequence' to include a vertical succession of facies that are 'organized in a coherent and predictable way' (Pettijohn, 1975), reflecting the natural evolution of a depositional environment. This idea was further perpetuated in landmark publications by Reading (1978) and Selley (1978a). Examples of facies sequences, in a sedimentological sense, would include coarsening-upward successions of deltaic facies (which many stratigraphers today would call 'parasequences'), or the repetition of channel fill, lateral accretion and overbank architectural elements that is typical of meandering river systems (which may be part of particular systems tracts in a stratigraphic sense). The development of seismic and sequence stratigraphy in the late 1970s and 1980s revitalized the use of the term 'sequence' in a stratigraphic context, which remained the dominant approach to date. It is therefore important to distinguish between the 'sequence' of sequence stratigraphy and the 'facies sequence' of sedimentology (see van Loon, 2000, for a full discussion).

The unconformity-bounded sequences promoted by Sloss (1963) and Wheeler (1964) in the pre-sequence stratigraphy era provided the geological community with informal mappable units that could be used for stratigraphic correlation and the subdivision of the rock record into genetically-related packages of strata. The concept of 'unconformity-bounded unit' (i.e., Sloss' 'sequence') was formalized by the European 'International Stratigraphic Guide' in 1994. The limitation of this method of stratigraphic analysis was imposed by the lateral extent of sequence-bounding

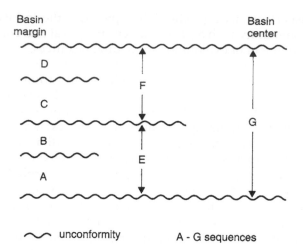

FIGURE 1.4 The concept of unconformity-bounded sequence of Sloss *et al.* (1949). As many unconformities are potentially restricted to the basin margins, the number of sequences mapped in the basin centre is often lower than the number of sequences present in an age-equivalent succession along the rim of the basin.

unconformities, which are potentially restricted to the basin margins. Hence, the number of sequences mapped within a sedimentary basin may significantly decrease along dip, from the basin margins towards the basin centre (Fig. 1.4). This limitation required a refinement of the early ideas by finding a way to extend sequence boundaries across an entire sedimentary basin. The introduction of 'correlative conformities,' which are extensions towards the basin center of basin-margin unconformities, marked the birth of modern seismic and sequence stratigraphy (Fig. 1.5) (Mitchum, 1977). The advantage of the modern sequence, bounded by a composite surface that may include a conformable portion, lies in its basin-wide extent — hence, the number of sequences mapped at the basin margin equals the number of sequences that are found in the basin center. Due largely to disagreements regarding the timing of the correlative conformity relative to a reference curve of base-level changes, this new sequence bounded by unconformities *or their correlative conformities* remains and informal designation insofar as has not yet been ratified by either the European or the North American commissions on stratigraphic nomenclature. Nonetheless, this usage has seen widespread adoption in the scientific literature of the past two decades.

Sequence Stratigraphy Era—Eustatic *vs.* Tectonic Controls on Sedimentation

Seismic stratigraphy emerged in the 1970s with the work of Vail (1975) and Vail *et al.* (1977). This new

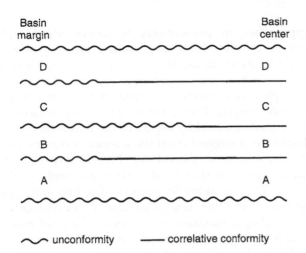

FIGURE 1.5 The concept of sequence as defined in seismic and sequence stratigraphy. The correlative conformities allow tracing sequences across an entire sedimentary basin. A–G—sequences.

method for analyzing seismic-reflection data stimulated a revolution in stratigraphy, with an impact on the geological community as important as the introduction of the flow regime concept in the late 1950s—early 1960s and the plate tectonics theory in the 1960s (Miall, 1995). The concepts of seismic stratigraphy were published together with a global sea-level cycle chart (Vail *et al.*, 1977), based on the underlying assumption that eustasy is the main driving force behind sequence formation at all levels of stratigraphic cyclicity. Seismic stratigraphy and the global cycle chart were thus introduced to the geological community as a seemingly inseparable package of new stratigraphic methodology. These ideas were then passed on to sequence stratigraphy in its early years, as seismic stratigraphy evolved into sequence stratigraphy with the incorporation of outcrop and well data (Posamentier *et al.*, 1988; Posamentier and Vail, 1988; Van Wagoner *et al.*, 1990). Subsequent publications (e.g., Hunt and Tucker, 1992; Posamentier and James, 1993; Posamentier and Allen, 1999) shift the focus away from eustasy and towards a blend of eustasy and tectonics, termed 'relative sea level.' Nonetheless, the global-eustasy model as initially proposed (Vail *et al.*, 1977) posed two challenges to the practitioners of 'conventional' stratigraphy: that sequence stratigraphy, as linked to the global cycle chart, constitutes a superior standard of geological time to that assembled from conventional chronostratigraphic evidence, and that stratigraphic processes are dominated by the effects of eustasy, to the exclusion of other allogenic mechanisms, including tectonism (Miall and Miall, 2001). Although the global cycle chart is now under intense scrutiny and criticism (e.g., Miall, 1992), the global-eustasy model is

still used for sequence stratigraphic analysis in some recent publications (e.g., de Graciansky *et al.*, 1998).

In parallel to the eustasy-driven sequence stratigraphy, which held by far the largest share of the market, other researchers went to the opposite end of the spectrum by suggesting a methodology that favored tectonism as the main driver of stratigraphic cyclicity. This version of sequence stratigraphy was introduced as 'tectonostratigraphy' (e.g., Winter, 1984). The major weakness of both schools of thought is that *a priori* interpretation of the main allogenic control on accommodation was automatically attached to any sequence delineation, which gave the impression that sequence stratigraphy is more of an interpretation artifact than an empirical, data-based method. This *a priori* interpretation facet of sequence stratigraphy attracted considerable criticism and placed an unwanted shade on a method that otherwise represents a truly important advance in the science of sedimentary geology. Fixing the damaged image of sequence stratigraphy only requires the basic understanding that base-level changes can be controlled by any combination of eustatic and tectonic forces, and that the dominance of any of these allogenic mechanisms should be assessed on a case by case basis. It became clear that sequence stratigraphy needed to be dissociated from the global-eustasy model, and that a more objective analysis should be based on empirical evidence that can actually be observed in outcrop or the subsurface. This realization came from the Exxon research group, where the global cycle chart originated in the first place: 'Each stratal unit is defined and identified only by physical relationships of the strata, including lateral continuity and geometry of the surfaces bounding the units, vertical stacking patterns, and lateral geometry of the strata within the units. Thickness, time for formation, and interpretation of regional or global origin are not used to define stratal units..., [which]... can be identified in well logs, cores, or outcrops and used to construct a stratigraphic framework regardless of their interpreted relationship to changes in eustasy' (Van Wagoner *et al.*, 1990).

The switch in emphasis from sea-level changes to relative sea-level changes in the early 1990s (e.g., Hunt and Tucker, 1992; Christie-Blick and Driscoll, 1995) marked a major and positive turnaround in sequence stratigraphy. By doing so, no interpretation of specific eustatic or tectonic fluctuations was forced upon sequences, systems tracts, or stratigraphic surfaces. Instead, the key surfaces, and implicitly the stratal units between them, are inferred to have formed in relation to a more 'neutral' curve of relative sea-level (base-level) changes that can accommodate any balance between the allogenic controls on accommodation.

Sequence Models

The concept of *sequence* is as good, or accepted, as the boundaries that define it. As a matter of principle, it is useless to formalize a unit when the definition of its boundaries is left to the discretion of the individual practitioner. The *sequence* defined by Sloss *et al.* (1949) as an unconformity-bounded unit, was widely embraced (and formalized in the 1994 International Stratigraphic Guide) because the concept of unconformity was also straightforward and surrounded by little debate. The modification of the original concept of *sequence* by the introduction of correlative conformities as part of its bounding surfaces triggered both progress and debates at the onset of the seismic and sequence stratigraphy era. The main source of contention relates to the nature, timing, and mappability of these correlative conformities, and as a result a number of different approaches to sequence definition and hence sequence models are currently in use, each promoting a unique set of terms and bounding surfaces. This creates a proliferation of jargon and concomitant confusion, and represents a barrier to communication of ideas and results. In time, many of these barriers will fade as the discipline matures and the jargon is streamlined. Likewise, the varying approaches to sequence delineation, also a cause for confusion, will become less contentious, and perhaps less important, as geoscientists focus more on understanding the origin of strata and less on issues of nomenclature or style of conceptual packaging. Some of the reasons for the variety of approaches in present-day sequence stratigraphy include: the underlying assumptions regarding primary controls on stratigraphic cyclicity; the type of basin from which models were derived; and the gradual conceptual advances that allowed for alternative models to be developed. The fact that controversy persists can be viewed as a healthy aspect in the maturation of the discipline; it suggests that the science is continuing to evolve, just as it should do. Present-day sequence stratigraphy can thus be described as a still-developing field that is taking the science of sedimentary geology in an exciting new direction of conceptual and practical opportunities, even though the road may be punctuated by disagreements and controversy.

The early work on seismic and sequence stratigraphy published in AAPG Memoir 26 (Payton, 1977) and SEPM Special Publication 42 (Wilgus *et al.*, 1988) resulted in the definition of the *depositional sequence*, as the primary unit of a sequence stratigraphic model. This stratigraphic unit is bounded by subaerial unconformities on the basin margin and their correlative conformities towards the basin center. The depositional sequence was subdivided into lowstand, transgressive, and highstand systems tracts on the basis of internal surfaces that correspond to changes in the direction of shoreline shift from regression to transgression and *vice versa* (Posamentier and Vail, 1988). Variations on the original depositional sequence theme resulted in the publication of several slightly modified versions of the depositional sequence model (Figs. 1.6 and 1.7).

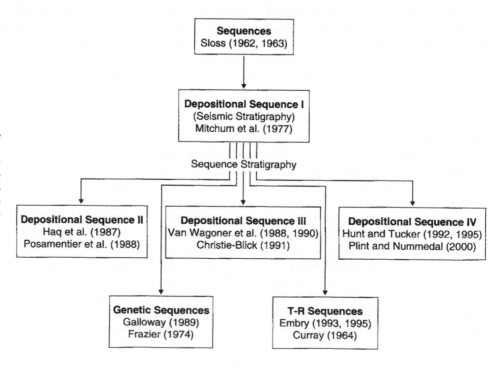

FIGURE 1.6 Family tree of sequence stratigraphy (modified from Donovan, 2001). The various sequence stratigraphic models mainly differ in the style of conceptual packaging of strata into sequences, i.e., with respect to where the sequence boundaries are picked in the rock record.

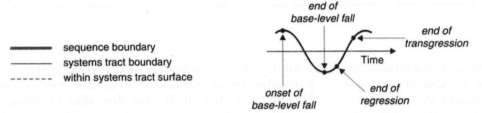

Sequence model / Events	Depositional Sequence II	Depositional Sequence III	Depositional Sequence IV	Genetic Sequence	T-R Sequence
end of transgression	HST	early HST	HST	HST	RST
end of regression	TST	TST	TST	TST	TST
end of base-level fall	late LST (wedge)	LST	LST	late LST (wedge)	
onset of base-level fall	early LST (fan)	late HST (fan)	FSST	early LST (fan)	RST
	HST	early HST (wedge)	HST	HST	

─── sequence boundary
─── systems tract boundary
------ within systems tract surface

FIGURE 1.7 Timing of system tracts and sequence boundaries for the sequence models currently in use (modified from Catuneanu, 2002). The conformable portion of the sequence boundary of the depositional sequence II was originally considered to form *during early* sea-level fall (Posamentier *et al.*, 1988), which was later revised to the *onset* of sea-level fall (Posamentier *et al.*, 1992b), as represented in this table. In addition to these classic models, other hybrid models are also in use, as for example the approach that recognizes the four systems tracts of the depositional sequence IV, but with a sequence boundary that conforms to the depositional sequence II (Coe, 2003). Abbreviations: LST—lowstand systems tract; TST—transgressive systems tract; HST—highstand systems tract; FSST—falling-stage systems tract; RST—regressive systems tract; T–R—transgressive-regressive.

Soon after the SEPM Special Publication 42, Galloway (1989), based on Frazier (1974), proposed that maximum flooding surfaces, rather than subaerial unconformities, be used as sequence boundaries. This unit was termed a *genetic stratigraphic sequence*, also referred to as a regressive–transgressive (R–T) sequence. Embry and Johannessen (1992) proposed a third type of stratigraphic unit, named a *transgressive–regressive (T–R) sequence*, corresponding to a full cycle of transgressive and regressive shoreline shifts (Figs. 1.6 and 1.7).

The various sequence models that are currently in use differ from each other mainly in the style of conceptual packaging of the stratigraphic record, using different timing for systems tract and sequence boundaries in relation to a reference cycle of base-level shifts (Figs. 1.6 and 1.7). Each sequence model may work best under particular circumstances, and no one model is universally preferable, or applicable to the entire range of case studies (Catuneanu, 2002). The dominant approaches, as reflected by the sequence stratigraphic literature, are those popularized by the Exxon school (Posamentier and Vail, 1988; Van Wagoner *et al.*, 1990; Posamentier and Allen, 1999) and to a somewhat lesser extent by Galloway (1989) and Embry and Johannessen (1992).

Nonetheless, the applicability and practical limitations of each approach are discussed in detail in this book.

SEQUENCE STRATIGRAPHIC APPROACH

Terminology

Figures 1.8 and 1.9 provide the most popular definitions for sequence stratigraphy and the main stratal units used in a sequence stratigraphic analysis. In contrast with all other types of stratigraphy (including allostratigraphy), and in spite of having been widely accepted in the geologic literature, sequence stratigraphy has not yet been formally incorporated into the North American Code of Stratigraphic Nomenclature, nor into the International Stratigraphic Guide. The reason for this is the lack of consensus on some basic principles, including the definition of a *sequence* (i.e., which surfaces should constitute the sequence boundaries), and also the proliferation of a complex jargon that is difficult to standardize.

FIGURE 1.8 Definitions of sequence stratigraphy. In the simplest sense, sequence stratigraphy deals with the sedimentary response to base-level changes, which can be analyzed from the scale of individual depositional systems to the scale of entire basins.

Sequence stratigraphy (Posamentier et al., 1988; Van Wagoner, 1995): the study of rock relationships within a time-stratigraphic framework of repetitive, genetically related strata bounded by surfaces of erosion or nondeposition, or their correlative conformities.

Sequence stratigraphy (Galloway, 1989): the analysis of repetitive genetically related depositional units bounded in part by surfaces of nondeposition or erosion.

Sequence stratigraphy (Posamentier and Allen, 1999): the analysis of cyclic sedimentation patterns that are present in stratigraphic successions, as they develop in response to variations in sediment supply and space available for sediment to accumulate.

Sequence stratigraphy (Embry, 2001a): the recognition and correlation of stratigraphic surfaces which represent changes in depositional trends in sedimentary rocks. Such changes were generated by the interplay of sedimentation, erosion and oscillating base level and are now determined by sedimentological analysis and geometric relationships.

Note that <u>sedimentation</u> is separated from <u>base-level</u> changes. Also note important keywords:
- "cyclicity": a sequence is a cyclothem, i.e. it corresponds to a stratigraphic cycle;
- "time framework": age-equivalent depositional systems are correlated across a basin. This provides the foundation for the definition of systems tracts. In the early days of sequence stratigraphy, bounding surfaces were taken as time lines, in the view of the global-eustasy model. Today, independent time control is required for large-scale correlations;
- "genetically related strata": no major hiatuses are assumed within a sequence.

The fact that several different sequence models are currently in use does not make the task of finding a common ground easy, even for what a *sequence* should be. A key aspect of the problem lies in the fact that the position of the sequence boundary (in both space and time) varies from one model to another, to the extent that any of the sequence stratigraphic surfaces may become a sequence boundary or at least a part of it. Nevertheless, all versions of sequence boundaries regardless of which model is employed include both unconformable and conformable portions, which means that the original definition of *sequence* by Mitchum (1977) (Fig. 1.9), which incorporates the notion of a correlative conformity, still satisfies most of the current approaches.

Jargon is a potential distraction that can make sequence stratigraphy a difficult undertaking for those embarking on the application of this approach. All sequence models purport to describe the same rocks, though they often use different sets of terms. Beyond this terminology barrier and beyond the issue of which surfaces constitute the sequence boundaries, sequence stratigraphy is, in fact, a relatively easy method to use. A careful analysis of the different models reveals a lot of common ground between the various approaches with much of the terminology synonymous or nearly so. Again, the main differences between these approaches lie in the conceptual packaging of the same succession of strata. Once these differences are understood, the geoscientist has the

flexibility of using whatever model works best for the particular circumstances of a specific case study. Having said that, it is also desirable to proceed towards a unified sequence stratigraphic approach, which is the only way that can lead to the formal standardization of sequence stratigraphic concepts. The differences highlighted in Fig. 1.7 show that (1) a significant part of the 'disagreement' is in fact a matter of semantics, hence it can be easily overcome; and (2) the position of the sequence boundary, especially its conformable portion, varies with the model. Beyond these issues, all models are bridged by the fact that the subdivisions of each type of sequence are linked to the same reference curve of base-level changes, and hence they are conceptual equivalents. It is therefore conceivable that a basic set of principles may ultimately be accepted as the formal backbone of the discipline by all practitioners of stratigraphic analysis. Such acceptance would not preclude divergence of analytical styles as a function of case study and/or the data available for analysis.

This book attempts to demonstrate that, irrespective of the model of choice, and its associated timing of sequence boundaries, the 'heartbeat' of sequence stratigraphy is fundamentally represented by shoreline shifts, whose nature and timing control the formation of all systems tracts and bounding surfaces. Beyond nomenclatural preferences, each stage of shoreline shift (normal regression, forced regression, transgression) corresponds to the formation of a

Depositional systems (Galloway, 1989): three-dimensional assemblages of process-related facies that record major paleo-geomorphic elements.

Depositional systems (Fisher and McGowan, 1967, in Van Wagoner, 1995): three-dimensional assemblages of lithofacies, genetically linked by active (modern) processes or inferred (ancient) processes and environments.

Depositional systems represent the sedimentary product of associated depositional environments. They grade laterally into coeval systems, forming logical associations of paleo-geomorphic elements (cf., systems tracts).

Systems tract (Brown and Fisher, 1977): a linkage of contemporaneous depositional systems, forming the subdivision of a sequence.

A systems tract includes all strata accumulated across the basin during a particular stage of shoreline shifts.

Systems tracts are interpreted based on stratal stacking patterns, position within the sequence, and types of bounding surfaces. The timing of systems tracts is inferred relative to a curve that describes the base-level fluctuations at the shoreline.

Sequence (Mitchum, 1977): a relatively conformable succession of genetically related strata bounded by unconformities or their correlative conformities.

Sequences and systems tracts are bounded by key stratigraphic surfaces that signify specific events in the depositional history of the basin. Such surfaces may be conformable or unconformable, and mark changes in the sedimentation regime across the boundary.

Sequences correspond to full stratigraphic cycles of changing depositional trends. The conformable or unconformable character of the bounding surfaces is not an issue in the process of sequence delineation, nor the degree of preservation of the sequence.

The concepts of sequence, systems tracts, and stratigraphic surfaces are independent of scale, i.e. time for formation, thickness, or lateral extent. Same sequence stratigraphic terminology can be applied to different orders of cyclicity, via the concept of hierarchy. Well-log signatures are not part of the definition of sequence stratigraphic concepts, although general trends may be inferred from the predictable stacking patterns of systems tracts. The magnitude of the log deflections will vary with the magnitude/importance of the mapped surfaces and stratal units.

FIGURE 1.9 Main building blocks of the sedimentary record from a sequence stratigraphic prospective. With an increasing scale of observation, these units refer to depositional systems, systems tracts, and sequences.

systems tract with unique stratal stacking patterns. Surfaces that can serve, at least in part, as systems tract boundaries constitute surfaces of sequence stratigraphic significance. These fundamental principles are common to all models, and ultimately provide the basis for a unified sequence stratigraphic approach.

Concept of Scale

It is important to note that the application and definition of sequence stratigraphic concepts is independent of scale (Figs. 1.8 and 1.9). This means that the same terminology can and should be applied for sequences, systems tracts, and surfaces that have developed at different temporal and spatial scales. The general sequence stratigraphic approach thus applies to features as small as those produced in an experimental flume, formed in a matter of hours (e.g., Wood *et al.*, 1993;

Koss *et al.*, 1994; Paola, 2000; Paola *et al.*, 2001), as well as to those that are continent wide and formed over a period of millions of years. Nonetheless a distinction must be made between larger- and the smaller-scale sequences, systems tracts, and stratigraphic surfaces. This is addressed through a hierarchy based on the use of modifiers such as first-order, second-order, third-order, etc., commonly in a relative rather than an absolute sense. Although this terminology is often associated with specific time ranges (Vail *et al.*, 1977, 1991; Krapez, 1996), this has not always been common practice in the scientific literature (see discussions in Embry, 1995; Posamentier and Allen, 1999; Catuneanu *et al.*, 2004, 2005). One reason for this is that we often do not know the scale (especially duration, but also lateral extent or thickness changes across a basin) of the stratal units we deal with within a given study area, so the use of specific names for specific scales may become quite subjective. Another advantage of using a consistent terminology regardless of scale is that jargon is kept to a minimum, which makes sequence stratigraphy more user-friendly and easier to understand across a broad spectrum of readership. These issues are tackled in more detail in Chapter 8, which deals with the hierarchy of sequences and sequence boundaries.

Among the key concepts shown in Fig. 1.9, the term *depositional system* is a general (conventional) notion defined on the basis of depositional setting and environment. The terms *systems tract* and *sequence* are specific sequence stratigraphic terms, defined in relationship to the base-level and the transgressive–regressive curves. A systems tract includes a sum of laterally correlative depositional systems (hence, the use of plural: *systems*). A sequence includes two or more systems tracts, depending on the model of choice (Fig. 1.7). The actual scale for sequence stratigraphic work is highly variable, depending on the problem in hand, ranging from depositional system scale (also highly variable) to the entire fill of the basin, and beyond. When applied to the analysis of a depositional system, e.g., an ancient delta (Fig. 1.10), sequence stratigraphy is mainly used to resolve the nature of contacts and the details of facies relationships. Such studies are often performed to describe the degree of reservoir compartmentalization in the various stages of oil field exploration and production. When applied to the scale of depositional system associations, the issue of stratigraphic correlation

becomes a primary objective, and provides the framework for the larger scale distribution of facies.

The principles outlined above provide a general idea about the range of potential outcomes and objectives of sequence stratigraphy as a function of scope and scale of analysis. There is a common misconception that sequence stratigraphy is always related to regional, continental, or even global scales of observation (sub-basins, basins, and global cycles)—this does not need to be the case, as sequence stratigraphy can be applied virtually to any scale. A good example of this is the study of the 'East Coulee Delta' (Posamentier *et al.*, 1992a), where an entire range of sequence stratigraphic elements (including 'classic' systems tracts) have been documented at a centimeter to meter scale (Fig. 1.11). In recent years there have been numerous flume-based studies where sequences have been created under controlled laboratory conditions (e.g., Wood *et al.*, 1993; Koss *et al.*, 1994; Paola, 2000; Paola *et al.*, 2001). Such studies have provided valuable insight as to variations on the general sequence model.

Sequence Stratigraphy *vs.* Lithostratigraphy and Allostratigraphy

Almost any type of study of a sedimentary basin fill requires the construction of cross sections. The lines we draw on these two-dimensional representations are of

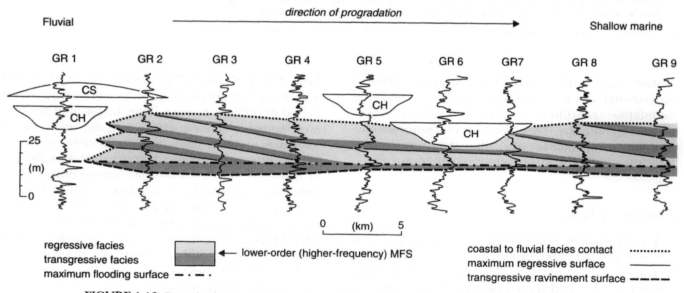

FIGURE 1.10 Example of sequence stratigraphy applied to understand the reservoir compartmentalization of a deltaic depositional system (case study illustrating the regression of the Late Cretaceous Bearpaw seaway, central Alberta). Abbreviations: GR—gamma ray logs; CH—fluvial channel fill; CS—crevasse splay; MFS—maximum flooding surface. Note that maximum flooding surfaces are associated with the finest-grained sediments, and their position reveals the overall progradation and geometry of the delta. The reservoir includes at least five separate hydrodynamic units, each corresponding to a stage of delta front progradation.

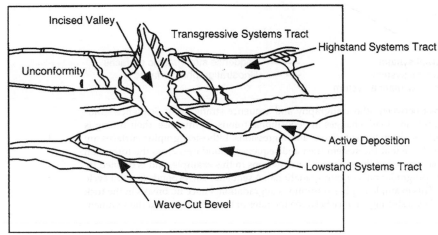

FIGURE 1.11 East Coulee Delta (approximately 1 m wide; modified from Posamentier *et al.*, 1992a; image courtesy of H.W. Posamentier), demonstrating the applicability of sequence stratigraphic concepts at virtually any scale. In this example, the highstand systems tract was left behind, and it was subsequently incised as a result of the fall in the local (pond) base level during the progradation of the lower elevation lowstand delta. See Posamentier *et al.* (1992a) for a more detailed interpretation.

two main types: (1) lines that build the chronostratigraphic or time framework of the studied interval, and (2) lines that illustrate lateral changes of facies or lithology. The chronostratigraphic framework is constructed by the correlation of surfaces of sequence stratigraphic significance, or true time markers such as bentonites or magnetic polarity boundaries. This is where some confusion can arise. Strictly speaking, sequence stratigraphic surfaces commonly are not true time lines but in fact are to some degree time transgressive, or diachronous. However, because true time lines are not commonly observed, the geoscientist is relegated to using these surfaces as proxies for time lines, being pragmatic and accepting the notion that in most instances, within the confines of most study areas they are at least very close to being time lines and therefore, are fundamentally useful. The degree of diachroneity of sequence stratigraphic surfaces, as well as of other types of stratigraphic surfaces, is discussed in more detail in Chapter 7.

Sequence stratigraphic surfaces are not necessarily easier to observe than the more diachronous contacts that mark lateral and vertical changes of facies. Consequently the practitioner can be faced with the dilemma of where to begin a stratigraphic interpretation; in other words, what lines should go first on a cross-section. The sequence stratigraphic approach yields a genetic interpretation of basin fill, which clarifies by time increment how a basin has filled with sediment. To accomplish this, a chronostratigraphic framework is first established, and sequence stratigraphic surfaces are interpreted. Subsequently, the sections between sequence stratigraphic surfaces are interpreted by recognizing facies contacts. These two types of surfaces (i.e., 'time lines' and 'facies contacts') define sequence stratigraphy and lithostratigraphy, respectively (Fig. 1.12).

The inherent difference between lithostratigraphy and sequence stratigraphy is important to emphasize, as both analyze the same sedimentary succession but with the focus on different stratigraphic aspects or rock properties. Lithostratigraphy deals with the lithology of strata and with their organization into units based on lithological character (Hedberg, 1976).

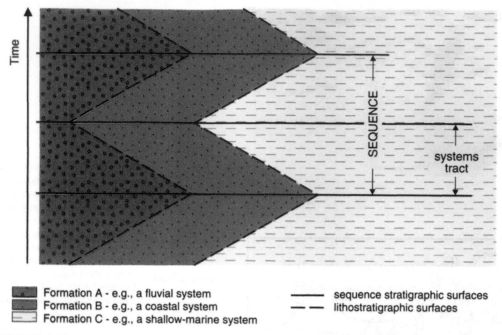

Formation A - e.g., a fluvial system
Formation B - e.g., a coastal system
Formation C - e.g., a shallow-marine system

—————— sequence stratigraphic surfaces
— — — lithostratigraphic surfaces

FIGURE 1.12 Conceptual contrast between lithostratigraphy and sequence stratigraphy. Sequence stratigraphic surfaces are event-significant, and mark changes in depositional trends. In this case, their timing is controlled by the turnaround points between transgressions and regressions. Lithostratigraphic surfaces are highly diachronous facies contacts. Note that the system tract and sequence boundaries cross the formation boundaries. Each systems tract is composed of three depositional systems in this example, and is defined by a particular depositional trend, i.e., progradational or retrogradational. A sequence corresponds to a full cycle of changes in depositional trends. This example implies continuous aggradation, hence no breaks in the rock record, with the cyclicity controlled by a shifting balance between the rates of base-level rise and the sedimentation rates.

The boundaries between lithostratigraphic units are often highly diachronous facies contacts, in which case they develop *within* the sedimentary packages bounded by sequence stratigraphic surfaces. Sequence stratigraphy deals with the correlation of coeval stratal units, irrespective of the lateral changes of facies that commonly occur across a basin, and which are bounded by low diachroneity (i.e., nearly synchronous) surfaces (Fig. 1.12). It is also important to note that facies analyses leading to the interpretation of paleoenvironments are much more critical for sequence stratigraphy than for lithostratigraphy, as illustrated in Figs. 1.13 and 1.14. These figures show that even along 1D vertical profiles, sequence stratigraphic units are often offset relative to the lithostratigraphic units due to their emphasis on different rock attributes. Understanding what constitutes a reasonable vertical and lateral relationship between facies within a time framework assists in correlating the same time lines through varying lithologies.

An example of a sequence stratigraphic—as contrasted with a lithostratigraphic—interpretation based on the same data set is illustrated in Fig. 1.15.

The interpretation of sequence stratigraphic surfaces is based on two fundamental observations: the type of stratigraphic contact, conformable or unconformable; and the nature of facies (depositional systems) which are in contact across each particular surface. The reconstruction of paleodepositional environments is therefore a critical pre-requisite for a successful sequence stratigraphic interpretation. In contrast, the lithostratigraphic cross-section does not require knowledge of paleoenvironments, but only mapping of lithological contacts. Some of these contacts may coincide with sequence stratigraphic surfaces, others may only reflect diachronous lateral changes of facies. As a result, the lithostratigraphic units (e.g., formations A, B, and C in Fig. 1.15) provide only descriptive information of lithologic distribution, which in some instances could combine the products of sedimentation of various depositional environments. Thus a simple map of lithologic distribution may give little insight as to the general paleogeography, and as a result be of little use in predicting lithologies away from known data points.

Allostratigraphy is a stratigraphic discipline that is intermediate in scope between lithostratigraphy

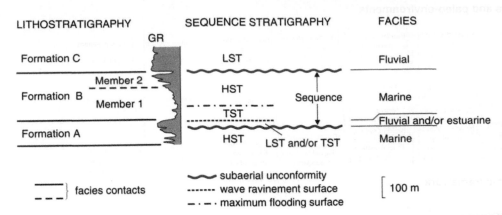

FIGURE 1.13 Lithostratigraphic and sequence stratigraphic interpretations of a gamma ray (GR) log (modified from Posamentier and Allen, 1999). Lithostratigraphy defines rock units on the basis of lithology, often irrespective of the depositional environment. Sequence stratigraphy defines rock units based on the event-significance of their bounding surfaces. Abbreviations: LST—lowstand systems tract; TST—transgressive systems tract; HST—highstand systems tract.

and sequence stratigraphy. The North American Commission on Stratigraphic Nomenclature (NACSN) introduced formal allostratigraphic units in the 1983 North American Stratigraphic Code to name discontinuity-bounded units. As currently amended, 'an allostratigraphic unit is a mappable body of rock that

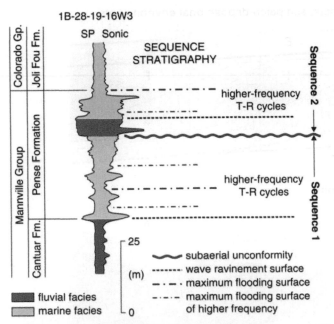

FIGURE 1.14 Relationship between depositional environments, lithostratigraphy, and sequence stratigraphy (wireline logs from the Western Canada Sedimentary Basin). Note that facies analysis (interpretation of paleodepositional environments) is more critical to sequence stratigraphy than to lithostratigraphy. Several higher frequency transgressive–regressive cycles can be noted within each sequence. The most prominent maximum flooding surface of each sequence, corresponding to the peak of finest sediment, belongs to the same hierarchical order as the sequence itself. *These* maximum flooding surfaces separate the transgressive and highstand systems tracts of sequences 1 and 2. Abbreviations: SP—spontaneous potential; T–R—transgressive–regressive.

is defined and identified on the basis of its bounding discontinuities' (Article 58). Allostratigraphic units, in order of decreasing rank, are allogroup, alloformation, and allomember—a terminology that originates and is modified from lithostratigraphy. The fundamental unit is the alloformation (NACSN, 1983, Art. 58). The bounding discontinuities which define the allostratigraphic approach are represented by any mappable lithological contact, with or without a stratigraphic hiatus associated with it. Basically, any type of stratigraphic contact illustrated in Fig. 1.16 may qualify as an allostratigraphic boundary. In this approach, all lithostratigraphic *and* sequence stratigraphic surfaces that are associated with a lithological contrast may be used for allostratigraphic studies (e.g., Bhattacharya and Walker, 1991; Plint, 2000).

Whereas allostratigraphy provides the means to take lithostratigraphy to a higher level of genetic interpretation of paleodepositional histories, because of the use of time-significant surfaces, its pitfall rests with the vague definition of 'discontinuities.' NACSN deliberately left the definition of 'discontinuity' to the practicing geologist who wishes to define or use allostratigraphic units, so the actual meaning of such units is largely equivocal. Because a stratigraphic unit is as well or poorly defined as its bounding surfaces, the formalization of allostratigraphic units in the North American Stratigraphic Code remains a half realized achievement until discontinuity surfaces are also defined and formalized. Between the European and the North American commissions on stratigraphic nomenclature, efforts are being made to clarify both the degree of overlap and the outstanding differences between the 'unconformity-bounded units' of the 1994 International Stratigraphic Guide (i.e., the pre-sequence stratigraphy 'sequences' of Sloss *et al.*, 1949) and the 'discontinuity-bounded units' of the 1983 NACSN (i.e., allostratigraphic units). Because the

1. Data: vertical profiles and paleo-environments

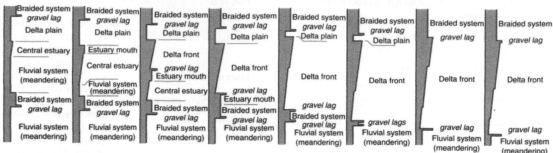

2. Sequence stratigraphic framework

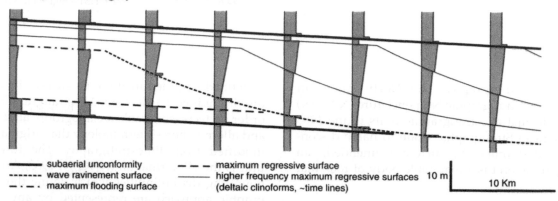

— subaerial unconformity
----- wave ravinement surface
—·—· maximum flooding surface

— — — maximum regressive surface
——— higher frequency maximum regressive surfaces
(deltaic clinoforms, ~time lines)

10 m

10 Km

3. Sequence stratigraphic framework, facies contacts, and paleo-depositional environments

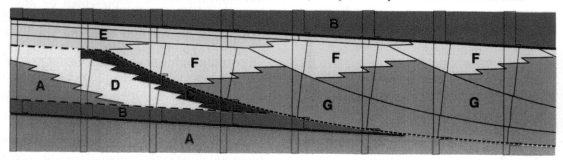

4. Cross-section emphasizing lithostratigraphic units

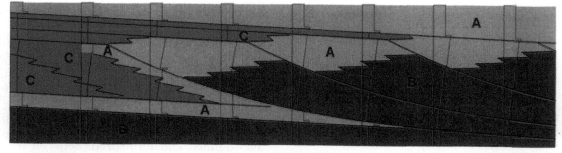

FIGURE 1.15 Sequence stratigraphic *vs.* lithostratigraphic frameworks, starting from the same set of facies data. 1. The reconstruction of paleodepositional environments *via* facies analysis is an important pre-requisite for sequence stratigraphic interpretations. The nature of stratigraphic contacts (scoured, conformable) also needs to be assessed *via* sedimentological analysis. 2. The sequence stratigraphic framework is constructed by correlating the key sequence stratigraphic surfaces. All sequence stratigraphic surfaces shown on the cross section are good chronostratigraphic markers (low diachroneity), with the exception of the transgressive wave-ravinement surface which is highly diachronous. 3. Sequence stratigraphic cross section, showing key surfaces, within-trend facies contacts, and paleodepositional environments. Within-trend facies contacts, marking lateral changes of facies, are placed on the cross-section *after* the sequence stratigraphic framework is constructed. Facies codes: A—meandering system; B—braided system; C—estuary-mouth complex; D—central estuary; E—delta plain; F—upper delta front; G—lower delta front—prodelta. 4. Lithostratigraphic cross-section. Three main lithostratigraphic units (e.g., formations) may be defined: A—a sandstone-dominated unit; B and C – mudstone-dominated units, with silty and sandy interbeds. Formations B and C are separated by Formation A. Additional lithostratigraphic units (e.g., members—subdivisions of units A, B, C) may be defined as a function of variations in lithology and color.

STRATIGRAPHIC CONTACTS

A. **Unconformity** = significant hiatus ± erosion (usually with erosion)

A substantial break or gap in the geological record ... It normally implies uplift and erosion with loss of the previously formed record. ... Relationship between rock strata in contact, characterized by a lack of continuity in deposition, and corresponding to a period of nondeposition, weathering, or esp. erosion (either subaerial or subaqueous) prior to the deposition of the younger beds.

   1. **Disconformity** = hiatus + erosion

   An unconformity in which the bedding planes above and below the break are essentially parallel, indicating a significant interruption in the orderly sequence of sedimentary rocks, generally by a considerable interval of erosion ..., and usually marked by a visible and irregular or uneven erosion surface of appreciable relief.

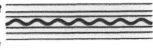

   2. **Paraconformity** = hiatus ± erosion (no discernable erosion)

   An obscure or uncertain unconformity in which no erosion surface is discernable ..., and in which the beds above and below the break are parallel.

   3. **Angular unconformity** = hiatus, erosion, and tilt

   An unconformity between two groups of rocks whose bedding planes are not parallel or in which the older, underlying rocks dip at a different angle (usually steeper) than the younger, overlying strata.

   4. **Nonconformity** = top of basement rocks

   An unconformity developed between sedimentary rocks and older igneous or metamorphic rocks that had been exposed to erosion before the overlying sediments covered them.

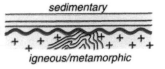

sedimentary

igneous/metamorphic

B. **Diastem** = short hiatus ± erosion (a minor paraconformity)

A relatively short interruption in sedimentation, involving only a brief interval of time, with little or no erosion before deposition is resumed; a depositional break of lesser magnitude than a paraconformity, or a paraconformity of very small time value.

C. **Conformity** = no hiatus

Undisturbed relationship between adjacent sedimentary strata that have been deposited in orderly sequence... True stratigraphic continuity in the sequence of beds.

FIGURE 1.16 Types of stratigraphic contacts (definitions from Bates and Jackson, 1987). Note that any of these stratigraphic contacts may qualify as an allostratigraphic unit boundary, i.e., a 'discontinuity,' as long as it is associated with a lithological contrast.

(lithological) 'discontinuity' is a much less specific term, including both unconformities and conformities (Fig. 1.16), 'unconformity-bounded units' remain only a special case of allostratigraphic units. In this context, the currently informal concepts of sequence stratigraphy may ultimately provide the framework that will allow previously defined types of stratigraphic units and surfaces to obtain a clear status in relation to each other and within the bigger picture of genetic stratigraphy. Formalizing sequence stratigraphic concepts is thus an important next task for all international commissions on stratigraphy.

2

Methods of Sequence Stratigraphic Analysis

INTRODUCTION

The roots of sequence stratigraphy can be traced far back in the classic principles of sedimentary geology, which established the fundamental guidelines of sedimentological and stratigraphic analyses. These 'first principles', as referred to by Posamentier and Allen (1999), set up the ground rules for the physics of flow and sediment motion, and the processes of sediment accumulation, bypass or erosion in relation to a shifting balance between relative sea-level changes, sediment supply and the energy of the transporting agent (Fig. 2.1). These principles still represent the scientific background of sequence stratigraphy, that allows old and modern concepts to blend into an evolving new way of looking at the sedimentary rock record.

It is therefore recognized that sequence stratigraphy is a fresh approach to analysis of sedimentary successions rather than a brand new method on its own. One cannot stress enough that a successful sequence stratigraphic study requires *integration* of various data sets and methods of data analysis into a unified, interdisciplinary approach (Fig. 1.1). This is not to say that sequence stratigraphy simply re-sells old concepts in a new package—in fact, the sequence stratigraphic approach allows for new insights into the genesis and architecture of sedimentary basin fills, which were not possible prior to the introduction of seismic stratigraphic concepts in the 1970s. The issues of facies formation and predictability in both mature and frontier hydrocarbon exploration basins are good examples of such new insights that were made possible by the sequence stratigraphic approach, and which are highly significant on both academic and economic grounds.

This chapter presents a brief account of the main methods that need to be integrated into a comprehensive sequence stratigraphic analysis, including facies analysis of ancient deposits (outcrops, core) and modern environments; analysis of well-log signatures; analysis

of seismic data; and the achievement of time control *via* relative and absolute age determinations. Each of these methods forms the core of a more conventional and dedicated discipline, so this presentation only reiterates aspects that are particularly relevant to sequence stratigraphy. Following the introduction to the various methods, a general guideline for a step-by-step sequence stratigraphic workflow is provided as a practical approach to the generation of geological models.

FACIES ANALYSIS: OUTCROPS, CORE, AND MODERN ANALOGUES

Facies analysis is a fundamental sedimentological method of characterizing bodies of rocks with unique lithological, physical, and biological attributes relative to all adjacent deposits. This method is commonly applied to describe the sediments and/or sedimentary rocks observed in outcrops, core, or modern environments. Facies analysis is of paramount importance for any sequence stratigraphic study, as it provides critical clues for paleogeographic and paleoenvironmental reconstructions, as well as for the definition of sequence stratigraphic surfaces. As such, facies analysis is an integral part of both sedimentology and sequence stratigraphy, which explains the partial overlap between these disciplines (Fig. 1.2). In the context of sequence stratigraphy, facies analysis is particularly relevant to the study of cyclic changes in the processes that form individual depositional systems in response to base-level shifts.

Concepts of Depositional System, Facies, and Facies Models

A depositional system (Fig. 1.9) is the product of sedimentation in a particular depositional environment;

Principles of flow and sediment motion

All natural systems tend toward a state of equilibrium that reflects an optimum use of energy. This state of equilibrium is expressed as a graded profile in fluvial systems, or as a base level in coastal to marine systems. Along such profiles, there is a perfect balance between sediment removal and accumulation.

Fluid and sediment gravity flows tend to move from high to low elevations, following pathways that require the least amount of energy for fluid and sediment motion.

Flow velocity is directly proportional to slope magnitude.

Flow discharge (subaerial or subaqueous) is equal to flow velocity times cross-sectional area.

Sediment load (volume) is directly proportional to the transport capacity of the flow, which reflects the combination of flow discharge and velocity.

The mode of sediment transport (bedload, saltation, suspension) reflects the balance between grain size/weight and flow competence.

Principles of sedimentation

Walther's Law: within a relatively conformable succession of genetically related strata, vertical shifts of facies reflect corresponding lateral shifts of facies.

The direction of lateral facies shifts (progradation, retrogradation) reflects the balance between sedimentation rates and the rates of change in the space available for sediment to accumulate.

Processes of aggradation or erosion are linked to the shifting balance between energy flux and sediment supply: excess energy flux leads to erosion, excess sediment load triggers aggradation.

The bulk of clastic sediments is derived from elevated source areas and is delivered to sedimentary basins by river systems.

As environmental energy decreases, coarser-grained sediments are deposited first.

FIGURE 2.1 Key 'first principles' of sedimentary geology that are relevant to sequence stratigraphy (modified from Posamentier and Allen, 1999).

hence, it includes the three-dimensional assemblage of strata whose geometry and facies lead to the interpretation of a specific paleodepositional environment. Depositional systems form the building blocks of systems tracts, the latter representing an essential concept for stratigraphic correlation and the genetic interpretation of the sedimentary basin fill. The study of depositional systems is intimately related to the concepts of facies, facies associations, and facies models, which are defined in Fig. 2.2.

Facies analysis is an essential method for the reconstruction of paleodepositional environments, as well as for the understanding of climatic changes and subsidence history of sedimentary basins. The understanding of facies and their associations are also essential for the correct interpretation of sequence stratigraphic surfaces, as is explained in more detail in Chapter 4.

Facies analysis is therefore a prerequisite for any sequence stratigraphic studies.

Classification of Depositional Environments

Depositional settings may be classified into three broad categories, as follows (Fig. 2.3): nonmarine (beyond the reach of marine flooding), coastal (intermittently flooded by marine water), and marine (permanently covered by marine water). An illustration of the subenvironments that encompass the transition from nonmarine to fully marine environments is presented in Fig. 2.4. Note that in coastal areas, the river-mouth environments (i.e., sediment entry points to the marine basin) are separated by stretches of open shoreline where the beach environment develops. The glacial environment is not included in the classification

Facies (Bates and Jackson, 1987): the aspect, appearance, and characteristics of a rock unit, usually reflecting the conditions of its origin; esp. as differentiating the unit from adjacent or associated units.

Facies (Walker, 1992): a particular combination of lithology, structural and textural attributes that defines features different from other rock bodies.

Facies are controlled by sedimentary processes that operate in particular areas of the depositional environments. Hence, the observation of facies helps with the interpretation of syn-depositional processes.

Facies Association (Collinson, 1969): groups of facies genetically related to one another and which have some environmental significance.

The understanding of facies associations is a critical element for the reconstruction of paleo-depositional environments. In turn, such reconstructions are one of the keys for the interpretation of sequence stratigraphic surfaces (see more details in Chapter 4).

Facies model (Walker, 1992): a general summary of a particular depositional system, involving many individual examples from recent sediments and ancient rocks.

A facies model assumes predictability in the morphology and evolution of a depositional environment, inferring "standard" vertical profiles and lateral changes of facies. Given the natural variability of allocyclic and autocyclic processes, a dogmatic application of this idealization introduces a potential for error in the interpretation.

FIGURE 2.2 Concepts of facies, facies associations, and facies models.

1. Nonmarine environments

  • **Colluvial and alluvial fans**
  • **Fluvial environments**
  • **Lacustrine environments**
  • **Aeolian environments**

2. Coastal (marginal marine) environments

  • **River mouth environments**
    - regressive river mouths: Deltas
    - transgressive river mouths: Estuaries

  • **Open shoreline (beach) environments**
    - foreshore
    - backshore

3. Marine environments

  • **Shallow marine environments**
    - shoreface
    - inner and outer shelf

  • **Deep marine environments**
    - continental slope
    - abyssal plain (basin floor)

FIGURE 2.3 Classification of depositional environments, based on the relative contributions of nonmarine and marine processes. The coastal/marginal-marine environments, also known as 'transitional', are intermittently flooded by marine water during tidal cycles and storms. Note that both types of coastal environments (river-mouth or open shoreline) may be transgressive or regressive. Depositional *systems* refer to products (bodies of rock in the stratigraphic record), whereas depositional *environments* refer to active processes in modern areas of sediment accumulation. This is similar to the conceptual difference between *cycle* and *cyclothem*, or between *period* and *system*, etc. The boundaries between the various coastal and shallow-marine environments are defined in Fig. 2.4.

scheme in Fig. 2.3 because it is climatically controlled and may overlap on any nonmarine, coastal, or marine setting. Within the nonmarine portion of the basin, a distinction can be made between the steeper-gradient *alluvial plain*, which captures the upstream reaches of fluvial systems, and the gently sloping *coastal plain* that may develop within the downstream reaches of the fluvial environment (Fig. 2.5). 'Coastal plain' is a geomorphological term that refers to a relatively flat area of prograded or emerged seafloor, bordering a coastline and extending inland to the nearest elevated land (Bates and Jackson, 1987; Fig. 2.5). Figure 2.5 illustrates the situation where the coastal plain forms by processes of progradation of the seafloor, rather than emergence. In this case, the sediments that accumulate on the coastal plain during the progradation of the shoreline are part of the so-called 'coastal prism', which includes fluvial to shallow-water deposits (Posamentier *et al.*, 1992b; Fig. 2.5). The coastal prism is wedge shaped, and expands landward from the coastal environment by onlapping the pre-existing topography in an upstream direction. The landward limit of the coastal prism was termed 'bayline' by Posamentier *et al.* (1992b), and it may shift upstream when the progradation of the shoreline is accompanied by aggradation.

Coastal environments are critical for sequence stratigraphy, as they record the history of shoreline shifts and are most sensitive in providing the clues for

FIGURE 2.4 Transition from marine to nonmarine environments. The large arrows indicate the direction of shoreline shift in the two river-mouth environments (R—regressive; T—transgressive). Between the river-mouth environments, the coastline is an open shoreline. Note that the character of the shoreline (transgressive *vs.* regressive) may change along strike due to variations in subsidence and sedimentation rates.

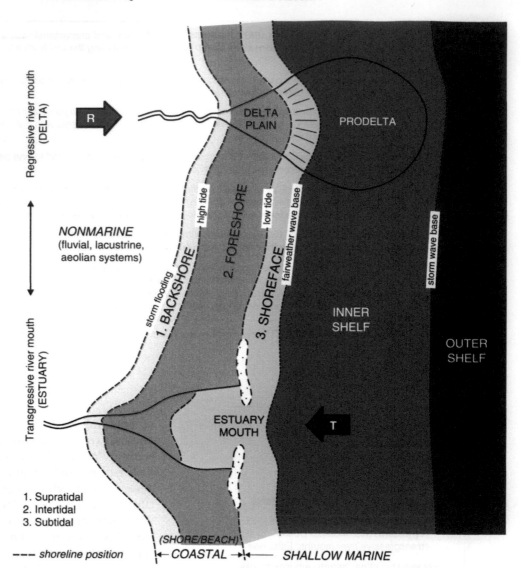

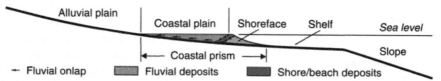

FIGURE 2.5 Dip-oriented profile illustrating the main geomorphic and depositional settings of a continental shelf: alluvial plain, coastal plain, coastline (including the intertidal and supratidal environments; Fig. 2.4), and shallow-marine (shoreface and shelf) environments (modified from Posamentier *et al.*, 1992b). Note that coastal plains may form by either the progradation or the emergence of the seafloor. This diagram illustrates the former situation, when a coastal prism of fluvial to shoreface deposits accumulates in the coastal plain to shallow-water settings (see text for details). For scale, coastal plains may be tens to hundreds of kilometers wide, depending on sediment supply and the gradient of the onlapped floodplain surface (e.g., the coastal plain of the Nueces River in Texas is approximately 40 km wide: Blum and Tornqvist, 2000; the coastal plain of the River Po in Italy is approximately 200 km wide: Hernandez-Molina, 1993; the coastal plain of the Mississippi River is at least 300–400 km wide: Blum and Tornqvist, 2000). Coastal prisms are typically associated with lowstand and highstand normal regressions (systems tracts). A lowstand coastal prism may be scoured by tidal- and/or wave-ravinement processes during subsequent transgression, whereas a highstand coastal prism is typically incised by rivers during subsequent base-level fall. Both lowstand and highstand coastal prisms may be preserved in the rock record where the original thickness of the coastal prism exceeds the amount of subsequent erosion.

the reconstruction of the cyclic changes in depositional trends. In fact, the development of sequence stratigraphic concepts started in the first place with the study of the transition zone between marine and nonmarine environments, where the relationship of facies and stratigraphic surfaces is easier to observe. From the shoreline, the application of sequence stratigraphy was gradually expanded in both landward and basinward directions, until a coherent basin-wide model that includes the stacking patterns expected in both fully fluvial and deep-marine successions was finally established. The importance of the coastline, as the link between the marine and nonmarine portions of the basin, is also reflected by the fact that the reference curve of base-level changes that is used to define the four main events of a stratigraphic cycle, and implicitly the timing of all systems tracts and stratigraphic surfaces (Fig. 1.7), is centered around the fluctuations in accommodation *at the shoreline*—this issue, which is the key to understanding sequence stratigraphic principles, is elaborated in subsequent chapters.

A reality that is commonly overlooked is that coastlines may change their transgressive *vs.* regressive character along strike, as a function of the fluctuations in subsidence and sedimentation rates (Fig. 2.4). This means that the predictable architecture and age relationships of depositional systems and systems tracts presented in 2D cross-sections along dip may be altered in a 3D view, due to the high diachroneity that may potentially be imposed on systems tract boundaries by the strike variability in subsidence and sedimentation. One should therefore keep an open mind when trying to extrapolate the reality of one dip-oriented profile to other locations along the strike. Autocyclic shifts in the distribution of energy and sediment within depositional environments, which could affect all settings in Fig. 2.3, are another reason why variations in stratigraphic geometry should be expected along strike from one dip-oriented profile to another.

Walther's Law

The connection between the vertical and lateral changes of facies observed in outcrop and subsurface is made by Walther's Law (Fig. 2.6). This is a fundamental principle of stratigraphy, which allows the geologist to visualize predictable lateral changes of facies based on the vertical profiles observed in 1D sections such as small outcrops, core, or well logs. As discussed by Miall (1997), vertical changes in litho- and biofacies have long been used to reconstruct paleogeography and temporal changes in depositional environments and, with the aid of Walther's Law, to interpret lateral shifts of these environments. As a note of caution, however, such interpretations are only valid within

Walther's Law (Middleton, 1973): in a conformable succession, the only facies that can occur together in vertical succession are those that can occur side by side in nature.

Walther's Law (Bates and Jackson, 1987): only those facies and facies-areas can be superimposed which can be observed beside each other at the present time.

Walther's Law (Posamentier and Allen, 1999): the same succession that is present vertically also is present horizontally *unless there is a break in sedimentation.*

In other words, a vertical change of facies implies a corresponding lateral shift of facies within a relatively conformable succession of genetically related strata.

FIGURE 2.6 Walther's Law: the principle that connects the lateral and vertical shifts of facies within a sequence (i.e., a relatively conformable succession of genetically related strata).

relatively conformable successions of genetically related strata. Vertical changes across sequence-bounding unconformities potentially reflect major shifts of facies between successions that are genetically unrelated, and therefore such changes should not be used to reconstruct the paleogeography of one particular time slice in the stratigraphic record.

A prograding delta is a good illustration of the Walther's Law concept. The deltaic depositional system includes prodelta, delta front, and delta plain facies, 'which occur side by side in that order and the products of which occur together in the same order in vertical succession. Use of the depositional system concept enables predictions to be made about the stratigraphy at larger scales, because it permits interpretations of the rocks in terms of broad paleoenvironmental and paleogeographic reconstructions. This technique has now become part of sequence stratigraphy, where sequences are regionally correlatable packages of strata that record local or regional changes in base level' (Miall, 1990, p. 7).

Beyond the scale of a depositional system, Walther's Law is equally valuable when applied to systems tracts, as the internal architecture of each systems tract involves progradational or retrogradational shifts of facies which translate into corresponding facies changes along vertical profiles. Figure 1.15 provides examples of how vertical profiles integrate and help to reconstruct the lateral facies relationships along dip-oriented sections.

Sedimentary Petrography

The observation of sedimentary facies in outcrops or core is often enough to constrain the position of sequence-bounding unconformities, where such contacts juxtapose contrasting facies that are genetically unrelated (Fig. 2.7). The larger the stratigraphic hiatus associated with sequence boundaries, the better the chance of mapping these surfaces by simple facies observations. There are however cases, especially in proximal successions composed of coarse, braided fluvial deposits, where subaerial unconformities are 'cryptic', difficult to distinguish from any other channel-scour surface (Miall, 1999). Such cryptic sequence boundaries may occur within thick fluvial successions consisting of unvarying facies, and may well be associated with substantial breaks in sedimentation. In the absence of abrupt changes in facies and paleocurrent directions across these sequence boundaries, petrographic studies of cements and framework grains may provide the only solid criteria for the identification and mapping of sequence-bounding unconformities. The Late Cretaceous Lower Castlegate Sandstone of the Book Cliffs (Utah) provides an example where a nonmarine sequence boundary was mapped updip into a continuous braided-fluvial sandstone succession only by plotting the position of subtle changes in the detrital petrographic composition, interpreted to reflect corresponding changes in provenance in relation to tectonic events in the Sevier highlands (Miall, 1999).

Besides changes in provenance and the related composition of framework grains, subaerial unconformities may also be identified by the presence of secondary minerals that replace some of the original sandstone constituents *via* processes of weathering under subaerial conditions. For example, it has been documented that subaerial exposure, given the availability of sufficient amounts of K, Al, and Fe that may be derived from the weathering of clays and feldspars, may lead to the replacement of calcite cements by secondary glauconite (Khalifa, 1983; Wanas, 2003). Glauconite-bearing sandstones may therefore be used to recognize sequence-bounding unconformities, where the glauconite formed as a replacement mineral. Hence, a distinction needs to be made between the syndepositional glauconite of marine origin (framework grains in sandstones) and the secondary glauconite that forms under subaerial conditions (coatings, cements), which can be resolved *via* petrographic analysis.

The distribution pattern of early diagenetic clay minerals such as kaolinite, smectite, palygorskite, glaucony, and berthierine, as well as of mechanically infiltrated clays, may also indicate changes in accommodation and the position of sequence stratigraphic surfaces (Ketzer *et al.*, 2003a, b; Khidir and Catuneanu, 2005; Figs. 2.8–2.10). As demonstrated by Ketzer *et al.* (2003a), 'changes in relative sea-level and in sediment supply/sedimentation rates, together with the climatic conditions prevalent during, and immediately after deposition of sediments control the type, abundance, and spatial distribution of clay minerals by influencing the pore-water chemistry and the duration over which the sediments are submitted to a certain set of geochemical conditions' (Figs. 2.8 and 2.9). The patterns of change in the distribution of early diagenetic clay minerals across subaerial unconformities may be preserved during deep-burial diagenesis, when late diagenetic minerals may replace the early diagenetic ones (e.g., the transformation of kaolinite into dickite with increased burial depth; Fig. 2.10).

Petrographic studies may also be used to emphasize grading trends (fining- *vs.* coarsening-upward) in vertical successions (outcrops, core). Vertical profiles are an integral part of sequence stratigraphic analyses,

FIGURE 2.7 Subaerial unconformity (arrows) at the contact between the Burgersdorp Formation and the overlying Molteno Formation (Middle Triassic, Dordrecht–Queenstown region, Karoo Basin). The succession is fluvial, with an abrupt increase in energy levels across the contact. Note the change in fluvial styles from meandering (with lateral accretion) to amalgamated braided systems. The unconformity is associated with an approximately 7 Ma stratigraphic hiatus (Catuneanu *et al.*, 1998a), and hence separates fluvial sequences that are genetically unrelated.

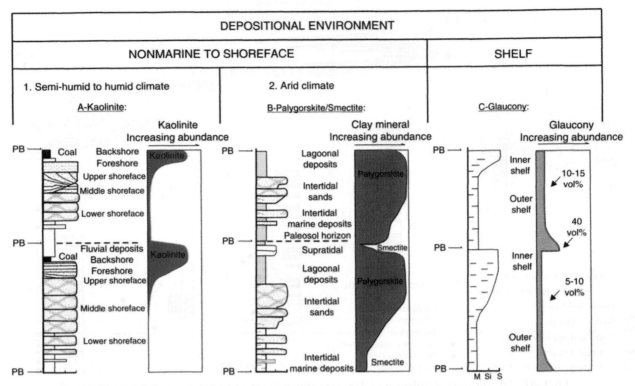

FIGURE 2.8 Predictive distribution of early-diagenetic clay minerals in a succession of fluvial to shallow-water regressive lobes ('parasequences') separated by flooding surfaces (redrafted and modified from Ketzer *et al.*, 2003a). A—kaolinite content increases toward the top of parasequences where continental facies are exposed to extensive meteoric water flushing under semi-humid to humid climatic conditions. Kaolinite content increases in the presence of unstable silicates and organic matter, as the degradation of the latter facilitates the formation of acidic fluids; B—palygorskite content increases toward the top of parasequences capped by evaporitic deposits, under arid climatic conditions; C—in fully marine successions, autochthonous glauconite is most abundant at the parasequence boundary, and decreases gradually toward the top of the parasequence. Abbreviation: PB—parasequence boundary.

and are commonly used to discern between progradational and retrogradational trends in marine successions, or to outline fluvial depositional sequences in nonmarine deposits. Fluvial sequences, for example, often show overall fining-upward trends that reflect aggradation in an energy-declining environment (e.g., Eberth and O'Connell, 1995; Hamblin, 1997; Catuneanu and Elango, 2001). From a sedimentological perspective, sequence boundaries (subaerial unconformities) in such fluvial successions are commonly picked at the base of the coarsest units, usually represented by amalgamated channel fills. This interpretation is generally correct in proximal settings, close to source areas, where renewed subsidence is closely followed by the onset of fluvial sedimentation. In more distal settings, however, independent time control may be required to find the actual position of unconformities, which are not necessarily placed at the base of the fining-upward successions but rather within the underlying fine-grained facies (Sweet *et al.*, 2003, 2005; Catuneanu and Sweet, 2005).

In spite of the potential limitations, the observation of grading trends remains a fundamental and useful method of emphasizing cyclicity in the stratigraphic record. As long as data are available, i.e., access to outcrops or core, plots reflecting vertical changes in grain size can be constructed by careful logging and textural analysis. The actual vertical profiles may reflect the absolute, bed-by-bed changes in grain size, or smoothed out curves that show the overall statistical changes in grain size (e.g., moving averages of overlapping intervals). The latter method is often preferred because it eliminates abnormal peaks that may only have local significance. The technique of constructing vertical profiles can also be adapted as a function of case study. The grain size logs may be plotted using an arithmetic horizontal scale, where fluctuations in grain size are significant, or on logarithmic scales where the succession is monotonous and the differences in grain size are very small. The latter technique works best in fine-grained successions, where logarithmic plots enhance the differences in grain size, but is less efficient in coarser deposits (D. Long, pers. comm., 2004).

The construction of grain size logs is generally a viable method of identifying cycles in individual

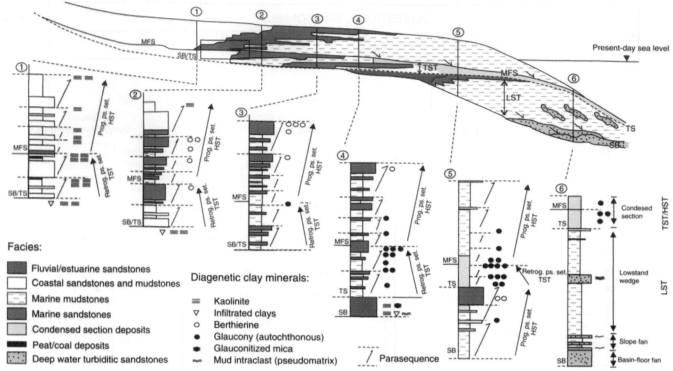

FIGURE 2.9 Predictive distribution of diagenetic clay minerals in a sequence stratigraphic framework (redrafted and modified from Ketzer *et al.*, 2003a). Abbreviations: MFS—maximum flooding surface; TS—transgressive surface; SB—sequence boundary; HST—highstand systems tract; TST—transgressive systems tract; LST—lowstand systems tract.

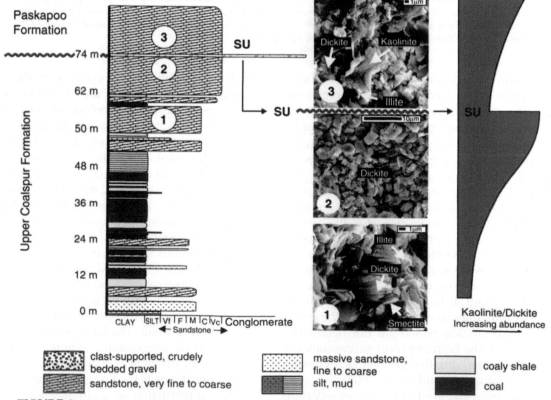

FIGURE 2.10 Pattern of change in the distribution of kaolinite/dickite in a fluvial sequence stratigraphic framework (from Khidir and Catuneanu, 2005). Kaolinite/dickite content increases gradually toward the top of the sequence, and decreases abruptly across the sequence boundary. Abbreviation: SU—subaerial unconformity.

outcrops or core, but matching such trends across a basin, solely based on the observed grading trends, is not necessarily a reliable correlation technique. Changes in sedimentation patterns across a basin due to variations in subsidence and sediment supply make it difficult to know which cyclothems are age equivalent when comparing vertical profiles from different sections. Under ideal circumstances, the availability of age data (biostratigraphic, magnetostratigraphic, radiometric, marker beds) represents the perfect solution to this problem. Often, however, such age data are missing, especially in the study of older successions, and in the absence of time control other sedimentological observations have to be integrated with the petrographic data in order to constrain geological interpretations. Paleocurrent measurements, derived from unidirectional flow-related bedforms, are particularly useful as a complement to petrographic data, as they provide a record of the tectonic tilt in the basin and changes thereof. The documentation of such changes helps us to infer events in the evolution of the basin, commonly reflected by sequence-bounding unconformities in the rock record, providing additional criteria to enhance correlations across the basin.

Paleocurrent Directions

The major breaks in the stratigraphic record are potentially associated with stages of tectonic reorganization of sedimentary basins, and hence with changes in tilt direction across sequence boundaries. This is often the case in tectonically active basins, such as grabens, rifts, or foreland systems, where stratigraphic cyclicity is commonly controlled by cycles of subsidence and uplift triggered by various tectonic, flexural, and isostatic mechanisms. Other basin types, however, such as 'passive' continental margins or intracratonic sag basins, are dominated by long-term thermal subsidence, and hence they may show little change in the tilt direction through time. In such cases, stratigraphic cyclicity may be mainly controlled by fluctuations in sea level, and paleocurrent measurements may be of little use to constrain the position of sequence boundaries.

In the case of tectonically active basins, where fluctuations in tectonic stress regimes match the frequency of cycles observed in the stratigraphic record (e.g., Cloetingh, 1988; Cloetingh et al., 1985, 1989; Peper et al., 1992), paleocurrent data may prove to provide the most compelling evidence for sequence delineation, paleogeographic reconstructions, and stratigraphic correlations, especially when dealing with lithologically monotonous successions that lack any high-resolution time control. A good example is the case study of the Early Proterozoic Athabasca Basin of Canada, where the basin fill is composed of dominantly siliciclastic deposits that show little variation in grain size in any given area. In this case, vertical profiles are equivocal, the age data to constrain correlations are missing, and the only reliable method to outline genetically related packages of strata is the measurement of paleocurrent directions. Based on the reconstruction of fluvial drainage systems, the Athabasca basin fill has been subdivided into four second-order depositional sequences separated by subaerial unconformities across which significant shifts in the direction of tectonic tilt are recorded (Ramaekers and Catuneanu, 2004).

Overfilled foreland basins represent a classic example of a setting where fluvial sequences and bounding unconformities form in isolation from eustatic influences, with a timing controlled by orogenic cycles of thrusting (tectonic loading) and unloading (Catuneanu and Sweet, 1999; Catuneanu and Elango, 2001; Catuneanu, 2004a). In such foredeep basins, fluvial aggradation takes place during stages of differential flexural subsidence, with higher rates towards the center of loading, whereas bounding surfaces form during stages of differential isostatic rebound. As the thrusting events are generally shorter in time relative to the intervening periods of orogenic quiescence, foredeep fluvial sequences are expected to preserve the record of less than half of the geological time (Catuneanu et al., 1997a; Catuneanu, 2004a). Renewed thrusting in the orogenic belt marks the onset of a new depositional episode. Due to the strike variability in orogenic loading, which is commonly the norm rather than the exception, abrupt changes in tilt direction are usually recorded across sequence boundaries (Fig. 2.11). In the absence of other unequivocal criteria (see for example the case of the Athabasca Basin discussed above), such changes in tectonic tilt may be used to outline fluvial sequences with distinct drainage patterns, and to map their bounding surfaces.

Pedology

Pedology (soil science) deals with the study of soil morphology, genesis, and classification (Bates and Jackson, 1987). The formation of soils refers to the physical, biological, and chemical transformations that affect sediments and rocks exposed to subaerial conditions (Kraus, 1999). Paleosols (i.e., fossil soils) are buried or exhumed soil horizons that formed in the geological past on ancient landscapes. Pedological studies started with the analysis of modern soils and Quaternary paleosols, but have been vastly expanded to the pre-Quaternary record in the 1990s due to their multiple

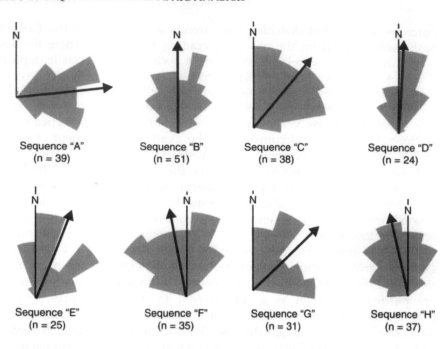

FIGURE 2.11 Paleoflow directions for the eight third-order depositional sequences of the Koonap-Middleton fluvial succession in the Karoo foredeep (from Catuneanu and Bowker, 2001). The succession spans a time interval of 5 Ma during the Late Permian, and measures a total thickness of 2630 m. 'n' represents the number of paleoflow measurements used to construct the rose diagram for each sequence. In this case study, sequence boundaries are marked not only by a change in tectonic tilt, but also by an abrupt change in fluvial styles and associated lithofacies.

geological applications. Notably, some of these geological applications include (1) interpretations of ancient landscapes, from local to basin scales; (2) interpretations of ancient surface processes (sedimentation, nondeposition, erosion), including sedimentation rates and the controls thereof; (3) interpretations of paleoclimates, including estimations of mean annual precipitation rates and mean annual temperatures; and (4) stratigraphic correlations, and the cyclic change in soil characteristics in relation to base-level changes (Kraus, 1999). All these applications, and particularly the latter, have relevance to sequence stratigraphy.

The complexity of soils, and thus of paleosols, can only begin to be understood by looking at the diversity of environments in which they may form; the variety of surface processes to which they can be genetically related; and the practical difficulties to classify them. Paleosols have been described from an entire range of nonmarine settings, including alluvial (Leckie *et al.*, 1989; Wright and Marriott, 1993; Shanley and McCabe, 1994; Aitken and Flint, 1996), palustrine (Wright and Platt, 1995; Tandon and Gibling, 1997) and eolian (Soreghan *et al.*, 1997), but also from coastal settings (e.g., deltaic: Fastovsky and McSweeney, 1987; Arndorff, 1993) and even marginal-marine to shallow-marine settings, where stages of base-level fall led to the subaerial exposure of paleo-seafloors (Lander *et al.*, 1991; Webb, 1994; Wright, 1994).

Irrespective of depositional setting, soils may form in conjunction with different surface processes, including sediment aggradation (as long as sedimentation rates do not outpace the rates of pedogenesis), sediment

bypass (nondeposition), and sediment reworking (as long as the rate of scouring does not outpace the rate of pedogenesis). Soils formed during stages of sediment aggradation occur within conformable successions, whereas soils formed during stages of nondeposition or erosion are associated with stratigraphic hiatuses, marking diastems or unconformities in the stratigraphic record. These issues are particularly important for sequence stratigraphy, as it is essential to distinguish between paleosols with the significance of sequence boundaries, playing the role of subaerial unconformities, and paleosols that occur within sequences and systems tracts. Theoretical and field studies (e.g., Wright and Marriott, 1993; Tandon and Gibling, 1994, 1997) show that the paleosol types observed in the rock record change with a fluctuating base level, thus allowing one to assess their relative importance and significance from a sequence stratigraphic perspective. For example, sequence boundaries of the Upper Carboniferous cyclothems in the Sydney Basin of Nova Scotia are marked by mature calcareous paleosols (calcretes; Fig. 2.12) formed during times of increased aridity and lowered base level, whereas vertisols and hydromorphic paleosols occur within sequences, being formed in aggrading fluvial floodplains during times of increased humidity and rising base level (Fig. 2.13; Tandon and Gibling, 1997).

The classification of soils and paleosols has been approached from different angles, and no universal scheme of pedologic systematics has been devised yet. The classification of modern soils relies on diagnostic horizons that are identified on the basis of properties

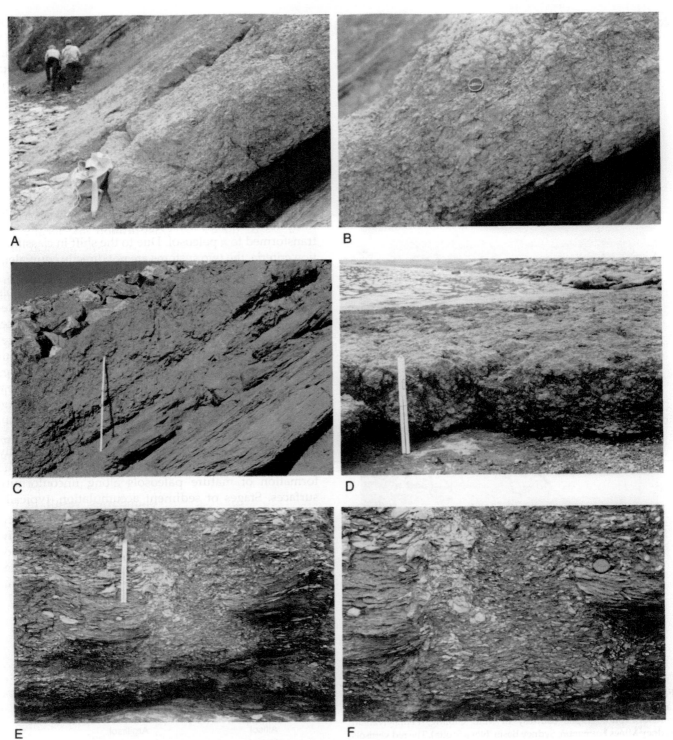

FIGURE 2.12 Calcareous paleosols and associated facies, formed during base-level fall and subaerial exposure (photographs courtesy of M.R. Gibling; Pennsylvanian Sydney Mines Formation, Sydney Basin, Nova Scotia; for more details, see Gibling and Bird, 1994; Gibling and Wightman, 1994; Tandon and Gibling, 1994, 1997). A—calcrete, marking a 'subaerial unconformity' (depositional sequence boundary) within coastal plain deposits. The carbonate soil implies a semi-arid climatic period, suggesting that lowstands in base level were relatively more arid than the peat-forming periods that represent the overlying transgressive and highstand systems tracts; B—close up of calcrete in image A, showing well-developed vertic and nodular fabric; C—calcrete in image A, with strong nodular texture. Note the non-disrupted nature of the siltstone below; D—calcrete exposed on wave-cut platform, with strong vertic fabric (scale 50 cm); E—upright tree cast, partially replaced by carbonate beneath a 'lowstand' calcrete layer. This occurrence suggests that carbonate-rich groundwaters caused local cementation through conduits below the main soil level; F—close up of carbonate-cemented tree in image E.

FIGURE 2.13 Coastal plain successions showing calcrete horizons (arrows—depositional sequence boundaries) overlain by red calcic vertisols (photographs courtesy of M.R. Gibling; Pennsylvanian Sydney Mines Formation, Sydney Basin, Nova Scotia). The red vertisols (dryland clastic soils) are interpreted as being formed within the transgressive systems tract under conditions of abundant sediment supply (Tandon and Gibling, 1997). A—'lowstand' carbonates (calcrete paleosols/sequence boundary – arrow) pass upward into dryland clastic soils, probably marking the renewal of clastic supply to the coastal plain as accommodation is made available by base-level rise; B — close up of concave-up, slickensided joints (mukkara structure) in red vertisols of image A; C—grey coastal-plain siltstones at lower left pass upward in meter-thick calcrete (arrows). Siltstones immediately below the calcrete are calcite cemented. Calcrete is overlain by red vertisols and thin splay sandstones, as sedimentation resumed on the dryland coastal plain, possibly as transgression allowed sediment storage on the floodplain.

such as texture, color, amount of organic matter, mineralogy, cation exchange capacity, and pH (Soil Survey Staff, 1975, 1998; Fig. 2.14). The main pitfalls of this approach, when applied to paleosols, are two-fold: (1) the taxonomic approach does not emphasize the importance of hydromorphic soils (i.e., 'gleysols', common in aggrading fluvial floodplains, defined on the basis of soil saturation; Fig. 2.14); and (2) it is dependent on soil properties, some of which (e.g., cation exchange capacity, or amount of organic matter) are not preserved in paleosols. For these reasons, Mack *et al.* (1993) devised a classification specifically for paleosols (Fig. 2.14), based on mineralogical and morphological properties that are preserved as a soil is transformed to a paleosol. Due to the shift in classification criteria, the two systems are not directly equivalent with respect to some soil/paleosol groups (Fig. 2.14).

From a sequence stratigraphic perspective, paleosols may provide key evidence for reconstructing the syndepositional conditions (e.g., high *vs.* low water table, accommodation, and sedimentation rates, paleoclimate) during the accumulation of systems tracts, or about the temporal significance of stratigraphic hiatuses associated with sequence-bounding unconformities. The types of paleosols that may form in relation to the interplay between surface processes (sedimentation, erosion) and pedogenesis are illustrated in Fig. 2.15. Stages of nondeposition and/or erosion, typically associated with sequence boundaries, result in the formation of mature paleosols along unconformity surfaces. Stages of sediment accumulation, typically

Soil systematics (Soil Survey Staff, 1975, 1998)	Paleosol systematics (Mack et al., 1993)
Entisol	Protosol
Inceptisol	
Vertisol	Vertisol
Histosol	Histosol
sub-class	Gleysol
Andisol	-
Oxisol	Oxisol
Spodosol	Spodosol
Alfisol	Argillisol
Ultisol	
-	Calcisol
-	Gypsisol
Aridisol	-
Mollisol	-
Gelisol	-

FIGURE 2.14 Comparison between the soil and paleosol classification systems of the United States Soil Taxonomy (Soil Survey Staff, 1975, 1998) and Mack *et al.* (1993). Due to differences in the classification criteria, not all soil or paleosol groups have equivalents in both systems.

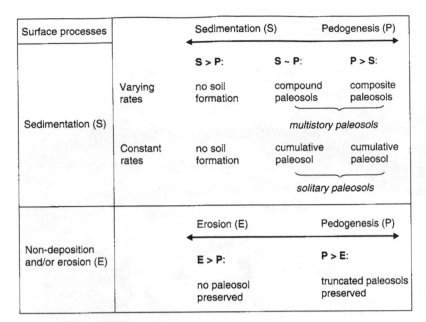

Surface processes		Sedimentation (S)		Pedogenesis (P)
		S > P:	S ~ P:	P > S:
Sedimentation (S)	Varying rates	no soil formation	compound paleosols	composite paleosols
			multistory paleosols	
	Constant rates	no soil formation	cumulative paleosol	cumulative paleosol
			solitary paleosols	
Non-deposition and/or erosion (E)		Erosion (E)		Pedogenesis (P)
		E > P:		P > E:
		no paleosol preserved		truncated paleosols preserved

FIGURE 2.15 Interplay of pedogenesis and surface processes (modified from Morrison, 1978; Bown and Kraus, 1981; Marriott and Wright, 1993; Kraus, 1999). Compound, composite and cumulative paleosols occur within conformable successions, hence within depositional sequences. 'Truncated' paleosols are associated with stratigraphic hiatuses, and therefore mark diastems or unconformities.

associated with the deposition of sequences, result in the formation of less mature and generally aggrading paleosols of compound, composite, or cumulative nature, whose rates of aggradation match the sedimentation rates (see Kraus, 1999, for a comprehensive review of these paleosol types).

Paleosols associated with sequence boundaries are generally strongly developed and well-drained, reflecting prolonged stages of sediment cut-off and a lowered base level (low water table in the nonmarine portion of the basin; Fig. 2.12). Besides base level, climate may also leave a strong signature on the nature of sequence-bounding paleosols (e.g., a drier climate would promote evaporation and the formation of calcic paleosols). Base level and climate are not necessarily independent variables, as climatic cycles driven by orbital forcing (e.g., eccentricity, obliquity, and precession cycles, with periodicities in a range of tens to hundreds of thousands of years; Fig. 2.16; Milankovitch, 1930, 1941; Imbrie and Imbrie, 1979; Imbrie, 1985; Schwarzacher, 1993) are a primary control on sea-level changes at the temporal scale of Milankovitch cycles. In such cases, stages of base-level fall may reflect times of increased climatic aridity (e.g., see Tandon and Gibling, 1997, for a case study). On the other hand, base-level changes may also be driven by tectonism, independent of climate changes, in which case base-level cycles may be offset relative to the climatic fluctuations. A more comprehensive discussion of the relationship between base-level changes, sea-level changes, tectonism, and climate is provided in Chapter 3.

Irrespective of the primary force behind a falling base level, the cut-off of sediment supply is an important parameter that defines the conditions of formation of sequence-bounding paleosols. Stages of sediment cut-off during the depositional history of a basin may be related to either autogenic or allogenic controls. In the case of sequence boundaries, the fall in base level and the sediment cut-off are intimately related, and are both controlled by allogenic mechanisms. The stratigraphic hiatus associated with a sequence-bounding unconformity/paleosol varies greatly with the rank (importance) of the sequence and the related allogenic controls, and it is generally in a range of 10^4 years (for the higher-frequency Milankovitch cycles) to 10^5–10^7 years for the higher-order sequences (Summerfield, 1991; Miall, 2000). Sequence-bounding unconformities are commonly regional in scale, as opposed to the more

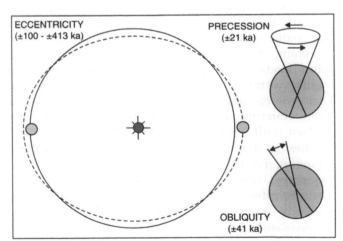

FIGURE 2.16 Main components of orbital forcing, showing the causes of Milankovitch-band (10^4–10^5 years) cyclicity (modified from Imbrie and Imbrie, 1979, and Plint *et al.*, 1992).

FIGURE 2.17 'Wet' and immature paleosol of gleysol type, formed in close association with a coal seam during an overall stage of base-level rise. This example comes from the Castlegate Formation in Utah, which consists of amalgamated braided fluvial channel fills interpreted to form a lowstand systems tract (positive but low rates of creation of accommodation). Such immature paleosols develop *within* depositional sequences, commonly over short time scales of 10^3 years or less (Fig. 2.18). The formation of wet and immature soils *vs.* coal seams is most likely a function of fluctuations in climatic conditions and fluvial discharge (subaerial exposure *vs.* flooding of overbank environments) rather than marine base-level changes.

localized diastems related to autogenic processes, and depending on paleo-landscape, can be surfaces with highly irregular topographic relief along which the amount of missing time may vary considerably (Wheeler, 1958). Accordingly, the paleosol associated with a sequence-bounding unconformity can show lateral changes that may be used to interpret lateral variations in topography and missing time (Kraus, 1999).

Paleosols that form within sequences may be weakly to well-developed, but are generally less mature than the sequence-bounding paleosols (Figs. 2.13 and 2.17). They form during stages of base-level rise (higher water table in nonmarine environments), when surface processes are dominated by sediment aggradation. As a result, these paleosols tend to be 'wetter' relative to the sequence-bounding paleosols, to the extent of becoming hydromorphic (gleysol type) around maximum flooding surfaces which mark the timing of the highest water table in the nonmarine environment. Such 'wetter' and immature paleosols form over relatively short time scales, and are often seen in close association with coal seams (Fig. 2.17). Figure 2.18 synthesizes the main contrasts between the sequence-bounding paleosols and the paleosols that form within sequences. The latter type may show aggradational features, often with a multistory architecture due to unsteady sedimentation rates (Fig. 2.15), but may also be associated with hiatuses where autogenic processes such as channel avulsion lead to a cut-off of sediment supply in restricted overbank areas. As the periodicity

of avulsion is estimated to be in a range of 10^3 years (Bridge and Leeder, 1979), the stratigraphic hiatuses that are potentially associated with paleosols developed within sequences are in general at least one order of magnitude less significant than the hiatuses associated with sequence-bounding paleosols (Fig. 2.18).

Figure 2.19 illustrates a generalized model of paleosol development in relation to a cycle of base-level changes. As a matter of principle, the higher the sedimentation rates the weaker developed the paleosol is. Hence, the most mature paleosols are predicted along sequence boundaries (zero or negative sedimentation rates), and the least developed paleosols are expected to form during transgressions, when aggradation rates and the water table are highest. Due to the high water table in the nonmarine environments during

Features \ Paleosol type	Sequence-bounding paleosols	Paleosols within sequences
maturity	strongly developed	weakly to well-developed
soil saturation	well-drained	wetter
hiatus	10^4 yr or more	0–10^3 yr
hiatus controls	allogenic	autogenic (e.g., avulsion)
hiatus extent	regional	local
significance	unconformity	diastem
accommodation	negative	positive
surface process	bypass or erosion	aggradation
water table	low	higher
architecture	solitary	commonly multistory

FIGURE 2.18 Comparison between sequence-bounding paleosols and the paleosols developed within sequences.

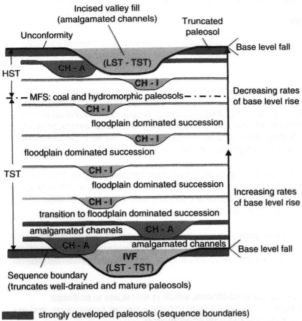

FIGURE 2.19 Generalized model of paleosol development in relation to a base-level cycle (modified from Wright and Marriott, 1993). In this model, the rates of fluvial aggradation (and implicitly the degree of channel amalgamation and the paleosol maturity) are directly linked to the rates of base-level rise. Note that low sedimentation rates (early and late stages of base-level rise) allow for channel amalgamation and the formation of well-developed paleosols; high sedimentation rates favor the formation of weakly developed paleosols within a succession dominated by floodplain deposits. Abbreviations: LST—lowstand systems tract; TST—transgressive systems tract; HST—highstand systems tract; IVF—incised-valley fill; CH-A—amalgamated (multistory) channels; CH-I—isolated channels; MFS—maximum flooding surface.

transgression, hydromorphic paleosols are often associated with regional coal seams (Fig. 2.19; Tandon and Gibling, 1994). It can be concluded that paleosols are highly relevant to sequence stratigraphy, complementing the information acquired *via* different methods of data analysis. Pedologic studies are routinely performed on outcrops and core (Leckie *et al.*, 1989; Lander *et al.*, 1991; Platt and Keller, 1992; Caudill *et al.*, 1997), and to a lesser extent on well logs (Ye, 1995), and may be applied to a wide range of stratigraphic ages, including strata as old as the Early Proterozoic (Gutzmer and Beukes, 1998).

Ichnology

General Principles

Ichnology is the study of traces made by organisms, including their description, classification and interpretation (Pemberton *et al.*, 2001). Such traces may be ancient

(trace fossils–the object of study of paleoichnology) or modern (recent traces—the object of study of neoichnology), and generally reflect basic *behavior patterns* (e.g., resting, locomotion, dwelling, or feeding—all of which can be combined with escape or equilibrium structures; Ekdale *et al.*, 1984; Frey *et al.*, 1987; Pemberton *et al.*, 2001) that can be linked to a number of *ecological controls* (e.g., substrate coherence, water energy, sedimentation rates, nutrients, salinity, oxygenation, light or temperature), and implicitly to particular *depositional environments* (Seilacher, 1964, 1978).

Trace fossils include a wide range of biogenic structures where the results of organism activities are preserved in sediments or sedimentary rocks, but not the organisms themselves or any body parts thereof. Ichnofossils also exclude molds of the body fossils that may form after burial, but include imprints made by body parts of active organisms (Pemberton *et al.*, 2001). Trace fossils are often found in successions that are otherwise unfossiliferous, and bring a line of evidence that can be used towards the reconstruction of paleoecological conditions and paleodepositional environments. As with any independent research method, the information brought by ichnology may be equivocal in some cases (e.g., when two or more different organisms contribute to the formation of one trace, or when one organism generates different structures in the same substrate due to changes in behavior; Fig. 2.20), so it is best that ichnological data be used in conjunction with other clues provided by classical paleontology and sedimentology. Integration of all these complementary techniques is therefore the best approach to facies analysis, which allows one to better constrain paleoenvironmental interpretations. A list of basic principles of ichnology is provided in Fig. 2.20.

The fossil record of an ichnocoenose, which is an association of environmentally related traces, is defined as an *ichnofacies* (e.g., Seilacher, 1964, 1967; Pemberton and MacEachern, 1995). Furthermore, besides the actual types of trace fossils, their abundance and disposition are also used to characterize the texture and internal structure of a deposit, which defines the concept of *ichnofabric* (Bromley and Ekdale, 1984). Lateral and vertical shifts in ichnofacies and ichnofabrics are generally used to interpret changes in space as well as through time in paleodepositional environments, based on the inferred shifts in paleoecological conditions.

The concept of ichnofacies, which is central to ichnology, was developed originally based on the observation that many of the environmental factors that control the distribution of traces change progressively with increased water depth (Seilacher, 1964, 1967). It is important to realize, however, that the ecology of an environment reflects the interplay of a multitude of factors (Fig. 2.20), and therefore the types and number

FIGURE 2.20 Basic principles of ichnology (compiled from Seilacher, 1964, 1978; Ekdale *et al.*, 1984; Frey *et al.*, 1987; Pemberton *et al.*, 2001).

Basic principles of Ichnology:

1. Trace fossils generally reflect the activity of *soft-bodied organisms*, which commonly lack hard (preservable) body parts. In many environments, such organisms represent the dominant component of the biomass.

2. Trace fossils may be classified into structures reflecting *bioturbation* (disruption of original stratification or sediment fabric: e.g., tracks, trails, burrows); *biostratification* (stratification created by organism activity: e.g., biogenic graded bedding, biogenic mats); *biodeposition* (production or concentration of sediments by organism activity: e.g., fecal pellets, products of bioerosion); or *bioerosion* (mechanical or biochemical excavation by an organism into a substrate: e.g., borings, gnawings, scrapings, bitings).

3. Trace fossils reflect *behavior patterns*, and so they have *long temporal ranges*. This hampers biostratigraphic dating, but facilitates paleoecological comparisons of rocks of different ages. Basic behavior patterns include resting, locomotion, dwelling and feeding, all of which can be combined with escape or equilibrium structures.

4. Trace fossils are sensitive to water energy (hence, they may be used to recognize and correlate event beds), substrate coherence, and other *ecological parameters* such as salinity, oxygen levels, sedimentation rates, luminosity, temperature, and the abundance and type of nutrients.

5. Behavior patterns depend on ecological conditions, which in turn relate to particular depositional environments. Hence, trace fossils tend to have a *narrow facies range*, and can be used for interpretations of *paleo-depositional environments*.

6. Trace fossils tend to be *enhanced by diagenesis*, as opposed to physical or chemical structures which are often obliterated by dissolution, staining or other diagenetic processes.

7. An individual trace fossil may be the product of *one organism* (easier to interpret), or the product of *two or more different organisms* (composite structures, more difficult to interpret).

8. An individual organism may generate *different structures* corresponding to different behavior in similar substrates, or to identical behavior in different substrates. At the same time, *identical structures* may be generated by different organisms with similar behavior.

of organisms that inhabit a particular area (and implicitly the resultant ichnofacies and ichnofabrics) do not necessarily translate into specific water depths, distance from shore, or tectonic or physiographic setting (Frey *et al.*, 1990; Pemberton and MacEachern, 1995). For example, the *Zoophycos* ichnofacies, typically formed under deeper-marine conditions, below the storm wave base, may also be found in other oxygen-poor settings such as restricted lagoons in coastal environments (Pemberton and MacEachern, 1995). This suggests that caution needs to be used when attempting to interpret absolute or relative paleobathymetry based on ichnofacies sequences, or to establish the syndepositional transgressive or regressive shifts of the shoreline.

Ichnofacies Classification

The classification of trace fossil assemblages (i.e., ichnofacies) is primarily based on substrate type and consistency, and has a direct bearing on paleoenvironmental interpretations (Fig. 2.21). The ichnofacies in Fig. 2.21 are listed in order of increasing marine influences, from fully nonmarine to marginal-, shallow-, and

deep-marine environments. The basic substrate types used in the classification of ichnofacies include *softgrounds* (either shifting or stable, but generally unconsolidated), *firmgrounds* (semi-consolidated substrates, which are firm but unlithified), *hardgrounds* (consolidated, or fully lithified substrates), and *woodgrounds* (*in situ* and laterally extensive carbonaceous substrates, such as peats or coal seams). Figure 2.21 shows that only three ichnofacies are substrate dependent (or 'substrate-controlled'; Ekdale *et al.*, 1984; Pemberton *et al.*, 2001), being associated with a specific substrate type (i.e., the *Teredolites* ichnofacies forms only on woodgrounds; the *Trypanites* ichnofacies is diagnostic for hardgrounds; and the *Glossifungites* ichnofacies indicates firmgrounds), whereas the rest of eight ichnofacies form on a variety of softground substrates, ranging from nonmarine to marginal-marine and fully marine, as a function of ecological conditions. On practical grounds, ichnofacies may therefore be broadly classified into two main groups, i.e., a softground-related group and a substrate-controlled group. As explained below, these two groups imply different genetic interpretations (e.g., conformities *vs.* unconformities), so the

Substrate	Ichnofacies	Environment		Trace fossils
Softground, *nonmarine*	Termitichnus	Subaerial	No flooding: paleosols developed on low watertable alluvial and coastal plains	*Termitichnus, Edaphichnium, Scaphichnium, Celliforma, Macanopsis, Ichnogyrus*
	Scoyenia	Freshwater	Intermittent flooding: shallow lakes or high watertable alluvial and coastal plains	*Scoyenia,* vertebrate tracks
	Mermia		Fully aquatic: shallow to deep lakes, fjord lakes	*Mermia, Gordia, Planolites, Cochlichnus, Helminthopsis, Palaeophycus, Vagorichnus*
Woodground	Teredolites	Marginal marine	Estuaries, deltas, backbarrier settings, incised valley fills	*Teredolites, Thalassinoides*
Softground, *marginal marine*	Psilonichnus		Backshore ± foreshore	*Psilonichnus, Macanopsis*
Hardground	Trypanites	Marginal marine to marine	Foreshore - shoreface - shelf	*Caulostrepsis, Entobia,* echinoid borings (unnamed), *Trypanites*
Firmground	Glossifungites			*Gastrochaenolites, Skolithos, Diplocraterion, Arenicolites, Thalassinoides, Rhizocorall.*
Softground, *marine*	Skolithos	Marine	Foreshore - shoreface	*Skolithos, Diplocraterion, Arenicolites, Ophiomorpha, Rosselia, Conichnus*
	Cruziana		Lower shoreface - inner shelf	*Phycodes, Rhizocorralium, Thalassinoides, Planolites, Asteriacites, Rosselia*
	Zoophycos		Outer shelf- slope	*Zoophycos, Lorenzinia, Spirophyton*
	Nereites		Slope - basin floor	*Paleodictyon, Helminthoida, Taphrhelminthopsis, Nereites, Cosmorhaphe, Spirorhaphe*

FIGURE 2.21 Classification of ichnofacies based on substrate type and consistency, as well as depositional environment (modified from Bromley *et al.*, 1984, and Pemberton *et al.*, 2001).

distinction is important for stratigraphic (and sequence stratigraphic) analyses.

Softground-related Ichnofacies

Softground substrates generally indicate active sediment accumulation (low to high rates) on moist to fully subaqueous depositional surfaces, and hence are associated with conformable successions. The only exception to this general trend is potentially represented by the mature paleosols of the *Termitichnus* ichnofacies, where pedogenesis on sediment-starved landscapes under low water table conditions may result in stratigraphic hiatuses in the rock record. All other softground ichnofacies are associated with the presence of water and active sediment aggradation. Softgrounds may be broadly classified into nonmarine, marginal-marine, and fully marine, as a function of location within the basin (Fig. 2.21).

Besides the *Termitichnus* ichnofacies, which forms under fully subaerial conditions, the other nonmarine softground ichnofacies require the presence of freshwater, at least to some degree. The *Scoyenia* ichnofacies is intermediate between subaerial and fully aquatic nonmarine environments, being indicative of a fluctuating water table (emergence—submergence cycles) such as in the case of floodplains, ephemeral lakes, or wet interdune areas in an eolian system. Under these conditions, the *Scoyenia* ichnofacies is associated with a moist to wet substrate consisting of argillaceous to sandy sediment (Pemberton *et al.*, 2001). The *Mermia* ichnofacies, the third and last in the nonmarine softground series (Fig. 2.21), forms on noncohesive and fine-grained substrates in fully aquatic (perennial) lacustrine environments (Pemberton *et al.*, 2001).

Marginal-marine softground substrates are represented by the *Psilonichnus* ichnofacies, which is typical for backshore (supratidal) environments. Such settings are subject to intermittent marine flooding and hence high fluctuations in energy levels, being dominated by marine processes during spring tides and storm surges, and by eolian processes during neap tides and fairweather. As a result, the sediment composition of the substrate also varies greatly, from muds, silts, and immature sands to mature, well-sorted sands with a variety of physical and biogenic sedimentary structures. Due to the occasional high energy levels, the marginal-marine softgrounds are considered as 'shifting substrates' (Pemberton and MacEachern, 1995), as clastic particles are often reworked by currents, waves or wind. The *Psilonichnus* ichnofacies may be intergradational with the *Scoyenia* ichnofacies, on the nonmarine side of its environmental range, and with the *Skolithos* ichnofacies towards the foreshore (Pemberton *et al.*, 2001).

Marine softgrounds indicate sediment aggradation on an unconsolidated seafloor, where sediments are shifting or are stable as a function of environmental energy. As a general trend, the water energy levels (as reflected by the action of waves and currents), as well as the grain size of sediment and the sedimentation rates, decrease from the shoreline towards the deep-sea basin floor in parallel with increasing water depths (Seilacher, 1964, 1967). However, exceptions to this trend may be caused by gravity-flow events, which may bring coarser sediment and increased energy levels to deep sea settings (slope, basin floor) that are otherwise dominated by low energy, pelagic sedimentation from suspension. Shifting softground substrates may therefore occur in any marine subenvironment, from shallow to deep, although statistically they are much more common in shoreface and adjacent (coastal and inner shelf) settings. Sedimentation rates on softground substrates may vary greatly, from very low to high, as a function of sediment supply and energy conditions. Condensed sections are at the lower end of the spectrum, and only some of them qualify as softground substrates, where the rates of submarine cementation do not outpace the sedimentation rate (Bromley, 1975). In many cases, however, condensed sections may be semilithified or even lithified (Loutit *et al.*, 1988), in which case they become firmgrounds or even hardgrounds. It can be concluded that softground substrates require a minimum rate of sediment accumulation, which needs to be higher than the rate of submarine seafloor cementation, and so they are indeed indicative of conformable successions.

Varying ecological conditions within a marine basin allow for the formation of four distinct ichnofacies on marine softgrounds (Fig. 2.21). The *Skolithos* ichnofacies commonly forms in foreshore to shoreface environments, where the energy level of waves and currents is relatively high, and the substrate consists of shifting particles of clean, well-sorted sand (Pemberton *et al.*, 2001; Fig. 2.22). The *Cruziana* ichnofacies is characteristic of the inner shelf, possibly extending into immediately adjacent subenvironments (lower shoreface and outer shelf), where energy levels are moderate to low and the sediment on the seafloor is generally poorly sorted, consisting of any relative amounts of mud, silt, and sand (Fig. 2.23). This ichnofacies forms on shifting to stable substrates, depending on water energy levels (Pemberton and MacEachern, 1995). Within the *Cruziana* environmental range, the highest energy and proportion of sand are recorded above the fairweather wave base, on a shifting particulate substrate, whereas both the energy and the sand content of the seafloor sediment decrease towards and below the storm wave base, where the substrate becomes more stable. The *Zoophycos* ichnofacies is typically seen, according to the bathymetric schemes, as intermediate between *Cruziana* and the deep-marine *Nereites*, on stable and poorly oxygenated seafloors that are below the storm wave base and free of gravity flows (Seilacher, 1967). Such environmental conditions often occur on outer shelves and continental slopes, where the substrate is composed mainly of fine-grained sediments (Figs. 2.21 and 2.24). While this view is generally valid, one must keep in mind that the *Zoophycos* ichnofacies has a much broader bathymetric range, extending basically to all quiet-water environments that are characterized by low oxygen levels and high organic content (reducing conditions) (Seilacher, 1978; Frey and Seilacher, 1980). In this context, *Zoophycos* traces may encompass a wide environmental range, from the deep sea settings illustrated in Fig. 2.21, to shallow-water epeiric basins and restricted coastal (back barrier) lagoons (Kotake, 1989, 1991; Frey *et al.*, 1990; Olivero and Gaillard, 1996; Uchman and Demircan, 1999). For this reason, Pemberton *et al.* (2001) speculate that the *Zoophycos* tracemaker was broadly adapted to a wide range of water depths and nutrient types, forming perhaps the most ecologically tolerant and environmentally versatile ichnofacies among all eleven shown in Fig. 2.21. This is also reflected by the fact that the *Zoophycos* ichnofacies is often intergradational with the *Cruziana* and *Nereites* traces assemblages (Crimes *et al.*, 1981). In contrast, the *Nereites* ichnofacies, the last in the marine softground series, has the least equivocal bathymetric implications, being indicative of deep sea environments ranging from slope to basin-floor settings, where suspension sedimentation alternates with the manifestation of gravity flows. This environment is characterized by mostly quiet but oxygenated water, periodically disrupted by the

FIGURE 2.22 *Skolithos* ichnofacies. A—*Skolithos* traces (Mississippian Etherington Formation, Jasper National Park, Alberta); B—*Ophiomorpha* traces on a bedding plane in shoreface to wave-dominated delta front deposits of Eocene age (Sunset Cove, Oregon; photo courtesy of M.K. Gingras); C—distal *Skolithos* ichnofacies: *Ophiomorpha* traces in core overprinting *Planolites*-dominated burrow mottling. Mud rip-up clasts are also present in the central part of the core. The core is deviated, and bedding is interpreted to be primarily horizontal (Cretaceous, east coast of Canada; photo courtesy of M.K. Gingras).

turbulence brought about by gravity-flow events, and by the scarcity of nutrients (Pemberton and MacEachern, 1995). Due to the potential diversity of sedimentary processes and sediment sources, the substrate lithology may also vary greatly, from pelagic and hemipelagic to turbidite silts and sands. Changes in the sand to mud ratio of the softground substrate in this deep-marine setting may vary significantly both laterally and vertically, as a function of a multitude of factors including sediment supply, basin physiography, and

base-level changes (a subject tackled in more detail in Chapters 5 and 6). Among all the defining features of the *Nereites* ichnofacies and environment, the water depth and the energy-related ecological factors seem to be more important than the manifestation of gravity-flow processes. This is argued by the fact that *Nereites* traces can be found not only within the confines of submarine fans, but also on distal basin floors, beyond the reach of gravity flows (Crimes *et al.*, 1981; Leszczynski and Seilacher, 1991; Miller, 1993).

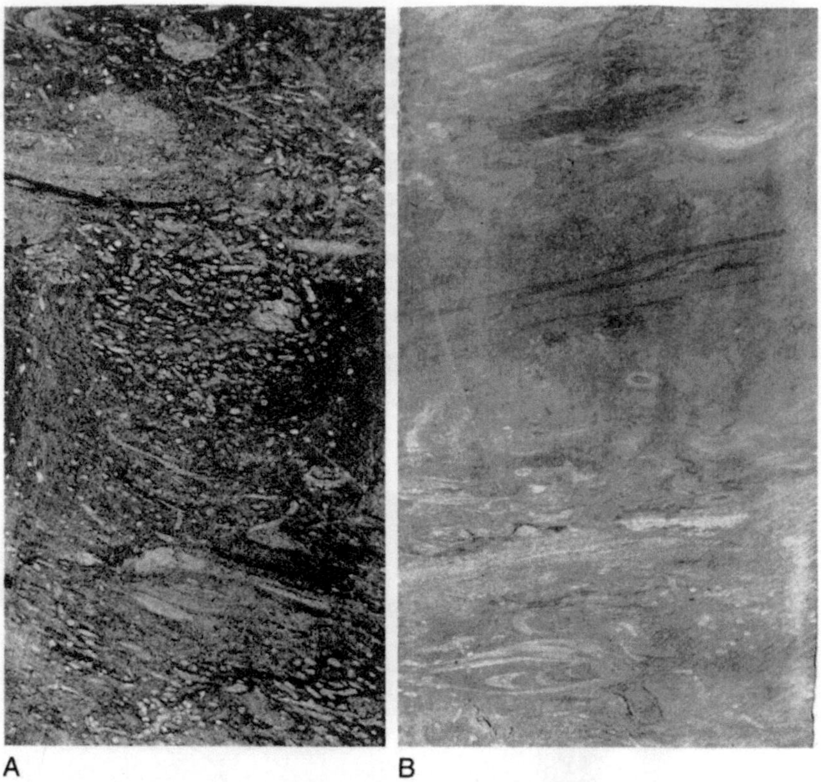

FIGURE 2.23 *Cruziana* to *Zoophycos* (A) and *Cruziana* (B) ichnofacies (Hibernia oilfield, eastern Canadian offshore; photos courtesy of M.K. Gingras). A—the core shows abundant *Chondrites*, *Zoophycos*, and reburrowed *Thalassinoides* traces; B—the core shows *Rhizocorallium* near the base, some white-rimmed *Terebellina*, and mottling due to *Planolites*. *Chondrites* traces are also present.

A   B

Where gravity-flow deposits are present, the *Nereites* ichnofacies may include distinct populations of pre-gravity flow traces, produced by stable communities adapted to low energy conditions, and also post-gravity flow traces formed under turbid conditions, generated by a less stable community that originates from shallower water. As turbidity gives way to a more 'normal', low energy environment, the pre-gravity flow community colonizes the softground substrate again (Frey and Seilacher, 1980).

Substrate-controlled Ichnofacies

The remaining three ichnofacies (*Glossifungites*, *Trypanites*, and *Teredolites*) are distinctly different from the softground-related group discussed above, in the sense that they are dependent on specific substrate types (firmgrounds, hardgrounds, and woodgrounds; Fig. 2.21). This substrate-controlled group is particularly important for stratigraphic analyses, as being most commonly associated with unconformities in the rock record. The substrate-controlled tracemakers populate *resilient* (as opposed to *soft*) substrates which are either erosionally exhumed (in the majority of cases) or simply the product of various processes during times of sediment starvation (nondeposition). In either case, firmgrounds, hardgrounds, and woodgrounds

mark the presence of stratigraphic hiatuses (± erosion) in the rock record. Such unconformities may practically be generated in any environment, from subaerial to subaqueous, but the actual colonization of the surface is regarded to reflect marine influence, particularly in pre-Tertiary times (Pemberton *et al.*, 2001). This fact has important implications for sequence stratigraphy, as far as the genetic interpretation of unconformities is concerned (MacEachern *et al.*, 1991, 1992, 1998, 1999; Pemberton *et al.*, 2001).

The *Glossifungites* ichnofacies (Figs. 2.25 and 2.26) develops on semi-cohesive (firm, but unlithified) substrates, best exemplified by dewatered muds. The process of dewatering takes place during burial, and subsequent erosional exhumation makes the substrate available to tracemakers (MacEachern *et al.*, 1992). Such erosion may occur in a variety of settings, from fluvial (e.g., caused by channel avulsion or valley incision) to shallow-water (e.g., tidal channels or wave erosion) and deeper-water (e.g., submarine channels eroding the seafloor) environments (Hayward, 1976; Fursich and Mayr, 1981; Pemberton and Frey, 1985). Despite the wide range of environments in which unconformities may form, firmground assemblages have only rarely been described from nonmarine successions (e.g., Fursich and Mayr, 1981), originating in their vast

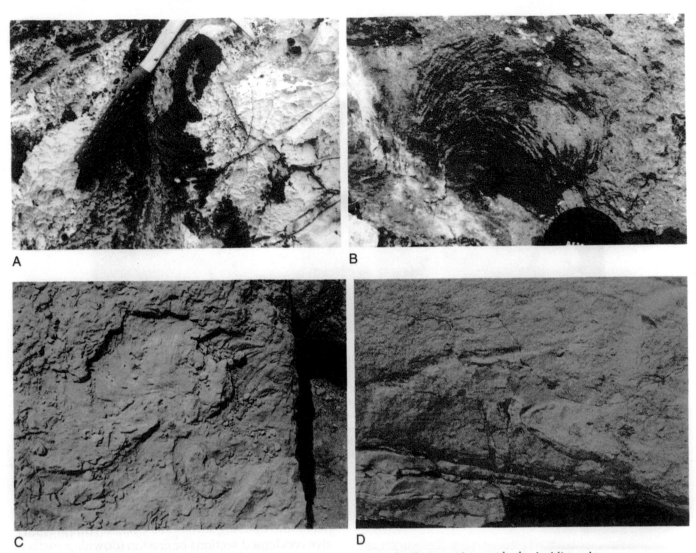

FIGURE 2.24 *Zoophycos* ichnofacies. A—*Zoophycos* trace fossil, concordant with the bedding plane (Mississippian Etherington Formation, Jasper National Park, Alberta); B—*Zoophycos* trace fossil, concordant with the bedding plane (Mississippian Shunda Formation, Jasper National Park, Alberta); C—*Zoophycos* trace fossil, concordant with the bedding plane (Cretaceous Wabiskaw Member of the Clearwater Formation, Fort McMurray area; photo courtesy of M.K. Gingras); *Zoophycos* ichnofacies, including *Zoophycos* and *Chondrites* in cross sectional view (Cretaceous Wabiskaw Member of the Clearwater Formation, Fort McMurray area; photo courtesy of M.K. Gingras).

majority in marine and marginal-marine settings, particularly in pre-Tertiary times (Pemberton *et al.*, 2001). Even though most of the firmgrounds are genetically linked to erosional processes, there are also cases where firmground assemblages form on semi-lithified condensed sections, where nondepositional breaks allow for early submarine cementation of seafloor sediments (Bromley, 1975). In such cases, the seafloor may be colonized by the *Glossifungites* ichnofacies tracemakers without the intervention of erosion. In the majority of studies, however, the *Glossifungites*

assemblage is found to be associated with erosionally exhumed substrates, indicating scour surfaces (MacEachern *et al.*, 1992; Gingras *et al.*, 2001). As indicated in Fig. 2.21, the *Glossifungites* ichnofacies has a relatively wide environmental spectrum, commonly ranging from marginal-marine to shallow-marine settings. From a sequence stratigraphic perspective, the *Glossifungites* ichnofacies may relate to scour surfaces cut by tidal currents in transgressive settings, waves in subtidal transgressive or forced regressive settings, incised valleys or submarine canyons, or

A   B

FIGURE 2.25 *Glossifungites* ichnofacies (photos courtesy of M.K. Gingras). A—*Glossifungites* ichnofacies at the base of a tidal channel fill (arrow). The photograph shows *Thalassinoides* burrows descending into underlying intertidal deposits. The firmground has the significance of a transgressive tidal-ravinement surface, and is overlain by tidal channel fill and estuary channel point bar deposits with inclined heterolithic strata (Pleistocene section, Willapa Bay, Washington); B—*Glossifungites* ichnofacies in a modern intertidal environment. The photograph shows burrows of *Upogebia pugettensis* (mud shrimp) descending into firm Pleistocene strata. The firmground is overlain by a thin veneer of unconsolidated (modern) mud, and has the significance of a transgressive tidal-ravinement surface (Goose Point at Willapa Bay, Washington).

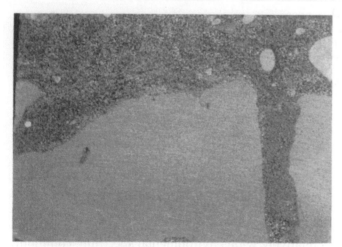

FIGURE 2.26 *Glossifungites* ichnofacies marking a transgressive wave-ravinement surface at the base of the Lower Albian Wilrich Member, Western Canada Sedimentary Basin (photo courtesy of M.K. Gingras). The firmground is associated with *Skolithos* and/or *Thalassinoides* burrows, and is overlain by transgressive glauconitic sand and chert pebble lag deposits. The wave-ravinement surface truncates the top of the dominantly nonmarine Gething Formation.

maximum flooding surfaces associated with transgressive condensed sections or erosion (downlap surfaces, cf. Van Wagoner *et al.*, 1990).

The *Trypanites* ichnofacies (Fig. 2.27) may form on a variety of fully lithified substrates, including rocky coasts, reefs, fully cemented condensed sections (hardgrounds), or any type of exhumed bedrock (Pemberton *et al.*, 2001). Most often, hardground substrates are associated with significant stratigraphic hiatuses (± erosion) and hence are important for the delineation of unconformities in the rock record, and implicitly for sequence stratigraphy. The generation or exposure of fully lithified substrates, such as the erosional exhumation of the bedrock for example, may take place in any environment, from subaerial to subaqueous. The colonization of such substrates, however, which leads to the formation of the *Trypanites* ichnofacies, is commonly the product of marine transgression, and therefore this trace fossil assemblage may be associated with transgressive tidal- or wave-ravinement surfaces, or with maximum flooding surfaces on the shelf. The environmental

FIGURE 2.27 *Trypanites* ichnofacies (photos courtesy of M.K. Gingras). A—large *Gastrochaenolites* traces, which are borings made by pholad bivalves into the base of a Pleistocene-age tidal channel. The channel fill is composed of organic-rich, unconsolidated sediment (dark color in the photograph). The underlying rock is a Miocene shoreface succession that belongs to the Empire Formation at Coos Bay, Oregon. The base of the channel corresponds to a transgressive tidal-ravinement surface; B—modern intertidal environment. The traces shown are *Gastrochaenolites*. The hardground occurs as a scour cut into Triassic bedrock by tidal currents, and has the significance of a transgressive tidal-ravinement surface. Boring density may locally exceed 1250 borings per square meter. Location is near Economy, Nova Scotia (Bay of Fundy, Minas Basin); C—modern intertidal environment (detail from B). The photograph shows the borings, the grooves cut by the bivalve (bioglyphs), and the tracemaker itself (*Zirfea pilsbyri*).

range of the *Trypanites* ichnofacies is thus relatively wide, similar to the *Glossifungites* assemblage (Fig. 2.21).

The *Teredolites* ichnofacies (Fig. 2.28) develops on woody substrates (woodgrounds), most commonly represented by driftwood pavements, peat or coal horizons (Bromley *et al.*, 1984; Savrda, 1991; Pemberton *et al.*, 2001). The woodgrounds themselves form in nonmarine to marginal-marine settings, and may or may not require erosional exhumation prior to colonization. The population of tracemakers that generates the *Teredolites* ichnofacies is distinctly different between freshwater (isopod borings) and marine-influenced

settings (wood-boring bivalves), with the latter being the dominant type of woodground assemblage. Woodground substrates are also resilient, as are the hardgrounds, but differ from the latter in terms of their organic nature. This characteristic makes them more flexible and readily biodegradable relative to the lithic substrates (Bromley *et al.*, 1984). In the majority of cases, the *Teredolites* ichnofacies is found in marginal-marine settings (Fig. 2.21), where shoreline transgression brings marine tracemakers on top of woodgrounds (e.g., peat or coal seams) previously formed in nonmarine environments. In this context,

FIGURE 2.28 *Teredolites* ichnofacies in a modern intertidal environment (Willapa Bay, Washington; photo courtesy of M.K. Gingras). The borings are sand-filled, which provides their typical mode of preservation, and are made by the terenid bivalve *Bankia*. The woodground has the significance of a transgressive tidal-ravinement surface. The association between *Teredolites* and transgressive coastlines is generally valid for both *in situ* and allochthonous woodgrounds.

and due to the resilient nature of woodground substrates, the *Teredolites* ichnofacies may be preserved below transgressive tidal- or wave-ravinement surfaces. Where the *Teredolites* ichnofacies is present at the base of an incised-valley fill, it provides evidence that the tidal- or wave-ravinement surface reworks the sequence boundary, and therefore that the valley-fill deposits are transgressive. Such analyses are important in sequence stratigraphy, as the nature of incised-valley fills (regressive *vs.* transgressive) has long been subject to debate (Embry, 1995; Emery and Myers, 1996; Posamentier and Allen, 1999). The proper identification of the *Teredolites* ichnofacies requires evidence that the woodground borings are *in situ*, as opposed to allochthonous, single pieces of xylic material (Arua, 1989; Dewey and Keady, 1987). Even in the latter case, however, recent work on modern coastline settings suggests that the most common occurrence of bored xylic clasts is from brackish to marine tidal channels, being thus associated with transgressive tidal-ravinement surfaces (Gingras *et al.*, 2004; Fig. 2.28).

Discussion

It is important to note that many individual trace fossils are common amongst different ichnofacies. For example, *Planolites* may be part of both *Mermia* (freshwater) and *Cruziana* (sea water) assemblages, *Thalassinoides* may populate softground, firmground, or woodground substrates, etc. (Fig. 2.21). Hence, the

context and the association of traces, coupled with additional clues provided by physical textures and structures, need to be used in conjunction for the proper interpretation of stratigraphic surfaces and paleodepositional environments.

In conclusion, the relevance of ichnology to sequence stratigraphy is two fold (Pemberton and MacEachern, 1995). *Softground-related ichnofacies*, which generally form in conformable successions, assist with the interpretation of paleodepositional environments and changes thereof with time. The vertical shifts in softground assemblages are governed by the same Walther's Law that sets up the principles of lateral and vertical facies variability in relatively conformable successions of strata, and therefore can be used to decipher paleodepositional trends (progradation *vs.* retrogradation) in the rock record. The recognition of such trends, which in turn relate to the regressions and transgressions of paleoshorelines, is central to any sequence stratigraphic interpretation. *Substrate-controlled ichnofacies*, which are genetically related to stratigraphic hiatuses, assist with the identification of unconformities in the rock record, and thus too have important applications for sequence stratigraphy. The actual type of unconformable sequence stratigraphic surface can be further evaluated by studying the nature and relative shift directions of the facies which are in contact across such omission surfaces. These aspects are presented in more detail in Chapter 4, which deals with stratigraphic surfaces. As stressed before, each individual method of facies analysis may be equivocal to some extent, so the integration of ichnology with conventional biostratigraphy and sedimentology provides an improved approach to facies analysis and sequence stratigraphy.

WELL LOGS

Introduction

Well logs represent geophysical recordings of various rock properties in boreholes, and can be used for geological interpretations. The most common log types that are routinely employed for facies analyses (lithology, porosity, fluid evaluation) and stratigraphic correlations are summarized in Fig. 2.29. Most of these log types may be considered 'conventional', as having been used for decades, but as technology improves, new types of well logs are being developed. For example, the new micro-resistivity logs combine the methods of conventional resistivity and dipmeter measurements to produce high-resolution images that simulate the

Log	Property measured	Units	Geological interpretation
Spontaneous potential	Natural electric potential (relative to drilling mud)	Millivolts	Lithology, correlation, curve shape analysis, porosity
Conventional resistivity	Resistance to electric current flow (1D)	Ohm-metres	Identification of coal, bentonites, fluid types
Micro resistivity	Resistance to electric current flow (3D)	Ohm-metres and degrees	Borehole imaging, virtual core
Gamma ray	Natural radioactivity (e.g., relater to K, Th, U)	API units	Lithology (including bentonites, coal), correlation, shape analysis
Sonic	Velocity of compressional sound wave	Microseconds/metre	Identification of porous zones, tightly cemented zones, coal
Neutron	Hydrogen concentration in pores (water, hydrocarbons)	Per cent porosity	Porous zones, cross plots with sonic and density for lithology
Density	Bulk density (electron density) (includes pore fluid in measurement)	Kilograms per cubic metre (g/cm^3)	Lithologies such as evaporites and compact carbonates
Dipmeter	Orientation of dipping surfaces by resistivity changes	Degrees (azimuth and inclination)	Paleoflow (in oriented core), stratigraphic, structural analyses
Caliper	Borehole diameter	Centimetres	Borehole state, reliability of logs

FIGURE 2.29 Types of well logs, properties they measure, and their use for geological interpretations (modified from Cant, 1992).

sedimentological details of an actual core. Such 'virtual' cores allow visualization of details at a millimeter scale, including sediment lamination, cross-stratification, bioturbation, etc., in three dimensions (Fig. 2.30).

Well logs have both advantages and shortcomings relative to what outcrops have to offer in terms of facies data. One major advantage of geophysical logs over outcrops is that they provide *continuous* information from relatively thick successions, often in a range of kilometers. This type of profile (log curves) allows one to see trends at various scales, from the size of individual depositional elements within a depositional system, to entire basin fills. For this reason, data provided by well logs may be considered more complete relative to the *discontinuous* information that may be extracted from the study of outcrops. Therefore, the subsurface investigations of facies relationships and stratigraphic correlations can usually be accomplished at scales much larger than the ones possible from the study of outcrops. On the other hand, nothing can replace the study of the actual rocks, hence the wealth of details that can be obtained from outcrop facies analysis cannot be matched by well-log analysis, no matter how closely spaced the boreholes may be (Cant, 1992).

FIGURE 2.30 New micro resistivity logs combine resistivity with dipmeter data to produce 'virtual cores' in three dimensions. Such detailed borehole imaging, with a vertical resolution of less than 8 mm, allows the observation of sedimentary structures in the absence of mechanical core (modified from data provided by Baker Atlas).

Well Logs: Geological Uncertainties

Well logs provide information on physical rock properties, but not a direct indication of lithology. Spontaneous potential and gamma ray logs are commonly used for the interpretation of siliciclastic successions in lithological terms, but one must always be aware of the potential pitfalls that may occur 'in translation'. Changes in rock porosity and pore-water chemistry (fresh *vs*. sea water) may induce different responses on spontaneous potential logs, including deflections in opposite directions, even if the lithology is the same. Similarly, gamma ray logs are often interpreted in grading terms (fining- *vs*. coarsening-upward), or worse, as it adds another degree of unconstrained interpretation, in bathymetric terms (deepening- *vs*. shallowing-upward trends). In reality, gamma ray logs simply indicate the degree of strata radioactivity, which is generally proportional to the shaliness of the rocks and/or the amount of organic matter.

Zones of high gamma ray response may correspond to a variety of depositional settings, from shelf and deeper-marine to coastal plains, backshore marshes and lacustrine environments. In fully subaqueous settings (marine or lacustrine), high gamma ray responses correlate to periods of restricted bottom-water circulation and/or with times of reduced sediment supply. Such periods favor the formation of condensed sections, which most commonly are associated with stages of shoreline transgression, and hence with maximum flooding surfaces (Galloway, 1989). However, due to the wide variety of environments which may result in the accumulation and preservation of organic matter and/or fine-grained sediment, the mere identification of high gamma ray zones is not sufficient to unequivocally identify condensed sections (Posamentier and Allen, 1999). At the same time, condensed sections may also be marked by a variety of chemical and biochemical precipitates formed during times of sediment starvation (e.g., siderite, glauconite, carbonate hardgrounds, etc.), thus exhibiting a wide range of log motifs which may not necessarily fit the classic high peaks on gamma ray logs (Posamentier and Allen, 1999).

The equivocal character of well logs, when it comes to geological interpretations, is also exemplified by the fact that fundamentally different depositional systems may produce similar log motifs. Figure 2.31 illustrates such an example, where comparable blocky log patterns formed in fluvial, estuarine, beach, shallow-marine, and deep-marine environments. Similarly, jagged log patterns are not diagnostic of any particular depositional system, and may be found all the way from fluvial to delta plain, inner shelf, and deep-water (slope to basin-floor) settings (Fig. 2.32). Such jagged log motifs simply indicate fluctuating energy conditions leading to the deposition of alternating coarser and finer sediments (heterolithic facies), conditions which can be met in many nonmarine, marginal-marine, shallow-marine, and deep-marine environments. Monotonous successions dominated by fine-grained sediments may also be common among different depositional systems, including deep-water 'overbanks' (areas of seafloor situated outside of channel-levee complexes or splay elements) and outer shelf settings (Fig. 2.33).

At the same time, one and the same depositional system may display different well-log signatures as a function of variations in depositional energy, sediment

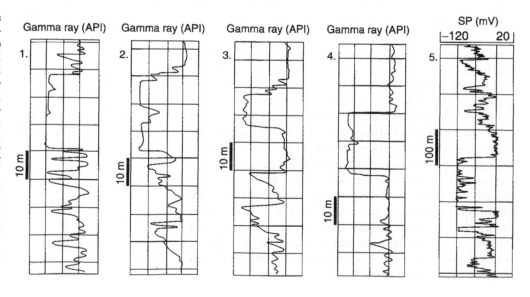

FIGURE 2.31 Well logs from five different siliciclastic depositional systems, each including a 'blocky' sandstone unit. 1—fluvial channel fill; 2—estuarine channel fill; 3—sharp-based shoreface deposits; 4—deep-water channel filled with turbidites; and 5—beach deposits (modified from Posamentier and Allen, 1999, and Catuneanu *et al.*, 2003a). Note the potentially equivocal signature of depositional systems on well logs. For this reason, the correct interpretation of paleodepositional environments requires integration of multiple data sets, including core, rock cuttings, biostratigraphy, and seismics. Abbreviation: SP—spontaneous potential.

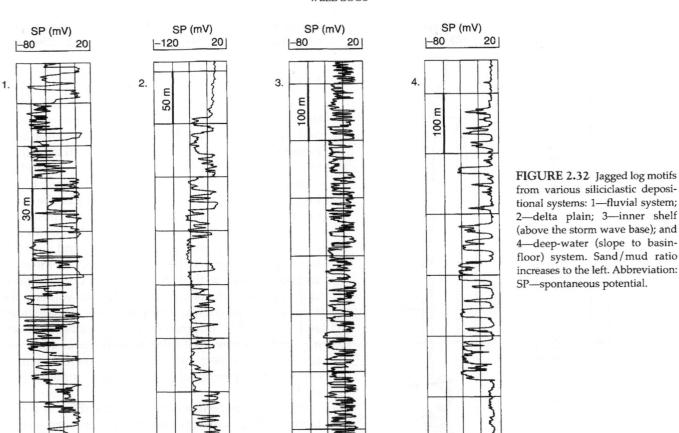

FIGURE 2.32 Jagged log motifs from various siliciclastic depositional systems: 1—fluvial system; 2—delta plain; 3—inner shelf (above the storm wave base); and 4—deep-water (slope to basin-floor) system. Sand/mud ratio increases to the left. Abbreviation: SP—spontaneous potential.

supply, accommodation, etc. For example, typical sand-bed meandering fluvial systems consist of an alternation of fining-upward channel fills and mud-dominated floodplain deposits; braided fluvial systems are often composed of amalgamated channel fills, which confer a blocky pattern to the well logs; in contrast, other types of rivers, including fine-grained meandering or flashy ephemeral, produce a more irregular, jagged type of motif on well logs (Fig. 2.34). Relatively thin (± meter scale) coarsening-upward trends may also be observed in fluvial successions in relation to crevasse splays, especially in low-energy and confined meandering-type rivers (Fig. 2.35). Similar to fluvial systems, slope to basin-floor deep-water systems may also generate a variety of log motifs, most commonly jagged or blocky, but also fining-upward and more rarely coarsening-upward, depending on sediment supply, type of sediment transport mechanism (contourites *vs.* gravity flows; types of gravity flows), and actual subenvironment penetrated by well (e.g., channels, levees, splays, etc.) (Fig. 2.36).

Log patterns are therefore diverse, generally indicative of changing energy regimes through time but not necessarily diagnostic for any particular depositional system or architectural element. An entire range of log motifs has been described in the past (e.g., Allen, 1975; Selley, 1978b; Anderson *et al.*, 1982; Serra and Abbott, 1982; Snedden, 1984; Rider, 1990; Cant, 1992; Galloway and Hobday, 1996), but the most commonly recurring patterns include 'blocky' (also referred to as 'cylindrical'), 'jagged' (also referred to as 'irregular' or 'serrated'), 'fining-upward' (also referred to as 'bell-shaped') and 'coarsening-upward' (also known as 'funnel-shaped') (Cant, 1992; Posamentier and Allen, 1999). The *blocky* pattern generally implies a constant energy level (high in clastic systems and low in carbonate environments) and constant sediment supply and sedimentation rates. The *jagged* motif indicates alternating high and low energy levels, such as seasonal flooding in a fluvial system, spring tides and storm surges in a coastal setting, storms *vs.* fairweather in an inner shelf setting, or gravity flows *vs.* pelagic fallout in deep-water environments. *Fining-upward* trends can again be formed in virtually any depositional environment, where there is a decline with time in energy levels or depocenters are gradually shifting away relative to the location under investigation.

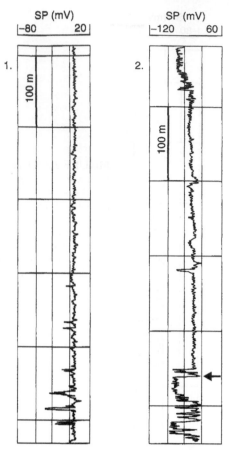

FIGURE 2.33 Examples of fine-grained-dominated successions from 1—deep-water basin-floor setting; and 2—outer shelf (below the storm wave base, but above the shelf edge) setting. Note that the outer shelf muds overlie prograding, shallower-marine (shoreface—inner shelf) deposits. The *association* of well-log facies thus provides important clues for the interpretation of paleodepositional environments. The arrow indicates a flooding event on the shelf.

Finally, the *coarsening-upward* pattern indicates a gradual shift towards the location under investigation of a progressively higher-energy depositional environment. It has been argued that this log motif is the least equivocal of all, especially for repeated sections 5–30 m thick (Posamentier and Allen, 1999). Typical examples of depositional elements that generate this log motif include prograding distributary mouth bars (deltaic settings) or prograding shoreface deposits (open shoreline settings) (Fig. 2.37). One needs to note, however, that a similar log pattern may also characterize crevasse splay deposits in fluvial settings (generally <5 m thick; Fig. 2.35), and even gravity-flow systems in deep-water environments, especially at the distal edge of a seaward-building turbidite lobe (Posamentier and Allen, 1999; Fig. 2.36). In the latter case though, the coarsening-upward trend is closely associated with a jagged

log motif, which provides an additional clue for the identification of the deep-water setting (Fig. 2.36).

Constraining Well-log Interpretations

The discussion in the previous section shows that the well-log interpretation of depositional systems, and implicitly of stratigraphic surfaces, is to a large extent speculative in the absence of actual rock data. Outcrop, core and well cuttings data (including sedimentologic, petrographic, biostratigraphic, ichnologic, and geochemical analyses) provide the most unequivocal 'ground truth' information on depositional systems (Posamentier and Allen, 1999). It follows that geophysical data, including well-log and seismic, which provide only indirect information on the solid and fluid phases in the subsurface, must be calibrated and verified with rock data in order to validate the accuracy of geological interpretations (Posamentier and Allen, 1999). Integration of all available data sets (e.g., outcrop, core, well cuttings, well-log, and seismic) is therefore the best approach to the correct identification of depositional systems and stratigraphic contacts.

Well logs are generally widely available, especially in mature hydrocarbon exploration basins, and so they are routinely used in stratigraphic studies. Seismic data are also available in most cases, as the seismic survey commonly precedes drilling. In the process of drilling, well cuttings are also routinely collected to provide information on lithology, porosity, fluid contents, and biostratigraphy (age and paleoecology). Core material is more expensive to collect, so it is generally restricted to the potentially producing reservoir levels (unless the borehole is drilled for research or exploration reference purposes, and continuous mechanical coring is performed). Nearby outcrops may be available when drilling is conducted in onshore areas, but are generally unavailable where drilling is conducted offshore. It can be concluded that, *to a minimum*, well logs can be analyzed in conjunction with seismic data and well cuttings, and to a lesser extent in combination with cores and outcrops. Constraining well-log interpretations with independent seismic and rock data is a fundamental step towards a successful generation of geological models. For example, two-dimensional seismic data provide invaluable insights regarding the tectonic setting (e.g., continental shelf *vs.* slope or basin floor) and the physiography of the basin. Three-dimensional seismic data add another level of constraint, by providing information about the plan-view morphology of the various depositional systems or elements thereof. Such information, combined with any available rock data, helps to place the well logs in

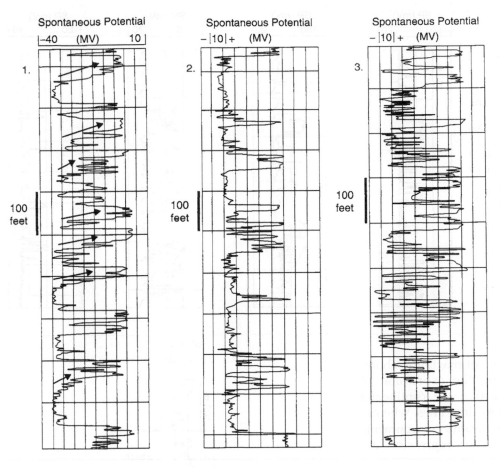

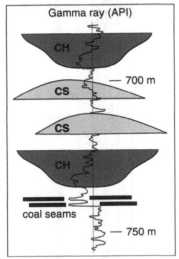

FIGURE 2.34 Log motifs in fluvial settings: 1—fining-upward patterns; 2—blocky patterns; and 3—jagged pattern (redrafted and modified from Posamentier and Allen, 1999). The sand/mud ratio increases to the left.

FIGURE 2.35 Log motifs of a low-energy fluvial system, showing both fining-upward (channel fills—CH) and coarsening-upward (crevasse splays—CS) trends. The example comes from the Horseshoe Canyon Formation (Maastrichtian) in south-central Alberta, Western Canada Sedimentary Basin.

the right context, from both tectonic and paleoenvironmental points of view.

The placement of a study area in the right tectonic and paleoenvironmental setting is crucial for the subsequent steps of well-log analysis. As noted by Posamentier and Allen (1999), 'correct identification of the depositional environment will guide which correlation style to use between wells. Thus, one style of correlation would be reasonable for prograding shoreface deposits, but a very different correlation style would be used for incised-valley-fill deposits, and still another style of correlation would be most reasonable for deep-water turbidites'. The reliability of well-log-based correlations is further improved by the presence of stratigraphic markers, which represent laterally extensive beds or groups of beds with a distinctive log response. Examples of such markers include bentonites and marine condensed sections, both of which are very useful as they (1) help to constrain correlations, and (2) are very close to time lines. Regional coal seams are also useful stratigraphic markers, although their chronostratigraphic significance needs to be assessed on a case-by-case basis.

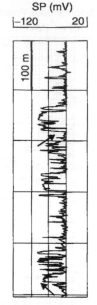

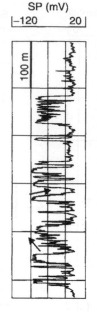
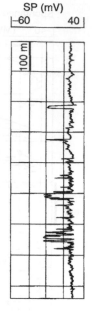

FIGURE 2.36 Log motifs in deep-water settings. Note the dominant jagged and blocky patterns, but occasional fining- and coarsening-upward trends may also be observed. The sand-dominated units in the deep-water systems are generally embedded within a thick succession of fine-grained (pelagic, hemipelagic) sediments. These examples depict a siliciclastic succession, with the logs showing an increased sand/mud ratio to the left.

Coal seams that form at the time of maximum shoreline transgression, representing the expression of maximum flooding surfaces within the continental portion of the basin, are close to time lines, whereas coals that originate from coastal swamp environments during shoreline transgression or regression are commonly time-transgressive.

An example of correlation style in a prograding shoreface setting, with the base of the underlying condensed section taken as the datum, is shown in Fig. 2.38. This correlation style accommodates some basic principles of stratal stacking patterns which are expected in such a setting, including the fact that clinoforms slope seawards and downlap the underlying transgressive shales (condensed section), and that the strata between clinoforms tend to thin and fine in an offshore direction. Without a good understanding of depositional environments and processes thereof, a 'blind' pattern matching exercise may easily lead to errors in interpretation by forcing correlations across depositional time lines (clinoforms in this example). Classic layer-cake models may still work in some cases, where depositional energy and sediment supply are

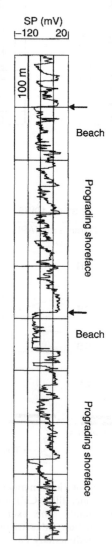

FIGURE 2.37 Coarsening-upward and blocky log motifs in a shallow-marine to coastal environment. Arrows indicate the most important flooding (transgressive) events, but many other less important flooding events are recorded at the top of each coarsening-upward prograding lobe. Sand/mud ratio increases to the left. Abbreviation: SP—spontaneous potential.

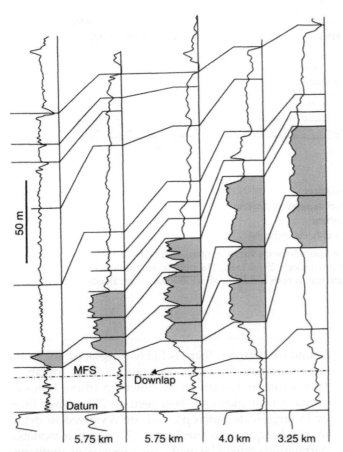

FIGURE 2.38 Gamma-ray cross section from the Upper Mannville Group in British Columbia, showing correlation by pattern matching in a shallow-marine setting (redrawn from Cant, 1992, with permission from the Geological Association of Canada). The correlation lines slope seaward (to the left), marking clinoforms which downlap onto the maximum flooding surface (MFS; top of transgressive shales). Shaded areas represent prograding shoreface sands.

constant over large distances (e.g., some distal basin-floor settings), but most environments tend to produce more complex stratigraphic architectures in response to variations along dip and strike in energy levels and sedimentation patterns. In a marginal-marine setting, for example, the selection of the right correlation approach is greatly facilitated by knowing the shoreline trajectory during that particular time interval—which, in turn, may be inferred from the more regional context constrained with seismic data. A cross section along the dip of a prograding delta would show clinoforms downlapping in a seaward direction (Fig. 2.39), whereas a strike-oriented cross-section may capture deltaic lobes wedging out in both directions (Fig. 2.40; e.g., Berg, 1982).

The analysis of well logs, therefore, may serve several interrelated purposes including, at an increasing scale of observation, the evaluation of rock and fluid phases in the subsurface, the interpretation of paleodepositional environments based on log motifs, and stratigraphic correlations based on pattern matching and the recognition of marker beds. Different scales of observation may therefore be relevant to different objectives. Details at the *smaller scale of individual depositional elements* are commonly used for the petrophysical analysis of reservoirs (lithology, porosity, and fluid evaluation), regardless of the depositional origin of that stratigraphic unit. Such analyses are usually performed by extracting information simultaneously from two or more log types. These 'cross-plots' work particularly well where the succession is relatively homogeneous, consisting of only two or three log types (e.g., mudstones,

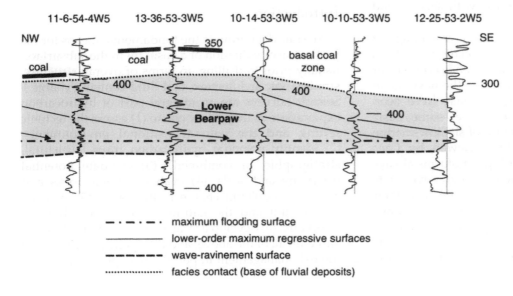

FIGURE 2.39 Dip-oriented gamma-ray cross section through the marine to marginal-marine facies of the Bearpaw Formation in central Alberta. The section is approximately 30 km long. The internal architecture of the formation (shaded area) shows clinoforms prograding to the right (seawards), and downlapping onto the maximum flooding surface which is placed at the top of the transgressive shales. Each prograding lobe corresponds to a lower-order (higher-frequency) transgressive–regressive cycle. The minor maximum flooding surfaces associated with each prograding lobe are not represented. The transgressive facies (fining-upward) are generally thinner than the regressive marine facies (coarsening-upward).

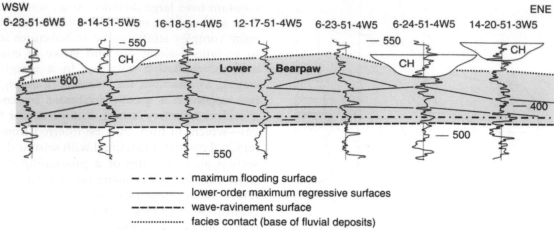

FIGURE 2.40 Strike-oriented gamma-ray cross section through the marine to marginal-marine facies of the Bearpaw Formation in central Alberta. The section is approximately 25 km long. The internal architecture of the formation (shaded area) shows deltaic lobes wedging out in both directions above the maximum flooding surface. Abbreviation: CH—fluvial channel fills.

siltstones, and sandstones) (Miall, 2000). As emphasized earlier, however, the log motifs of individual depositional elements are generally nondiagnostic of the paleoenvironment, and it is rather the *larger-scale context* within which these individual units are observed that allows one to infer the original depositional setting (Posamentier and Allen, 1999). For example, blocky patterns associated with fining-upward depositional elements may reflect a fluvial setting; similar blocky patterns associated with coarsening-upward trends may be indicative of a coastal environment (e.g., Fig. 2.37); and finally, blocky sandstones interbedded with shales may likely be the product of sedimentation in a deep-water setting. But again, all these interpretations based on overall log motifs need to be constrained with seismic and rock data.

The validation, within geological reasoning, of well-log-based cross-sections of correlation has been a fundamental issue for decades, and criteria for connecting the 'kicks' from one log to the next one in ways that make most geological sense have been developed accordingly. For example, some basic 'rules' that apply to the correlation of shallow-marine successions have been recently reviewed by Cant (2004), and include: (1) prograding clinoforms always slope seaward; (2) shallow-marine regressive units tend to have lateral continuity along dip, and their number may only change in the shoreline area; (3) units tend to fine and thin seaward; (4) unit thicknesses do not vary randomly; (5) where superimposed units show complementary thinning and thickening, the boundary between them is likely misplaced; (6) strata may terminate landward by onlap, offlap, toplap or truncation, and seaward by downlap (these types of stratal terminations are best seen on 2D seismic lines or in large-scale outcrops, and are reviewed in detail in a subsequent chapter); and (7) where reasonable correlations cannot be made, the presence of an unconformity may be inferred—such contacts exert an important control on clastic reservoirs, and may have a frequent occurrence in the rock record.

SEISMIC DATA

Introduction

Seismic data provide the fundamental means for the preliminary evaluation of a basin fill in the subsurface, usually prior to drilling, in terms of overall structure, stratigraphic architecture, and fluid content ('charge'). Seismic surveys are an integral part of hydrocarbon exploration, as they allow one to (1) assess the tectonic setting and the paleodepositional environments; (2) identify potential hydrocarbon traps (structural, stratigraphic, or combined); (3) evaluate potential reservoirs and seals; (4) evaluate source rocks and estimate petroleum charge in the basin; (5) evaluate the amount and the nature of fluids in individual reservoirs; (6) develop a strategy for borehole planning based on all of the above; and (7) significantly improve the risk management in petroleum exploration.

The development of seismic exploration techniques allowed for the transition from classical stratigraphy to seismic stratigraphy (the precursor of sequence stratigraphy—see Chapter 1 for a discussion) in the 1970s (Vail, 1975; Payton, 1977), and led to the establishment of the first criteria of interpreting seismic information in seismic stratigraphic and sequence stratigraphic terms (Mitchum and Vail, 1977; Mitchum et al., 1977; Vail et al., 1977). Seismic data have both advantages and shortcomings relative to the outcrop, core or well-log data, as emphasized below, so the integration of all these techniques is critical for mutual calibration and the development of reliable geological models.

In the initial steps of any seismic survey, seismic data are collected along a grid of linear profiles, resulting in the acquisition of two-dimensional (two-way travel time vs. horizontal distance) seismic lines. In modern seismic surveys, the information from this grid of two-dimensional seismic lines is integrated by computer interpolation to produce three-dimensional seismic volumes (Brown, 1991; Fig. 2.41). Following initial acquisition, the raw seismic information requires further processing (e.g., demultiplexing, gain recovery, static corrections, deconvolution, migration, etc.; Hart, 2000) before it is ready to be used for geological interpretations. Once available for analysis, the seismic lines provide *continuous* subsurface information over distances of tens of kilometers and depths in a range of kilometers. The continuous character of seismic data represents a major advantage of this method of stratigraphic analysis over well logs, core or outcrops, which only provide information from *discrete* locations in the basin. There are also shortcomings of the seismic data relative to well logs, core or outcrops, mainly in terms of vertical resolution (thinnest package of strata that can be recognized as such on seismic lines) and the nature of information (physical parameters as opposed to direct geology) that is represented on seismic lines.

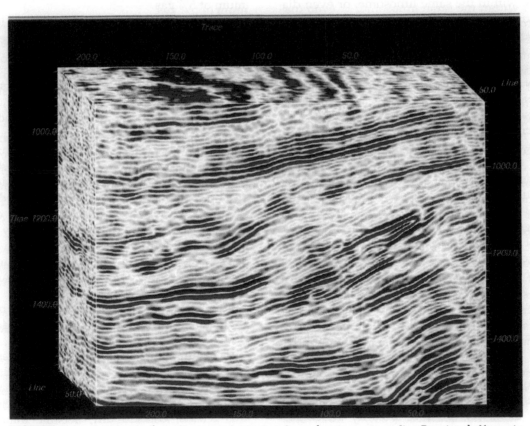

FIGURE 2.41 Sample of a three-dimensional seismic volume showing a prograding Permian shelf margin from the Delaware Basin (from Hart, 2000, reprinted by permission of the Society for Sedimentary Geology). This volume can be scrolled through in any direction to observe structural or stratigraphic changes through the study area.

Physical Attributes of Seismic Data

The makeup of a seismic image reflects the interaction between the substrate geology and the seismic waves traveling through the rocks, modulated by the physical properties of the rocks. The seismic waves emitted by a source at the surface are characterized by specific physical attributes, including shape (spatial form as depicted by a seismograph), polarity (direction of main deflection), frequency (number of complete oscillations per second), and amplitude (magnitude of deflection, proportional to the energy released by source). Excepting for frequency, which is a constant parameter that depends upon the source of the seismic signal, all other attributes may change as the waves travel through the geological substrate.

The physical properties which are most relevant to seismic data include the travel velocity of seismic waves, and the acoustic impedance (velocity multiplied by the rock's density) of the various layers and the contrasts thereof. Changes in acoustic impedance with depth are marked on seismic lines by reflections, which can signify changes in lithology, changes in fluid content within the same lithosome, or even diagenetic contrasts. Often, however, seismic reflections do not necessarily correspond to single lithological or fluid contacts, but may amalgamate a succession of strata that has a thickness less than the seismic resolution of that particular data set. As a general rule, a seismic reflection that preserves the polarity of the original seismic signal (i.e., 'positive polarity') indicates an increase in acoustic impedance with depth across that geological 'interface', whereas a change in the polarity of the seismic signal ('negative polarity') indicates a decrease in acoustic impedance with depth. The amplitude of the seismic reflection is usually proportional to the contrast in acoustic impedance across the geological 'contact'. Thus, high negative anomalies at the top of reservoir facies are commonly seen as a good 'sign' for petroleum exploration, as they suggest a sudden decrease in acoustic impedance inside the reservoir, which may potentially be related to the presence of porosity and low density fluids (i.e., hydrocarbons). For example, negative polarity reflections may mark a change from shales to underlying porous sandstones with hydrocarbons (ideal context of sealed reservoirs), but also a potential shift from compact sandstones (high acoustic impedance) to underlying shales (relatively lower acoustic impedance). Similarly, positive polarity reflections may also be equivocal, and indicative of various scenarios: shale overlying compact sandstones, porous sandstones overlying shale, or top of salt diapirs which are generally characterized by high acoustic impedance.

The nature of the seismic reflector (single contact *vs.* amalgamated package of strata) adds another degree of uncertainty to any attempts to interpret polarity data in terms of rock and fluid phases. Where the vertical distance between stratigraphic horizons is greater than the vertical resolution (i.e., seismic reflectors may correspond to single geological interfaces), the polarity of the reflections is more reliable in terms of geological interpretations. However, where seismic reflectors amalgamate closely spaced stratigraphic horizons, polarity interpretations become less reliable, as what we see on seismic lines is a composite signal. Therefore, besides simple polarity and amplitude studies, an entire range of additional techniques has been developed to assist with the fluid evaluation from seismic data, including the observation of bright spots (gas-driven high negative anomalies), flat spots (hydrocarbon/water contacts marked by horizontal high positive anomalies), and AVO (amplitude variance with offset) methods of computer data-analysis that increase the chances of locating natural gas or light petroleum with a minimum of 5% gas.

The vertical resolution of seismic data is primarily a function of the frequency of the emitted seismic signal. A high-frequency signal increases the resolution at the expense of the effective depth of investigation. A low-frequency signal can travel greater distances, thus increasing the depth of investigation, but at the expense of the seismic resolution. In practice, vertical resolution is generally calculated as a quarter of the wavelength of the seismic wave (Brown, 1991), so it also depends to some extent on travel velocity, which in turn is proportional to the rocks' densities. For example, the vertical resolution provided by a 30 Hz seismic wave traveling with a velocity of 2400 m/s is 20 m. This means that a sedimentary unit with a thickness of 20 m or less cannot be seen as a distinct package, as its top and base are amalgamated within a single reflection on the seismic line. Acquiring the optimum resolution for any specific case study requires therefore a careful balance between the frequency of the emitted signal and the desired depth of investigation (Fig. 2.42).

The limitation imposed by vertical resolution has been a main hindrance to the use of seismic data in resolving the details at the smaller scale of many individual reservoirs or depositional elements. For this reason, traditionally, seismic data have been regarded as useful for assessing the larger-scale structural and stratigraphic styles, but with limited applications when it comes to details at smaller-scale level.

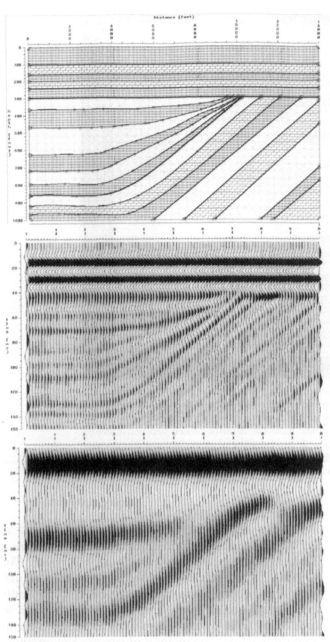

FIGURE 2.42. The effect of frequency on resolution and the observed stratigraphic geometry (from Hart, 2000, reprinted by permission of the Society for Sedimentary Geology). The real geometry is visible in the seismic model constructed with a 75 Hz wavelet (middle), but misleading in the model based on a 20 Hz frequency (bottom), where an onlap relationship is apparent.

However, as technology has improved, the limits of vertical resolution have been pushed from tens of meters down to meters, and spectacular three-dimensional seismic images can now be obtained from the geological substrate. In spite of this technological progress, seismic data still provide only indirect information on the solid and fluid phases in the subsurface, so calibration with

borehole data is essential for fine tuning the seismic facies—lithofacies relationship, for velocity measurements, or for time—depth conversions (Fig. 2.43).

Workflow of Seismic Data Analysis

The analysis of seismic data is facilitated by computer algorithms, and this is routinely performed by exploration geologists and geophysicists. The common routine, or workflow, includes an initial assessment of the large-scale structural and stratigraphic styles, followed by detailed studies in the smaller-scale areas that show features of potential economic interest. The following sections present the main steps of this routine, in workflow order.

Reconnaissance Studies

The reconnaissance analysis of a new seismic volume (e.g., Fig. 2.41) starts with an initial scrolling through the data (side to side, front to back, top to bottom) in order to assess the overall structural and stratigraphic styles (Hart, 2000). In this stage, as well as in all subsequent stages of data analysis, the interpreter must be familiar with a broad range of depositional and structural patterns in order to determine what working hypotheses are geologically reasonable for the new data set (Fig. 2.44). Following the reconnaissance scrolling, the seismic volume is 'sliced' in the areas that show the highest potential, where structural or stratigraphic traps may be present. The occurrence of such traps is often marked by seismic 'anomalies' (e.g., Fig. 2.45), which can be further highlighted and studied by applying a variety of techniques of data analysis. Slicing through the seismic volume is one of the most common techniques, and different slicing styles may be performed during the various phases of data handling (Fig. 2.46). The easiest slices that can be obtained in the early stages of data analysis are the *time slices* (horizontal or inclined planar slices through the volume; Fig. 2.46), which can be acquired before seismic reflections are mapped within the volume. The disadvantage of time slices is that they are usually time transgressive, as it is unlikely that a paleodepositional surface (commonly associated with some relief, and potentially affected by subsequent tectonism or differential compaction) corresponds to a perfect geometrical plane inside the seismic volume. For this reason, time slices are seldom true representations of paleo-landscapes or paleo-seafloors, unless the slice is obtained from very recent sediments at shallow depths. Once seismic reflections are interpreted and mapped throughout the volume, *horizon slices* can be obtained by flattening the seismic

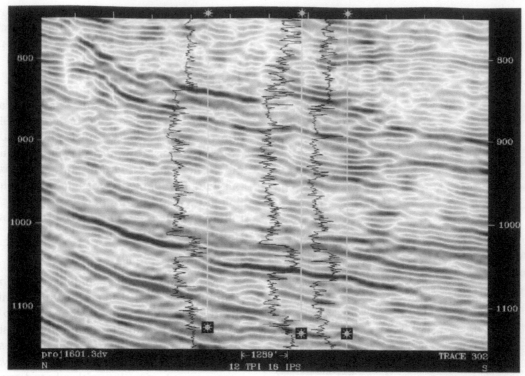

FIGURE 2.43 Example of a seismic line with well-log overlay (from Hart, 2000, reprinted by permission of the Society for Sedimentary Geology). The transect shows the basinward progradation (to the right) of a Permian mixed siliciclastic/carbonate continental slope in the Delaware Basin. The true location of the gamma ray curves is indicated by the white vertical lines. Note the correspondence between lithology contrasts (low GR—clean carbonates; higher GR—dolomitic sandstones and siltstones) and the location of prominent reflections. This type of display, only possible to view once time/depth relationships have been established, can be used to calibrate both the seismic and the well-log data.

horizon of interest (interpreted to correspond to a specific paleodepositional surface) and slicing the volume along it (Fig. 2.46). Such horizon slices may reveal astonishing geomorphologic details of past landscapes, seascapes, and depositional environments, and provide key evidence for the interpretation of paleodepositional settings and the calibration of well-log data. The role of horizon slices in the geological modeling of seismic volumes became more evident in recent years, as the seismic resolution improved in response to significant technological advances, to the extent that a new discipline is now emerging as 'seismic geomorphology' (e.g., Posamentier, 2000, 2004a; Posamentier and Kolla, 2003).

Still in the reconnaissance stage, the seismic anomalies emphasized by volume slicing can be further studied with additional techniques, such as voxel picking and opacity rendering, which can enhance geomorphologic interpretations. A voxel is a 'volume element,' similar with the concept of pixel ('picture element') in remote sensing, but with a third dimension ('z') that corresponds to time or depth. The other

two dimensions (measured along horizontal axes 'x' and 'y') of a voxel are defined by the bin size, which is the area represented by a single seismic trace. The vertical ('z') dimension is defined by the digital sampling rate of the seismic data, which is typically 2 or 4 milliseconds two-way travel time. Defined as such, each voxel is associated with a certain seismic amplitude value. The method of *voxel picking* involves auto-picking of connected voxels of similar seismic character, which can illuminate discrete depositional elements in three dimensions. Similarly, *opacity rendering*, which makes opaque only those voxels that lie within a certain range of seismic values, can also bring out features of stratigraphic interest (Posamentier, 2004b; B. Hart, pers. comm., 2004; Fig. 2.45).

Interval Attribute Maps

Once the stratigraphic objectives have been identified in the initial reconnaissance stages, the intervals bracketing sections of geologic interest can be evaluated in more detail by constructing interval attribute maps for those particular seismic 'windows' (Figs. 2.47

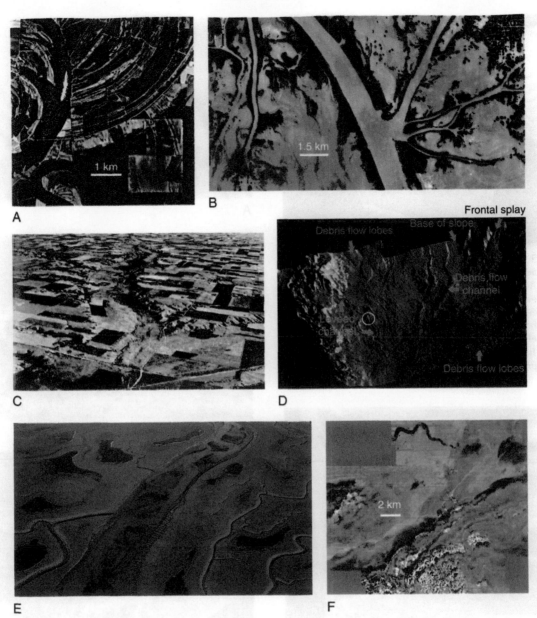

FIGURE 2.44 Analogs of modern and near modern depositional systems (images courtesy of H.W. Posamentier). A—aerial photograph of the Mississippi River, Louisiana, showing scroll bars associated with point bar development. B—aerial photograph of distributary channel and associated crevasse splays and crevasse channels in the Main Pass area of the Mississippi delta, Louisiana. C—oblique aerial photograph of a modern incised-valley system, Colorado; note the lateral tributary channels associated with drainage off the associated interfluve areas (for scale, note the roads and farm houses). D—seismically derived image of the modern seafloor in the ultra-deep waters of the DeSoto canyon area of the eastern Gulf of Mexico. Shown here are the base of slope (slope angle is approximately 1.8°) and the adjacent basin floor (slope angle is approximately 0.3°). Features such as debris flow channels and lobes, turbidite leveed channels and turbidite frontal splays are shown (for scale, the encircled channel is 300 m wide). E—oblique aerial photograph of an abandoning distributary channel, Mississippi delta, Louisiana. Note the thalweg and alternate bars within the channel (for scale, the main channel in the photograph is 1 km wide). The smaller channels shown constitute tidal creeks. F—seismic time slice through the Quaternary deposits of offshore eastern Java, Indonesia. The shelf edge is defined by slump scars; a small incised valley feeding a shelf edge delta is present on the outer shelf and presumably constitutes a forced regressive depositional system.

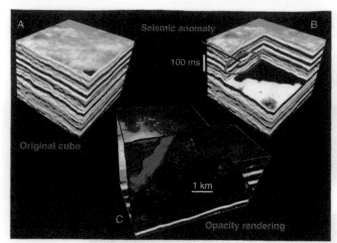

FIGURE 2.45 Reconnaissance interpretation of a seismic volume (Western Canada Sedimentary Basin; images courtesy of H.W. Posamentier). A—the original 3D seismic cube —— the image shows two section views and a plan view in the amplitude domain. B—chair slice through the 3D seismic cube. A seismic amplitude anomaly is highlighted. C—opacity rendered cube where only high amplitude voxels are rendered opaque; all other voxels are rendered transparent. This allows for visualization of a linear amplitude anomaly, interpreted as a channel.

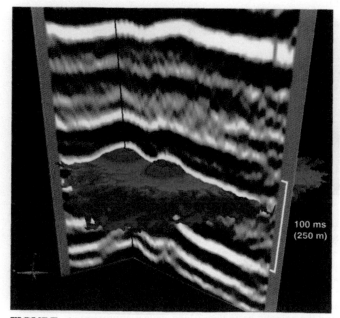

FIGURE 2.47 Two Devonian pinnacle reefs in the Western Canadian Sedimentary Basin, shown in section and three-dimensional view (image courtesy of H.W. Posamentier). Colors on the map view indicate time structure with reds/greens representing highs and purple representing lows. For scale, each reef is about 720 m wide. The two reefs are separated by a 200 m wide tidal channel.

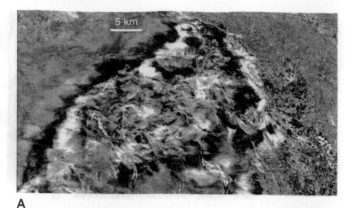

A

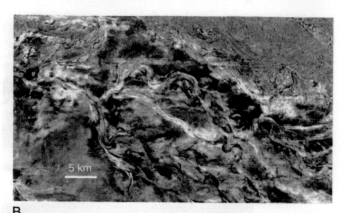

B

C

FIGURE 2.46 Reconnaissance interpretation of a 3D seismic volume using different slicing techniques (images courtesy of H.W. Posamentier). A—time slice: amplitude extraction from a planar horizontal slice. Shown here is part of a densely channeled deep-water turbidite system, eastern Gulf of Mexico. B—dipping planar slice: amplitude extraction from a planar surface dipping at approximately 2° to the east-southeast. C—horizon slice: amplitude extraction from a surface oriented parallel to a throughgoing mappable seismic reflection close to the section of interest. This type of slice yields the best image of the complete depositional system.

and 2.48). Most commonly, different types of amplitude extraction maps, seismic facies maps and seismic trace coherence maps are constructed, each with the potential of highlighting different features of the depositional systems under analysis (Figs. 2.48–2.52).

The *amplitude extraction maps* may display various amplitude attributes calculated over the selected interval (e.g., averages, positive polarity, negative polarity, cumulative amplitudes, amplitude peaks, square roots, etc.), and commonly reflect contrasts in acoustic impedance that may be interpreted in terms of lateral facies changes. Hence, such maps often enable the interpreter to visualize geomorphologic features that may be diagnostic for specific depositional systems, or even individual depositional elements within depositional systems (e.g., a fluvial channel fill in Fig. 2.49, or reef structures in Fig. 2.48).

The *seismic facies maps* also require the selection of an interval (e.g., 34 ms in Fig. 2.51), within which the shape of the seismic traces is analyzed by computer algorithms and classified into a number of waveforms.

The color codes used to differentiate between the different waveform classes enable the construction of maps that again can be interpreted in terms of facies and depositional elements (Figs. 2.50 and 2.51). This means that, as in the case with the lateral changes in amplitude attributes along the selected window, the change in seismic waveforms is also influenced by lateral shifts of facies, and hence each trace shape may be associated with a specific lithology-fluid 'package'. Of course, such a relationship needs to be calibrated with borehole data, although the overall geomorphology of depositional elements on the seismic facies maps may often allow one to infer with a high degree of confidence what lithofacies are expected in the various areas of a depositional system. For example, classes 9 and 10 in Fig. 2.50 (encircled area) are thought to indicate the location of the best reservoir sands within the channel fill. Once waveforms are interpreted in lithofacies terms, the visualization of particular depositional elements may be enhanced by highlighting only selected classes of trace shapes (Fig. 2.51).

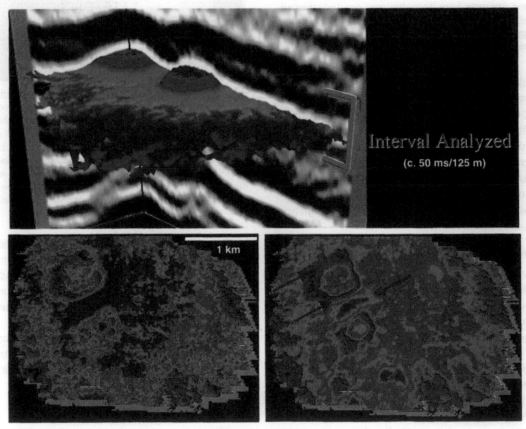

FIGURE 2.48 Interval attribute maps (maximum amplitude values to the left, and positive polarity total amplitude to the right) for the two reefs in Fig. 2.47 (images courtesy of H.W. Posamentier). The interval analyzed is approximately a 50 ms window. Note the amplitude asymmetry around the reef structures (blue arrows), possibly reflecting different patterns of current circulation around the reefs, with asymmetry suggesting a landward and a leeward side (prevailing wind direction is from the upper right). The amplitude anomaly between the reef structures (red arrow) indicates a different lithology, possibly associated with enhanced tidal scouring between the reefs.

FIGURE 2.49 Interval attribute of a Cretaceous distributary channel (Western Canada Sedimentary Basin; image courtesy of H.W. Posamentier). The attribute illustrates the amplitude strength within a 40 ms window. The lineaments within the 1.5 km wide channel represent alternate bars. Note the two, smaller, channels crosscutting the principal distributary channel. The crosscutting relationship suggests that the two small channels are younger than the larger channel.

The correlation, or lack thereof, of seismic traces within a chosen volume may be further emphasized by constructing *coherence maps*, which provide additional means for the study of geomorphological features (Fig. 2.52). Coherence is a volume attribute that emphasizes the correlation of seismic traces—light colors are assigned where seismic traces correlate, and dark colors indicate a lack of correlation. Coherence highlights seismic edges, which may correspond to structural or depositional elements.

Horizon Attribute Maps

Horizon attribute maps enhance the visualization of geomorphologic and depositional elements of specific paleodepositional surfaces (past landscapes or seascapes), by picking the geological horizon of interest within the seismic window studied in the previous step. If the interpretation of seismic reflections is correct, these horizon slices should be very close to time lines, providing a snapshot of past depositional environments. Horizon maps are constructed by extracting various seismic attributes along that particular reflection, such as dip azimuth, dip magnitude, roughness, or curvature (Fig. 2.53). Amplitude may also be extracted from a surface oriented parallel to a throughgoing mappable seismic reflection, as exemplified in Fig. 2.46C. Such horizon slices yield the best image of the complete depositional system.

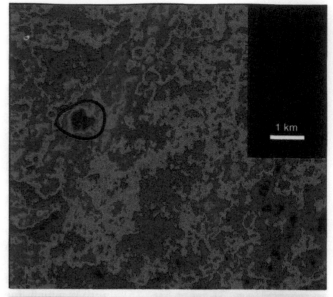

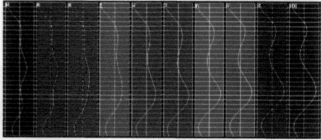

FIGURE 2.50 Seismic facies map based on a ten-fold classification of seismic traces (image courtesy of H.W. Posamentier). This example shows a channelized system in the Western Canada Sedimentary Basin. The black outline delineates a small structural high. Note that the deepest purple seismic class (i.e., Class 10) is confined to this outline, suggesting the possible presence of an accumulation of hydrocarbons within the channel at this location. Overall, the channel fill facies is dominated by Classes 7–10, whereas the interfluve area is dominated by Classes 1–6.

Time structure maps ('depth' in time below a surface datum) may also be obtained for a geological horizon mapped in three dimensions, and add important information regarding the subsidence history and the structural style of the studied area (Fig. 2.54). Interval or horizon attributes may be combined to enhance visualization effects, such as superimposing dip magnitude attributes on a time structure map (Fig. 2.55), or co-rendering coherence with amplitude data (Fig. 2.56).

3D Perspective Visualization

Three-dimensional perspective views add another degree of refinement to the information already available from the interval and horizon attribute maps. 3D perspective views illustrate surfaces extracted from 3D seismic data and depicted in x–y–z space. Interpreted horizons are then illuminated from a preferred direction

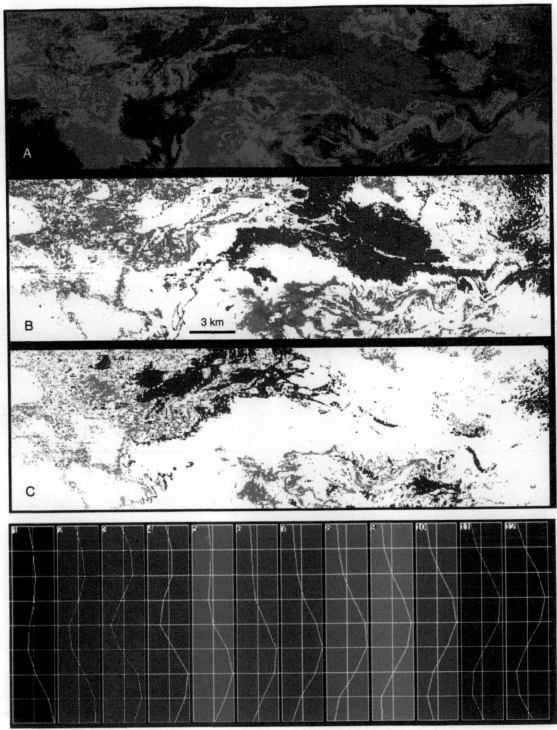

FIGURE 2.51 Seismic facies map of a deep-water mass transport complex, eastern Gulf of Mexico basin floor (images courtesy of H.W. Posamentier). The map is time transgressive, showing debris flow deposits (upper-left side of the image; proximal) overlying a channelized turbidity system (lower-right side of the image; distal). The analysis is based on a 34 ms interval, with twelve seismic classes defined. A—all classes are highlighted; a pattern of large-scale convolute deformation can be observed; B—only classes 2, 3, 4, and 9 are highlighted; this image reveals the more sheet-like portion of the mass transport complex in the more distal area; C—only classes 9 and 12 are highlighted; this image reveals the more convolute part of the mass transport complex, in the more proximal part of the system.

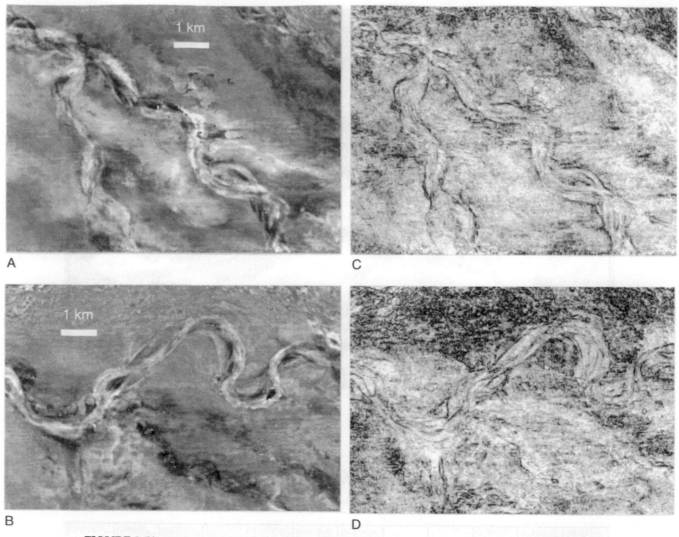

FIGURE 2.52 Interval attributes that characterize a deep-water Plio-Pleistocene channel system in the eastern Gulf of Mexico (images courtesy of H.W. Posamentier). A and B: Amplitude extraction from two horizon slices — these images capture successive positions of the channel thalweg and illustrates episodes of channel avulsion. The multiple thalweg images suggest meander loop migration towards the right and concomitant flow in that direction. C and D: Coherence slices of the same channel system shown in A and B. Coherence is a volume attribute that emphasizes the correlation of seismic traces—light colors are assigned where seismic traces correlate, and dark colors indicate a lack of correlation. Coherence highlights seismic edges (i.e., edges of depositional elements), and in this image enhances the channel margins also observed in the amplitude domain in A and B.

designed to highlight the relief and the depositional element morphology. Figures 2.57, 2.58, and 2.59 illustrate examples of such three-dimensional perspective images, which provide outstanding reconstructions of landscapes sculptured by fluvial systems (Fig. 2.57), seascapes of carbonate platforms (Fig. 2.58), or basin floors in deep-water settings dominated by gravity flows (Fig. 2.59). Such seismic data are of tremendous help in the reconstruction of paleodepositional environments and the calibration of borehole data. The examination of geological features in a three-dimensional perspective view may also be enhanced by

changing the angle of view, or by changing the angle of incidence of the light source that illuminates a particular image (Fig. 2.60).

AGE DETERMINATION TECHNIQUES

Age determinations refer to the evaluation of geological age by faunal or stratigraphic means, or by physical methods involving the relative abundance of radioactive parent/daughter isotopes (Bates and

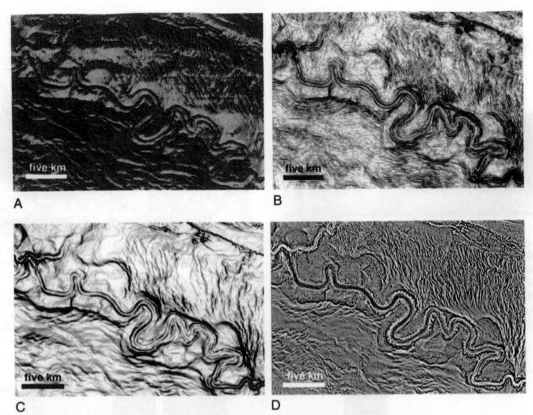

FIGURE 2.53 Horizon attributes that characterize the deep-water mid to late Pleistocene 'Joshua' channel in the northeastern Gulf of Mexico (modified from Posamentier, 2003; images courtesy of H.W. Posamentier). A—dip azimuth map: this map depicts the orientation of the surface such that north facing surfaces are assigned light colors, south facing are assigned dark colors, with intermediate orientations assigned colors between light and dark. This type of map creates a pseudo-3D image. Note the apparent knife-edge top of the small raised channel. Note also, the small sediment waves flanking the channel belt. B—surface roughness map: this map captures the roughness of a surface; rough areas are assigned dark colors, whereas smooth areas are assigned light colors. C—dip magnitude map: this map captures slope angles across the surface. Steep slope angles are indicated in black, whereas gentle slopes are depicted in white. In this display the raised channel is not imaged as a knife-edged feature. Rather, it is characterized by a flat to rounded feature, convex-upward. D — Curvature map: this map illustrates the curvature of the horizon, and outlines depositional elements by assigning dark colors for low-curvature (flat) areas and light colors for high-curvature edges of geomorphological features. Detailed morphology not as readily observed on the other attribute maps include small slump scars on the inner levee flanks adjacent to the raised channel, as well as sediment waves observed in the overbank areas.

Jackson, 1987). Time control may generally be achieved by means of biostratigraphy, magnetostratigraphy, isotope geochemistry, or by the mapping of lithological time-markers. Age data are always desirable to have, and are particularly useful to constrain correlations at larger scales.

The resolution of the various dating techniques varies with the method, as well as with the age of the deposits under investigation. For example, biostratigraphic determinations may provide resolutions of 0.5 Ma (Cretaceous ammonite zonation in the Western Canada Sedimentary Basin; Obradovich, 1993), 1 Ma (upper Cretaceous nonmarine palynology in the Western Canada Sedimentary Basin; A.R. Sweet, pers. comm., 2005), or 2 Ma (Permo-Triassic vertebrate fossils in the Karoo Basin; Rubidge, 1995). Biostratigraphy used in conjunction with magnetostratigraphy leads to even better results, increasing the resolution to about 0.4–0.5 Ma (the span of polarity chrons) for selected Cretaceous and Tertiary intervals. Geochronology produces results with an error margin of less than 0.5 Ma for the Phanerozoic, and more than 1 Ma for the Precambrian. In addition to these methods, lithological time markers, such as ash layers or widespread paleosol horizons, add to the available time control by providing excellent reference time-lines (Fig. 2.61).

The resolution of age determinations generally decreases with older strata due to a number of factors including facies preservation, postdepositional tectonics, diagenetic transformations, metamorphism, and

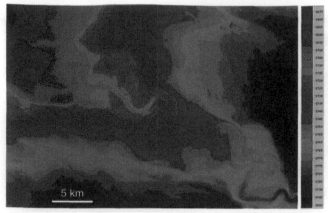

FIGURE 2.54 Time structure map on the channel shown in Fig. 2.53 (image courtesy of H.W. Posamentier). This image illustrates the elevated aspect of the thalweg as well as the entire channel belt, as a result of post-depositional differential compaction. The channel belt is elevated approximately 65 m above the adjacent overbank area. The direction of flow was from left to right. Red and orange indicate higher elevations relative to green, blue, and purple, with purple marking the greatest depth beneath the sea level. The scale to the right is in ms below the sea level.

FIGURE 2.56 Co-rendered or superimposed images from two seismic attribute maps of a Plio-Pleistocene deep-water leveed channel from the eastern Gulf of Mexico (image courtesy of H.W. Posamentier). The two attribute maps comprise amplitude and coherence. This image captures lithologic information inherent to the amplitude domain, and combines it with edge effects delineating the channel inherent in the coherence domain. Multiple channel thalwegs are observed, with meander loop migration verging to the right indicating flow from left to right across this area.

FIGURE 2.55 The base-Cretaceous unconformity in the Western Canada Sedimentary Basin, as depicted on four horizon attribute maps (images courtesy of H.W. Posamentier): A—dip magnitude map; B—dip azimuth map; C—time structure map; and D—co-rendered time structure and dip azimuth map. This surface is characterized by numerous fluvial channels at or near the basal Cretaceous boundary.

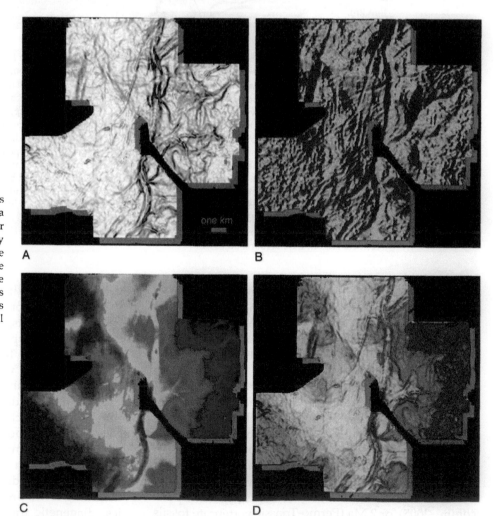

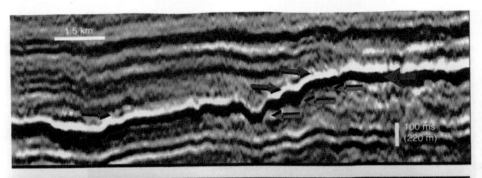

FIGURE 2.57 The base Cretaceous unconformity in the Western Canada Sedimentary Basin, as it appears on a two-dimensional seismic line and on a three-dimensional perspective view map (modified from Posamentier, 2004a; images courtesy of H.W. Posamentier). This is the same surface as shown in Fig. 2.55. The unconformity (red arrow on the seismic line) separates Cretaceous strata from the underlying Devonian deposits, and is associated with significant erosion (yellow arrows indicate truncation) and change in tectonic and depositional setting. The unconformity is onlapped by the Cretaceous strata (blue arrows), and corresponds to a first-order sequence boundary that marks a change from a divergent continental margin to the tectonic setting of a foreland system. The top of the Devonian deposits is incised by Cretaceous fluvial systems. Note the paleo tributary drainage network associated with inferred flow off the high area to the right of the perspective view. The white line on the three-dimensional perspective view map indicates the position of the two-dimensional seismic line.

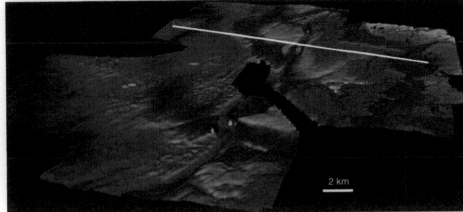

FIGURE 2.58 Three-dimensional perspective view of a Devonian channel in the Western Canada Sedimentary Basin (image courtesy of H.W. Posamentier). This channel is filled with bioclastic material and is interpreted to be a possible tidal channel on a carbonate platform. This feature is also illustrated in Fig. 2.48.

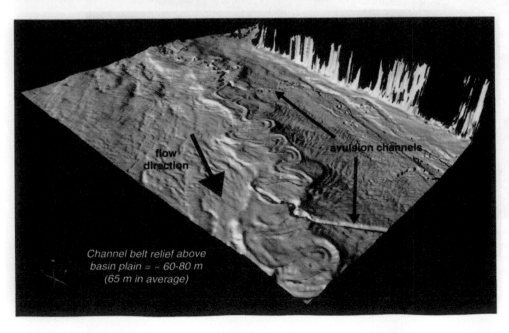

FIGURE 2.59 Three-dimensional perspective view of the Pleistocene 'Joshua' channel in the eastern Gulf of Mexico (modified from Posamentier, 2003; image courtesy of H.W. Posamentier). This deep-water channel is characterized by two avulsion events. The avulsion channels are mud filled as indicated by their concave-up transverse profiles, in contrast with the convex-up sand filled Joshua channel. This channel is also illustrated in Figs. 2.53 and 2.54. For scale, the channel fill is approximately 625 m wide.

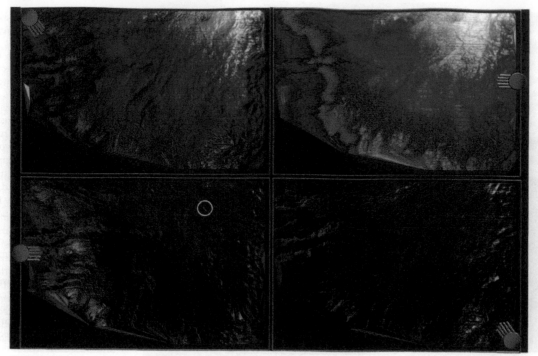

FIGURE 2.60 Illumination effects, such as changing the angle of incident light, may significantly enhance the geomorphologic features of the geological horizon of interest (images courtesy of H.W. Posamentier). This example shows the modern deep-water seascape in the DeSoto Canyon area of the eastern Gulf of Mexico (compare with Fig. 2.44-D). For scale, the encircled channel is 300 m wide.

FIGURE 2.61 Bentonite layers in the Bearpaw Formation (Late Campanian-Early Maastrichtian; St. Mary River, Alberta, Western Canada Sedimentary Basin). Such bentonites have a lateral extent of tens to hundreds of kilometers, in outcrop and subsurface. They may be dated with radiometric methods, and may also be tied against the biostratigraphic record of ammonite, palynological, or foraminiferal zonation.

evolution of life forms. At the lower end of the stratigraphic spectrum, the constraint of Precambrian rocks' ages is exclusively based on radiometric methods. However, even in the near-absence of chronological constraints, sequence stratigraphic models can still be constructed based on a good knowledge of the paleoenvironments and facies relationships within the basin (Christie-Blick *et al.*, 1988; Beukes and Cairncross, 1991; Krapez, 1993, 1996, 1997; Catuneanu and Eriksson, 1999, 2002; Eriksson and Catuneanu, 2004a).

WORKFLOW OF SEQUENCE STRATIGRAPHIC ANALYSIS

The accuracy of sequence stratigraphic analysis, as with any geological interpretation, is proportional to the amount and quality of the available data. Ideally, we want to integrate as many types of data as possible, derived from the study of outcrops, cores, well logs, and seismic volumes. Data are of course more abundant in mature petroleum exploration basins, where models are well constrained, and sparse in frontier regions. In the latter situation, sequence stratigraphic principles generate model-driven predictions, which enable the formulation of the most realistic, plausible, and predictive models for petroleum, or other natural resources exploration (Posamentier and Allen, 1999).

The following sections outline, in logical succession, the basic steps that need to be taken in a systematic sequence stratigraphic approach. These suggested steps by no means imply that the same rigid template has to be applied in every case study—in fact the interpreter must have the flexibility of adapting to the 'local conditions,' partly as a function of geologic circumstances (e.g., type of basin, subsidence, and sedimentation history) and partly as a function of available data.

The checklist provided below is based on the principle that a general understanding of the larger-scale tectonic and depositional setting must be achieved first, before the smaller-scale details can be tackled in the most efficient way and in the right geological context. In this approach, the workflow progresses at a gradually decreasing scale of observation and an increasing level of detail. The interpreter must therefore change several pairs of glasses, from coarse- to fine-resolution, before the resultant geologic model is finally in tune with all available data sets. Even then, one must keep in mind that models only reflect current data and ideas, and that improvements may always be possible as technology and geological thinking evolve.

Step 1—Tectonic Setting (Type of Sedimentary Basin)

The type of basin that hosts the sedimentary succession under analysis is a fundamental variable that needs to be constrained in the first stages of sequence stratigraphic research. Each tectonic setting is unique in terms of subsidence patterns, and hence the stratigraphic architecture, as well as the nature of depositional systems that fill the basin, are at least in part a reflection of the structural mechanisms controlling the formation of the basin. The large group of extensional basins for example, which include, among other types, grabens, half grabens, rifts and divergent continental margins, are generally characterized by subsidence rates which increase in a distal direction (Fig. 2.62). At the other end of the spectrum, foreland basins formed by the flexural downwarping of the lithosphere under the weight of orogens show opposite subsidence patterns with rates increasing in a proximal direction (Fig. 2.63). These subsidence patterns represent primary controls on the overall *geometry and internal architecture* of sedimentary basin

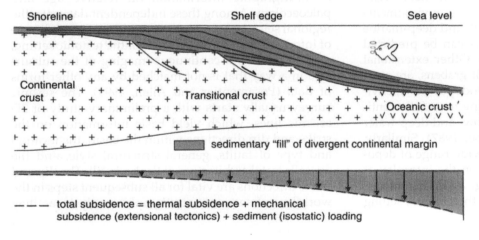

FIGURE 2.62 Generalized dip-oriented cross section through a divergent continental margin, illustrating overall subsidence patterns and stratigraphic architecture. Note that subsidence rates increase in a distal direction, and time lines converge in a proximal direction.

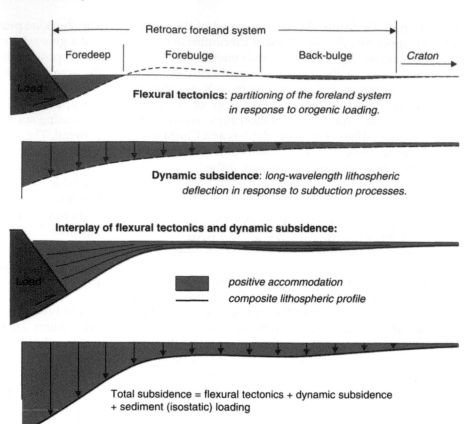

FIGURE 2.63 Generalized dip-oriented cross section through a retroarc foreland system showing the main subsidence mechanisms and the overall basin-fill geometry (modified from Catuneanu, 2004a). Note that subsidence rates generally increase in a proximal direction, and as a result time lines diverge in the same direction.

fills, as reflected by the converging or diverging trends displayed by time-line horizons in proximal or distal directions (Figs. 2.62 and 2.63). It is therefore imperative to acquire a good understanding of the tectonic setting before proceeding with the construction of stratigraphic models.

In addition to allowing an inference of syndepositional subsidence trends, the knowledge of the tectonic setting may also have bearing on the prediction of depositional systems that build the sedimentary succession, and their spatial relationships within the basin. In the context of a divergent continental margin, for example, fluvial to shallow-marine environments are expected on the continental shelf, and deep-marine (slope to basin-floor) environments can be predicted beyond the shelf edge (Fig. 2.62). Other extensional basins, such as rifts, grabens, or half grabens, are more difficult to predict in terms of paleodepositional environments, as they may offer anything from fully continental (alluvial, lacustrine) to shallow- and deep-water conditions (Leeder and Gawthorpe, 1987). Similarly, foreland systems may also host a wide range of depositional environments, depending on the interplay of subsidence and sedimentation (Fig. 2.64). This means that, even though a knowledge of the tectonic setting

narrows down the range of possible interpretations and provides considerable assistance with the generation of geological models, especially in terms of overall geometry and stratal architecture, the reconstruction of the actual paleodepositional environments represents another step in the sequence stratigraphic workflow, as suggested in this chapter.

The reconstruction of a tectonic setting must be based on regional data, including seismic lines and volumes, well-log cross-sections of correlation calibrated with core, large-scale outcrop relationships, and biostratigraphic information on relative age and paleoecology. Among these independent data sets, the regional seismic data stand out as the most useful type of information in the assessment of the tectonic setting, as they provide a continuous imaging of the subsurface in a way that is not matched by any other forms of data (Posamentier and Allen, 1999). The seismic survey usually starts with a preliminary study of 2D seismic lines, which yield basic information on the strike and dip directions within the basin, the location and type of faults, general structural style, and the overall stratal architecture of the basin fill. The dip and strike directions are vital for all subsequent steps in the workflow of sequence stratigraphic analysis, as they

1. Underfilled phase: deep marine environment in the foredeep

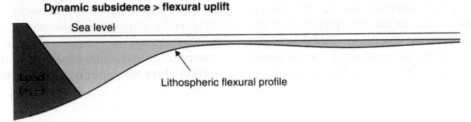

2. Filled phase: shallow marine environment across the foreland system

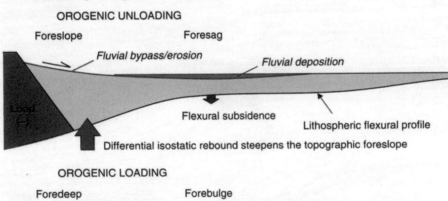

3. Overfilled phase: fluvial environment across the foreland system

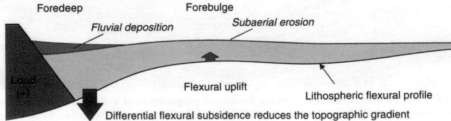

FIGURE 2.64 Patterns of sedimentation across a foreland system as a function of the interplay between accommodation and sedimentation (synthesized from Catuneanu *et al.*, 2002, and Catuneanu, 2004a, b; see Catuneanu, 2004a for full details and a review of case studies).

allow to infer shoreline trajectories, lateral relationships of depositional systems, and patterns of sediment transport within the basin. In addition to this, the converging or diverging character of seismic reflections, as long as they are considered to approximate time lines, reveal key information regarding the subsidence patterns along any given transect. Subsidence is differential in most cases, with rates varying mostly along dip (Figs. 2.62 and 2.63), although strike variability is also possible, to a lesser degree.

Figure 2.65 provides an example of a 2D seismic line which shows the overall progradation of a divergent continental margin. In this case, the position of the shelf edge can easily be mapped for different time slices, and the distribution of fluvial to shallow-marine (landward relative to the shelf edge) *vs.* deep-marine (slope to basin-floor) paleoenvironments can be assessed preliminarily with a high degree of confidence. Following the initial 2D seismic survey, 3D seismic data help to further enhance the interpretation of the physiographic elements of the basin under analysis (Fig. 2.66), thus providing the framework for the subsequent steps of the sequence stratigraphic workflow.

Step 2—Paleodepositional Environments

The interpretation of paleodepositional environments is another key step in the sequence stratigraphic workflow. Once the tectonic setting and the overall style of stratal architecture are elucidated, the interpreter needs to zoom in and constrain the nature of depositional systems that build the various portions of the basin fill. Paleoenvironmental reconstructions are important for several reasons, both inside and outside the scope of sequence stratigraphy. From a sequence stratigraphic perspective, the spatial and temporal relationships of depositional systems, including their shift directions through time, are essential criteria to validate the interpretation of sequence stratigraphic surfaces and systems tracts. Within this framework, the genesis, distribution and geometry of petroleum reservoirs, coal seams or mineral placers may be assessed in relation to the process-sedimentation principles that are relevant to each depositional environment. The identification of specific depositional elements (e.g., channel fills, beaches, splays, etc.) is also critical at this stage, as their morphology has a direct bearing on the economic evaluation of the stratigraphic units of interest.

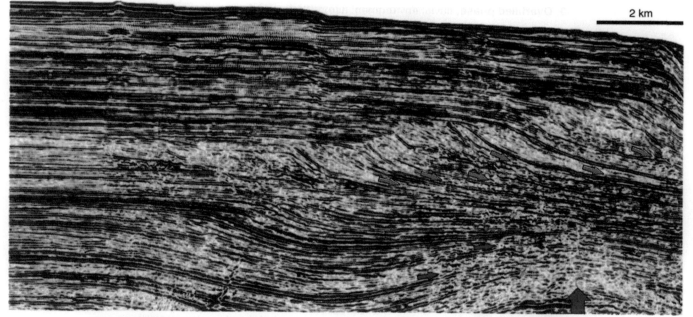

FIGURE 2.65 2D seismic transect showing the overall progradation of a divergent continental margin (from Catuneanu *et al.*, 2003a; image courtesy of PEMEX). The shelf edge position can easily be mapped for consecutive time slices, and hence a preliminary assessment of the paleodepositional environments can be performed with a high degree of confidence. In this case, fluvial to shallow-marine systems are inferred on the continental shelf (landward relative to the shelf edge), whereas deep-marine systems are expected in the slope to basin-floor settings. The prograding clinoforms downlap the seafloor (yellow arrows), but due to the rise of a salt diapir (blue arrow) some downlap type of stratal terminations may be confused with onlap (red arrows).

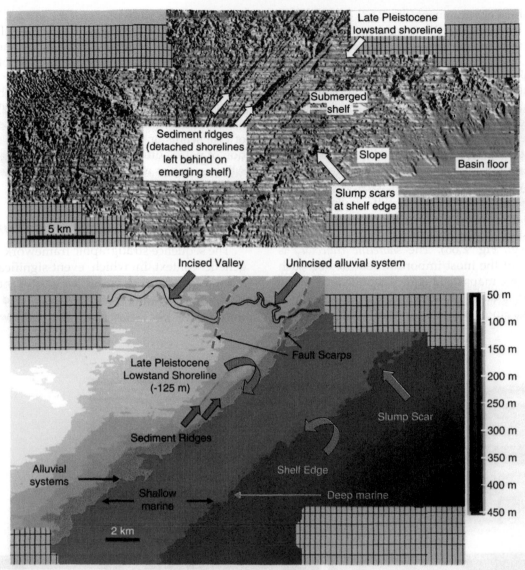

FIGURE 2.66 Azimuth map (top) and structure map (bottom) of the seafloor relief, offshore east Java, Indonesia, showing the tectonic and depositional settings during the Late Pleistocene relative sea-level lowstand (images courtesy of H.W. Posamentier). The detached shorelines (sediment ridges) on the continental shelf formed during relative sea-level fall, prior to the shoreline reaching its lowstand position. The slump scars indicate instability at the shelf edge, a situation that is common during times of relative fall. Note that the lowstand shoreline remained inboard of the shelf edge, which explains the presence of unincised fluvial systems on the outer continental shelf. In this example, the change from incised to unincised fluvial systems is controlled by a fault scarp—rivers incise into the more elevated footwall of the seaward-plunging normal fault.

The success of paleoenvironmental interpretations depends on the integration of multiple data sets (seismic, well-log, core, outcrop), as each type of data has its own merits and pitfalls. As discussed before, the geophysical data (seismic, well-log) provides more continuous, but indirect information on the subsurface geology. On the other hand, the rock data (core, outcrop) allow for a direct assessment of the geology, but most commonly from discrete locations within the basin. The mutual calibration of geophysical and rock data is therefore the best approach to the stratigraphic modeling of the subsurface geology. At this stage in the workflow, the 3D seismic data are significantly more useful than the 2D seismic transects. The latter are ideal to reveal structural styles and the overall stratal stacking patterns, as explained for step 1 above (e.g., Fig. 2.65), but fall short when it comes to the identification of depositional systems. In contrast, 3D horizon slices often provide outstanding geomorphologic details that help constrain the nature of the paleodepositional environment (Fig. 2.67). For unequiv-ocal results, however, the 3D seismic geomorphology needs to be combined with a knowledge of the tectonic setting (step 1), well-log motifs, and the direct informa-tion supplied by core and nearby outcrops, where such data are available. Paleoecology from palynology, pale-ontology, or ichnology, which can be inferred from the study of core and outcrops, may also assist consider-ably with the interpretation of the depositional setting.

The results of paleoenvironmental reconstructions may be presented in the form of paleogeographic maps (e.g., syntheses by Kauffman, 1984; Mossop and Shetsen, 1994; Long and Norford, 1997; Fielding *et al.*, 2001), which show the main physiographic and depo-sitional features of the studied area for a particular time slice (e.g., Fig. 2.66). The *shoreline trajectory* is arguably one of the most important features on such paleogeographic maps, as it shows the location of the sediment entry points into the marine basin relative to the basin margin or other important physiographic

elements of that particular tectonic setting. For example, in the context of a divergent continental margin, the position of the shoreline relative to the shelf edge represents a critical control on the type of terrigenous sediment (sand/mud ratio) that may be delivered to the slope and basin-floor settings, and hence on the development of deep-water reservoirs. The shoreline also exerts a critical control on the lateral development of coal seams or placer deposits, and on the distribu-tion of petroleum reservoirs of different genetic types. All these topics are discussed in more detail in subse-quent chapters of this book.

Step 3—Sequence Stratigraphic Framework

The sequence stratigraphic framework provides the genetic context in which event-significant surfaces, and the strata they separate, are placed into a coherent model that accounts for all temporal and spatial rela-tionships of the facies that fill a sedimentary basin.

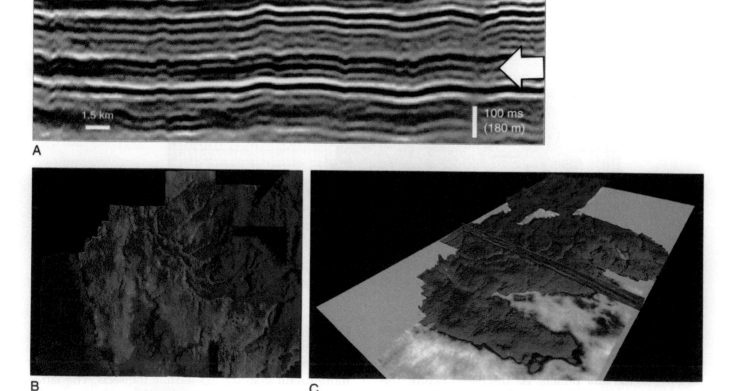

FIGURE 2.67 Devonian fluvial system in the Western Canada Sedimentary Basin (images courtesy of H.W. Posamentier). Note that the nature of the depositional system is difficult to infer from the 2D seismic transect (A) without having seen the three dimensional image (B and C). The interpretation of a fluvial drainage network becomes clear when the surface is viewed in three dimensions either as an illuminated and struc-turally color-coded plan view (B) or in a perspective view (C). Image C enhances the 3D aspect of this system by illustrating the key horizon along with a planar and cross-sectional slice. For scale, channels in images B and C are approximately 300 m wide.

Ultimately, this is the geological model that allows for the most efficient exploration approach for natural resources, as facies tend to develop following predictive patterns within this genetic framework.

As argued in Chapter 1, depositional trends, and changes thereof, represent the primary stratigraphic attribute that is used to develop a chronostratigraphic framework for the succession under analysis (Fig. 1.3). The recognition of depositional trends is based on lateral and vertical facies relationships, where the paleoenvironmental reconstructions of step 2 of the sequence stratigraphic workflow play a major part, and on observing the geometric relationships between strata and the surfaces against which they terminate. Some of these stratal terminations may be diagnostic for particular depositional trends, such as the coastal onlap for transgression (retrogradation), or the downlap pattern for regression (progradation). Only after the depositional trends are constrained, can the sequence stratigraphic surfaces that mark changes in such trends be mapped and labeled accordingly (e.g., the maximum flooding surface would be placed at the contact between retrograding strata and the overlying prograding deposits). This is why the construction of the sequence stratigraphic framework, in this third stage of the overall workflow, starts with the observation of stratal terminations, followed by the identification of sequence stratigraphic surfaces, which in turn allows for the proper labeling of the packages of strata between them in terms of systems tracts and sequences. The logical succession of steps in this routine is also adopted for the presentation of sequence stratigraphic concepts in the subsequent sections (Chapters 4–6) of this book.

Stratal Terminations

Stratal terminations refer to the geometric relationships between strata and the stratigraphic surfaces against which they terminate, and may be *observed* on continuous surface or subsurface data sets including large-scale outcrops and 2D seismic transects. The type of stratal termination (e.g., onlap, downlap, offlap, etc.) may provide critical information regarding the direction and type of syndepositional shoreline shift; this topic is dealt with in full detail in Chapter 4. Examples of downlap, indicating progradational depositional trends, are illustrated in Fig. 2.65 (2D seismic transect), 2.68, and 2.69 (large-scale outcrops). Stratal terminations may also be *inferred* on well-log cross-sections of correlation (Figs. 2.38 and 2.39), based on a knowledge of the depositional setting and the trends that are expected in that particular environment.

Stratigraphic Surfaces

Sequence stratigraphic surfaces help to build the chronostratigraphic framework for the sedimentary succession under analysis. Such surfaces can be identified on the basis of several criteria, including (1) the nature of the contact (conformable or unconformable); (2) the nature of depositional systems that are in contact across that surface; (3) types of stratal terminations associated with that surface; and (4) depositional trends that are recorded below and above that stratigraphic contact. The entire range of sequence stratigraphic surfaces, including the criteria used for their recognition, is discussed in detail in Chapter 4. What is worth mentioning at this point is that excepting for actual time-marker beds (e.g., ash layers, etc.), most of

FIGURE 2.68 Gilbert-type delta front, prograding to the left (Panther Tongue, Utah). The delta front clinoforms downlap the paleo-seafloor (arrows). Note person for scale.

FIGURE 2.69 River-dominated delta showing prodelta fine-grained facies at the base, delta front sands prograding to the left, and coal-bearing delta plain facies at the top (the Ferron Sandstone, Utah). The prograding delta front clinoforms dip at an angle of 5–7°, and downlap the underlying prodelta deposits (arrows). The outcrop is about 30 m high.

these surfaces provide the closest approximation to time lines that one can possibly have on a cross-section of correlation. This is the reason why sequence stratigraphic surfaces should always be mapped first, before lateral changes of facies and the associated facies contacts are marked on cross-sections.

In the case of continuous data (e.g., seismic transects, large-scale outcrops), tracing sequence stratigraphic surfaces may be straight forward, unless the basin is structurally complex. In the latter situation, independent time control (biostratigraphy, magnetostratigraphy, isotope geochemistry, or lithological time-markers) is needed to constrain stratigraphic correlations. Independent time control is also desirable to have where correlations are based on discontinuous data collected from discrete locations within the basin (small outcrops, core, well logs). Depending on data availability, all sources of direct and indirect geological information need to be integrated at this point to convey maximum credibility to the sequence stratigraphic model.

Systems Tracts and Sequences

This is the last step of the sequence stratigraphic workflow, when both event-significant surfaces and the packages of strata between them are interpreted in genetic terms. Identification of systems tracts on cross-sections is a straight forward procedure once the position and type of stratigraphic surfaces are at hand.

The terminology used to define systems tracts may vary with the model (Fig. 1.7), but beyond the trivial issue of semantics each systems tract is unequivocally characterized by specific stratal stacking patterns (depositional trends) and position within the framework of sequence stratigraphic surfaces. The types of systems tracts, as well as of sequences which they may build, are fully discussed in Chapters 5 and 6. Each data set may contribute with useful information towards the recognition of depositional trends in the rock record, but large-scale outcrops and seismic volumes stand out as particularly relevant for this purpose. The utility of different data sets for constraining the information required during the various steps of the sequence stratigraphic workflow is summarized in Figs. 2.70 and 2.71.

The observation of depositional trends allows for interpretations of syndepositional shoreline shifts, which in turn depend on the interplay of sedimentation and the space available for sediments to accumulate. All these aspects are an integral part of the sequence stratigraphic model, and ultimately allow an understanding of the logic of lithofacies distribution within the basin. Sediment is as important as the space that it requires to accumulate, so the assessment of extrabasinal sediment sources, weathering efficiency in relation to paleoclimates, distances and means of sediment transport, and the location of sediment entry points into the marine portion of the basin (river-mouth environments;

Key: √√√ good
√√ fair
√ poor

	Rock data			Geophysical data		
	Outcrops				Seismic data	
	Large-scale	Small-scale	Core	Well logs	2D	3D
Tectonic setting	√√	√	√	√√	√√√	√√√
Lithofacies	√√√	√√√	√√√	√√	√	√√
Depositional elements	√√√	√√	√√	√√	√	√√√
Depositional systems	√√√	√√	√√	√√	√	√√√
Depositional trends	√√√	√√	√√	√√	√√	√√√
Stratal terminations	√√√	√	√	√	√√√	√√√
Nature of contacts	√√√	√√√	√√√	√√	√√	√√

FIGURE 2.70 Utility of different data sets for constraining tectonic and stratigraphic interpretations. The seismic and large-scale outcrop data provide continuous subsurface and surface information, respectively. In contrast, small-scale outcrops, core, and well logs provide discontinuous data collected from discrete locations within the basin.

Data set	Main applications / contributions to sequence stratigraphic analysis
Seismic data	Continuous subsurface imaging; tectonic setting; structural styles; regional stratigraphic architecture; imaging of depositional elements; geomorphology
Well-log data	Vertical stacking patterns; grading trends; depositional systems; depositional elements; inferred lateral facies trends; calibration of seismic data
Core data	Lithology; textures and sedimentary structures; nature of stratigraphic contacts; physical rock properties; paleocurrents in oriented core; calibration of well-log and seismic data
Outcrop data	3D control on facies architecture; insights into process sedimentology; lithofacies; depositional elements; depositional systems; all other applications afforded by core data
Geochemical data	Depositional environment; depositional processes; diagenesis; absolute ages; paleoclimate
Paleontological data	Depositional environment; depositional processes; ecology; relative ages

FIGURE 2.71 Contributions of various types of data sets to the sequence stratigraphic interpretation. Integration of insights afforded by various data sets is the key to a reliable sequence stratigraphic model.

Figs. 2.3 and 2.4) provides critical insights regarding the validity of the model and the exploration potential of the study area. Intrabasinal sediment sources are also important, as they may explain the presence of potential reservoirs in areas that are seemingly unrelated to any extrabasinal sediment sources.

As depositional environments respond in a predictable way to changes in base level, a sequence stratigraphic model provides a first-hand interpretation of the base-level fluctuations in a basin, starting from the reconstructed depositional history. The predictable character of this relationship makes sequence stratigraphy a very efficient exploration tool in the search for natural resources, by allowing one to infer lateral changes of facies related to particular stages in the evolution of the basin. With a strong emphasis on the timing of depositional events, linked to the formation of the key bounding surfaces, sequence stratigraphy improves our understanding of the temporal and spatial development of economically important facies such as placer deposits, hydrocarbon reservoirs, source rocks, and seals. The emphasis on depositional processes also led to a shift in the focus of petroleum exploration from structural traps to combined or purely stratigraphic traps (Bowen et al., 1993; Posamentier and Allen, 1999). An entire range of new petroleum play types is now defined in the light of sequence stratigraphic concepts.

3

Accommodation and Shoreline Shifts

INTRODUCTION

This chapter introduces a set of core concepts relevant to sequence stratigraphy, including sediment accommodation, shoreline shifts, and the controls thereof, whose understanding is fundamental before approaching the more specialized topics related to sequence stratigraphic surfaces, systems tracts, and stratigraphic sequences. These basic concepts allow one to see why and how sequence stratigraphy works, and what the 'engine' is that unifies stratal stacking patterns across a basin into coherent models of stratigraphic architecture.

One of the key premises of sequence stratigraphy, which also served as a main incentive for its conceptual development, is that this approach allows for facies predictions from the confines of individual depositional systems to the scale of entire sedimentary basin fills (Fig. 1.2). This premise implies that depositional trends within all environments established within a sedimentary basin are synchronized to a large extent, being governed by external (allogenic) mechanisms that operate from basinal to global scales. This allogenic 'umbrella' controls regional depositional trends, and provides the basis for the definition of systems tracts and the development of sequence models of facies predictability.

Changes in depositional trends arguably represent the essence of sequence stratigraphic research (Fig. 1.3), and reflect the interplay between the space available for sediments to fill and the amount of sediment influx. The space available for sediments to fill (i.e., 'accommodation') is in turn modified by the basin-scale influence of allogenic controls, which thus provide the common thread that links the depositional trends across a sedimentary basin, from its fluvial to its marine reaches. At the limit between nonmarine and marine environments, the shoreline trajectory defines the type of depositional trend established at any given time. Shoreline trajectories are thus central to sequence stratigraphy, and their changes through time control the timing of all systems tracts and sequence stratigraphic surfaces. The effects of allogenic controls on sedimentation, the space available for sediments to fill, and shoreline trajectories and associated depositional trends are thus intricately related and form the foundation of the sequence stratigraphic approach.

ALLOGENIC CONTROLS ON SEDIMENTATION

Significance of Allogenic Controls

Sedimentation is generally controlled by a combination of autogenic and allogenic processes, which determine the distribution of depositional elements within a depositional system, as well as the larger-scale stacking patterns of depositional systems within a sedimentary basin.

Autogenic processes (e.g., self-induced avulsion in fluvial and deep-water environments) are particularly important at sub-depositional system scale, and are commonly studied using the methods of conventional sedimentology and facies analysis. Allogenic processes, on the other hand, are directly relevant to sequence stratigraphy, as they control the larger-scale architecture of the basin fill.

Allogenic controls provide the common platform that connects and synchronizes the depositional trends recorded at any given time in all environments established within a sedimentary basin, thus allowing for sequence stratigraphic models to be developed at the basin scale. This in turn is the key for the facies predictability applications of sequence stratigraphy, which are so valuable to both academic and

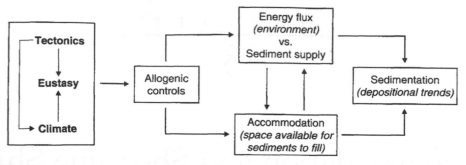

FIGURE 3.1 Allogenic controls on sedimentation, and their relationship to environmental energy flux, sediment supply, accommodation, and depositional trends (modified from Catuneanu, 2003). In any depositional environment, the balance between energy flux and sediment supply is key to the manifestation of processes of sediment accumulation or reworking. Besides tectonics, additional processes such as thermal subsidence (crustal cooling), sediment compaction, water-depth changes, isostatic, and flexural loading, also contribute to the total subsidence or uplift in the basin. Accommodation is affected by the balance between energy flux and sediment supply (i.e., increased energy 'erodes' accommodation; increased sediment supply adds to the amount of available accommodation), but it is also independently controlled by external factors such as eustasy and tectonism. At the same time, changes in accommodation controlled directly by external factors may alter the balance between energy flux and sediment supply at any location within the basin (e.g., deepening of the water as a result of sea-level rise lowers the energy flux at the seafloor). The interplay of all allogenic controls on sedimentation, as reflected by changes in accommodation and energy flux/sediment supply, ultimately determines the types of depositional trends established within the basin.

industry practitioners. The basic allogenic controls on sedimentation include the climate, tectonics, and sea-level changes, and their relationship with the environmental energy flux, sediment supply, accommodation, and depositional trends is summarized in Fig. 3.1. Tectonics is commonly equated with basin subsidence, but additional processes such as crustal cooling, crustal loading, water-depth changes and sediment compaction may also bring important contributions to the total subsidence in the basin. The dissolution and/or withdrawal of evaporites at depth have also been documented as possible subsidence mechanisms (e.g., Waldron and Rygel, 2005). Eustasy and tectonics both control directly the amount of space (accommodation) that is available for sediments to accumulate. Climate mainly affects accommodation *via* eustasy, as for example during glacio-eustatic falls and rises in sea level, but also by changing energy levels in continental to marine environments (e.g., seasonal fluvial discharge; wind regimes in eolian environments; fairweather *vs.* storm waves and currents in marine or lacustrine settings). The effect of climate is also reflected in the amount of sediment supply, by modifying the efficiency of weathering, erosion, and sediment transport processes.

It is important to note that the allogenic controls are 'external' relative to the sedimentary basin, but not necessarily independent of each other (Fig. 3.1).

Eustatic fluctuations of global sea level are controlled by both tectonic and climatic mechanisms, over various time scales (Fig. 3.2). Global climate changes are primarily controlled by orbital forcing (e.g., Milankovitch cycles with periodicities of 10^4–10^5 years; Fig. 2.16), but at more local scales may also be triggered by tectonic processes such as the formation of thrust-fold belts that may act as barriers for atmospheric circulation. Tectonism is primarily driven by forces of internal Earth dynamics, which are expressed at the surface by plume or plate tectonic processes. There is increasing

Hierarchical order	Duration (My)	Cause
First order	200-400	Formation and breakup of supercontinents
Second order	10-100	Volume changes in mid-oceanic spreading centers
Third order	1-10	Regional plate kinematics
Fourth and fifth order	0.01-1	Orbital forcing

FIGURE 3.2 Tectonic and orbital controls on eustatic fluctuations (modified from Vail *et al.*, 1977, and Miall, 2000). Local or basin-scale tectonism is superimposed and independent of these global sea-level cycles, often with higher rates and magnitudes, and with a wide range of time scales.

evidence that the tectonic regimes which controlled the formation and evolution of sedimentary basins in the more distant geological past were much more erratic in terms of origin and rates than formerly inferred solely from the study of the Phanerozoic record (e.g., Eriksson *et al.*, 2004; Eriksson *et al.*, 2005a, b). The more recent basin-forming processes seem to be largely related to a rather stable plate tectonic regime, whereas the formation of Precambrian basins reflects a combination of competing mechanisms, including magmatic-thermal processes ('plume tectonics') and a more erratic plate tectonic regime (Eriksson and Catuneanu, 2004b). These insights offered by the Precambrian record are critical for extracting the essence of how one should categorize the stratigraphic sequences that can be observed within a sedimentary succession at different scales. This issue is discussed in more detail in the chapter dealing with the sequence stratigraphic hierarchy (Chapter 8).

Signatures of Allogenic Controls

The signature of the eustatic control on sedimentation may be recognized from (1) the tabular geometry of sedimentary sequences, suggesting that accommodation was created in equal amounts across the entire basin; (2) the synchronicity of depositional and erosional events across the entire basin, and beyond; and (3) the lack of source area rejuvenation, as it may be suggested by the absence of conglomerates along the proximal rim of the basin. The sea-level control on sedimentation has been documented in numerous case studies, with a degree of confidence that improves with decreasing stratigraphic age (e.g., Suter *et al.*, 1987; Plint, 1991; Miller *et al.*, 1991, 1996, 1998, 2003, 2004; Long, 1993; Locker *et al.*, 1996; Stoll and Schrag, 1996; Kominz *et al.*, 1998; Coniglio *et al.*, 2000; Kominz and Pekar, 2001; Pekar *et al.*, 2001; Posamentier, 2001; Olsson *et al.*, 2002). Estimates of sea-level changes in the geological record have been obtained in recent years by back-stripping, accounting for water-depth variations, sediment loading, compaction, basin subsidence and foraminiferal $\delta^{18}O$ data. Studies of the 'ice-house world' of the past 42 Ma have demonstrated a relationship between depositional sequence boundaries and global $\delta^{18}O$ increases, linking stages of sequence-boundary formation with glacio-eustatic sea-level lowerings (e.g., Miller *et al.*, 1996, 1998). Even for the 'greenhouse world' of the Late Cretaceous—Early Cenozoic interval (prior to 42 Ma), backstripping studies on the New Jersey Coastal Plain, which was subject to minimal tectonic activity, indicate that sea-level fluctuations occurred

with amplitudes of > 25 m on time scales of < 1 Ma (Miller *et al.*, 2004). Such studies have questioned the assumption of a completely ice-free world during the Cretaceous interval, and have revamped the importance of sea-level changes on accommodation and sedimentation (e.g., Stoll and Schrag, 1996; Price, 1999; Miller *et al.*, 2004).

Tectonism is a common control in any sedimentary basin, and its manifestation leads to (1) a wedge-shaped geometry of sedimentary sequences, due to differential subsidence; (2) the accumulation of coarser-grained facies along the proximal rim of the basin in relation to the rejuvenation (uplift) of the source areas; (3) variations in the maximum burial depths of the sedimentary succession across the basin, as can be determined from the study of late diagenetic minerals, fluid inclusions, vitrinite reflection, apatite fission track, etc.; (4) changes in syndepositional topographic slope gradients, as inferred from the shift in fluvial styles through time; and (5) changes in the direction of topographic tilt, as inferred from paleocurrent measurements. The role of tectonic mechanisms in the development of stratigraphic cycles and unconformities has been documented for sedimentary basins spanning virtually all stratigraphic ages, from Precambrian to Phanerozoic and present-day depositories. Early assumptions indicated that tectonic processes may operate mainly on long time scales, of > 10^6 years (e.g., Vail *et al.*, 1977, 1984, 1991; Haq *et al.*, 1987; Posamentier *et al.*, 1988; Devlin *et al.*, 1993), leaving eustasy as the likely cause of higher-frequency cyclicity, at time scales of 10^6 years or less. Advances in our understanding of tectonic processes have led to the realization that tectonically-driven cyclicity may actually develop over a much wider range of time scales, both greater than and less than 1 Ma (e.g., Cloetingh *et al.*, 1985; Karner, 1986; Underhill, 1991; Peper and Cloetingh, 1992; Peper *et al.*, 1992, 1995; Suppe *et al.*, 1992; Karner *et al.*, 1993; Eriksson *et al.*, 1994; Gawthorpe *et al.*, 1994, 1997; Peper, 1994; Yoshida *et al.*, 1996, 1998; Catuneanu *et al.*, 1997a, 2000; Catuneanu and Elango, 2001; Davies and Gibling, 2003). Therefore, the eustatic and tectonic mechanisms may compete toward the generation of any order of stratigraphic cyclicity. The challenge in this situation is to evaluate their relative importance on a case by case basis. In this light, it has been noted that the amplitudes of sea-level changes reconstructed by means of backstripping (e.g., Miller *et al.*, 1991, 1996, 1998, 2003, 2004; Locker *et al.*, 1996; Stoll and Schrag, 1996; Kominz *et al.*, 1998; Coniglio *et al.*, 2000; Kominz and Pekar, 2001; Pekar *et al.*, 2001) are in many cases lower than those interpreted from seismic data (e.g., Haq *et al.*, 1987), questioning the accuracy of seismic data interpretations in terms of eustatic sea-level changes (Miall, 1986, 1992, 1994, 1997; Christie-Blick *et al.*, 1990; Christie-Blick and Driscoll, 1995). Field observations also indicate that the amount of erosion associated with many sequence-bounding unconformities in tectonically active basins was often greater than the inferred amplitude of eustatic fluctuations, suggesting that the basinward shifts of facies associated with stages of base-level fall are not necessarily related to changes in sea level (e.g., Christie-Blick *et al.*, 1990; Christie-Blick and Driscoll, 1995). All these insights re-emphasized the importance of tectonism as a control on accommodation and sedimentation, which, in tectonically active basins, may explain the observed cyclicity at virtually any time scale.

Climate changes within the 10^4–10^5 years Milankovitch band are attributed to several separate components of orbital variation, including orbital eccentricity, obliquity, and precession. Variations in orbital eccentricity, which refers to the shape (degree of stretching) of the Earth's orbit around the Sun, have major periods at around 100 and 413 ka. Changes of up to 3° in the obliquity (tilt) of the ecliptic have a major period of 41 ka. The precession of the equinoxes, which refers to the rotation (wobbling) of the Earth's axis as a spinning top, records an average period of about 21 ka (Fig. 2.16; Imbrie and Imbrie, 1979; Imbrie, 1985; Schwarzacher, 1993). In addition to Milankovitch-band processes, other astronomical forces may affect the climate over shorter time intervals, from a solar band (tens to hundreds of years range; e.g., sun-spot cycles), to a high-frequency orbital band (e.g., nutation cycles of the motion of the axis of rotation of the Earth about its mean position, with a periodicity of about 18.6 years) and a calendar band (cyclicity related to seasonal rhythms, such as freeze–thaw, varves, or fluvial discharge cycles, and other sub-seasonal effects driven by the Earth–Moon system interaction) (e.g., Fischer and Bottjer, 1991; Miall, 1997). Fluctuations in the syndepositional paleoclimate may be reconstructed by combining independent research methods such as (1) thin section petrography of the detrital framework constituents in sandstones, looking at the balance between stable and unstable grains; (2) the mineralogy of the early diagenetic constituents, assuming a short lag time between the deposition of the detrital grains and the precipitation of early diagenetic minerals; (3) the isotope geochemistry of early diagenetic cements; and (4) foraminiferal $\delta^{18}O$ data. Each of these techniques may potentially be affected by drawbacks when it comes to the unequivocal interpretation of syndepositional paleoclimates, so their use in conjunction allows for more reliable conclusions (e.g., Khidir and Catuneanu, 2003). The role of climate as a major control on sedimentation has been emphasized in numerous case studies,

including Blum (1994), Tandon and Gibling (1994, 1997), Miller *et al.* (1996, 1998), Blum and Price (1998), Heckel *et al.*, (1998), Miller and Eriksson (1999), Ketzer *et al.* (2003a, b) and Gibling *et al.* (2005).

Relative Importance of Allogenic Controls

The relative importance of climate, tectonism, and sea-level change on sediment accommodation is illustrated in Fig. 3.3. In marine environments, the balance between eustasy and subsidence changes according to the subsidence patterns that characterize each tectonic setting. For example, the rates of subsidence in extensional settings increase in a distal direction, and the opposite is true for foreland systems (Figs. 2.62 and 2.63). In fluvial environments, the effect of sea-level change

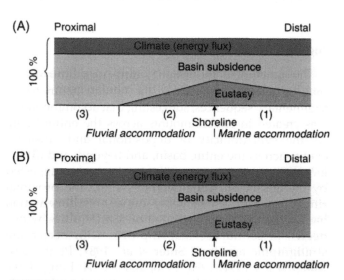

FIGURE 3.3 Relative importance of allogenic controls on accommodation in (A) extensional and (B) foreland basins. Subsidence patterns affect the balance between subsidence and eustasy in marine environments. Sedimentary basins may be subdivided into three distinct areas, based on the dominant controls on accommodation: (1) marine (or lacustrine, if eustasy is substituted with lake level) environments, where the amount of available accommodation is mainly controlled by subsidence and sea-level change; (2) downstream reaches of fluvial environments, which are still affected by sea-level change; and (3) upstream reaches of fluvial environments, unaffected by sea-level change. Note that the vertical scale suggests relative contributions of allogenic controls, and not actual amounts of accommodation. Accommodation increases in a distal direction in extensional basins, and in a proximal direction in foreland settings (Figs. 2.62 and 2.63). Variations in energy flux induced mainly (but not exclusively) by climate may affect accommodation in all environments. The boundaries that separate the relative contributions of eustasy, subsidence and climate may shift depending on local conditions. See also Fig. 3.4 for the actual processes that facilitate the climatic, subsidence and eustatic controls on fluvial and marine accommodation.

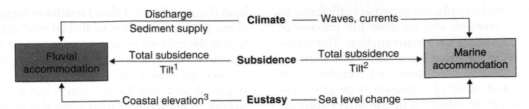

FIGURE 3.4 Processes that enable climatic, subsidence and eustatic controls on fluvial and marine accommodation. Notes: [1]—differential subsidence may modify the water velocity in fluvial systems; [2]—differential subsidence may influence the type of gravity flows that are manifest in marine/lacustrine environments; [3]—changes in coastal elevation may trigger shifts in slope gradients (and corresponding fluvial-energy flux) in the downstream reaches of fluvial systems. Ultimately, all allogenic controls modify the balance between sediment supply and energy flux in each depositional environment, leading to the manifestation of processes of erosion (negative accommodation) or sediment accumulation (positive accommodation).

diminishes in an upstream direction. Beyond the landward limit of eustatic influences, fluvial processes of aggradation or erosion are entirely controlled by climate and tectonism. Fluctuations in environmental energy flux, largely (but not exclusively) controlled by climate over various time scales, also have an impact on the amounts of accommodation that are available in each depositional environment (Figs. 3.3 and 3.4). Increases in energy flux result in losses of available accommodation, whereas decreases in energy flux allow for more sediment accumulation. Such fluctuations in environmental energy may occur from seasonal and sub-seasonal time scales (e.g., seasonal changes in mean precipitation rates and their impact on fluvial discharge, or the effect of fairweather *vs.* storm conditions on marine waves and currents) to longer-term time scales (e.g., Milankovitch cycles of glaciation and deglaciation, and their long-term effects on fluvial discharge).

The total amount of subsidence in the basin is arguably the most important control on accommodation, as the overall geometry of the basin fill ultimately reflects the pattern of basin subsidence (Figs. 2.62 and 2.63). As sea-level change commonly affects accommodation only in restricted portions of a basin (zones 1 and 2 in Fig. 3.3), subsidence also provides a common thread for the general patterns of accommodation changes across the entire basin. These overall trends are modified by fluctuations in energy flux, as explained above, and also by the superimposed effects of sea-level change. Figure 3.3 only provides a schematic illustration of these basic principles, and the boundaries that separate the relative contributions of the main allogenic controls on accommodation may shift as a function of local conditions in each sedimentary basin. These issues are discussed in more detail below, as well as in subsequent chapters of this book.

SEDIMENT SUPPLY AND ENERGY FLUX

Sediment Supply

Sediment supply is an important variable in sequence stratigraphic analyses, and it refers to the amount (or flux) and type (grain size) of sediment that is supplied from source areas to depositional areas by various transport agents, including gravity, water, and wind. The importance of sediment supply in stratigraphy, and especially on the manifestation of transgressions and regressions, was recognized at least since the eighteenth century, when Hutton attributed the migration of shorelines to the shifting balance between riverborne sediment supply and the marine processes of sediment reworking within the receiving basin (in Playfair, 1802). These early ideas have been subsequently refined in landmark publications by Lyell (1868), who related the progradation of deltas to an excess of sediment supply; Grabau (1913), who linked transgressions and regressions to the interplay of sediment supply and the 'depression' caused by subsidence within the receiving basin (precursor of what we call today 'accommodation'); and Curray (1964), who reiterated the role of sediment supply and relative sea level as the primary controls on transgressions and regressions. Following the birth of seismic and sequence stratigraphy in the 1970s and 1980s, the integration of sediment supply in modern stratigraphic analyses has become the norm (e.g., Jervey, 1988; Flemings and Jordan, 1989; Jordan and Flemings, 1991; Swift and Thorne, 1991; Thorne and Swift, 1991; Schlager, 1992, 1993; Johnson and Beaumont, 1995; Helland-Hansen and Martinsen, 1996; Catuneanu *et al.*, 1998b; Cross and Lessenger, 1999; Paola *et al.*, 1999; etc.)

Sediment supply is primarily a by-product of climate and tectonism. A wetter climate increases the amount of sediment supply, *via* increased efficiency of weathering and erosion, and so does the process of tectonic uplift *via* source area rejuvenation. The transport capacity of the transport agents may also increase under wetter climatic conditions (e.g., higher river discharge) and as a result of increased slope gradients due to tectonic tilt. In addition to the direct controls exerted by climate (e.g., *via* precipitation rates, temperature fluctuations) and source area tectonism, the substrate lithology and the vegetation cover of the sediment source areas also influence the flux and grain size of the sediment transported by rivers or wind (Blum, 1990; Einsele, 1992; Miall, 1996).

Sediment supply is critical to the stratigraphic architecture of any sedimentary basin, as it is one of the fundamental variables that determine the type of depositional trends in all fluvial to marine environments (Fig. 3.1). Once accommodation is made available by subsidence or sea-level change, the lithology, location, and stacking patterns of depositional elements are largely a function of the volume and type of sediment supply. At the same time, as a consequence of sediment accumulation, more accommodation is created as a result of isostatic sediment loading (Matthews, 1984; Schlager, 1993). The relationship between sedimentation and accommodation is thus a two-way process/response interaction, as sedimentation does not only consume accommodation made available by other mechanisms, but may also create additional space as sediment aggradation/loading proceeds. This fact is valid for all fluvial to marine environments, as isostatic sediment loading contributes to the total subsidence in the basin that is otherwise caused by tectonic, thermal, or sediment compaction processes.

Sediment Supply *vs.* Environmental Energy Flux

Variations in sediment supply may also be conducive to the manifestation of depositional processes of aggradation or erosion, but the significance of such variations is relative to the energy flux of each particular environment. In marine basins, sediment is transported by a variety of subaqueous currents, including wave-induced (longshore, rip), tidal, contour, or gravity flows, and the nature of processes at the seafloor (sediment accumulation *vs.* erosion) is dictated by the balance between the energy (transport capacity) of the current and its sediment load. A marine current that has more energy than that required to transport its sediment load (i.e., underloaded flow) erodes the seafloor, whereas a current that drops its energy below the level that is required to transport its entire sediment

load (i.e., overloaded flow) results in aggradation. The same principle applies to fluvial and eolian systems, where the balance between the energy of the transport agent (water, wind) and its sediment load controls surface processes of aggradation or downcutting (Figs. 3.5 and 3.6). Even though the role of sediment supply in reducing or increasing the amount of available accommodation is not captured in Fig. 3.3, it is implied that the 'energy flux' factor stands for this dynamic energy/sediment balance, as an increase in energy flux *relative to sediment supply* leads to a loss of accommodation, and a decrease in energy flux *relative to sediment supply* results in a gain of accommodation.

Shifts in the balance between energy flux and sediment supply may be caused by each of the allogenic controls on accommodation (climate, subsidence/uplift, or sea-level change; Figs. 3.1 and 3.4), either independently or in any combination thereof. In the early days of sequence stratigraphy it was generally implied that sea-level change exerts the main control on stratigraphic architecture, and implicitly on processes of aggradation or erosion (Vail *et al.*, 1977; Posamentier *et al.*, 1988). In the 1990s, tectonism was emphasized as an equally important control, and the combination of eustatic and tectonic processes was invoked as the key driving force behind surface processes of deposition or sediment reworking (Hunt and Tucker, 1992; Posamentier and Allen, 1999). Climate was generally left out of sequence stratigraphic models, as it was the most difficult allogenic mechanism to quantify, but its effect on sediment aggradation or erosion was proven to be as important as the control exerted by eustasy or tectonism (Blum, 1994; Blum and Price, 1998; Gibling *et al.*, 2005). Syndepositional surface processes of aggradation or erosion ultimately reflect the interplay of all three allogenic controls, whose effects may enhance or cancel each other out depending on local circumstances. The Late Cenozoic fluvial record of the U.S. Gulf Coast provides an example where climate and sea-level change promoted opposite depositional trends during stages of glaciation and interglaciation. In this case study, the climatic control on fluvial discharge outpaced the effects of sea-level change, leading to fluvial aggradation during glacial periods (driven by a drop in fluvial discharge/energy flux, in spite of the coeval glacio-eustatic fall) and fluvial erosion during interglacial stages (as a result of increased fluvial discharge due to ice melting, and despite the rise in sea level) (Blum, 1990, 1994). Similar examples of fluvial incision triggered by climate-controlled increases in discharge during times of glacial melting and global sea-level rise are also found in western Canada (Fig. 3.7).

Ultimately, *all processes of aggradation or erosion are linked to the shifting balance between environmental energy*

FIGURE 3.5 Surface processes that reflect the dynamic interplay of sediment supply and wind energy in eolian environments. Sediment supply exceeding the transport capacity (energy) of winds results in the accumulation of sand as sheets or dunes, depending on flow regimes. Winds stronger relative to their sediment load lead to erosion and the formation of deflation surfaces. A—sand dunes in the Namib Desert (Namibia), formed as a result of abundant sediment supply (sediment supply > wind energy; photo courtesy of Roger Swart); B—deflation surface on Mars (wind energy > sediment supply; photo courtesy of NASA); C—deflation surface in the Namib Desert, Namibia (wind energy > sediment supply); D—deflation surface in the Namib Desert, Namibia (detail showing the concentration of heavy minerals as lag deposits on top of the Precambrian dolomites basement rocks).

flux and sediment supply (i.e., aggradation occurs only where sediment supply outpaces energy flux, and erosion occurs only where energy outpaces sediment load). In turn, accommodation is closely related to the shifting balance between energy flux and sediment supply, both as a control but also as a controlled variable (see the two-way relationship indicated in Fig. 3.1). On the one hand, the balance between energy flux and sediment supply affects the amounts of available accommodation, although accommodation is also independently controlled by other factors as well (Figs. 3.1, 3.3, and 3.4). As a general rule, accommodation is inversely proportional to energy flux (i.e., an increase

in energy 'erodes' accommodation) and directly proportional to sediment supply (i.e., an increase in sediment supply adds to the amount of available accommodation; Fig. 3.6). On the other hand, changes in accommodation controlled directly by allogenic mechanisms may also affect the balance between energy flux and sediment supply within the basin. For example, an increase in accommodation, such as in response to subsidence or sea-level rise tends to reduce the energy level at the seafloor, thus promoting sediment aggradation. This explains why, in virtually any situation, depositional trends may ultimately be related to shifts in the balance between energy flux

FIGURE 3.6 Satellite image of southern Arabian Peninsula showing a gradual shift in the balance between sediment supply and wind energy from the upper-left corner of the image (sediment supply dominant) to the lower-right corner of the image (wind energy dominant). Accommodation is positive where sediment supply dominates, leading to the accumulation of sand on top of the basement rocks. Accommodation is negative where energy is in excess, leading to the exposure and erosion of the basement rocks. The longitudinal dunes shown in this image are parallel to the prevailing northeasterly winds, and are the equivalent of parting lineation of the upper flow regime of subaqueous bedforms. The wind regime in this case is higher energy relative to the wind regime that generated the transversal dunes shown in Fig. 3.5A, which are the equivalent of dunes of the lower flow regime of subaqueous bedforms.

FIGURE 3.7 Aerial photograph showing the modern incised valley of the Red Deer River (Alberta). Note farm houses for scale. Tributaries are also incised, which is one of the diagnostic features of incised valleys. The incision of the Red Deer River valley was climate-controlled, and caused by the significant increase in fluvial discharge associated with the rapid glacial melting during the Late Pleistocene.

and sediment supply. The two-way process/response relationship between energy flux/sediment supply, on the one hand, and accommodation, on the other, is illustrated in Fig. 3.1.

A simple illustration of how a shifting balance between sediment supply and environmental energy flux may affect accommodation and depositional processes in a shallow-marine setting is presented in Fig. 3.8. The scenario in Fig. 3.8 assumes that sediment is supplied by a river that flows along its graded profile, to a stable coastline that is not affected by subsidence or sea-level changes. The elimination of the effects of subsidence and sea-level change on accommodation allows for a direct evaluation of the depositional processes that take place in this shallow-marine environment in response to the interplay of sediment supply and wave energy. If sediment supply and environmental energy flux are in perfect balance (case A in Fig. 3.8), all sediments will bypass this area, without erosion or aggradation, being removed by

longshore drift. In this case accommodation is zero, in spite of the available water column in the marine environment, and the base level is superimposed on the seafloor—in other words, the seafloor corresponds to a graded profile. If sediment supply outpaces the capacity of the environment to remove it, sediment aggradation and progradation will occur (case B in Fig. 3.8). In this case, base level is above the seafloor and accommodation is positive. Where the energy of the environment outpaces sediment supply, erosion of the seafloor will occur (case C in Fig. 3.8). In this case, base level is below the seafloor, accommodation is negative, and coastline erosion may lead to the retrogradation of the shoreline. An important lesson from this diagram is that the amount of available *accommodation is not measured to the sea level*, but rather to a graded profile (base level) that may be in any spatial relationship with the sea level and the seafloor. The situation depicted in Fig. 3.8 is a simplification of the common reality, which is that other factors, such as subsidence and sea-level change, may also affect accommodation in parallel with (and independent of) fluctuations in energy and/or sediment supply (Fig. 3.8).

This discussion indicates that accommodation and sediment supply are not independent variables, as they are often in a process/response relationship that is modulated by environmental energy flux. Consequently, the axiom that the sequence stratigraphic architecture

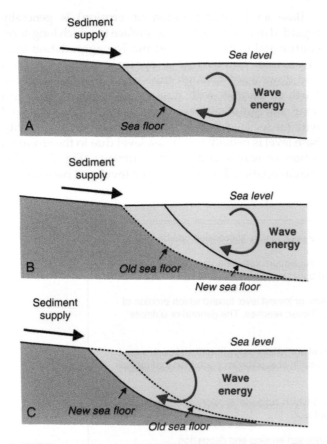

FIGURE 3.8 Relationship between energy flux, sediment supply, base level and accommodation in a shoreface environment that is not affected by subsidence or sea-level change. A—sediment supply is perfectly balanced by wave energy. In this case, all sediment bypasses the area, base level is superimposed on the seafloor, and accommodation is zero; B—sediment supply outpaces wave energy. In this case, sediment aggradation and progradation take place, base level is above the seafloor (superimposed on the sea level, as drafted in the diagram), and accommodation is positive; C—sediment supply is outpaced by wave energy. In this case, coastal and seafloor erosion take place, base level is below the seafloor, and accommodation is negative. Note that *accommodation is not measured to the sea level*, but rather to a graded profile (base level) that may be in any spatial relationship with the sea level and the seafloor. Where accommodation is not affected by subsidence or sea-level change, it is entirely controlled by fluctuations in energy flux and sediment supply. Also note that no confusion should be made between accommodation (space available for *sediments* to fill, measured from the seafloor to the base level) and water depth (space available for *water* to fill, measured from the seafloor to the sea level). For example, more volume is made available for water to fill in case C, but accommodation is negative due to the exceedingly strong wave energy.

is controlled by the interplay between the rate of change in accommodation and the rate of sediment supply (e.g., Schlager, 1993) is only valid as an approximation, since the two variables depend on each other. For this axiom to be true, the approximations being made are that accommodation is measured to the sea level, rather

than to the base level (in which case accommodation becomes independent of sediment supply), and that sediment supply is proportional to the sedimentation rates. In reality, none of these approximations are entirely accurate, as discussed above and also in more detail in the following section of this chapter. One has to keep in mind the difference between *sediment supply*, which is measured as a flux, and *sedimentation* (rate), which is measured as a change in vertical distance at any location. Depending on energy flux conditions, a high sediment supply does not necessarily translate into high rates of sedimentation. While accommodation depends on sediment supply, it is measured independently of sedimentation. Therefore, the correct relationship in terms of the controls on stratigraphic architecture is portrayed by the interplay between the rate of change in accommodation and the rate of sedimentation, as both are measured, independently of each other, in units that reflect changes in vertical distance at any particular location. Further discussion on this topic is provided in the following section of this chapter.

The amounts of available marine accommodation may be modified by all three allogenic controls, whose relative importance varies with the basin (Fig. 3.3). Fluvial processes of aggradation or erosion (positive or negative fluvial accommodation, respectively) are increasingly influenced by sea-level change towards the shoreline and by climate and tectonism towards the source areas (Blum, 1990; Posamentier and James, 1993). In nonmarine regions, eustasy is therefore a more important downstream factor, whose importance diminishes in a landward direction, whereas climate and tectonism compensate this trend by becoming increasingly important upstream (Fig. 3.3). More details about the intricate process/response relationship between the allogenic controls, accommodation, and sedimentation are provided in the following sections of this chapter, as well as throughout the book.

SEDIMENT ACCOMMODATION

Definitions—Accommodation, Base Level, and Fluvial Graded Profiles

The concept of sediment 'accommodation' describes the amount of space that is available for sediments to fill, and it is measured by the distance between base level and the depositional surface (Jervey, 1988). This concept was initially applied to marine environments, as a tool to enable mathematical simulations of progradational basin-filling on divergent continental margins (Jervey, 1988). In this context, base level was equated, at first approximation, with sea level, and hence the original definition of 'accommodation' did not require further explanations of what the meaning of 'base level' may be in continental environments. It is now widely agreed that accommodation may be made available in both fluvial and marine environments by the combined effects of climate, tectonism, and sea-level change (Fig. 3.3). The expansion of the concept of accommodation into the nonmarine portion of sedimentary basins brought about further scrutiny of the concept of base level, which led to conflicting ideas and terminology (Shanley and McCabe, 1994; Fig. 3.9).

Base level (of deposition or erosion) is generally regarded as a global reference surface to which long-term continental denudation and marine aggradation tend to proceed. This surface is dynamic, moving upward and downward through time relative to the center of Earth in parallel with eustatic rises and falls in sea level. For simplicity, base level is often approximated with the sea level (Jervey, 1988; Schumm, 1993). In reality, base level is usually below sea level due to the erosional action of waves and marine currents (Fig. 3.8). This spatial relationship between sea level and base level is

---

Base level (Twenhofel, 1939): highest level to which a sedimentary succession can be built.

Base level (Sloss, 1962): an imaginary and dynamic equilibrium surface above which a particle cannot come to rest and below which deposition and burial is possible.

Base level (Bates and Jackson, 1987): theoretical limit or lowest level toward which erosion of the Earth's surface constantly progresses but rarely, if ever, reaches. The general or ultimate base level for the land surface is sea level.

Base level (Jervey, 1988): ... is controlled by sea level and, at first approximation, is equivalent to sea level ... although, in fact, a secondary marine profile of equilibrium is attained that reflects the marine-energy flux in any region.

Base level (Schumm, 1993): the imaginary surface to which subaerial erosion proceeds. It is effectively sea level, although rivers erode slightly below it.

Base level (Cross, 1991): a surface of equilibrium between erosion and deposition.

Base level (Cross and Lessenger, 1998): a descriptor of the interactions between processes that create and remove accommodation space and surficial processes that bring sediment or that remove sediment from that space.

Base level (Posamentier and Allen, 1999): the level that a river attains at its mouth (i.e., either sea level or lake level), and constitutes the surface to which the equilibrium profile is anchored.

There are two schools of thought regarding the concept of base level:

(1) Base level is more or less the sea level, although usually below it due to the action of waves and currents. The extension of this surface into the subsurface of continents defines the ultimate level of continental denudation. On the continents, processes of aggradation versus incision are regulated via the concept of graded (equilibrium) fluvial profile. Graded fluvial profiles meet the base level at the shoreline.

(2) The concept of base level is generalized to define the surface of balance between erosion and sedimentation within both marine and continental areas (the "stratigraphic" base level of Cross and Lessenger, 1998). In this acceptance, the concept of graded fluvial profile becomes incorporated within the concept of base level. The stratigraphic base level will thus include a continental portion (fluvial base level = graded fluvial profile) and a marine portion (marine base level ~ sea level).

The drawback of the second approach is that fluvial base-level shifts are controlled by marine base-level shifts, especially in the downstream reaches of the river system, and hence the two concepts are in a process/response relationship. This suggests that it is preferable to keep these two concepts separate as opposed to incorporating them into one "stratigraphic base level". This is the approach adopted in this book, where the fluvial base level is referred to as the fluvial graded profile, and the marine base level is simply referred to as the base level.

FIGURE 3.9 Definitions of the concept of base level.

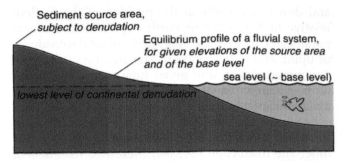

FIGURE 3.10 The concept of base level, defined as the lowest level of continental denudation (modified from Plummer and McGeary, 1996). Graded (equilibrium) fluvial profiles meet the base level at the shoreline. As the elevation of source areas changes in response to denudation or tectonic forces, graded fluvial profiles adjust accordingly. Graded profiles also respond in kind to changes in base level. See also Fig. 3.9 for alternative definitions of base level.

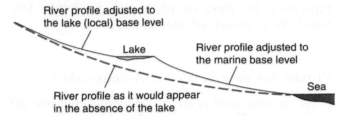

FIGURE 3.11 Marine and local base levels as illustrated by a river flowing into a lake and from the lake into the sea (modified from Press and Siever, 1986). In each river segment, the graded profile adjusts to the lowest level it can reach.

also supported by the fact that rivers meeting the sea erode below sea level (Schumm, 1993), i.e., to the base level (Fig. 3.9).

Figure 3.10 shows a marine to continental area, in which base level is approximated with sea level. The base level may be projected into the subsurface of the continents, marking the lowest level of subaerial erosion (Plummer and McGeary, 1996). The surface topography tends to adjust to base level by long-term continental denudation. Between the source areas that are subject to denudation and the marine shorelines, processes of nonmarine aggradation may still take place when the amount of sediment load exceeds the transport capacity (energy flux) of any particular transport agent (gravity-, air-, or water-flows).

Coupled with the concept of base level, fluvial equilibrium (graded) profiles are particularly important to understanding processes of sedimentation in continental areas. For any given elevations of the source area and of the body of water into which the river debouches, fluvial systems tend to develop a dynamic equilibrium in the form of a graded longitudinal profile (Miall, 1996, p. 353). This equilibrium profile is achieved when the river is able to transport its sediment load without aggradation or degradation of the channels (Leopold and Bull, 1979). Rivers that are out of equilibrium will aggrade or incise in an attempt to reach the graded profile (Butcher, 1990, p. 376). In this context, fluvial systems start adjusting to new equilibrium profiles as soon as the elevation of source areas, the level of the body of water into which the river debouches, and/or any shifts in the balance between fluvial-energy flux and sediment load that these changes may trigger, are modified due to factors such as tectonism, climate, or sea-level change. An equilibrium profile may be below or above the land surface

(triggering incision or aggradation, respectively), and it merges with the base level at the marine shoreline (Fig. 3.10). In a more general sense, the base level for fluvial systems is represented by the level of any body of water into which a river debouches, including sea level, lake level, or even another river (Posamentier and Allen, 1999; Fig. 3.11). Surface processes in inland basins dominated by eolian processes may also be related to local base levels, which are represented by deflation surfaces associated with the level of the groundwater table (Kocurek, 1988).

The marine base level (~ sea level) and the fluvial graded profiles are sometimes used in conjunction to define a composite ('stratigraphic') base level, which is the surface of equilibrium between erosion and deposition within both marine and continental areas (Cross, 1991; Cross and Lessenger, 1998; Fig. 3.9). At any given location, the position of this irregular 3D surface is determined by the competing forces of sedimentation and erosion, and it may be placed either above the land surface/seafloor (where aggradation occurs), or below the land surface/seafloor (where subaerial/submarine erosion occurs).

The debate regarding the relationship between base level and the fluvial graded profile still persists in current sequence stratigraphic terminology. One school of thought argues that the term 'base level' should apply to both concepts, as the same definition can describe them both (i.e., a dynamic surface of equilibrium between deposition and erosion; Barrell, 1917; Sloss, 1962; Cross, 1991; Cross and Lessenger, 1998). A second school of thought restricts the term 'base level' to the level of the body of water into which the river debouches, where an abrupt decrease in fluvial-energy flux is recorded (Powell, 1875; Davis, 1908; Bates and Jackson, 1987; Schumm, 1993; Posamentier and Allen, 1999; Catuneanu, 2003). Terminology is trivial to some extent, but there seems to be value in keeping the concepts of graded fluvial profile and base level separate, as they are in a process/response relationship—i.e., the position in space of the fluvial graded profile is in part a function of the elevation of the base level (Fig. 3.9). This is the approach adopted in this book.

Proxies for Base Level and Accommodation

As the base level is an imaginary and dynamic 4D surface of equilibrium between deposition and erosion, largely dependent on fluctuations in environmental energy and sediment supply (Fig. 3.8), the precise quantification of accommodation at any given time and in any given location is rather difficult. For this reason, proxies may be used for an easier visualization of the available accommodation. At first approximation, sea level is a proxy for base level (Jervey, 1988; Schumm, 1993), and so the available accommodation in a marine environment may be measured as the distance between the sea level and the seafloor. Both the sea level and the seafloor may independently change their position with time relative to the center of Earth in response to various controls, and therefore the amount of available accommodation fluctuates accordingly. Sea level is one of the primary allogenic controls on sedimentation, and it is in turn controlled by climate and tectonism, as discussed in the previous sections (Fig. 3.1). The upward

and downward shifts in the position of the seafloor relative to the center of Earth depend on two main parameters, namely the magnitude of total subsidence or uplift, and sedimentation. The amount of available accommodation at any given time and in any given location therefore equals the balance between how much accommodation is created (or destroyed) by factors such as tectonism and sea-level change, and how much of this space is consumed by sedimentation at the same time. The distinction between these two members of the accommodation equation (creation/destruction *vs.* consumption) is one of the key themes of sequence stratigraphy, which allows one to understand the fundamental mechanisms behind the formation of systems tracts and sequence stratigraphic surfaces.

Figure 3.12 helps to define some of the basic concepts involved in the accommodation equation, such as eustasy (sea level relative to the center of Earth), relative sea level (sea level relative to a datum that is independent of sedimentation), and water depth (sea level relative to the seafloor). A change in relative sea level is a proxy for how much accommodation was created or lost during a period of time, independent of sedimentation, whereas water depth is a proxy for how much accommodation is still available after the effect of sedimentation is

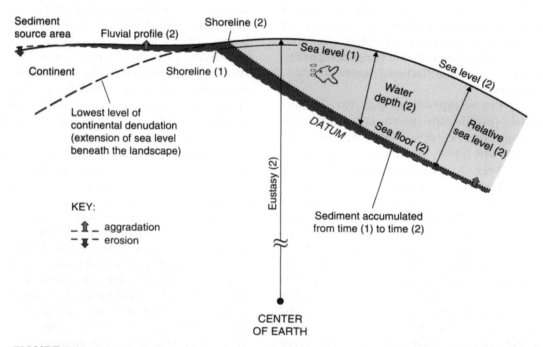

FIGURE 3.12 Eustasy, relative sea level, and water depth as a function of sea level, seafloor, and datum reference surfaces (modified from Posamentier *et al.*, 1988). The datum is a subsurface reference horizon that monitors the amount of total subsidence or uplift relative to the center of Earth. In this diagram, the datum corresponds to the ground surface (subaerial and subaqueous) at time (1). Sedimentation (from time 1 to time 2 in this diagram) buries the datum, which, at any particular location, may be visualized as a G.P.S. that monitors changes in elevation through time (i.e., distance relative to the center of Earth).

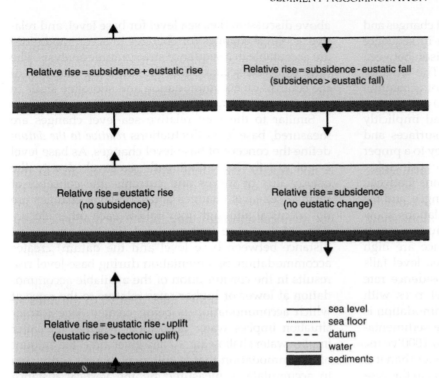

Relative rise = subsidence + eustatic rise

Relative rise = subsidence - eustatic fall
(subsidence > eustatic fall)

Relative rise = eustatic rise
(no subsidence)

Relative rise = subsidence
(no eustatic change)

Relative rise = eustatic rise - uplift
(eustatic rise > tectonic uplift)

—— sea level
········· sea floor
– – – datum
░ water
▓ sediments

FIGURE 3.13 Scenarios of relative sea-level rise. If base level is equated with sea level for simplicity (by neglecting the energy of waves and currents), then relative sea-level rise becomes synonymous with base-level rise. Note that the newly created accommodation may be consumed by sedimentation at any rates, resulting in the shallowing or deepening of the water. The length of the arrows is proportional to the rates of vertical tectonics and eustatic changes.

also taken into account. The datum in Fig. 3.12 monitors the total amount of subsidence or uplift (including the effects of sediment loading and compaction) recorded in any location within the basin relative to the center of Earth. This datum reference horizon is taken as close to the seafloor as possible in order to capture the entire subsidence component related to sediment compaction, but its actual position is not as important as the change in the distance between itself and the

sea level. This is because we are more interested in the *changes* in relative sea level (i.e., changes in the distance between the datum and the sea level), which reflect how much accommodation is created or lost during a period of time, rather than the actual *amount* of relative sea level (i.e., the actual distance between the datum and the sea level) at any given time. Different scenarios for rises and falls in relative sea level are illustrated in Figs. 3.13 and 3.14.

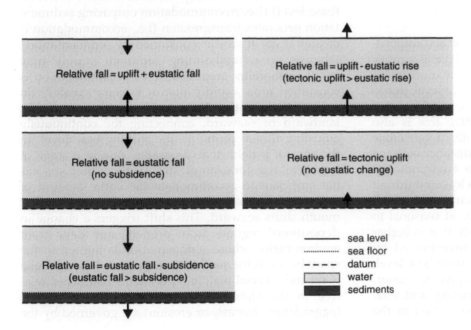

Relative fall = uplift + eustatic fall

Relative fall = uplift - eustatic rise
(tectonic uplift > eustatic rise)

Relative fall = eustatic fall
(no subsidence)

Relative fall = tectonic uplift
(no eustatic change)

Relative fall = eustatic fall - subsidence
(eustatic fall > subsidence)

—— sea level
········· sea floor
– – – datum
░ water
▓ sediments

FIGURE 3.14 Scenarios of relative sea-level fall. If base level is equated with sea level for simplicity (by neglecting the energy of waves and currents), then relative sea-level fall becomes synonymous with base-level fall. Falling base level results in loss of available accommodation, and almost invariably in the shallowing of the water. The length of the arrows is proportional to the rates of vertical tectonics and eustatic changes.

The separation between relative sea-level changes and sedimentation is a fundamental approach in sequence stratigraphy, which allows for the comparison between their rates as independent variables. The balance between these rates (creation/destruction of accommodation *vs.* consumption of accommodation) controls the direction and type of shoreline shifts, and implicitly the timing of all sequence stratigraphic surfaces and systems tracts. This approach is therefore key to a proper understanding of sequence stratigraphic principles. Failure to do so may result in confusions between relative sea-level changes, water-depth changes, and the directions of shoreline shift. Simple calculations show that the relative sea level may rise even during stages of sea-level fall, if the rates of subsidence are high enough (Fig. 3.13). For example, if the sea level falls at a rate of 5 m/1000 years but the subsidence rate is 9 m/1000 years, the relative sea level rises with 4 m/1000 years, which means that accommodation is created at a rate of 4 m/1000 years. If the sedimentation rate in that particular location is 3 m/1000 years, it means that accommodation is created faster than it is consumed, and hence the water is deepening, in this case at a rate of 1 m/1000 years. If the location in this example is placed in the vicinity of the shoreline, then the increase in water depth is likely to be associated with a shoreline transgression. As shown by numerical modeling, the correlation between water-depth changes and the direction of shoreline shift (i.e., water shallowing = regression, and water deepening = transgression) is only truly valid for shallow-marine areas, and it may be distorted offshore (Catuneanu *et al.*, 1998b).

Changes in Accommodation

The above discussion on the controls on accommodation is based on the assumption that sea level is a proxy for base level. This is true at first approximation, but in reality base level is commonly *below* the sea level, due to the energy flux brought about by waves and currents (Fig. 3.8). As noted by Schumm (1993), this is also supported by the fact that at their mouths, rivers erode slightly below the sea level. The actual distance between base level and sea level depends on environmental energy, as for example the base level is lowered during storms relative to its position during fairweather. Such energy fluctuations usually take place at seasonal to sub-seasonal time scales, at a frequency that is higher than most highest-frequency cycles investigated by sequence stratigraphy. Longer-term shifts in base level, at scales relevant to sequence stratigraphy, are generally controlled by the interplay of eustasy and total subsidence. In other words, the proxies used in the

above discussion (i.e., sea level for base level, and relative sea-level changes for changes in accommodation) are acceptable in a sequence stratigraphic analysis. The most complete scenario that illustrates the interplay of the controls on accommodation and shoreline shifts in a marine environment is presented in Fig. 3.15.

Similar to the way relative sea-level changes are measured, base-level fluctuations *relative to the datum* define the concept of base-level changes. As base level is not exactly coincident with sea level, due to the energy flux of waves and currents, the concepts of relative sea-level changes and base-level changes are not identical although they follow each other closely (Fig. 3.15). A rise in base level (increasing vertical distance between base level and the datum) creates accommodation. Sedimentation during base-level rise results in the consumption of the available accommodation at lower or higher rates relative to the rates at which accommodation is being created. The former situation implies water deepening, whereas the latter implies water shallowing. At any given time, the amount of accommodation that is still available for sediments to accumulate is measured by the vertical distance between the seafloor and the base level. Similarly, a fall in base level (decreasing vertical distance between base level and the datum) destroys accommodation. Almost invariably, such stages result in water shallowing in that particular location, irrespective of the depositional processes.

The contrast between the rates of change in accommodation and the sedimentation rates in locations placed in the vicinity of the shoreline allows one to understand why the shoreline may shift either landward or seaward during times of relative sea-level (base-level) rise. Accommodation outpacing sedimentation generates transgression (i.e., accommodation is created faster than it is consumed by sedimentation), whereas an overwhelming sediment supply may result in shoreline regression (i.e., accommodation is consumed more rapidly than it is being created). In either situation, the river mouth moves accordingly, landward or seaward, connecting the continuously adjusting fluvial profile to the shifting base level. In the case of a delta that progrades during a stage of base-level rise, for example, the newly created space is not sufficient to accommodate the entire amount of sediment brought by the river, and as a result the river mouth shifts seaward. This shift triggers a change in depositional regimes from prodelta and delta front environments, where sedimentation is limited to the space between the seafloor and the base level, to delta plain and alluvial plain environments (landward relative to the shoreline), where depositional trends (aggradation, bypass, or erosion) are governed by the

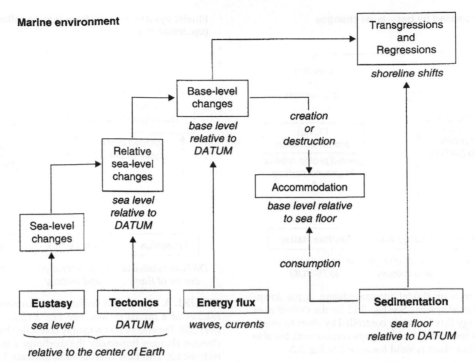

FIGURE 3.15 Controls on accommodation and shoreline shifts in a marine environment (modified from Catuneanu, 2003). This diagram also applies to lacustrine environments by substituting sea level with lake level. See Fig. 3.12 for the definition of the *DATUM*. The energy flux lowers the base level *via* the effects of waves, wave-generated currents, tidal currents, contour currents, or gravity flows. Short-term climatic changes (seasonal to sub-seasonal time scales) are accounted for under energy flux, whereas the longer-term climatic changes (e.g., Milankovitch type) are built into eustasy. The 'energy flux' box stands for the dynamic balance between environmental energy and sediment supply, as an increase in energy *relative to sediment supply* leads to base-level fall (loss of accommodation), and a decrease in energy *relative to sediment supply* leads to base-level rise (gain of accommodation). Note the difference between 'sediment supply' (load moved by a transport agent) and 'sedimentation' (amount of vertical aggradation). For example, depending on energy flux conditions, high sediment supply does not necessarily result in high sedimentation rates. Base-level changes depend on sediment supply, but are measured independently of sedimentation. In contrast, relative sea-level changes are independent of both sediment supply and sedimentation. This flow chart is valid for zone 1 in Fig. 3.3.

relative position between the fluvial graded profile and the actual fluvial profile.

The fluvial graded profile is the conceptual equivalent of the marine base level in the nonmarine realm, as it describes the imaginary and dynamic surface of equilibrium between deposition and erosion in the fluvial environment. In this context, the amount of *fluvial accommodation* is defined as the space between the graded profile and the actual fluvial profile (Posamentier and Allen, 1999). If we compare this definition with the concept of *marine accommodation*, discussed above, the graded profile is the equivalent of the base level, and the actual fluvial profile is the counterpart of the seafloor in the marine environment. If we follow this comparison even farther, we notice that the sea level, which is used as a proxy for base level, does not have an equivalent in the fluvial realm, which makes the visualization of fluvial accommodation rather difficult as there is no physical proxy for

the fluvial graded profile. The only observable surface is the actual fluvial landscape, whose position relative to an independent datum changes in response to surface processes of aggradation or erosion (Fig. 3.12). In turn, these surface processes are triggered by an attempt of the river to reach its graded profile.

The graded profile is 'anchored' to the base level at the river mouth, and as the base level rises and falls, this anchoring point moves either landward or seaward, or up or down, triggering an in-kind response of the graded profile (Posamentier and Allen, 1999). Therefore, base-level changes exert an important control on graded profiles, and implicitly on fluvial accommodation, especially in the downstream reaches of the fluvial system (Shanley and McCabe, 1994; Fig. 3.16). The position of graded profiles also depends on fluctuations in energy flux, which are mainly attributed to the effects of climate on a river's transport capacity (Blum and Valastro, 1989; Blum, 1990; Fig. 3.16). Such energy fluctuations may

Fluvial system influenced by base-level changes
(downstream end)

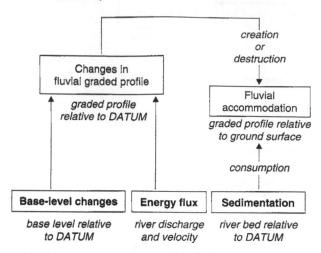

FIGURE 3.16 Controls on fluvial accommodation in the downstream reaches of a fluvial system. See Fig. 3.12 for the definition of the *DATUM*. The energy flux is mainly controlled by short to longer term climatic changes (especially the discharge component), but also by tectonic tilt. This flow chart is valid for zone 2 in Fig. 3.3.

Fluvial system isolated from marine influences
(upstream end)

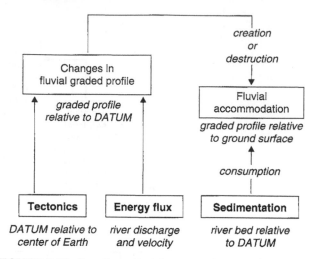

FIGURE 3.17 Controls on fluvial accommodation in the upstream reaches of a fluvial system. See Fig. 3.12 for the definition of the *DATUM*. The energy flux is mainly controlled by short- to longer-term climatic changes (especially the discharge component), but also by tectonic tilt. This flow chart is valid for zone 3 in Fig. 3.3.

be recorded over different time scales, from seasonal climatic changes that may occur with a frequency higher than the highest-frequency cycles studied by sequence stratigraphy, to Milankovitch-scale orbital forcing.

The effect of base-level changes on fluvial processes (aggradation *vs.* erosion) is only 'felt' by rivers within a limited distance upstream relative to the river mouth, which is usually in a range of less than 200 km (Miall, 1997). Beyond the landward limit of base-level influences, rivers respond primarily to a combination of tectonic and climatic controls (Fig. 3.17). Tectonism dictates the overall geometry of fluvial sequences, as the creation of fluvial accommodation follows the patterns of regional subsidence. For example, the rates of subsidence induced by flexural loading in a foreland basin increase in a proximal direction, toward the center of loading, whereas the rates of thermal and mechanical subsidence in an extensional basin increase in a distal direction. Superimposed on these general trends, the climatic control on runoff and discharge also affect the position of graded profiles, as discussed above (Fig. 3.17).

The role of climate as a control on accommodation is always difficult to quantify, as it operates *via* other variables such as eustasy and environmental energy flux. In a marine environment, the short-term climatic changes (seasonal to sub-seasonal time scale) translate into fluctuations in energy flux, whereas the longer-term changes are accounted for under eustasy (Fig. 3.15). In the case of fluvial environments, both short- and

longer-term climatic changes are reflected in the fluctuations in energy flux, as there is no physical proxy for the graded profile that could be related to the longer term climate shifts (Figs. 3.16 and 3.17). Climate is also relevant to the 'sedimentation' box in all cases (Figs. 3.15–3.17), as the amount of sediment supply transferred from source areas to the sedimentary basin depends on the efficiency of weathering and sediment transport processes, both partly dependent on climate.

Changes in accommodation, in conjunction with the rates of sedimentation, represent a key control on depositional trends, which are in turn reflected by specific shoreline shifts (e.g., progradation is associated with shoreline regression, and retrogradation relates to shoreline transgression). Quantitative modeling of the interplay between subsidence, sea-level change and sedimentation shows that even though the shoreline may only shift in one direction along a dip oriented profile at any given time, accommodation may change with different rates, and even in opposite directions, along the same cross-sectional profile (Jervey, 1988; Catuneanu *et al.*, 1998b). This coeval change in the rates and sign of accommodation shifts is caused by differential subsidence, which is usually the norm in any sedimentary basin. The higher the contrasts in the rates of differential subsidence between various areas in the basin, the more pronounced the difference between the amounts of available accommodation will be. For example, during a stage of sea-level fall, accommodation may be negative in slowly subsiding areas (i.e., the rate of eustatic fall exceeds the rate of subsidence), but positive in areas where rapid subsidence prevails over the rates of sea-level fall (Jervey, 1988; Catuneanu et al., 1998b).

As sedimentation rates also vary along dip oriented sections, the interplay of accommodation and sedimentation results in even more complex water-depth trends characterized by different rates of change (e.g., slow vs. rapid deepening or shallowing), or direction of change (shallowing vs. deepening) between various areas in the basin (Jervey, 1988; Catuneanu et al., 1998b). Despite this variability in accommodation and water-depth trends within a basin at any given time, sequence stratigraphic models account for only one reference curve of base-level changes relative to which all systems tracts and sequence stratigraphic surfaces are defined (Fig. 1.7). This reference curve describes changes in accommodation at the shoreline. The interplay between sedimentation and this curve of base-level changes controls the transgressive and regressive shifts of the shoreline, which are referred to in the nomenclature of systems tracts (e.g., 'transgressive systems tracts', or 'regressive systems tracts'; Fig. 1.7). These issues of numerical modelling, and their consequences for the timing of specific events during the evolution of the basin, are dealt with in more detail in Chapter 7.

The success of sequence stratigraphic analyses depends on the understanding of the basic principles. Common sources of confusion are related to the concepts of (1) base-level changes vs. (2) water-depth changes vs. (3) shoreline shifts (transgressions, regressions) vs. (4) grading trends (fining- and coarsening-upward). Keeping these concepts separate is as important as separating data from interpretations. Water shallowing is often confused with base-level fall, and similarly, water deepening may be confused with base-level rise. Base-level changes are measured independent of the sediment that accumulates on the seafloor (i.e., base level relative to datum; Figs. 3.12 and 3.15), whereas water-depth changes include the sedimentation component (i.e., sea level relative to the seafloor; Fig. 3.12). For example, either water deepening or shallowing may occur during a stage of base-level rise, as a function of the balance between the rates of creation and consumption of accommodation. Grading is a characteristic of facies that can be directly observed in outcrops, core, or well logs. Describing the rocks in terms of fining- and coarsening-upward trends is always objective, and does not necessarily translate in terms of specific base-level or water-depth changes. Grading indicates a consistent change through time in sediment supply across the area of observation, such as the progradation of the sediment entry points associated with shoreline

regression. The trend associated with this lateral shift of facies, coarsening-upward in this example, may occur during base-level rise, base-level fall, water shallowing, or water deepening at the point of observation. The correlation between grain size and marine water depth is only safely valid for nearshore areas, where changes with depth in depositional energy are more predictable, but it may be altered offshore where the balance between wave, tide, gravity, and contour currents is less predictable. In the latter situation, the sediment transport energy may fluctuate independently of water-depth changes, and hence no linear correlation between water depth and grain size can be established. Other possible confusions, between base-level changes and shoreline shifts, or between water-depth changes and shoreline shifts, are addressed in the following section of this chapter. These issues are also examined in more detail, using numerical models, in Chapter 7.

SHORELINE TRAJECTORIES

Definitions

The interplay between base-level changes and sedimentation controls the fluctuations in water depth, as well as the transgressive and regressive shifts of the shoreline (Fig. 3.15). The types of shoreline shifts are critical in a sequence stratigraphic framework, as they determine the formation of packages of strata associated with particular depositional trends and hence characterized by specific stacking patterns, known as systems tracts.

A transgression is defined as the landward migration of the shoreline. This migration triggers a corresponding landward shift of facies, as well as a deepening of the marine water in the vicinity of the shoreline. Transgressions result in retrogradational stacking patterns, e.g., marine facies shifting towards and overlying nonmarine facies (Fig. 3.18). Within the nonmarine side of the basin, the transgression is commonly indicated by the appearance of tidal influences in the fluvial succession, e.g., sigmoidal cross-bedding, tidal (heterolithic wavy, flaser, and lenticular) bedding, oyster beds and brackish to marine trace fossils (Shanley et al., 1992; Miall, 1997). Retrogradation is the diagnostic depositional trend for transgressions, and is defined as the backward (landward) movement or retreat of a shoreline or of a coastline by wave erosion; it produces a steepening of the beach profile at the breaker line (Bates and Jackson, 1987). As defined by Bates and Jackson (1987), the terms 'shoreline' and 'coastline' are often used synonymously, especially when referring to processes that occur over geological (Milankovitch band and

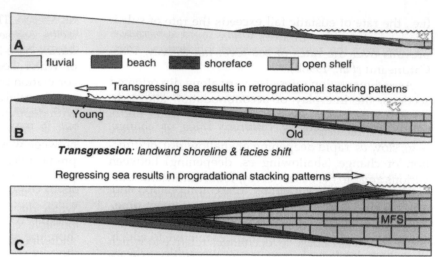

FIGURE 3.18 Transgressions and regressions. Note the retrogradation and progradation (lateral shifts) of facies, as well as the surface that separates retrogradational from overlying progradational geometries. This surface is known as the maximum flooding surface (MFS).

larger) time scales. In the solar to calendar band of time (hundreds of years and less), however, there is a tendency to regard 'coastline' as a limit fixed in position for a relatively long time, and 'shoreline' as a limit constantly moving across the intertidal area (i.e., the intersection of a plane of water with the beach, which migrates with changes of the tide or of the water level) (Bates and Jackson, 1987). In the context of this book, reference is made mainly to processes that operate over geological time scales, above the solar-band range, and therefore the terms 'shoreline' and 'coastline' are used interchangeably.

A regression is defined as the seaward migration of the shoreline. This migration triggers a corresponding seaward shift of facies, as well as a shallowing of the marine water *in the vicinity of the shoreline*. Regressions result in progradational stacking patterns, e.g., nonmarine facies shifting towards and overlying marine facies (Fig. 3.18). Progradation is the diagnostic depositional trend for regressions, and is defined as *the building forward or outward toward the sea of a shoreline or coastline (as of a beach, delta, or fan) by nearshore deposition of riverborne sediments or by continuous accumulation of beach material thrown up by waves or moved by longshore drifting* (Bates and Jackson, 1987).

The direct relationship between transgressions and regressions, on the one hand, and water deepening and shallowing, on the other hand, is only safely valid for the shallow areas adjacent to the shoreline (see *italics* in the definitions of transgressions and regressions). In offshore areas, the deepening and shallowing of the water may be out of phase relative to the coeval shoreline shifts, as subsidence and sedimentation rates vary along the dip of the basin (Catuneanu *et al.*, 1998b). For example, the Mahakam delta in Indonesia

(Verdier *et al.*, 1980) provides a case study where the progradation (regression) of the shoreline is accompanied by a deepening of the water offshore, due to the interplay between sedimentation and higher subsidence rates. Also, the progradation of submarine fans during the rapid regression of the shoreline often occurs in deepening waters due to the high subsidence rates in the central parts of many extensional basins.

Transgressions, as well as two types of regressions may be defined as a function of the ratio between the rates of base-level changes and the sedimentation rates at the shoreline (Fig. 3.19). The top sine curve in Fig. 3.19 idealizes the cyclic rises and falls of base level through time, allowing for equal periods of time of base-level fall and rise. This symmetry is often distorted in real case studies, but the principles remain the same regardless of the shape of the reference base-level curve. During the falling leg of the base-level cycle, accommodation is reduced by external controls (primarily the interplay of subsidence and sea-level change), and the shoreline is forced to regress irrespective of the sedimentation factor. This type of regression driven by base-level fall is known as 'forced' regression (Posamentier *et al.*, 1992b). During the rising leg of the base-level cycle, accommodation is created and consumed at the same time, so the actual direction of shoreline shift depends on the interplay of these two competing forces. Sedimentation tends to dominate in the early and late stages of base-level rise, when the rates of rise are low, whereas rising base level tends to be the dominant factor around the inflexion point of the reference curve, when the rates of rise are highest.

To better understand the changes in the direction of shoreline shift that may occur during base-level rise, the bottom sine curve in Fig. 3.19 displays the *rates* of

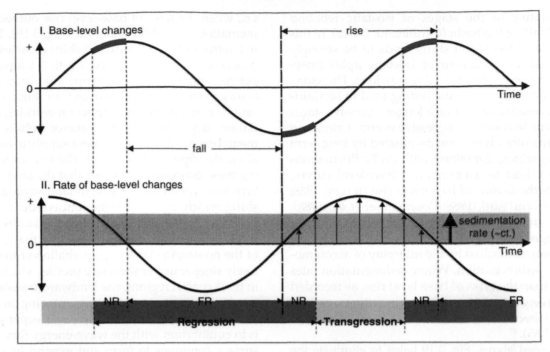

FIGURE 3.19 Concepts of transgression, normal regression, and forced regression, as defined by the interplay between base-level changes and sedimentation. The top sine curve shows the magnitude of base-level changes through time. The thicker portions on this curve indicate early and late stages of base-level rise, when the rates of base-level rise (increasing from zero and decreasing to zero, respectively) are outpaced by sedimentation rates. The sine curve below shows the rates of base-level changes. Note that the rates of base-level change are zero at the end of base-level rise and base-level fall stages (the change from rise to fall and from fall to rise requires the motion to cease). The rates of base-level change are the highest at the inflection points on the top curve. Transgressions occur when the rates of base-level rise outpace the sedimentation rates. For simplicity, the sedimentation rates are kept constant during the cycle of base-level shifts. The reference base-level curve is shown as a symmetrical sine curve for simplicity, but no inference is made that this should be the case in the geological record. In fact, asymmetrical shapes are more likely, as a function of particular circumstances in each case study (e.g., glacio-eustatic cycles are strongly asymmetrical, as ice melts quicker than it builds up), but this does not change the fundamental principles illustrated in this diagram. Abbreviations: FR—forced regression; NR—normal regression.

base-level change (first derivative of the top sine curve), which may be compared directly with the rates of sedimentation. In this diagram, sedimentation rates are assumed to be constant during a full cycle of base-level shifts, for simplicity, but other mathematical functions can be used as well to reflect more realistic fluctuations in sedimentation rates through time. What is important to emphasize is that in the early stages of base-level rise, when the rates of rise are low as increasing from zero, sedimentation rates are most likely to outpace the rates of creation of accommodation, leading to a *'normal' regression* of the shoreline, thus continuing the regressive trend of the falling leg. The timing of the end of shoreline regression is therefore not the end of base-level fall at the shoreline, but rather during the early stages of base-level rise. Once the increasing rates of base-level rise outpace the rates of sedimentation, a *transgression* of the shoreline begins (Fig. 3.19).

In the late stages of base-level rise, when the rates of rise are low as decreasing to zero, sedimentation takes over once again triggering a second *'normal' regression* of the base-level cycle. The timing of the end of shoreline transgression is therefore not the onset of base-level fall, but rather during the late stages of base-level rise (Fig. 3.19).

The discussion above implies that transgressive stages may be shorter in time (less than half of a cycle) relative to the regressive stages (normal plus forced), given a symmetrical curve of base-level changes. The actual balance between the temporal duration of transgressive and regressive stages changes with the basin, depending on the dominant allogenic controls on accommodation, as well as on sediment supply. In *foreland basins* for example, where flexural tectonics is the main control on accommodation, stages of flexural subsidence (and base-level rise) are significantly shorter in time relative to the stages of isostatic rebound (base-level fall) in the basin (Catuneanu, 2004a). In this case, a cycle of base-level shifts tends to be strongly asymmetrical, in the favour of isostatic uplift (base-level fall) and associated forced regressions. Therefore, transgressions in this tectonic setting tend to be short-lived events relative to the much longer regressive stages that intervene between transgressive events. *Extensional basins*, on the other hand, are dominated by long-term subsidence, which, combined with cyclic fluctuations in sea level, lead to asymmetrical base-level curves, this time in the favour of base-level rise (Jervey, 1988; Posamentier and Vail, 1988; Posamentier *et al.*, 1988). In this case, transgressions may potentially last longer than the regressive stages, but their relative durations are ultimately controlled by the interplay of accommodation and sedimentation. Where sedimentation rates are higher than the rates of base-level rise, as recorded in many divergent continental margin settings, normal regressions become the dominant type of shoreline shift (Fig. 2.65).

As explained above, Fig. 3.19 helps to eliminate the confusion between base-level changes and shoreline shifts. A common misconception is that base-level fall equates with shoreline regression, and base-level rise signifies shoreline transgression, by neglecting the effect of sedimentation. In reality, the turnaround point from base-level fall to subsequent base-level rise in the shoreline area is temporally offset relative to the turnaround point from shoreline regression to subsequent transgression with the duration of the early rise normal regression. Similarly, the onset of shoreline regression is separated in time from the onset of base-level fall at the shoreline by the duration of late rise normal regression (Fig. 3.19).

The succession of transgressive and regressive shoreline shifts illustrated in Fig. 3.19 represents the most complete scenario of stratigraphic cyclicity, where one forced regression, two normal regressions and one transgression manifest during a full cycle of base-level changes. In practice, simplified versions of stratigraphic cyclicity may also be encountered, such as: (1) repetitive successions of transgressive and normal regressive facies, where *continuous base-level rise* in the basin outpaces and is outpaced by sedimentation in a cyclic manner; and (2) repetitive successions of forced and normal regressions, where the *high sediment input* consistently outpaces the rates of base-level rise (hence, no transgressions). The stratal geometries associated with these basic types of shoreline shifts are presented below.

Transgressions

Transgressions occur when accommodation is created more rapidly than it is consumed by sedimentation,

i.e., when the rates of base-level rise outpace the sedimentation rates at the shoreline (Fig. 3.19). This results in a retrogradation (landward shift) of facies. The main processes that take place in the transition zone between nonmarine and marine environments during transgression are summarized in Fig. 3.20. These processes involve both sediment reworking and aggradation, depending on the balance between environmental energy flux and sediment supply in each location along the dip-oriented profile. The key for understanding these processes is the fact that the shoreline trajectory involves a combination of landward and upward shifts, which implies that the concave-up, wave-carved shoreface profile gradually migrates landward on top of fluvial or coastal facies. Assuming that the gradient of the nonmarine landscape is shallower than the relatively steeper upper shoreface profile, which is the case in most coastal regions, the landward translation of the shoreline triggers active wave scouring in the upper shoreface, in an attempt to carve a steeper profile that is in equilibrium with the wave-energy flux. This scour surface continues to form and expand in a landward direction for as long as the shoreline transgresses, and it is one of the sequence stratigraphic surfaces, diagnostic for transgression.

The scour surface cut by waves during the shoreline transgression (wave-ravinement surface) is onlapped by the aggrading and retrograding lower shoreface and shelf deposits (Fig. 3.20). The combination of wave scouring in the upper shoreface and deposition in the lower shoreface is required to preserve the concave-up shoreface profile that is in equilibrium with the wave energy during transgression (Bruun, 1962; Dominguez and Wanless, 1991). The onlapping deposits that accumulate in the lower shoreface and shelf environments 'heal' the bathymetric profile of the seafloor which, following shoreline transgression, has a gradient that is too steep relative to the new, lower energy conditions. These onlapping shallow-marine sediments form a transgressive wedge known as 'healing-phase' deposits (Posamentier and Allen, 1993; Fig. 3.20). The patterns of sediment redistribution as a result of wave-ravinement erosion in the upper shoreface during transgression are illustrated in Fig. 3.21. Note that the sediment eroded in the upper shoreface is transported both in landward and seaward directions. The portion of the sediment carried towards the coast may form backstepping beaches or estuary-mouth complexes, whereas the sediment carried offshore generates healing-phase wedges. Healing-phase deposits are relatively easy to recognize on seismic lines, as they form a package of convex-up reflections that onlap the last (youngest) regressive clinoform (Fig. 3.22).

The rise in base level at the shoreline promotes coastal aggradation in estuarine (river-mouth) or

Transgressive shorelines:

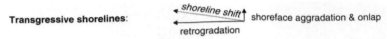

1. Coastal aggradation

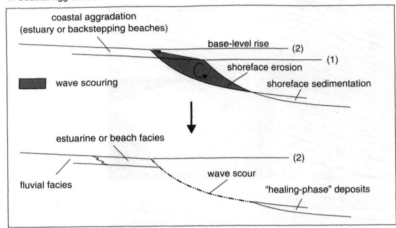

2. Coastal erosion

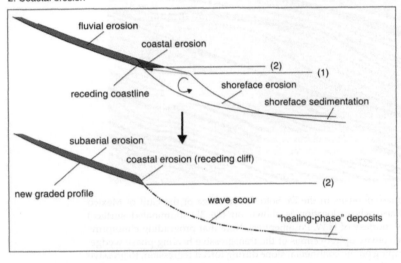

FIGURE 3.20 Shoreline trajectory in transgressive settings (from Catuneanu, 2003). Transgressions are driven by base-level rise, where the rates of base-level rise outpace the sedimentation rates in the shoreline area. The balance between the opposing trends of aggradation (in front of the shoreline) and wave scouring (behind the shoreline) determines the type of transgressive coastline. Irrespective of the overall nature of coastal processes (aggradation *vs.* erosion), the scour cut by waves in the upper shoreface is onlapped by transgressive lower shoreface and shelf ('healing phase') deposits. Low-gradient coastal plains are prone to coastal aggradation, whereas steeper coastal plains are prone to coastal erosion. In both cases, the gradients may be shallower than the average shoreface profile (approximately 0.3°).

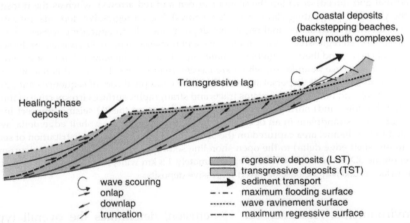

FIGURE 3.21 Patterns of sediment redistribution during shoreline transgression (modified from Posamentier and Allen, 1993, and Willis and Wittenberg, 2000). Some sediment is carried landward as backstepping beaches (open shorelines) or backstepping estuary-mouth complexes (river-mouth settings), while the coarser fraction typically mantles the ravinement surface as a transgressive lag. The transgressive coastal deposits may or may not be preserved as a function of the balance between the rates of coastal aggradation and the rates of wave-ravinement erosion. In addition, some sediment is transported seaward of the last clinoform of the underlying progradational deposits (LST) and forms a wedge-shaped deposit referred to as the healing-phase unit. Abbreviations: LST—lowstand systems tract; TST—transgressive systems tract. The definition of sequence stratigraphic surfaces follows in Chapter 4.

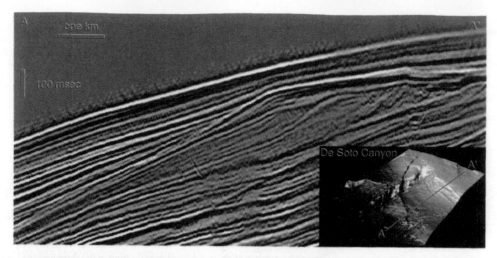

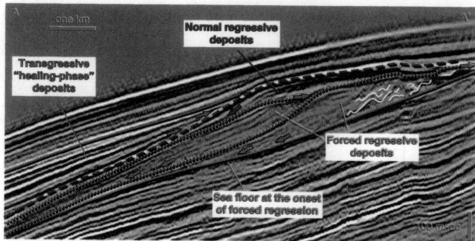

FIGURE 3.22 Shelf-edge and healing-phase deposits in the De Soto Canyon area of the Gulf of Mexico (uninterpreted and interpreted seismic lines, whose location is shown on the 3D illuminated surface) (modified from Posamentier, 2004a; images courtesy of H.W. Posamentier). Note that prograding clinoforms tend to be concave-up, in contrast with the convex-up reflections of the transgressive healing-phase wedge. The white wavy lines indicate possible slumping on the continental slope during forced regression. Regressive deposits (both normal and forced) downlap the seafloor (green and red arrows), whereas the transgressive deposits onlap the youngest prograding clinoform (blue arrows). Forced regressive deposits are associated with offlap (yellow arrows), whereas normal regressive deposits include an aggrading topset. These three genetic types of strata (forced regressive, normal regressive, and transgressive) are *independent* of the sequence stratigraphic model of choice, and their recognition is more important than the nomenclature of systems tracts or even the position of sequence boundaries, which are *model-dependent* (Fig. 1.7). For this reason, shoreline shifts, and their associated sediment dispersal systems, form the conceptual core of sequence stratigraphy as they control the formation and timing of all systems tracts and stratigraphic surfaces irrespective of the model of choice. Note that the (lowstand) normal regressive deposits shown on the 2D seismic transect include a prograding and aggrading strandplain in an open shoreline setting rather than a shelf edge delta, which is small and restricted to the channel area captured on the 3D illuminated surface. The distribution of sediment from the river mouth (shelf edge delta) to the open shoreline setting is attributed to longshore currents. For scale, the channel on the 3D illuminated surface is approximately 1.8 km wide, and 275 m deep at shelf edge. The illuminated surface is taken at the base of forced regressive deposits.

beach (open shoreline) environments. However, the tendency of coastal aggradation is counteracted by the wave scouring in the upper shoreface, as the latter gradually shifts in a landward direction. The balance between these two opposing forces, of sedimentation *vs.*

erosion, determines the overall type of transgressive coastline (Fig. 3.20). Coastlines dominated by aggradation lead to the preservation of estuarine or backstepping beach facies in the rock record (Fig. 3.23). Coastlines dominated by erosion are associated with

A

B

FIGURE 3.23 Estuarine facies preserved in the rock record, showing tidally-influenced inclined heterolithic strata (Dinosaur Park Formation, Belly River Group, Alberta). A—estuary-channel point bar (the section is approximately 4 m thick); B—amalgamated estuary channel fills (the section is approximately 6 m thick). The preservation of estuary facies in the rock record indicates coastal aggradation during shoreline transgression, which means that the rates of aggradation in the estuary outpaced the rates of wave scouring in the upper shoreface. This scenario is conducive to the preservation of underlying lowstand normal regressive fluvial deposits, which are protected from transgressive wave scouring by the estuary strata.

A

B

FIGURE 3.24 Coastal erosion in a transgressive open shoreline setting (Canterbury Plains, New Zealand). A—wave-ravinement erosion outpaces coastal aggradation in spite of rising base level. As a result, a receding cliff forms instead of backstepping beaches. Beyond the cliff face, the coastal plain is subject to subaerial erosion. B—note the gravel beach, indicating high energy upper shoreface-shoreline systems. The gravel is supplied by (1) coastal erosion of the wave-cut cliff, which consists of gravel-rich Quaternary deposits, and (2) rivers (Fig. 3.25). In this open shoreline setting, the riverborne gravel is redistributed along the coastline by strong longshore currents.

unconformities in the nonmarine part of the basin, whose stratigraphic hiatuses are age-equivalent with the transgressive marine facies. Regardless of the overall nature of coastal processes, the wave-ravinement surface is onlapped by transgressive shallow-marine ('healing-phase') deposits, which provides a clue for understanding the transgressive nature of some subaerial unconformities.

A modern example of an erosional transgressive coastline is represented by the shore of the Canterbury Plains in the Southern Island of New Zealand (Leckie, 1994). In this wave-dominated setting, the rates of wave erosion outpace the rates of coastal aggradation in both open shoreline and river-mouth settings. As a result, estuaries are incised into the coastal plain, and the open shorelines are marked by receding cliffs (Figs. 3.24–3.26). The extreme wave energy that leads to overall coastal erosion is caused by oceanic swell originating as far

FIGURE 3.25 Shallow gravel-bed braided system, supplying coarse-grained sediment to the Canterbury Plains shoreline. From the sediment entry points (river mouths), the gravel is reworked and redistributed along the open coastline by strong longshore currents. Southern Alps, New Zealand.

FIGURE 3.26 Coastal erosion in a transgressive river-mouth setting (A—panoramic view and B—close up). Wave-ravinement erosion outpaces coastal aggradation in spite of rising base level. As a result, the estuary is incised, about 20 m into the coastal plain. The width of the incised estuary is about 1 km. Ashburton River, Canterbury Plains, New Zealand.

away as 2000 km. The wave-cut cliffs, which may be up to 25 m high, recede at a rate of approximately 1 m per year. Coastal erosion lowers the fluvial graded profile below the topographic profile (Fig. 3.20), causing the rivers to incise 1.5–4.2 mm per year in the vicinity of the coastline. The amount of incision gradually decreases inland from the coast, until it becomes minimal 8–15 km upstream (Leckie, 1994).

Forced Regressions

Forced regressions occur during stages of base-level fall, when the shoreline is forced to regress by the falling base level irrespective of sediment supply (Fig. 3.19). A variety of processes may accompany the forced regression of the shoreline in the transition zone between marine and nonmarine environments, including erosion, aggradation, or a combination of both. These processes affect both fluvial and marine environments, and the manifestation of one over the other (erosion vs. aggradation) in any region depends on the relative position between the energy flux equilibrium profile (fluvial graded profile or base level) and the ground surface (subaerial or subaqueous).

In shallow-marine settings, equilibrium profiles are generally concave-up and reflect the energy flux of fairweather waves. These profiles are dynamic, being sensitive to any changes in marine-energy flux that may occur during storms or due to the activity of marine currents. The dominant processes that manifest during forced regression in a shallow-marine environment are therefore a function of the relative position between the wave equilibrium profile and the seafloor. Low-gradient seafloors are more susceptible to wave erosion during a fall in base level, whereas steeper seafloors (with a gradient higher than the gradient of the wave equilibrium profile) are less affected by the wave-energy flux, being rather prone to aggradation (Fig. 3.27). Seafloor gradients in coastal regions are in turn controlled by the basin physiography, as well as by the dominant process of sediment distribution in the subtidal areas adjacent to the coastline.

In wave-dominated coastal settings, such as open shorelines or wave-dominated deltas, the preservation of the concave-up seafloor profile that is in equilibrium with the wave energy requires coeval deposition and erosion in the upper and lower parts of the subtidal area, respectively (Bruun, 1962; Plint, 1988; Dominguez and Wanless, 1991; Fig. 3.27). As the shoreline shifts basinward, the upper subtidal forced regressive deposits downlap the scour generated in the lower subtidal zone (Fig. 3.27). At the same time, the subaerially exposed area is commonly subject to sediment starvation, pedogenesis, or fluvial and wind degradation. The amount of nonmarine downcutting is generally proportional to the magnitude of base-level fall, but it also depends on the changes in slope gradients of the ground surface exposed by the fall in base level (see Posamentier, 2001, for a discussion of incised vs. unincised fluvial bypass systems).

In the case of river-dominated deltas, the angle of repose of delta front clinoforms is generally steeper than the gradient required to balance the energy of the waves, so there is no reason for wave scouring in the lower delta front area (Fig. 3.27). Therefore, the marine scour surface that forms in shallow-marine wave-dominated settings during forced regression is missing from the stratal architecture of forced regressive river-dominated deltas. In the former case, a vertical profile through the shallow-marine forced regressive succession shows an abrupt shift of facies from offshore muds to upper subtidal sands (Figs. 3.28 and 3.29), whereas this facies shift is gradational in the latter situation (Fig. 3.30).

Landward relative to the shoreline, processes of fluvial erosion or aggradation reflect changes in fluvial-energy flux that are in part controlled by the contrast between the gradients of the fluvial and seafloor profiles at the onset of forced regression. As the shoreline regresses and the seafloor becomes subaerially exposed, steeper seafloor gradients (relative to the fluvial profile at the onset of forced regression) lead to increased fluvial-energy flux and incision, whereas shallower seafloor gradients trigger a decrease in fluvial-energy flux and sediment aggradation (cases A and C in Fig. 3.31, respectively). Both processes of fluvial incision or aggradation propagate gradually from the shoreline upstream through a series of landward-migrating knickpoints (Figs. 3.31 and 3.32). Each knickpoint represents an abrupt shift in slope gradients along the fluvial profile at a particular time, and it is the change in fluvial-energy flux induced by such shifts in slope gradients that triggers aggradation or fluvial incision. A downstream increase in valley slope is prone to fluvial incision (case A in Fig. 3.31; Fig. 3.32), whereas a downstream decrease in valley slope promotes fluvial aggradation (case C in Fig. 3.31) (Pitman and Golovchenko, 1988; Butcher, 1990; Posamentier and Allen, 1999). The fluvial response to such changes in valley slope is in fact much more complex than depicted in Fig. 3.31, as rivers may internally adjust their flow parameters (e.g., the degree of channel sinuosity) in order to adapt to changing topographic gradients without aggradation or incision (Schumm, 1993).

The diagrams in Fig. 3.27 illustrate a scenario where the gradient of the seafloor in the subtidal zone is steeper than the gradient of the downstream fluvial profile,

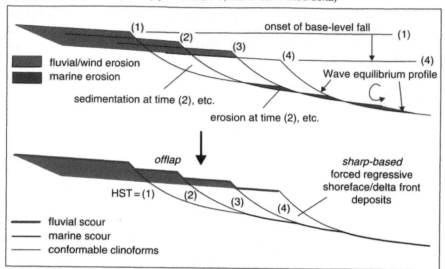

FIGURE 3.27 Shoreline trajectory in forced regressive settings (modified from Catuneanu, 2003). Forced regressions are driven by base-level fall, irrespective of sediment supply, and the rates of progradation are generally high. Wave-dominated subtidal settings are characterized by low gradients of the seafloor, which is subject to wave scouring in order to preserve a profile that is in equilibrium with the wave energy. River-dominated deltas generally have delta front clinoforms that are steeper than the wave equilibrium profile, and therefore no wave scouring takes place during forced regression. HST—highstand systems tract.

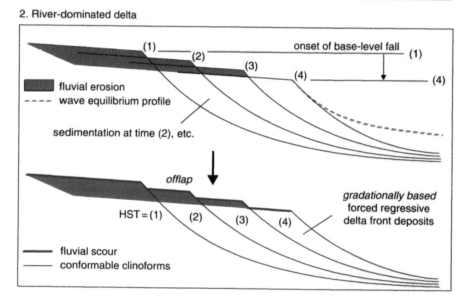

which is the case in the majority of coastal regions. Other situations may, however, occur as well, as illustrated in Fig. 3.31. These three possible scenarios may explain why rivers do not always incise during stages of base-level fall, as commonly inferred in the sequence stratigraphic literature (case A in Fig. 3.31; Fig. 3.32), but they may also bypass (case B in Fig. 3.31) or even aggrade (case C in Fig. 3.31) during the forced regression of the shoreline. As discussed earlier in this chapter, however, changes in base level controlled by tectonism and sea-level change, which are accounted for in Fig. 3.31, may be overprinted by the effect of

climate change to the extent that processes of fluvial incision or aggradation may proceed in a fashion that is opposite to what is normally expected from relative sea-level changes (Blum, 1990, 1994). All these aspects of fluvial sedimentation are detailed more in Chapter 4 (discussion on subaerial unconformities), Chapter 5 (discussion of the falling-stage systems tract) and Chapter 6 (discussion of fluvial processes in a sequence stratigraphic framework).

Stages of forced regression are generally characterized by a significant increase of sediment supply to the deep-water depositional systems. This is due to (1) a lack

FIGURE 3.28 Forced regressive, wave-dominated shoreface sands (with A—swaley cross-stratification–Fig. 3.29) abruptly overlying inner shelf interbedded sands and muds (B). The upper shoreface sands (A) are 'sharp-based' due to wave scouring in the lower shoreface during base-level fall. The exposed section below the wave scour is approximately 2 m thick. Blackhawk Formation, Utah.

of accommodation in the fluvial to shallow-marine environments, and therefore the terrigenous sediment tends to bypass these settings and be delivered to the deep-water environment; and (2) additional sediment may be supplied by erosional processes in the fluvial and lower shoreface environments.

The stratal architecture of shallow-marine forced regressive deposits is a function of sediment supply, rates of base-level fall, and gradient of the seafloor (Ainsworth and Pattison, 1994; Posamentier and Morris, 2000). The interplay of these variables controls the character of the forced regressive prograding lobes,

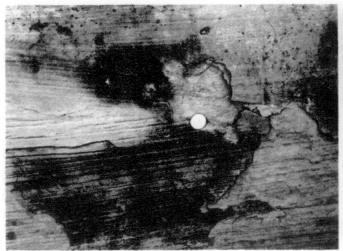

FIGURE 3.29 Swaley cross-stratification in wave-dominated, upper shoreface sandstones. Blackhawk Formation, Utah.

A

B

FIGURE 3.30 Forced regressive, river-dominated deltaic succession (Panther Tongue, Utah). A—conformable shift of facies from prodelta to the overlying delta front deposits. The delta front sands are 'gradationally based', as no wave scouring took place during the progradation of the delta; B—relatively steep delta front clinoforms (dipping to the right in the photograph, at an angle of 5–15°). As the clinoforms are steeper than the wave equilibrium profile (approximately 0.3°), no wave scouring took place during the progradation of the delta. The delta front succession is topped by a transgressive lag (sandstone layer—see arrow), which in turn is overlain by transgressive shale. Hence, no delta plain deposits are present.

which may be attached vs. detached, stepped-topped vs. smooth-topped, and spread over short or long distances (Fig. 3.33). Criteria for the recognition of shallow-marine forced regressive deposits in outcrop, core, well logs and seismic data are also provided by Posamentier and Morris (2000). Perhaps the most important defining signature of coastal to shallow-marine forced regressive deposits is their offlapping (seaward downstepping) character, which is caused by the fall in relative sea level (Fig. 3.27). This stratal stacking pattern may be observed on seismic lines (Fig. 3.22), and it is particularly significant for the exploration of age-equivalent deep-water

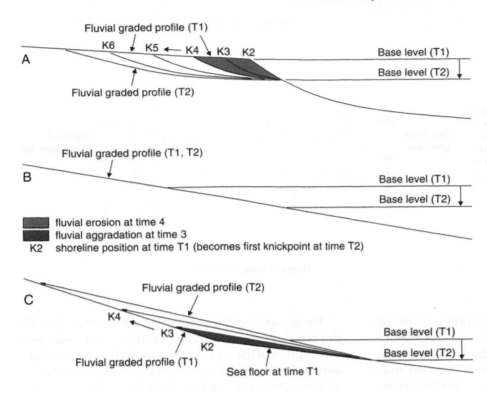

FIGURE 3.31 Fluvial responses to base-level fall, as a function of the contrast in slope gradients between the fluvial and the seafloor profiles at the onset of forced regression (modified from Summerfield, 1985; Pitman and Golovchenko, 1988; Butcher, 1990; Schumm, 1993; Posamentier and Allen, 1999; Blum and Tornqvist, 2000). A—fluvial incision; B—fluvial bypass; C—fluvial aggradation. Knickpoints (K) mark abrupt changes in the gradient of fluvial profiles. A downstream increase in slope gradient (and corresponding fluvial-energy flux) is prone to fluvial erosion (case A). A downstream decrease in slope gradient (and corresponding fluvial-energy flux) is prone to fluvial aggradation (case C). Knickpoints migrate upstream with time, resulting in a landward expansion of the subaerial unconformity (case A) or in a backfill of the landscape to the level of the new graded profile, accompanied by fluvial onlap of the old graded profile (case C). Case A is most likely, case C is least likely. Case B may describe the forced regression across a continental shelf, where minor fluvial incision (or aggradation) may still occur below the seismic resolution.

reservoirs (more details on this topic are presented in Chapters 5 and 6). Offlapping forced regressive deposits may also be observed in modern environments, such as for example in areas that are currently subject to post-glacial isostatic rebound at a rate that exceeds the present day rate of sea-level rise (Fig. 3.34).

Normal Regressions

Normal regressions occur during early and late stages of base-level rise, when sedimentation rates outpace the low rates of base-level rise at the shoreline (Fig. 3.19). In this case, the newly created accommodation is totally

FIGURE 3.32 Upstream-migrating fluvial knickpoint (arrow) along a small-scale, actively incising 'valley'. Note the decrease in the elevation of the 'coastal plain' as a result of base-level fall. The older coastal plain, which existed during the early stage of incision, is now preserved as a stranded terrace.

FIGURE 3.33 Stratal architecture of shallow-marine forced regressive deposits, as a function of sediment supply, rates of base-level fall and gradient of the seafloor. The interplay of these variables may result in a variety of possibilities, with the prograding forced regressive lobes being attached or detached, stepped-topped or smooth-topped, and spread over short or long distances (see Posamentier and Morris, 2000, for a more detailed discussion).

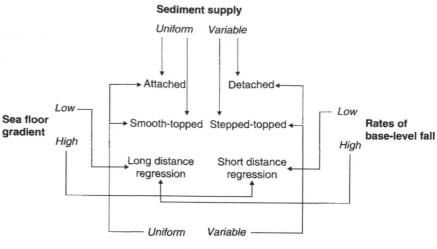

consumed by sedimentation, aggradation is accompanied by sediment bypass (the surplus of sediment for which no accommodation is available), and a progradation of facies occurs (Fig. 3.35). Such seaward shifts of facies result in the formation of conformable successions, which consist typically of coarsening-upward shallow-marine deposits topped by coastal to fluvial facies (Fig. 3.36). Normal regressive successions may develop in both river-mouth (deltaic) and open coastline settings. In the former case, the vertical profile records a shift from prodelta, to delta front and delta plain facies (Fig. 3.36), whereas in the latter setting the change is from shelf to shoreface and overlying beach and alluvial facies (Figs. 3.37 and 3.38).

FIGURE 3.34 Modern forced regressive delta showing offlapping stratal stacking patterns (photo courtesy of J. England). In this case, the fall in base level is triggered by post-glacial isostatic rebound in the Canadian arctics, at a rate that exceeds the rate of present day sea-level rise.

The dip angle of the prograding clinoforms (Fig. 3.35) depends on the dominant controls on sediment distribution in the subtidal area, as well as on sediment supply. In the case of wave-dominated open coastlines, or wave-dominated deltas, the angle of repose is very low, averaging 0.3° (mean gradient of the wave equilibrium profile). This angle is steeper in the case of river-dominated deltas, ranging from less than a degree (where rivers bring a significant amount of fine-grained suspension load, and the sediment transport in the delta front environment is primarily attributed to low-density turbidity flows) to approximately 30° (Gilbert-type deltas, where the riverborne sediment is dominantly sandy and its transport within the delta front environment is largely linked to the manifestation of grain flows). In either case, the creation of accommodation in the coastal and adjacent fluvial and shallow-marine regions is prone to aggradation along the entire nearshore profile, and hence no significant fluvial or wave scouring are expected to be associated with this type of shoreline shift (Fig. 3.35). As a result, normal regressive shoreface or delta front deposits are *gradationally based* (Fig. 3.36), in contrast with the forced regressive shoreface or wave-dominated delta front facies which are *sharp-based* (Figs. 3.27 and 3.28).

The process of coastal aggradation, in response to rising base level, also confers another important diagnostic feature that separates normal regressive from forced regressive deposits (Figs. 3.27 and 3.35). As accommodation is positive in the coastal region, a *topset* of intertidal to supratidal deposits (delta plain in river-mouth settings, Fig. 3.36; or beach/strandplain sediments in open shoreline settings, Fig. 3.38) accumulates and progrades on top of the shallow-marine delta front/shoreface facies (Fig. 3.35). Such a topset is absent in the case of forced regressions, where the subtidal facies

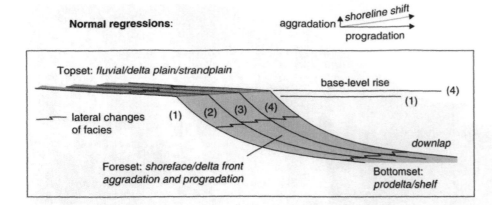

Normal regressions:

Topset: *fluvial/delta plain/strandplain*

lateral changes of facies

Foreset: *shoreface/delta front aggradation and progradation*

base-level rise

downlap

Bottomset: *prodelta/shelf*

FIGURE 3.35 Shoreline trajectory in normal regressive settings, defined by a combination of progradation and aggradation in fluvial to shallow-marine systems. Normal regressions are driven by sediment supply, where the rates of base-level rise at the shoreline are outpaced by sedimentation rates. Normal regressions occur during early and late stages of base-level rise, when the rates of creation of accommodation are low (Fig. 3.19). Progradation rates are generally low. Normal regressions are prone to aggradation in fluvial, coastal (delta plains in river-mouth settings, or strandplains along open shorelines), and marine environments.

FIGURE 3.36 Normal regressive deltaic succession (river-dominated delta), showing a conformable transition from shallow-marine muds and sands (shelf, prodelta, delta front) to coastal and fluvial deposits (Ferron Sandstone, Utah). The arrow points at the conformable facies contact between delta front sands and the overlying coal-bearing delta plain and fluvial facies. This facies contact marks the base of the deltaic topset (Fig. 3.35).

FIGURE 3.37 Aggrading upper shoreface sandstones in a wave-dominated open coastline setting. These wave ripple-marked strata are interpreted as part of a late rise (highstand) normal regressive systems tract (Rubidge *et al.*, 2000). Waterford Formation (Late Permian), Ecca Group, Karoo Basin.

FIGURE 3.38 Aggrading beach deposits in a normal regressive setting. The sands are massive, with low-angle stratification, typical of foreshore open-shoreline systems. The beach sands overlie coarsening-upward shelf to shoreface deposits (in subsurface in this particular location), and are overlain by fluvial floodplain facies. The latter contact is sharp but conformable. Uppermost Bearpaw Formation sands (Early Maastrichtian), Castor area, Western Canada Sedimentary Basin.

offlap and are truncated by processes of subaerial erosion (Fig. 3.27). The thickness of topset successions varies with the case study, depending on the duration of normal regression, the rates of coastal aggradation, and available sediment supply. The topset may be identified in core or outcrop based on facies analysis, but its recognition on seismic lines as a distinct unit may or may not be possible, depending on seismic resolution relative to the unit's thickness (Fig. 3.22).

The surface that separates the topset package from the underlying subtidal deposits is always represented by a conformable (and diachronous, with the rate of shoreline regression) facies contact (dotted line in Fig. 3.35; Fig. 3.36). The upper boundary of the topset unit may also be conformable, where no subsequent erosion reworks it (e.g., in the case of early rise 'lowstand' normal regressions, where the topset is overlain by transgressive fluvial and/or estuarine strata), but often it is scoured by subaerial erosion (e.g., late rise 'high-stand' topsets truncated by subaerial unconformities) or transgressive reworking (e.g., early rise 'lowstand' topsets truncated by tidal- or wave-ravinement surfaces). The preservation potential of topset packages is higher in the case of early rise ('lowstand') normal regressive deposits, as the creation of accommodation continues following the maximum regression of the shoreline, and lower in the case of late rise ('highstand') normal regressive successions which are followed by stages of base-level fall and potential subaerial erosion.

4

Stratigraphic Surfaces

INTRODUCTION

Stratigraphic surfaces mark shifts through time in depositional regimes (i.e., changes in depositional environments, sediment load and/or environmental energy flux), and are created by the interplay of base-level changes and sedimentation. Such shifts in depositional regimes may or may not correspond to changes in depositional trends, may or may not be associated with stratigraphic hiatuses, and may or may not place contrasting facies in contact across a particular surface. The correct identification of the various types of stratigraphic surfaces is key to the success of the sequence stratigraphic approach, and the criteria used for such identifications are explored in this chapter.

Stratigraphic surfaces provide the fundamental framework for the genetic interpretation of any sedimentary succession, irrespective of how one may choose to name the packages of strata between them. For this reason, stratigraphic surfaces in conjunction with shoreline trajectories, which are core concepts *independent* of the sequence stratigraphic model of choice, are more important than the nomenclature of systems tracts or even the position of sequence boundaries, which are *model-dependent* (Fig. 1.7). Across the spectrum of existing sequence stratigraphic models, the significance of stratigraphic surfaces may change from sequence boundaries to systems tract boundaries or even within systems tract facies contacts (Figs. 1.6 and 1.7).

Stratigraphic surfaces may be identified based on a number of criteria, including the nature of contact (conformable or unconformable), the nature of facies which are in contact across the surface, depositional trends recorded by the strata below and above the contact (forced regressive, normal regressive, or transgressive), ichnological characteristics of the surface or of the facies which are in contact across the surface, and strata terminations associated with each particular surface. It can be noted that most of these criteria involve preliminary facies analyses and an understanding of the environments in which the stratigraphic contact and the juxtaposed facies that it separates, originated. The reconstruction of the depositional setting therefore enables the interpreter to apply objective criteria for the recognition, correlation, and mapping of stratigraphic surfaces.

Depending on the type of data available for analysis, some contacts that separate packages of strata characterized by contrasting stacking patterns may be mapped solely on the basis of how strata terminate against the contact being mapped, without independent constraints on paleodepositional environments. This is often the case where only 2D seismic lines are available for the preliminary screening of the subsurface stratigraphy. In such cases, truncation, toplap, onlap, offlap or downlap surfaces may be identified from local to regional scales, simply based on the geometric relationship of the underlying and/or overlying strata with the contact that separates them. Integration of additional data, such as 3D seismic volumes, well logs and core, provides additional constraints on depositional setting and the genesis of stratal termination in an environmental context, thus allowing for a proper identification of the stratigraphic contact(s) under investigation.

Stratigraphic surfaces may generally be classified in *environment-dependent surfaces*, which have specific environments of origin and hence a specific stratigraphic context (e.g., surfaces of fluvial incision, transgressive wave scouring, regressive wave scouring), *geometric surfaces*, defined by stacking patterns and stratal terminations (e.g., onlap surface, downlap surface), and *conceptual surfaces*, which are environment-dependent and/or geometric surfaces that carry a specific significance (e.g., systems tract or sequence boundary) within the context of sequence stratigraphic models (e.g., subaerial unconformities, correlative conformities, maximum flooding or maximum regressive surfaces) (Galloway, 2004). In an empirical, rather than model-driven approach, the designation of conceptual surfaces should only be done at the end of a sequence stratigraphic study, once the environment-dependent and geometric surfaces are properly identified, mapped, and tested for their chronostratigraphic reliability. Once this observational framework is in place, the selection of the most useful and geologically meaningful conceptual surfaces for defining regional genetic units, such as systems tracts and sequences, may be performed (Galloway, 2004). The selection of conceptual surfaces depends on the particular circumstances of each case study, and therefore should not follow any rigid templates to which all data sets must conform in order to fit the predictions of any particular model.

Stratigraphic surfaces may also be classified as a function of their relevance to sequence stratigraphy. Surfaces that can serve at least in part as systems tract or sequence boundaries are *sequence stratigraphic surfaces*. Depending on scope and scale of observation, such surfaces are used to build the chronostratigraphic framework of a sedimentary succession, from the scale of individual depositional systems to entire basin fills. Once this sequence stratigraphic framework is established, additional surfaces may be traced within the genetic units (i.e., systems tracts) bounded by sequence stratigraphic surfaces. Such internal surfaces have been defined as *within-trend facies contacts* (Embry and Catuneanu, 2001, 2002), and help to illustrate the patterns of facies shifts within individual systems tracts. The following sections of this chapter present the types of stratal terminations that are used to interpret geometric surfaces and associated depositional trends and shoreline trajectories, followed by a discussion of all types of stratigraphic surfaces that have relevance to sequence stratigraphy.

TYPES OF STRATAL TERMINATIONS

Stratal terminations are defined by the geometric relationship between strata and the stratigraphic surface against which they terminate, and are best observed at larger scales, particularly on 2D seismic lines and in large-scale outcrops (Figs. 2.65, 2.68, 2.69, and 3.22). The main types of stratal terminations are described by truncation, toplap, onlap, downlap, and offlap (Fig. 4.1). Excepting for truncation, which is a term stemming from classical geology, the other concepts have been

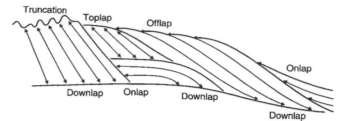

FIGURE 4.1 Types of stratal terminations (modified from Emery and Myers, 1996). Note that tectonic tilt may cause confusion between onlap and downlap, due to the change in ratio between the dip of the strata and the dip of the stratigraphic surface against which they terminate.

introduced with the development of seismic stratigraphy in the 1970s to define the architecture of seismic reflections (Mitchum and Vail, 1977; Mitchum *et al.*, 1977). These terms have subsequently been incorporated into sequence stratigraphy in order to describe the stacking patterns of stratal units and to provide criteria for the recognition of the various surfaces and systems tracts (e.g., Posamentier *et al.*, 1988; Van Wagoner *et al.*, 1988; Christie-Blick, 1991). The definitions of the key types of stratal terminations are provided in Fig. 4.2.

Stratal terminations form in relation to specific depositional trends, and therefore allow one to infer the type of syndepositional shoreline shifts and implicitly to reconstruct the history of base-level changes at the shoreline (Fig. 4.3). In some instances, the interpretation of stratal terminations in terms of shoreline shifts is unequivocal, as for example *coastal onlap* indicates transgression, and *offlap* is diagnostic for forced regressions. In other cases, stratal terminations may allow for alternative interpretations, as for example *downlap* may form in relation to either normal or forced regressions. In such cases, additional criteria have to be used in order to cut down the number of choices and arrive at unequivocal conclusions. In this example, the differentiation between normal and forced regressions that can be associated with *downlap* may be performed by studying the depositional trends (aggradation or erosion) in the syndepositional coastal setting. Evidence of scouring, as indicated by an uneven erosional relief, lag deposits, or the presence of offlap at the top of the prograding package would point towards forced regression, whereas coastal aggradation would suggest base-level rise and hence normal regression.

The process of coastal aggradation during normal regressions results in the formation of *topset* packages of delta plain (in a prograding river-mouth environment; Figs. 2.3 and 2.4), strandplain (a wide beach characterized by subparallel ridges and swales, in places with associated dunes, which forms by processes

Truncation: termination of strata against an overlying erosional surface. *Toplap* may develop into truncation, but truncation is more extreme than toplap and implies either the development of erosional relief or the development of an angular unconformity.

Toplap: termination of inclined strata (clinoforms) against an overlying lower angle surface, mainly as a result of nondeposition (sediment bypass), ± minor erosion. Strata lap out in a landward direction at the top of the unit, but the successive terminations lie progressively seaward. The toplap surface represents the proximal depositional limit of the sedimentary unit. In seismic stratigraphy, the *topset* of a deltaic system (delta plain deposits) may be too thin to be "seen" on the seismic profiles as a separate unit (thickness below the seismic resolution). In this case, the topset may be confused with toplap (i.e., *apparent toplap*).

Onlap: termination of low-angle strata against a steeper stratigraphic surface. Onlap may also be referred to as *lapout*, and marks the lateral termination of a sedimentary unit at its depositional limit. Onlap type of stratal terminations may develop in marine, coastal, and nonmarine settings:

- marine onlap: develops on continental slopes during transgressions (*slope aprons*, Galloway, 1989; *healing-phase deposits*, Posamentier and Allen, 1993), when deep-water transgressive strata onlap onto the maximum regressive surface.

- coastal onlap: refers to transgressive coastal to shallow-water strata onlapping onto the transgressive (tidal, wave) ravinement surfaces.

- fluvial onlap: refers to the landward shift of the upstream end of the aggradation area within a fluvial system during base-level rise (normal regressions and transgression), when fluvial strata onlap onto the subaerial unconformity.

Downlap: termination of inclined strata against a lower-angle surface. Downlap may also be referred to as *baselap*, and marks the base of a sedimentary unit at its depositional limit. Downlap is commonly seen at the base of prograding clinoforms, either in shallow-marine or deep-marine environments. It is uncommon to generate downlap in nonmarine settings, excepting for lacustrine environments. Downlap therefore represents a change from marine (or lacustrine) slope deposition to marine (or lacustrine) condensation or nondeposition.

Offlap: the progressive offshore shift of the updip terminations of the sedimentary units within a conformable sequence of rocks in which each successively younger unit leaves exposed a portion of the older unit on which it lies. Offlap is the product of base-level fall, so it is diagnostic for forced regressions.

FIGURE 4.2 Types of stratal terminations (definitions from Mitchum, 1977; Galloway, 1989; Emery and Myers, 1996).

of coastal aggradation and progradation in an open shoreline setting; Figs. 2.3 and 2.4) and/or coastal plain deposits (Fig. 2.5). The *topset* is not a type of stratal termination, but rather a unit consisting of nearly horizontal layers of sediments deposited on the top surface of a prograding coastline, which covers the edge of the seaward-lying foreset beds and is continuous with the landward alluvial plain (Bates and Jackson, 1987). The thickness of the *topset* package depends on the duration of normal regression, and the rates of base-level rise and sediment supply. The concept of *toplap*, as a stratal termination that forms in relation to a regressive coastline during base-level stillstand (i.e., neither normal nor forced regression; Fig. 4.3) is, in reality, often associated with the formation of topsets, especially where the topset thickness is less than the vertical seismic resolution. Ideally, the formation of *toplap* requires progradation of foreset beds (delta front or shoreface clinoforms) coeval with perfect sediment bypass in the coastal environments (delta plain, strandplain, or coastal plain). This means an ideal case where the base level at the shoreline does not change with time, as a base-level rise would result in topset, and a base-level fall would result in offlap. Such a situation may only happen for relatively short periods of time, as the base level (controlled by the interplay of several independent factors) is hardly, if ever, stable. The concept of *toplap* was developed from the analysis of seismic data, where the thickness of topset packages often falls below the seismic resolution, being reduced to a seismic interface. The toplap type of stratal terminations is therefore *apparent* in most cases (Fig. 4.4). Apparent toplaps may also develop during stages of base-level fall (forced regressions) associated with minimum erosion, where the evidence for erosion is undetectable on seismic lines (Fig. 4.3).

Stratal termination	Shoreline shift	Base level
Truncation, fluvial	FR	Fall
Truncation, marine	FR, T	Fall, Rise
Toplap	R	Stillstand
Apparent toplap	NR, FR	Rise, Fall
Offlap	FR	Fall
Onlap, fluvial	NR, T	Rise
Onlap, coastal	T	Rise
Onlap, marine	T	Rise
Downlap	NR, FR	Rise, Fall

FIGURE 4.3 Interpretation of stratal terminations in terms of syndepositional shoreline shifts and base-level changes. Exceptions from these general trends are, however, known to occur, as for example fluvial incision (truncation) may also take place during base-level rise and transgression (Fig. 3.20). Abbreviations: R—regression; FR—forced regression; NR—normal regression; T—transgression.

In terms of the inferred relationship between stacking patterns and base-level changes, some stratal terminations are generally considered to form only during stages of base-level rise (i.e., all types of onlap), some are specific for a falling base level (e.g., fluvial

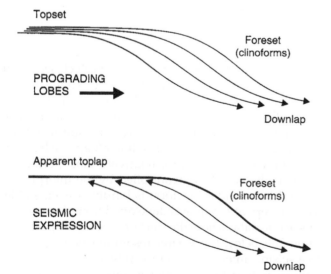

FIGURE 4.4 Seismic expression of a topset package that is thinner relative to the seismic resolution. The top diagram shows the stratal architecture of a deltaic system in a normal regressive setting. Note the possible confusion between topset and toplap on low-resolution seismic data.

incision/truncation and offlap), whereas others may be associated with either falling or rising base level (i.e., truncation related to processes of marine erosion, apparent toplap, or downlap) (Fig. 4.3). Exceptions to these general rules are, however, known to occur, as for example fluvial incision may also take place during stages of base-level rise and transgression (Fig. 3.20).

Additional general principles may be formulated with respect to the nature of stratigraphic surfaces (conformable *vs.* unconformable) and the type of stratal terminations recorded by the surface itself or by the underlying and overlying strata against it. For example, *strata below* a conformable surface do not terminate against it, as conformities tend to parallel the bedding of the underlying deposits, but may terminate against a younger unconformity (i.e., truncation or toplap). At the same time, both types of surfaces, conformable or unconformable, may be offlapped, onlapped, or downlapped by the *strata above*. As for the stratigraphic contacts themselves, they may terminate by onlap, offlap, or downlap against *older* stratigraphic horizons.

A good knowledge of the tectonic and depositional settings is often critical for the proper identification of specific stratal terminations. For example, the *marine onlap* describes deep-water gravity-flow deposits onlapping onto the continental slope, whereas *fluvial* and *coastal onlaps* develop on continental shelves, in nonmarine and coastal to shallow-marine environments, respectively. The differentiation between fluvial, coastal, and marine onlap is therefore important for paleogeographic reconstructions, and requires knowledge of the types of facies that onlap onto the steeper landscape or seafloor surfaces. Another example is offered by *truncation* surfaces, which may be caused by erosional processes in either fluvial or marine environments (Fig. 4.3). Here too, knowledge of the facies that are in contact across the scour surface, as well as of the overall stratal stacking patterns, are critical for the proper identification of the truncation type. In wave-dominated forced regressive coastal settings, truncation is produced by wave scouring in the shallow-marine environment as the base-level falls, and the juxtaposed facies below and above the scour surface are both marine in nature. In this case, the truncation surface is downlapped by prograding forced regressive subtidal deposits. At the same time, another erosional surface is cut by fluvial systems adjusting to a lower-elevation graded profile, landward relative to the shoreline (Fig. 3.27). Truncation surfaces may also be formed by processes of wave scouring in the subtidal environment during shoreline transgression, but this time the scour is onlapped by 'healing-phase' shallow-marine strata (coastal onlap; Fig. 3.20).

Where seismic data provide the only source of geological information, as is often the case in frontier hydrocarbon basins, one must be aware that most stratigraphic units thinner than several meters, depending on seismic resolution, are generally amalgamated within single seismic reflections. For this reason, as noted by Posamentier and Allen (1999), '... because of limited seismic resolution, the location of stratal terminations, imaged on seismic data as reflection terminations, will, in general, not be located where the reflection terminations are observed. Coastal onlap as well as downlap terminations, in particular, can, in fact, be located a considerable distance landward and seaward, respectively, of where they appear on seismic data, because of stratal thinning.' Another potential artefact of limited seismic resolution is that reflection geometries observed on seismic transects (i.e., stratal terminations as imaged on seismic data) may not always be representative of true stratal stacking patterns. For example, *apparent* onlap may be inferred on seismic lines along which stratigraphic units drape, and not terminate against a pre-existing topography, particularly where the thickness of those units is less than the seismic resolution (Hart, 2000; Fig. 2.42).

Postdepositional tectonic tilt may add another level of difficulty to the recognition and interpretation of stratal terminations, both in outcrop and on seismic data. In particular, onlap and downlap may easily be affected by differential subsidence or tectonic uplift, which may change the syndepositional slope gradients of strata and of the surfaces against which they terminate. For example, the upward motion of salt diapirs during the evolution of a basin may modify the original inclination of pre-existing strata, turning depositional downlap into apparent onlap, or *vice versa* (e.g., see red arrows in Fig. 2.65, which resemble onlap geometries, but correspond in fact to depositional downlap related to the progradation of the divergent continental margin).

The correct interpretation of stratal terminations is of paramount importance for the success of the sequence stratigraphic method, as it provides critical evidence for the reconstruction of syndepositional shoreline shifts, and implicitly for the identification of systems tracts and sequence stratigraphic surfaces. Shoreline trajectories, as inferred from stratal terminations and stacking patterns, are also important for understanding sediment distribution and dispersal systems within a sedimentary basin. This, in turn, has important ramifications for the effort of locating facies with specific economic significance, such as petroleum reservoirs, coal-bearing successions, or mineral placers. Offlapping prograding lobes, for example, are a promising 'sign' for the exploration of deep-water systems, because the inferred base-level fall at the shoreline is one of the main controls that facilitates the transfer of coarser-grained sediment from fluvial and coastal systems into the deep-water environment. Evidence for normal regressions or transgressions is equally important for designing exploration strategies, because the depocenters for sediment accumulation, and implicitly the distribution of economically-significant facies, shift accordingly as a function of shoreline trajectory, shoreline location in relation to the main physiographic elements of the basin, available accommodation, and sediment supply. All these issues are explored in more detail in the subsequent chapters of this book.

SEQUENCE STRATIGRAPHIC SURFACES

Surfaces that can serve, at least in part, as systems tract or sequence boundaries, are surfaces of sequence stratigraphic significance. Sequence stratigraphic surfaces are defined relative to two curves; one describing the base-level changes at the shoreline, and one describing the associated shoreline shifts (Figs. 4.5 and 4.6). The two curves are offset relative to one another by the duration of normal regressions, whose timing is controlled by the interplay of base level and sedimentation *at the shoreline* (Fig. 4.5). As explained in Chapter 3, normal regressions most likely occur in the early ('lowstand') and late ('highstand') stages of base-level rise, when the rates of rise are very low (starting from zero and approaching zero, respectively), being outpaced by the rates of sedimentation at the shoreline.

Base-level changes in Figs. 4.5 and 4.6 are idealized, being defined by symmetrical sine curves. This may not necessarily be the case in reality. Pleistocene examples from the Gulf of Mexico suggest longer stages of base-level fall relative to base-level rise in relation to glacio-eustatic climatic fluctuations, as it takes more time to build ice caps (base-level fall) than to melt the ice (Blum, 2001). The tectonic control on base-level changes may also generate asymmetrical base-level curves. The case study of the Western Canada foreland system shows that stages of thrusting in the adjacent orogen, responsible for subsidence in the foredeep, were shorter in time relative to the stages of orogenic quiescence that triggered isostatic rebound and uplift in the foredeep (Catuneanu et al., 1997a). Given the likely asymmetrical nature of the reference curve of base-level changes, the associated transgressive–regressive curve is bound to display an even more

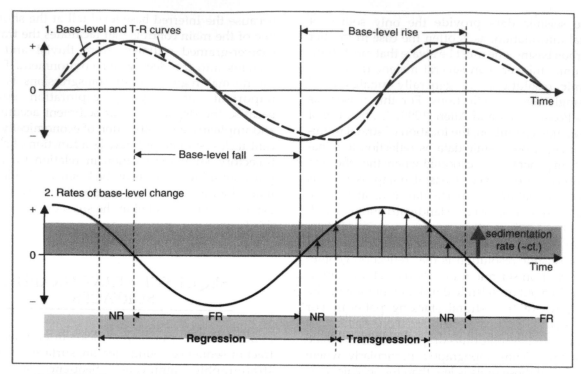

FIGURE 4.5 Base-level and transgressive–regressive (T–R) curves. Sequence stratigraphic surfaces, and systems tracts, are all defined relative to these curves (Fig. 4.6). The T–R curve, describing the shoreline shifts, is the result of the interplay between sedimentation and base-level changes *at the shoreline*. Sedimentation rates during a cycle of base-level change are considered constant, for simplicity. Similarly, the reference base-level curve is shown as a symmetrical sine curve for simplicity, but no inference is made that this should be the case in the geological record. In fact, asymmetrical shapes are more likely, as a function of particular circumstances in each case study (e.g., glacio–eustatic cycles are strongly asymmetrical, as ice melts more rapidly than it builds up), but this does not change the fundamental principles illustrated in this diagram. Abbreviations: FR—forced regression; NR—normal regression.

asymmetrical shape, with much shorter transgressions relative to the regressive stages, within the context of the examples above.

As a function of the interplay between sedimentation and base-level fluctuations at the shoreline, four main events associated with changes in depositional trends are recorded during a complete cycle of base-level shifts (Figs. 1.7, 4.5, and 4.7):

1. *Onset of forced regression* (onset of base-level fall at the shoreline): this is accompanied by a change from sedimentation to erosion/bypass in the fluvial to shallow-marine environments;
2. *End of forced regression* (end of base-level fall at the shoreline): this marks a change from degradation to aggradation in the fluvial to shallow-marine environments;
3. *End of regression* (during base-level rise at the shoreline): this marks the turnaround point from shoreline regression to subsequent transgression;

4. *End of transgression* (during base-level rise at the shoreline): this marks a change in the direction of shoreline shift from transgression to subsequent regression.

These four events control the formation of all sequence stratigraphic surfaces, as outlined below. In addition to the seven surfaces of sequence stratigraphy (Fig. 4.7), which can serve at least in part as *systems tract boundaries*, additional stratigraphic surfaces may be mapped *within* systems tracts. These within-trend facies contacts are lithological discontinuities that may have a strong physical expression in outcrop, core, or subsurface, but are more suitable for lithostratigraphic or allostratigraphic analyses (Fig. 4.8). The nomenclature and definition of systems tracts differ among the various sequence models (Figs. 1.6 and 1.7), but invariably, the timing of each systems tract boundary corresponds to one of the four main events of the base-level cycle (Figs. 1.7 and 4.7).

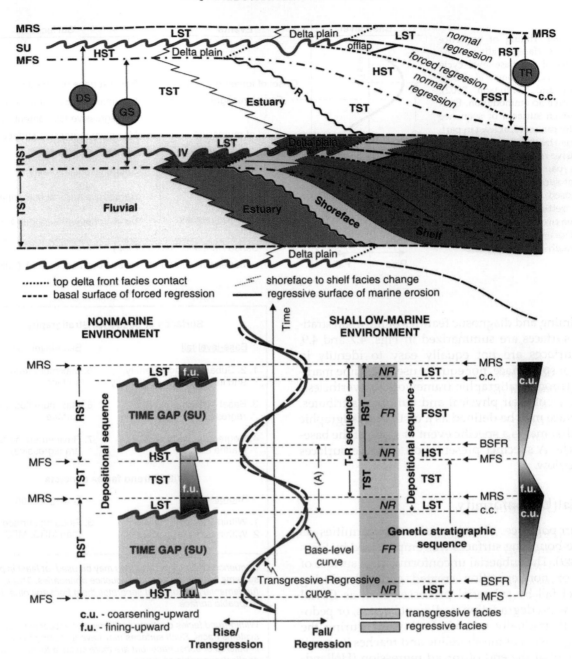

FIGURE 4.6 Sequences, systems tracts, and stratigraphic surfaces defined in relation to the base-level and the transgressive–regressive curves (modified from Catuneanu *et al.*, 1998b). Abbreviations: SU—subaerial unconformity; c.c.—correlative conformity (*sensu* Hunt and Tucker, 1992); BSFR—basal surface of forced regression (= correlative conformity *sensu* Posamentier *et al.*, 1988); MRS—maximum regressive surface; MFS—maximum flooding surface; R—transgressive wave-ravinement surface; IV—incised valley; (A)—positive accommodation (base-level rise); NR—normal regression; FR—forced regression; LST—lowstand systems tract (*sensu* Hunt and Tucker, 1992); TST—transgressive systems tract; HST—highstand systems tract; FSST—falling-stage systems tract; RST—regressive systems tract; DS—depositional sequence; GS—genetic stratigraphic sequence; TR—transgressive–regressive sequence.

FIGURE 4.7 Timing of sequence stratigraphic surfaces relative to the main events of the base-level cycle (modified from Catuneanu *et al.*, 1998b, and Embry and Catuneanu, 2002). (-A)—negative accommodation. Each of these seven surfaces of sequence stratigraphy can serve, at least in part, as systems tract boundaries. The 'transgressive ravinement surfaces' include a pair of *wave-* and *tidal-*ravinement surfaces, which are often superimposed, especially in open shoreline settings. In river-mouth settings, the two transgressive ravinement surfaces may be separated by estuary-mouth complex deposits.

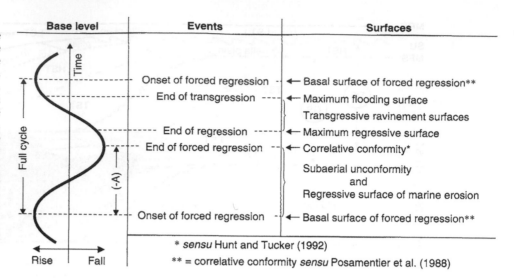

Base level	Events	Surfaces
	Onset of forced regression	← Basal surface of forced regression**
	End of transgression	← Maximum flooding surface
		Transgressive ravinement surfaces
	End of regression	← Maximum regressive surface
	End of forced regression	← Correlative conformity*
		Subaerial unconformity and Regressive surface of marine erosion
	Onset of forced regression	← Basal surface of forced regression**

* *sensu* Hunt and Tucker (1992)

** = correlative conformity *sensu* Posamentier et al. (1988)

The timing and diagnostic features of the main stratigraphic surfaces are summarized in Figs. 4.7 and 4.9. These surfaces are not equally easy to identify in outcrop or subsurface, nor equally useful as time markers in a chronostratigraphic framework. Nevertheless, irrespective of their physical and temporal attributes, each surface may be defined as a distinct stratigraphic contact that marks a specific event or stage of the base-level cycle. A succinct presentation of these surfaces follows below.

Subaerial Unconformity

The importance of subaerial unconformities as sequence-bounding surfaces was emphasized by Sloss *et al.* (1949). The subaerial unconformity is a surface of erosion or nondeposition created generally during base-level fall by subaerial processes such as fluvial incision, wind degradation, sediment bypass, or pedogenesis. It gradually extends basinward during the forced regression of the shoreline and reaches its maximum extent at the end of forced regression (Helland-Hansen and Martinsen, 1996: 'seaward, the subaerial unconformity extends to the location of the shoreline at the end of fall'). Owing to their timing and mode of formation, subaerial unconformities correspond to the largest stratigraphic hiatuses in the sedimentary rock record (Fig. 4.6), separate strata that are genetically unrelated (i.e., which belong to different cycles of base-level change), and mark abrupt basinward shifts of facies (e.g., Fig. 4.10). The subaerial unconformity has a marine correlative conformity whose timing corresponds to the end of base-level fall at the shoreline (*sensu* Hunt and Tucker, 1992; Figs. 4.6 and 4.7). Criteria for the recognition of subaerial unconformities

---

Surfaces of Sequence Stratigraphy

Base-level <u>fall</u>	Base-level <u>rise</u>
1, 2. Subaerial unconformity, and its correlative conformity*	5. Maximum regressive surface
3. Basal surface of forced regression**	6. Maximum flooding surface
4. Regressive surface of marine erosion	7. Ravinement surfaces (transgressive)

Within-trend facies contacts

Regression	Transgression
1. Within-trend NR surface	3. Flooding surface (other than MRS, MFS, or RS)
2. Within-trend FR surface	

Sequence stratigraphic surfaces may be used, at least in part, as systems tract boundaries or sequence boundaries. This is their fundamental attribute that separates them from any other type of mappable surface.

Within-trend facies contacts are lithological discontinuities within systems tracts. Such surfaces may have a strong physical expression in outcrop or subsurface, but are more suitable for lithostratigraphic or allostratigraphic analyses.

FIGURE 4.8 Types of stratigraphic surfaces (modified from Embry, 2001b and Catuneanu, 2002). The top seven surfaces are proper sequence stratigraphic surfaces that may be used, at least in part, as *systems tract or sequence boundaries*. The bottom three represent facies contacts developed *within* systems tracts. Such within-trend facies contacts may be marked on a sequence stratigraphic cross-section only after the sequence stratigraphic framework has been constructed. The transgressive ravinement surfaces include a pair of *wave-* and *tidal-*ravinement surfaces, which are often superimposed, especially in open shoreline settings. Notes: *—sensu* Hunt and Tucker, 1992; **—correlative conformity *sensu* Posamentier *et al.*, 1988. Abbreviations: MRS—maximum regressive surface; MFS—maximum flooding surface; RS—transgressive ravinement surfaces; NR—normal regressive; FR—forced regressive.

Stratigraphic surface	Nature of contact	Facies		Depositional trends[3]		Substrate-controlled ichnofacies	Stratal terminations	Temporal attributes[8]
		below	above	below	above			
Subaerial unconformity	Scoured or bypass	Variable (where marine, c-u)	Nonmarine	NR, FR	NR, T	N/A	Above: fluvial onlap Surface: offlap Below: truncation, toplap	Variable hiatus
Correlative conformity[1]	Conformable	Marine, c-u	Marine (c-u on shelf)	FR	NR	N/A	Above: downlap Surface: downlap Below: N/A	Low diachroneity
Basal surface of forced regression[2]	Conformable or scoured	Marine (c-u on shelf)	Marine, c-u	NR	FR	Glossifungites, where reworked by the RWR	Above: downlap Surface: downlap Below: N/A, truncation	Low diachroneity
Regressive wave ravinement	Scoured	Shelf, c-u	Shoreface, c-u	NR, FR	FR, NR	Glossifungites	Above: downlap Surface: N/A Below: truncation	High diachroneity
Maximum regressive surface	Conformable[7]	Variable[5]	Variable (where marine, f-u)	NR	T	N/A	Above: marine onlap Surface: onlap, downlap Below: N/A	Low diachroneity
Maximum flooding surface	Conformable or scoured	Variable (where marine, f-u)	Variable (where marine, c-u)	T	NR	Glossifungites, Trypanites, Teredolites	Above: downlap Surface: onlap, downlap[4] Below: N/A, truncation	Low diachroneity
Transgressive wave ravinement	Scoured	Variable (where marine, c-u)	Marine, f-u	NR, T	T	Glossifungites, Trypanites, Teredolites	Above: coastal onlap Surface: N/A Below: truncation	High diachroneity
Transgressive tidal ravinement	Scoured	Variable (where marine, c-u)	Estuary mouth complex	NR, T	T	Glossifungites, Trypanites, Teredolites	Above: coastal onlap Surface: N/A Below: truncation	High diachroneity
Within-trend NR surface	Conformable	Delta front or beach	Delta plain or fluvial	NR	NR	N/A	N/A	High diachroneity
Within-trend FR surface[6]	Conformable	Prodelta	Delta front	FR	FR	N/A	Above: downlap Surface: N/A Below: N/A	High diachroneity
Flooding surface	Conformable or scoured	Variable	Marine, f-u or c-u	T, NR	T, NR	Glossifungites, Trypanites, Teredolites	Above: onlap, downlap Surface: onlap, downlap[4] Below: truncation	Low to high diachroneity

FIGURE 4.9 Diagnostic features of the main stratigraphic surfaces (modified from Catuneanu, 2002, 2003, and Embry and Catuneanu, 2002). These contacts include seven *sequence stratigraphic surfaces* (by grouping the transgressive wave- and tidal-ravinement surfaces into 'transgressive ravinement surfaces'; Figs. 4.7 and 4.8), and three *within-trend facies contacts* (Fig. 4.8). Notes: [1]—*sensu* Hunt and Tucker (1992); [2]—correlative conformity *sensu* Posamentier *et al.* (1988); [3]—where all systems tracts are preserved; [4]—in a transgressive setting, downlap may only be apparent as it may mark the base of a sedimentary unit at its erosional rather than depositional limit; [5]—where marine, coarsening-upward in shallow water and fining-upward in deep water; [6]—this facies contact may only develop in the case of river-dominated deltas; [7]—see text for a discussion of possible exceptions; [8]—the temporal attributes listed in this table are valid for *dip-oriented* sections (see Chapter 7 for a full discussion of temporal attributes, both along dip and strike). Note that *conformable* stratigraphic contacts may onlap or downlap the depositional surface, but no stratal terminations against them are recorded by the facies below. *Unconformable* stratigraphic contacts truncate the strata below, and are commonly associated with substrate-controlled ichnofacies where the overlying strata are marine. The substrate-controlled ichnofacies refer to the *Glossifungites*, *Trypanites*, and *Teredolites* trace fossil assemblages, and do not include the softground ichnofacies (see Chapter 2 for more details). Both conformable and unconformable stratigraphic contacts are commonly onlapped or downlapped by the strata above. Abbreviations: c-u—coarsening-upward; f-u—fining-upward; RWR—regressive wave ravinement (= regressive surface of marine erosion); NR—normal regression; FR—forced regression; T—transgression.

FIGURE 4.10 Outcrop photograph of a subaerial unconformity (arrow) at the contact between swaley cross-stratified shoreface deposits and the overlying fluvial strata (Bahariya Formation, Lower Cenomanian, Bahariya Oasis, Western Desert, Egypt). In this example, the subaerial unconformity marks the base of an incised valley. Owing to their timing and mode of formation, subaerial unconformities correspond to the largest stratigraphic hiatuses in the sedimentary rock record (Fig. 4.6), separate strata that are genetically unrelated (i.e., which belong to different cycles of base-level change), and mark abrupt basinward shifts of facies. Preserved subaerial unconformities are always overlain by fluvial deposits (Fig. 4.9; see text for details).

in the field have been reviewed by Shanmugam (1988), and are synthesized in Fig. 4.9.

Forced regressions generally require fluvial systems to adjust to new (lower) graded profiles, especially in the downstream reaches where fluvial processes are primarily controlled by base-level changes (Figs. 3.3, 3.16, and 3.31A). The response of fluvial systems to base-level fall is complex and depends, among other parameters, on the magnitude of fall and the contrast in slope gradients between the seafloor exposed to subaerial processes and the fluvial landscape at the onset of forced regression. A small base-level fall at the shoreline may be accommodated by changes in channel sinuosity, roughness and width, with only minor incision (Schumm, 1993; Ethridge *et al.*, 2001). The subaerial unconformity generated by such unincised fluvial systems is mainly related to the process of sediment bypass (Posamentier, 2001). A larger base-level fall at the shoreline, such as the lowering of the base level below a major topographic break (e.g., the shelf edge) results in fluvial downcutting and the formation of incised valleys (Schumm, 1993; Ethridge *et al.*, 2001; Posamentier, 2001; Fig. 4.11). The interfluve areas are generally subject to sediment starvation and soil development. The subaerial unconformity can thus be traced at the top of paleosol horizons that are correlative to the unconformities generated in the channel subenvironment (Wright and Marriott, 1993; Gibling and Bird, 1994; Gibling and Wightman, 1994;

Tandon and Gibling, 1994, 1997; Kraus, 1999; Figs. 2.12, 2.13, and 4.12).

The subaerial unconformity may be placed at the top of any type of depositional system (fluvial, coastal, or marine), but it is always overlain by nonmarine deposits (Figs. 4.9, 4.10, and 4.13). The preservation of the overlying nonmarine deposits is thus required for the recognition and labeling of a subaerial unconformity as such. The underlying fluvial to shallow-marine strata may be either normal regressive (landward from the shoreline position at the onset of base-level fall) or forced regressive (within the area of forced regression). The overlying fluvial deposits may be either normal regressive (lowstand) or transgressive, depending on landscape gradients and the degree of development of lowstand normal regressive strata (Fig. 4.9). Low landscape gradients coupled with extended periods of time of lowstand normal regression are prone to the development of normal regressive fluvial topsets on top of the subaerial unconformity. The subaerial unconformity may be subsequently reworked (and replaced) by younger stratigraphic surfaces, in which cases the contact should be described using the name of the youngest preserved surface, which imposes its attributes on that particular stratigraphic contact. For example, subaerial unconformities may be reworked by transgressive ravinement surfaces, in which case the unconformable contact is directly overlain by transgressive marine facies (Fig. 4.14).

FIGURE 4.11 Subaerial unconformity at the base of the Early Cretaceous Mannville Group, where fluvial deposits overlie Devonian carbonates (Western Canada Sedimentary Basin), on 3D seismic data (images courtesy of H.W. Posamentier). This illuminated horizon is characterized by high-sinuosity fluvial channels incised into the underlying carbonate section.

Besides sedimentological methods of documenting the seaward shift of facies that accompanies the fall in base level, the observation of ichnofacies and ichnofabrics may provide additional clues for the identification of subaerial unconformities. The process of subaerial erosion may result in the formation of firmgrounds, by the exhumation of semi-cohesive deposits, but in the absence of marine or marginal-marine conditions no substrate-controlled ichnofacies may form (Fig. 4.9). Instead, subaerial unconformities may be associated with nonmarine softgrounds, particularly the paleosol-related *Termitichnus* ichnofacies, and also with abrupt shifts from marine to overlying nonmarine ichnofabrics. In a case study from the Ebro Basin in Spain, Siggerud and Steel (1999) document subaerial unconformities on the basis of ichnofabric transitions, from intertidal and subtidal deposits with *Ophiomorpha* burrows, to overlying *Taenidium*, *Scoyenia*, and *Planolites* trace assemblages that formed in fluvial environments. In the absence of nonmarine ichnofacies, subaerial unconformities may still be identified based on other evidence of subaerial exposure, such as the presence of rooted paleosols cross-cutting marginal to shallow-marine ichnofabrics (Taylor and Gawthorpe, 1993). Where

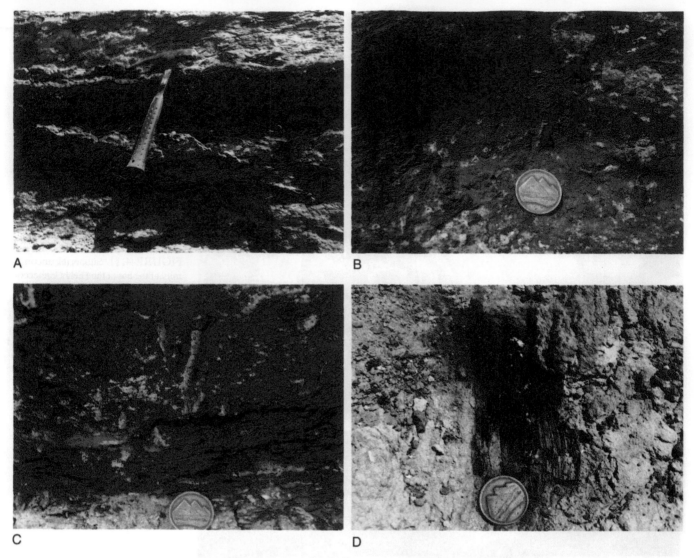

FIGURE 4.12 Outcrop photographs of a subaerial unconformity (top of ferruginous paleosol horizon in image A) and associated facies, which formed within a fully nonmarine succession of fine-grained fluvial overbank deposits (Bahariya Formation, Lower Cenomanian, Bahariya Oasis, Western Desert, Egypt). Floodplain claystones are present both above and below the paleosol horizon. Plant roots are abundant, and present within both the paleosol and the underlying claystones (images B and C). Dessication cracks and wood fragments filled with iron oxides are also present within the claystone intervals (image D). Concretions are occasionally associated with the paleosol horizon, and rip up clasts are found above the subaerial unconformity, at the base of the overlying depositional sequence.

subaerial unconformities are replaced by subsequent transgressive ravinement surfaces, the composite stratigraphic contact may be marked by substrate-controlled ichnofacies (commonly *Glossifungites*, but also *Trypanites* and *Teredolites*), as the return of marine conditions allows the colonization of the formerly exposed surface by marine tracemakers (Pemberton and MacEachern, 1995). In this case, the contact may no longer be referred to as a subaerial unconformity, as it takes over the attributes of a transgressive ravinement surface.

The stratigraphic hiatus associated with the subaerial unconformity is variable, due to differential fluvial incision and the gradual expansion of subaerial erosion in a basinward direction during the stage of base-level fall. The mechanics of formation of subaerial unconformities are suggested in Figs. 3.27 ('fluvial erosion' associated with forced regressions) and 3.31 (case A, where the subaerially exposed seafloor is steeper than the fluvial landscape at the onset of forced regression). Note that the subaerial unconformity not

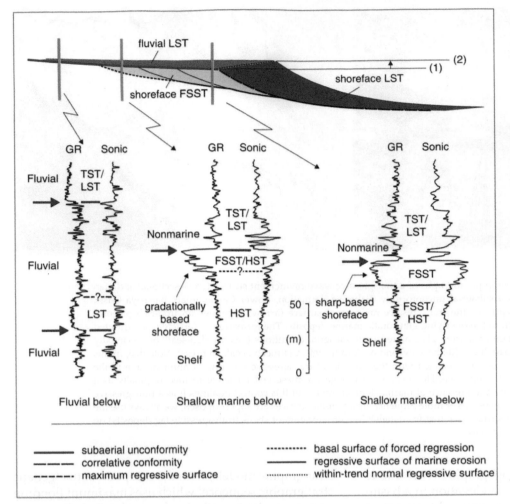

FIGURE 4.13 Well-log expression of the subaerial unconformity (arrows; modified from Catuneanu, 2002, 2003). See Fig. 4.9 for a summary of diagnostic features of the subaerial unconformity. Log examples from the Scollard and Paskapoo formations (left), and Cardium Formation (center and right), Western Canada Sedimentary Basin. Note that the subaerial unconformity may top either gradationally based (highstand or earliest forced regressive) or sharp-based (forced regressive) shoreface deposits. Abbreviations: GR—gamma ray log; LST—lowstand systems tract; TST—transgressive systems tract; HST—highstand systems tract; FSST—falling-stage systems tract.

only expands in a seaward direction as the seafloor is gradually exposed by the falling base level, but at the same time it also expands in a landward direction as well *via* the upstream migration of fluvial knickpoints (Figs. 3.31 and 3.32).

One should note that the generally inferred genetic relationship between subaerial unconformities and forced regressions reflects a 'most likely' scenario, and that exceptions do occur. For example, subaerial unconformities may also form during shoreline transgression, where extreme wave energy results in coastal erosion (Leckie, 1994; Fig. 3.20). In such cases, the (transgressive) subaerial unconformity is reworked by the transgressive wave-ravinement surface, which is onlapped by transgressive shallow-marine deposits, and no intervening fluvial to coastal deposits accumulate during shoreline transgression (Fig. 3.20). On the other hand, forced regressions may also be accompanied by fluvial aggradation where the gradient of the exposed seafloor is less than that of the fluvial landscape at the onset of base-level fall (case C in Fig. 3.31, most likely in the case of fault-bounded basins), or

where the climate-induced decrease in fluvial discharge during stages of glaciation (prone to fluvial aggradation) outpaces the influence of glacio-eustatic fall (prone to fluvial erosion). At the same time, subaerial unconformities may form during stages of glacial melting and global sea-level rise, due to climate-controlled increases in fluvial discharge (Fig. 3.7). All these departures from the prediction of standard sequence stratigraphic models need to be kept in mind and considered on a case-by-case basis.

Subaerial unconformities may be identified with any kind of data (outcrop, core, seismic, and well-log), as afforded by their physical and geometric attributes. An examination of the actual rock facies in outcrop and/or core allows one to observe the evidence for scouring, the nature of juxtaposed facies and depositional trends, and the abrupt seaward shift of facies across the unconformity. The indirect geophysical information afforded by seismic data provides more details about the regional geometric attributes of this type of stratigraphic contact, including offlapping stratal terminations along the unconformity, truncation

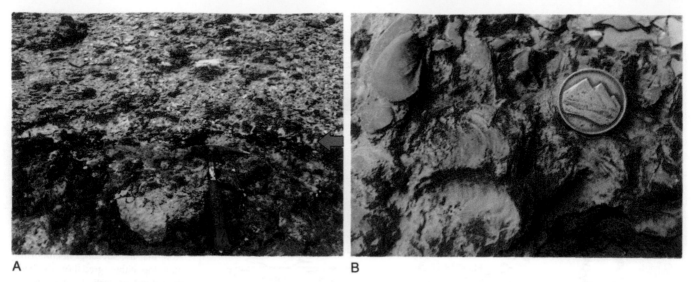

A   B

FIGURE 4.14 Outcrop photographs of a transgressive wave-ravinement surface (cross sectional and plan views) that replaces a subaerial unconformity (Bahariya Formation, Lower Cenomanian, Bahariya Oasis, Western Desert, Egypt). A—the transgressive ravinement surface (arrow) separates an iron-rich paleosol horizon (ferricrete) from the overlying glauconitic marine deposits. The formation of ferricrete is attributed to the *in situ* alteration of marine glauconite under subaerial conditions (i.e., a paleo-seafloor subaerially exposed by a fall in base level; El-Sharkawi and Al-Awadi, 1981; Catuneanu *et al.*, in press). Note that, in this case, the amount of erosion associated with the subsequent transgressive scouring is minimal, due to the indurated nature of the ferricrete. However, even though the preserved ferricrete formed originally as a subaerial unconformity, the presence of marine deposits on top of this contact qualifies it as a transgressive ravinement surface (where two or more sequence stratigraphic surfaces are superimposed, we always use the name of the younger surface; see text for details); B—concentration of shells (transgressive lag deposits) on top of the ravinement surface.

of subjacent strata, irregular topographic relief due to differential erosion, and a loss in elevation in a basinward direction (Fig. 4.15). The basinward termination of the subaerial unconformity indicates the shoreline position at the end of forced regression, which is an important inference for the construction of paleogeographic maps. The position of the shoreline during late stages of forced regression relative to the major physiographic elements of the basin (e.g., the shelf edge in a divergent continental margin setting) is also critical for the evaluation of sediment distribution between the shallow- and deep-water depositional systems. Subsequent to the end of base-level fall at the shoreline, the subaerial unconformity may be onlapped by fluvial lowstand normal regressive or transgressive strata (Fig. 4.9), as the area of fluvial aggradation gradually expands upstream during base-level rise, or may be draped by a normal regressive topset (Fig. 4.15).

The subaerial unconformity is arguably the most important type of stratigraphic contact, as it corresponds to the most significant breaks in the rock record and hence it separates the sedimentary succession into relatively conformable packages of genetically related strata (Fig. 4.6). For this reason, subaerial unconformities are adopted as sequence boundaries in most sequence

stratigraphic models, with the exception of the 'genetic stratigraphic sequence' which uses maximum flooding surfaces as its boundaries (more detailed discussion on this topic follows in Chapter 6). The alternative use of maximum flooding surfaces as sequence boundaries stems from the fact that they are usually the easiest to be identified on well logs, at the heart of condensed sections that form in shallow-marine environments during shoreline transgression (Galloway, 1989). In contrast, subaerial unconformities may be more difficult to pick on well logs because of the variety of facies that can be associated with them (Fig. 4.9), depending on the location within the basin.

Within incised-valley systems, subaerial unconformities may be easy to identify at the base of coarse-grained valley-fill deposits, which may directly overlie finer-grained shallow-marine strata (Figs. 4.16A and 4.10). The identification of subaerial unconformities as such, at the base of incised-valley fills, requires the preservation of fluvial strata above the basal unconformity of the incised valley. Sometimes, however, the fluvially-cut surface at the base of the incised valley may be modified during subsequent transgression, where no fluvial deposits are preserved above the unconformity, and the valley fill is represented by

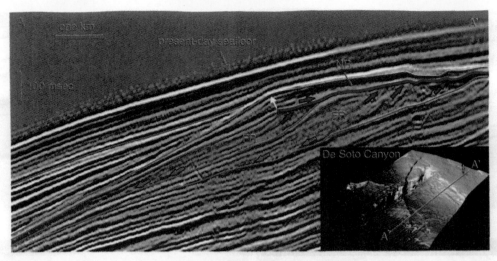

FIGURE 4.15 Subaerial unconformity (red line) on a dip-oriented, 2D seismic transect (location shown on the 3D illuminated surface) (De Soto Canyon area, Gulf of Mexico; image courtesy of H.W. Posamentier). Red arrows indicate truncation of underlying forced regressive shallow-marine strata. The deep-water forced regressive deposits downlap the prograding continental slope (yellow arrows). Thinner yellow lines provide a sense of the overall stratal stacking patterns. Note that the subaerial unconformity is associated with offlap, decrease in elevation in a basinward direction, and irregular topographic relief (differential erosion). The basinward termination of the subaerial unconformity indicates the shoreline position at the end of forced regression. The subaerial unconformity is onlapped (fluvial onlap; green arrow) and overlain by a topset of lowstand normal regressive strata. The white arrow indicates the shoreline trajectory during the subsequent lowstand normal regression. For scale, the channel on the 3D illuminated surface is approximately 1.8 km wide, and 275 m deep at shelf edge. The illuminated surface is taken at the base of forced regressive deposits. Abbreviations: FR—forced regressive deposits; NR—normal regressive deposits.

tidally-influenced estuarine deposits. In such cases, the subaerial unconformity is replaced by a younger transgressive surface of erosion at the contact between normal regressive highstand and overlying transgressive deposits (e.g., Ainsworth and Walker, 1994).

Subaerial unconformities may also be marked by sharp facies contacts in fully fluvial successions, where abrupt shifts in fluvial styles are recorded across the contacts (Fig. 4.13; cases B and C in Fig. 4.16). In such cases, the contrast in fluvial styles commonly reflects an increase in fluvial energy levels associated with a basinward shift of facies. In some interfluve areas, however, the facies and log expression of subaerial unconformities may be much more cryptic, as they may occur within fine-grained successions of overbank deposits (Fig. 4.16D). Well-drained and mature paleosols may also mark the position of subaerial unconformities, being formed during times of base-level fall and lowering of the water table in the nonmarine portion of the basin (Figs. 2.12, 2.13, and 4.12). Synonymous terms for the subaerial unconformity include the *'lowstand unconformity'* (Schlager, 1992), the *'regressive surface of fluvial erosion'* (Plint and Nummedal, 2000) and the *'fluvial entrenchment/incision surface'* (Galloway, 2004).

Correlative Conformity

The correlative conformity forms within the marine environment at the end of base-level fall at the shoreline (*sensu* Hunt and Tucker, 1992; Figs. 4.6 and 4.7). This surface approximates the paleo-seafloor at the end of forced regression, which is the *youngest clinoform associated with offlap*, and it correlates with the seaward termination of the subaerial unconformity (Figs. 4.17 and 4.18). The correlative conformity separates forced regressive deposits below from lowstand normal regressive deposits above, and, as with any clinoform, it downlaps the underlying succession. In turn, the end-of-fall paleo-seafloor is downlapped by the overlying prograding clinoforms, but no termination is recorded by the strata below against this conformable surface (Fig. 4.9).

A different 'correlative conformity' was defined by Posamentier *et al.* (1988), and subsequently refined by Posamentier and Allen (1999) as the paleo-seafloor at the onset of forced regression; that surface is dealt with under its synonymous term of 'basal surface of forced regression'. The distinction between these two types of correlative conformities is necessary because they are physically separated by the prograding and offlapping forced regressive deposits. The end-of-fall and the

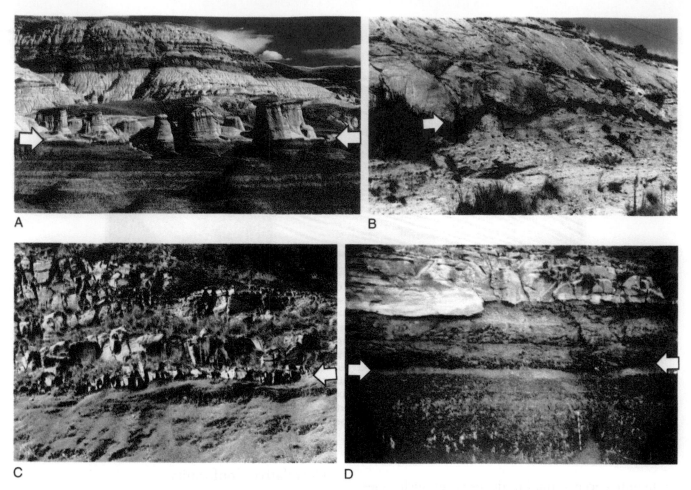

FIGURE 4.16 Outcrop examples of subaerial unconformities (arrows). A—subaerial unconformity at the contact between shallow-marine shales (Bearpaw Formation) and the overlying incised-valley-fill fluvial sandstones (Horseshoe Canyon Formation) (Late Cretaceous, Red Deer Valley, Western Canada Sedimentary Basin; facies interpretations from Ainsworth, 1994). Note that accurate paleoenvironmental reconstructions are crucial for the correct identification of sequence stratigraphic surfaces. For example, the basal sandstones of the Horseshoe Canyon Formation were previously interpreted as deltaic (Shepheard and Hills, 1970), which would make this contact a regressive surface of marine erosion. This subaerial unconformity may have been modified into a transgressive surface of erosion, if the fluvial sandstones are attributed to an estuarine environment (Ainsworth and Walker, 1994). B—subaerial unconformity at the contact between the Bamboesberg and Indwe members of the Molteno Formation (Late Triassic, Dordrecht region, Karoo Basin). The succession is entirely fluvial, with an abrupt increase in energy levels across the contact. Note the irregular character of this surface, due to differential fluvial erosion. C—subaerial unconformity at the contact between the Balfour Formation and the overlying Katberg Formation (Early Triassic, Nico Malan Pass, Karoo Basin). The succession is fully fluvial, with an abrupt increase in energy levels across the contact. Note the change in fluvial styles from a floodplain-dominated meandering system to the overlying amalgamated braided stream channels. D—subaerial unconformity at the top of a paleosol horizon (Burgersdorp Formation, Early-Middle Triassic, Queenstown region, Karoo Basin). The paleosol (with rootlets) is overlain by meandering-stream floodplain deposits. The scale is 1.4 m long. Note that in all cases, the strata overlying the subaerial unconformity are nonmarine.

onset-of-fall correlative conformities also have different preservation potentials in the rock record. The end-of-fall paleo-seafloor has a high preservation potential because it is followed by a stage of base-level rise, when aggradation is the prevalent depositional trend. The onset-of-fall paleo-seafloor, on the other hand, is potentially subject to erosion in both shallow and deep-water environments due to the subsequent fall in base level that may trigger wave scouring on the shelf, shelf-edge instability, and the onset of significant gravity flows in the deep-water environment. This 'correlative conformity' has therefore less potential to be preserved

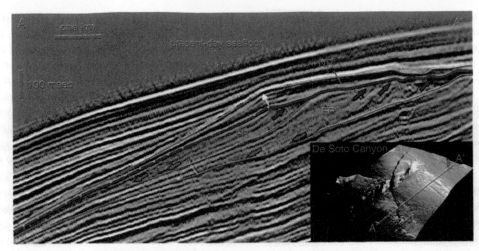

FIGURE 4.17 Correlative conformity (*sensu* Hunt and Tucker, 1992; red dashed line) on a dip-oriented, 2D seismic transect (location shown on the 3D illuminated surface) (De Soto Canyon area, Gulf of Mexico; image courtesy of H.W. Posamentier). The solid red line shows the subaerial unconformity, whose basinward termination meets the correlative conformity at the point that corresponds to the position of the shoreline at the end of forced regression. The correlative conformity is the *youngest clinoform associated with offlap*. Red arrows indicate truncation of shallow-marine forced regressive strata by the subaerial unconformity. The deep-water forced regressive deposits downlap the prograding continental slope (yellow arrows). The white arrow indicates the shoreline trajectory during the subsequent lowstand normal regression. For scale, the channel on the 3D illuminated surface is approximately 1.8 km wide, and 275 m deep at shelf edge. The illuminated surface is taken at the base of forced regressive deposits. Abbreviations: FR—forced regressive deposits; NR—normal regressive deposits.

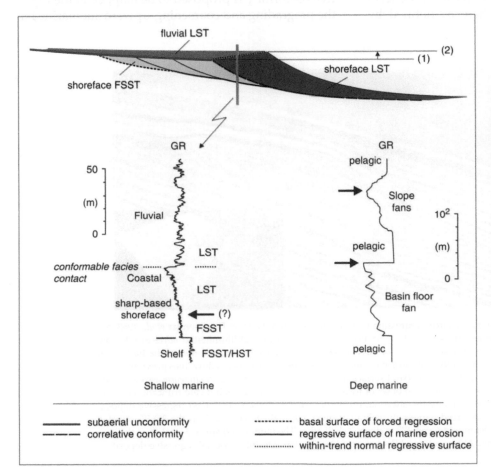

FIGURE 4.18 Well-log expression of the correlative conformity (arrows; modified from Catuneanu, 2002, 2003). See Fig. 4.9 for a summary of diagnostic features of the correlative conformity. In a shoreface succession, the correlative conformity is the clinoform that correlates with the basinward termination of the subaerial unconformity, but this surface is difficult to pinpoint on individual 1D logs (see question mark) because it is part of a continuous coarsening-upward trend. Log examples from the Lea Park Formation, Western Canada Sedimentary Basin (left), and modified from Vail and Wornardt (1990) and Kolla (1993) (right). Abbreviations: GR—gamma ray log; LST—lowstand systems tract; FSST—falling-stage systems tract; HST—highstand systems tract.

as a conformable surface in the rock record. The factors and the circumstances which diminish the preservation potential of the onset-of-fall paleo-seafloor ('basal surface of forced regression') are discussed in more detail in the next section of this chapter.

Even though the correlative conformity of Posamentier *et al.* (1988) has historical priority, the use of the end-of-fall marine surface as the conformable portion of the sequence boundary has been adopted in more recent models (i.e., depositional sequences III and IV; Figs. 1.6 and 1.7) because the onset-of-fall choice allows a portion of the subaerial unconformity, and the correlative conformity, to be both intercepted along the same vertical profile within the area of forced regression (Hunt and Tucker, 1992). In this case, the correlative conformity (*sensu* Posamentier *et al.*, 1988) does not correlate with the seaward termination of the subaerial unconformity, the two surfaces being separated by forced regressive deposits (Fig. 4.19). In addition to this, the depositional sequence II model (Posamentier *et al.*, 1988; Posamentier and Allen, 1999; Fig. 1.7) does not provide a name for the surface that separates forced regressive from overlying lowstand normal regressive strata, even though the end of base-level fall at the shoreline is one of the key events of the base-level cycle (Fig. 4.7). For these reasons, the term 'correlative conformity' is used here as defined by

Hunt and Tucker (1992) (end-of-fall marine surface), whereas the original correlative conformity of Posamentier *et al.* (1988) (onset-of-fall marine surface) is referred to as the 'basal surface of forced regression'.

The correlative conformity turned out to be a problem surface in sequence stratigraphy, surrounded by controversies regarding its timing and physical attributes. The main problem relates to the difficulty of recognizing it in most outcrop sections, core, or wireline logs (Fig. 4.18), although at the larger scale of seismic data one can infer its approximate position as the clinoform that correlates with the basinward termination of the subaerial unconformity (Fig. 4.17). The latter method of mapping the correlative conformity is limited by the relatively low seismic resolution, which makes it possible that a number of discrete clinoforms may be amalgamated as one seismic horizon.

The shallow-marine portion of the correlative conformity develops within a conformable prograding package (coarsening-upward trends below and above; Fig. 4.9), lacking lithofacies and grading contrasts (Fig. 4.18). As such, no substrate-controlled ichnofacies can be associated with the correlative conformity, and the juxtaposed deposits display no contrast in ichnofabrics. In the deep-marine environment, the correlative conformity is proposed to be mapped at the top of the prograding and coarsening-upward submarine fan

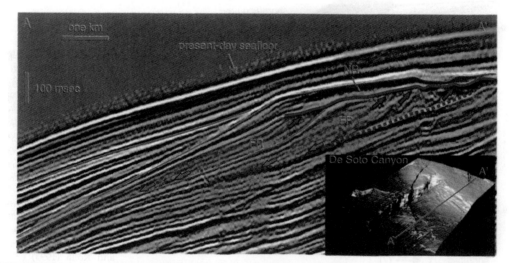

FIGURE 4.19 Basal surface of forced regression (= correlative conformity *sensu* Posamentier *et al.*, 1988; red dotted line) on a dip-oriented 2D seismic transect (location shown on the 3D illuminated surface) (De Soto Canyon area, Gulf of Mexico; image courtesy of H.W. Posamentier). The solid red line shows the basinward portion of the subaerial unconformity that formed during forced regression. Thinner yellow lines provide a sense of the overall stratal stacking patterns. The basal surface of forced regression is the *oldest clinoform associated with offlap*, and corresponds to the seafloor at the onset of forced regression. Red arrows indicate truncation of shallow-marine forced regressive strata by the subaerial unconformity. The deep-water forced regressive deposits downlap the basal surface of forced regression (yellow arrows). For scale, the channel on the 3D illuminated surface is approximately 1.8 km wide, and 275 m deep at shelf edge. The illuminated surface is taken at the base of forced regressive deposits. Abbreviations: FR—forced regressive deposits; NR—normal regressive deposits.

complex (the 'basin floor component' of Hunt and Tucker, 1992; Fig. 4.18). The overlying gravity-flow deposits tend to display a fining-upward trend due to the gradual cut-off of sediment supply to the deep-water environment during rising base level, as terrigenous sediment starts to be trapped in aggrading fluvial, coastal, and shallow-marine systems (Posamentier and Walker, 2002; Posamentier and Kolla, 2003). Beyond these models, the mapping of the end-of-fall surface within deep-water facies is in fact much more difficult because the manifestation of gravity flows, sediment supply and the associated vertical profiles, depend on a multitude of factors, some of which are independent of base-level changes. In addition to this, the idea of coeval changes along strike from coarsening- to fining-upward trends is based on the assumption that there is a uniform linear source of sediment to the outer shelf, slope, and basin floor. This is generally untrue in most clastic basins, where sediment entry points are restricted to river-mouth systems, and the clastic sediment influx to the basin is rarely enough to affect deposition in more than a small region at any one time (Frazier, 1974). Considering the autogenic shifts in the *locus* of sediment accumulation, both within a submarine fan complex and in the deep-water environment in general, there is little likelihood that changes from coarsening- to fining-upward are synchronous along strike, or even that the succession is conformable, as inferred by the term correlative 'conformity'.

The correlative conformity is implied to be a time line, i.e., 'the time surface that is correlative with the "collapsed" unconformity' (Posamentier and Allen, 1999). At the same time, the correlative conformity is also defined in relation to general stacking patterns, at 'a change from rapidly prograding parasequences to aggradational parasequences' (Haq, 1991) or at the top of submarine fan deposits (Hunt and Tucker, 1992). The latter definitions imply a diachronous correlative conformity, younger basinward, with a rate that matches the rate of offshore sediment transport (Fig. 4.9; Catuneanu et al., 1998b; Catuneanu, 2002).

Basal Surface of Forced Regression

The term 'basal surface of forced regression' was introduced by Hunt and Tucker (1992) to define the base of all deposits that accumulate in the marine environment during the forced regression of the shoreline. This corresponds to the correlative conformity of Posamentier et al. (1988), and it approximates the *paleo-seafloor at the onset of base-level fall at the shoreline* (Figs. 4.6 and 4.7). Where preserved from subsequent erosion, the basal surface of forced regression occurs within a fully marine succession, separating highstand

normal regressive strata below from forced regressive strata above (Fig. 4.9). On the shelf, both underlying and overlying deposits record progradational trends, and, within this overall coarsening-upward succession, the onset-of-fall surface is a clinoform that downlaps the preexisting strata. In turn, the basal surface of forced regression is downlapped by the younger forced regressive prograding clinoforms. As with all other conformable stratigraphic contacts, strata below do not terminate against this surface. Where the basal surface of forced regression is reworked by marine waves or currents, the scoured contact truncates the underlying strata (Fig. 4.9).

It is generally inferred that the onset-of-fall marine surface is (1) conformable, and (2) a time surface. The chances of this stratigraphic interface being preserved as a conformity in the rock record are discussed in more detail in the following paragraphs of this section. Regarding its temporal attributes, the chronohorizon status of the basal surface of forced regression, as with any other candidate for a sequence-bounding 'correlative conformity' (see Chapter 7 for further discussion), is acceptable relative to the resolution of available biostratigraphic and geochronologic age-dating techniques. Nevertheless, as at least portions of this marine surface on the shelf and on the continental slope are represented by prograding clinoforms, a low diachroneity is recorded in relation to the rates of offshore sediment transport, as it takes time for the terrigenous sediment supplied at the shoreline to reach any depozone in the deeper portions of the marine basin (Fig. 4.9; Catuneanu, 2002).

In seismic stratigraphic terms, the basal surface of forced regression is the *oldest clinoform associated with offlap* (i.e., the youngest clinoform of the underlying normal regressive deposits that is offlapped by forced regressive lobes; Fig. 4.19). This onset-of-fall marine surface is positioned below the subaerial unconformity within the area of forced regression of the shoreline (Fig. 4.19), and, providing that there is a good preservation of the earliest forced regressive deposits, the two surfaces meet at a point that marks the shoreline position at the onset of forced regression. The potential pitfall of this approach is that the subaerial unconformity and/or the subsequent transgressive wave-ravinement erosion may remove the earliest offlapping sandstone strata, so one cannot always determine where the offlapping deposits actually begin on the seismic section. This shortcoming is even more pronounced where the pattern of stratal offlap is obliterated by subsequent subaerial or transgressive ravinement erosion.

In shallow-marine (shoreface to shelf) environments, the fall in base level lowers the wave base, which may expose the seafloor to wave scouring processes, depending on the seafloor gradient (shallower or steeper relative to the wave equilibrium profile; Fig. 3.27) and the magnitude of base-level fall. High magnitude falls in base level may result in the subaerial exposure of the entire shallow-marine seafloor, which reduces significantly the chances of preservation of shallow-marine forced regressive deposits, and implicitly of their basal surface. For lower magnitude falls in base level, the preservation potential of the basal surface of forced regression within shallow-marine successions increases accordingly. The nature of scouring *vs.* aggradational processes that affect the shallow-marine seafloor during forced regression depends largely on the *angle of repose* of the prograding clinoforms relative to the wave graded profile, which in turn reflects the influence of sediment supply and of the processes that control sediment redistribution in the subtidal and inner shelf environments. A differentiation is therefore required between wave-dominated shallow-marine environments, where the seafloor gradient is small (commonly < 1°) and in balance with the wave energy, and river-dominated settings where the angle of repose of clinoforms (generally > 1°) is steeper than the wave equilibrium profile.

Wave-dominated settings, such as subtidal environments in front of open coastlines or wave-dominated deltas, are particularly prone to wave scouring during forced regression in an attempt to maintain the seafloor graded profile that is in balance with the wave energy (Fig. 4.20). In such settings, the preservation potential of the basal surface of forced regression as a

FIGURE 4.20 Stratigraphic surfaces that form in response to forced regression in a *wave-dominated* coastal to shallow-marine setting (modified from Bruun, 1962; Plint, 1988; Dominguez and Wanless, 1991). The shoreface profile that is in equilibrium with the wave energy is preserved during forced regression by a combination of coeval sedimentation and erosion processes in the upper and lower shoreface, respectively. The onset-of-fall paleo-seafloor (basal surface of forced regression) is preserved at the base of the earliest forced regressive shoreface lobe, but it is reworked by the regressive surface of marine erosion seaward relative to a lever point of balance between sedimentation and erosion. As a result, the earliest falling-stage shoreface deposits are gradationally based, whereas the rest of the offlapping lobes are sharp-based.

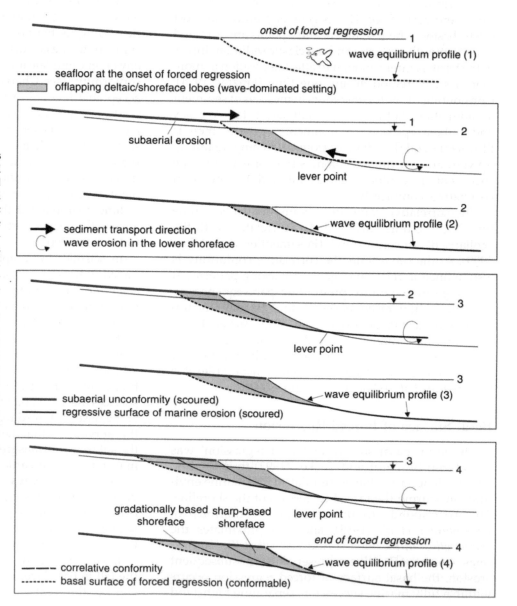

conformable paleo-seafloor is relatively low. Maintaining the wave equilibrium profile during base-level fall requires coeval sediment accumulation in the upper subtidal area and wave scouring in the lower subtidal environment (Bruun, 1962; Plint, 1988; Dominguez and Wanless, 1991; Fig. 4.20). As a result, the onset-of-fall paleo-seafloor may be preserved adjacent to the shoreline position at the onset of forced regression, at the base of the earliest prograding forced regressive lobe, but it is reworked by the regressive surface of marine erosion offshore relative to a *lever point* of balance between sedimentation and erosion (Fig. 4.20). The actual location of this lever point depends on the balance between sediment supply and wave energy, moving seaward as sediment supply increases relative to wave energy, and *vice-versa*. Landward from the initial lever point at the onset of base-level fall, the forced regressive shoreface deposits are *gradationally based*, whereas seaward from the same lever point the forced regressive shoreface deposits are *sharp-based* (Figs. 4.20 and 4.21). This onset-of-fall lever point therefore marks the place where the basal surface of forced regression and the regressive surface of marine erosion m*eet* al*ong* a dip-oriented cross-sectional profile (Figs. 4.20 and 4.22). The forced regressive shoreface deposits, either gradationally or sharp-based, are commonly truncated at the top, as being subject to subsequent subaerial or transgressive ravinement erosion. Where preserved from such subsequent erosion, the forced regressive shoreface deposits are always *thinner* than the depth of the fairweather wave base, with thicknesses most often in a range of meters, and they are generally represented by swaley cross-stratified upper shoreface facies (Fig. 4.21).

The onset-of-fall paleo-seafloor preserved at the base of the earliest forced regressive prograding lobe may be subject to subsequent erosion by the subaerial unconformity, as fluvial graded profiles adjust to the successively lower elevations of the forced regressive shoreline (Fig. 4.20). The preservation of this portion of the conformable basal surface of forced regression is therefore possible where the fall in base level and the associated subaerial erosion in the shoreline area are *less* than the depth of the fairweather wave base. As the base level falls and the shoreline is forced to regress, the regressive surface of marine erosion generated by wave scouring in the lower shoreface continues to expand in a basinward direction (Fig. 4.20), forming a highly diachronous unconformity (Fig. 4.9). At the same time, basinward relative to the scouring area, sediments accumulate in the deeper inner and outer shelf environments, allowing the preservation of the basal surface of forced regression at their base (Plint, 1991; Plint and Nummedal, 2000; Figs. 4.23 and 4.24).

These forced regressive shelf deposits may be truncated at the top by the seaward-expanding regressive surface of marine erosion (profiles D and E in Fig. 4.24), or, beyond the seaward termination of this scour surface, they may be conformably overlain by normal regressive lowstand deposits (profile F in Fig. 4.24).

It can be concluded that in wave-dominated shallow-marine successions, the conformable basal surface of forced regression may be preserved in two distinct areas separated by a zone of wave scouring of the onset-of-fall paleo-seafloor: at the base of early-fall gradationally based shoreface deposits, and at the base of forced regressive shelf deposits (Figs. 4.23 and 4.24). Where preserved, either within shoreface or shelf successions, the basal surface of forced regression poses the same recognition problems as the correlative conformity (coarsening-upward strata below and above, and a lack of lithofacies contrast across the contact; Figs. 4.9 and 4.25). As in the case of the correlative conformity, the *conformable* basal surface of forced regression may not be recognized based on ichnological criteria, because no substrate-controlled ichnofacies are associated with it, and no contrast in ichnofabrics is recorded between the strata below and above. Where reworked by wave scouring, the basal surface of forced regression is replaced by the regressive surface of marine erosion, and the composite surface may be delineated by the *Glossifungites* ichnofacies (Fig. 4.9).

In contrast to the wave-dominated settings, the preservation potential of the basal surface of forced regression within shallow-marine successions is much greater in front of river-dominated deltas, where the angle of repose of the prograding clinoforms is steeper than the wave equilibrium profile. As a result, the fall in base level does not trigger wave scouring in the lower subtidal environment, for as long as the water remains deeper that the fairweather wave base (Fig. 4.26). In such settings, no regressive surface of marine erosion forms during the forced regression of the shoreline, and the forced regressive shoreface deposits are gradationally based, being conformably underlain by normal and forced regressive shelf facies (Fig. 3.30).

In the deep-water environment, the basal surface of forced regression is taken at the base of the prograding submarine fan complex (Hunt and Tucker, 1992), as the scour cut by the *earliest* gravity flows associated with the forced regression of the shoreline (Fig. 4.25). In this case, the basal surface of forced regression separates pelagic sediments below from gravity-flow deposits above (Fig. 4.27). The pitfall of this approach is that the arrival of the first gravity-flow deposits in the deep-water environment may not necessarily coincide with the start of base-level fall, but may in fact happen any time during

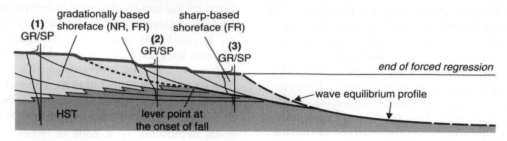

A. Stepped-topped forced regressive shoreface deposits

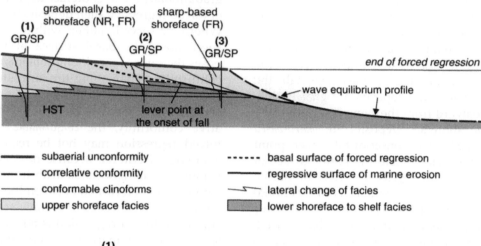

B. Smooth-topped forced regressive shoreface deposits

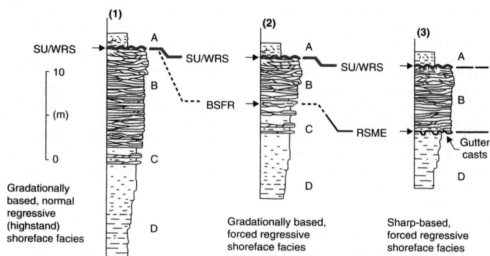

FIGURE 4.21 Forced regressive shoreface deposits in a wave-dominated setting, showing the nature of facies (dominantly swaley cross-stratified upper shoreface sands), top contacts (stepped- *vs.* smooth-topped), and basal contacts (gradationally *vs.* sharp-based). The geometry of the subaerial unconformity (stepped *vs.* smooth) depends primarily on the interplay of sediment supply and the rates of base-level fall (see Chapter 3 for more details). The basal surface of forced regression and the regressive surface of marine erosion meet at the onset-of-fall lever point of balance between upper shoreface sedimentation and lower shoreface wave scouring (Fig. 4.20). Stratal offlap may be difficult or even impossible to recognize (see the smooth-topped forced regressive shoreface deposits), but the pattern of truncation of the underlying normal regressive clinoforms, as well as the seaward dipping trend of the top unconformity, provide additional criteria to recognize the forced regressive nature of the prograding shoreface deposits. Abbreviations: GR/SP—synthetic gamma ray/spontaneous potential logs; HST—highstand systems tract (underlying normal regressive deposits); NR—normal regressive; FR—forced regressive; SU—subaerial unconformity; WRS—transgressive wave-ravinement surface; BSFR—basal surface of forced regression; RSME—regressive surface of marine erosion. Facies: A—nonmarine or transgressive marine; B—upper shoreface (swaley cross-stratified); C—lower shoreface to inner shelf (hummocky cross-stratified); D—outer shelf (bioturbated silts and muds).

FIGURE 4.22 Wave-dominated shallow-marine succession showing the transition between gradationally based (A) and sharp-based (B) upper shoreface forced regressive facies (Blackhawk Formation, Utah). The dashed line represents the inferred basal surface of forced regression (preserved onset-of-fall paleo-seafloor), and the solid line marks the regressive surface of marine erosion which separates upper shoreface sands (above) from inner shelf interbedded sands and muds (below). The direction of progradation is from left to right. Compare this field example with the diagrams in Figs. 4.20, 4.21 and 4.23.

fall, depending on physiography and sediment supply. Therefore, the base of the submarine fan complex may potentially be (much) younger than the onset of fall, depending on when the first gravity flows arrive in any particular area of the deep-water environment.

Regressive Surface of Marine Erosion

The regressive surface of marine erosion (referred to as the 'regressive wave ravinement' in Fig. 4.9) forms during forced regression in *wave-dominated shelf settings*, where seafloor gradients are low and in balance with the wave energy. This ravinement surface is a scour cut by waves in the lower shoreface during base-level fall at the shoreline, as the shoreface attempts to preserve its concave-up profile that is in equilibrium with the wave energy (Bruun, 1962; Plint, 1988; Dominguez and Wanless, 1991; Fig. 4.20). The process of wave scouring is only possible where the seafloor gradient beyond the toe of the shoreface is *lower* than the gradient of the wave equilibrium profile, which is approximated by the seafloor gradient of the shoreface. This condition is fulfilled in most wave-dominated shelf settings, where the shelf gradient averages approximately 0.01–0.03°, and the shoreface gradient is an order of magnitude steeper, of approximately 0.1–0.3° (Elliott, 1986; Cant, 1991; Walker and Plint, 1992; Hampson and Storms, 2003). Due to this contrast in seafloor gradients, the lowering of the fairweather wave base during base-level fall results in the erosion of the formerly aggrading lower shoreface to inner shelf areas, which enables the progradation of swaley cross-stratified upper to middle shoreface sandstones directly over a scour surface cut in inner to outer shelf mudstone-dominated

facies (Plint, 1991). In settings where the seafloor beyond the fairweather wave base is *steeper* than the wave equilibrium profile, such as in front of river-dominated deltas (clinoforms commonly steeper than 1°) or on continental slopes (averaging a gradient of approximately 3°), the fall in base level is not accompanied by wave scouring and the formation of a wave-ravinement surface (Fig. 3.27). In such settings, the forced regressive shoreface deposits are gradationally based (Figs. 3.30 and 4.26).

The amount of erosion that affects the seafloor of shallow-marine wave-dominated settings during forced regression is highest in the *lower shoreface* environment, close to the fairweather wave base, and is commonly in a range of meters (Plint, 1991). Seaward of the toe of the shoreface, erosion is replaced by sediment bypass and eventually by uninterrupted deposition in the deeper shelf environment (Plint, 1991). During base-level fall, the *inner shelf* is generally an area of sediment bypass, up to tens of kilometers wide, although meter-thick hummocky cross-stratified sands may still accumulate above the storm wave base (Plint, 1991). The preservation potential of these hummocks, however, is relatively low because, as the base level falls, the wave-scoured lower shoreface area shifts across the former inner shelf environment, and as a result the hummocky cross-stratified beds are truncated by the regressive surface of marine erosion (Figs. 4.23 and 4.24). Beyond the storm wave base, the *outer shelf* environment may record continuous aggradation, providing that the fall in base level does not subaerially expose the entire continental shelf (Plint, 1991). Given the low preservation potential of forced regressive inner shelf facies, it is therefore common to

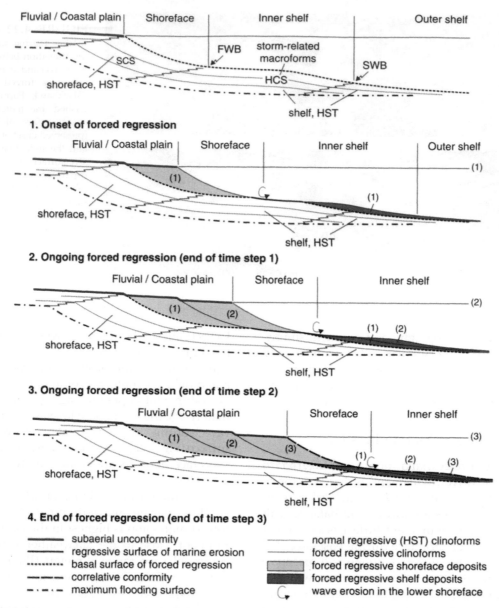

1. Onset of forced regression

2. Ongoing forced regression (end of time step 1)

3. Ongoing forced regression (end of time step 2)

4. End of forced regression (end of time step 3)

——— subaerial unconformity	·········· normal regressive (HST) clinoforms
——— regressive surface of marine erosion	——— forced regressive clinoforms
·········· basal surface of forced regression	░░░ forced regressive shoreface deposits
— — — correlative conformity	▓▓▓ forced regressive shelf deposits
— · — · — maximum flooding surface	ᵧ wave erosion in the lower shoreface

FIGURE 4.23 Shallow-marine deposits of the falling stage, in a *wave-dominated* shelf setting (modified from Catuneanu, 2003). The shallow-marine forced regressive deposits may include: gradationally based shoreface (underlain by the basal surface of forced regression), sharp-based shoreface (underlain by the regressive surface of marine erosion) and shelf facies (gradationally based, underlain by the basal surface of forced regression). The basal surface of forced regression may in parts be eroded by the subaerial unconformity and by the regressive surface of marine erosion. Where preserved, the basal surface of forced regression is a systems tract boundary. The regressive surface of marine erosion may become a systems tract boundary where it reworks the basal surface of forced regression. Note that the inner shelf environment widens during forced regression in response to falling base level and shelf aggradation, in order to maintain the same depth of the SWB. The inner shelf accumulates hummocky cross-stratified deposits, which aggrade during storm events forming positive-relief features on the seafloor (Arnott *et al.*, 2004). As a result, the seafloor does not necessarily describe the commonly inferred smooth concave-up profile, but rather displays inner shelf macroforms (meter-scale height to hundreds of meters wide) above the average concave-up seafloor profile (Catuneanu, 2003; Arnott *et al.*, 2004). Abbreviations: HST—highstand systems tract; HCS—hummocky cross-stratification; SCS—swaley cross-stratification; FWB—fairweather wave base; SWB—storm wave base.

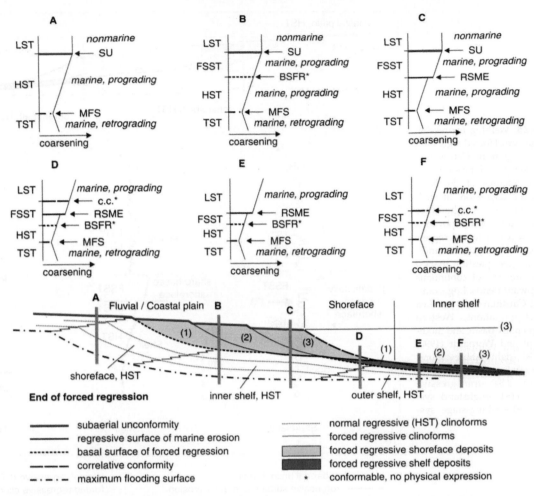

FIGURE 4.24 Architecture of sequence stratigraphic surfaces in a wave-dominated, shallow-marine setting (continued from Fig. 4.23) (modified from Catuneanu, 2003). Vertical profiles are not to scale, and it is assumed that the stratigraphy shown on the cross-section is overlain by (lowstand) normal regressive deposits preserved from subsequent transgressive ravinement erosion. The basal surface of forced regression may be preserved at the base of either shoreface or shelf deposits as the youngest clinoform of the underlying normal regressive succession. This surface may in parts be replaced (reworked) by the regressive surface of marine erosion, as well as by the subaerial unconformity. Note that the regressive surface of marine erosion and the basal surface of forced regression may both occur in the same location (e.g., vertical profile D), separated by falling-stage shelf deposits. Abbreviations: TST—transgressive systems tract; HST—highstand systems tract; FSST—falling-stage systems tract; LST—lowstand systems tract; SU—subaerial unconformity; c.c.—correlative conformity; BSFR—basal surface of forced regression; RSME—regressive surface of marine erosion; MFS—maximum flooding surface.

find the sharp-based swaley cross-stratified upper to middle shoreface deposits directly overlying falling-stage outer shelf mudstones (Plint and Nummedal, 2000; Fig. 4.28). Where no forced regressive shelf deposits are preserved, the sharp-based shoreface may prograde directly on top of normal regressive (high-stand) shelf facies, which have a much better preservation potential than their forced regressive equivalents (e.g., vertical profile C in Fig. 4.24).

The preservation potential of forced regressive shelf sediments depends on the balance between the thickness of the succession that accumulated prior to the fairweather wave base approach and the amount of subsequent wave scouring. Whether or not forced regressive shelf deposits are preserved as a result of this scouring, the regressive surface of marine erosion is always placed between *shelf facies below* (either normal or forced regressive) and *shoreface facies above* (again, either forced or normal regressive; Figs. 4.9, 4.24 and 4.29). The origin of the underlying shelf facies (highstand normal regressive *vs.* forced regressive) is difficult to establish especially when working with well-log data, because the basal surface of forced regression, where preserved as a conformable paleo-seafloor, has

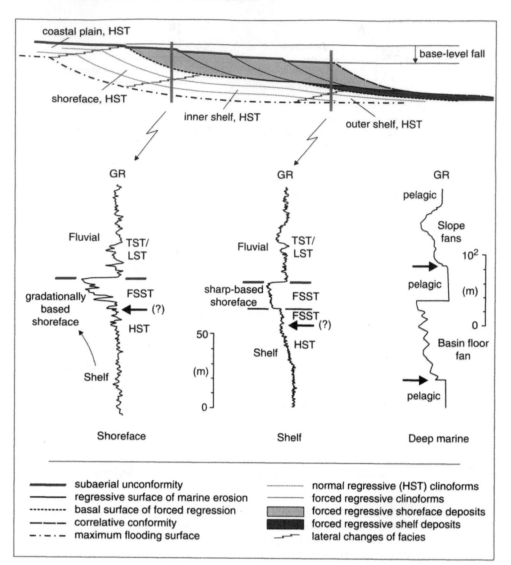

FIGURE 4.25 Well-log expression of the basal surface of forced regression (arrows; modified from Catuneanu, 2002, 2003). See Fig. 4.9 for a summary of diagnostic features of the basal surface of forced regression. In shallow-marine successions (shoreface and shelf), the conformable portions of the basal surface of forced regression are difficult to recognize on individual 1D logs (see question marks) because they are part of continuous coarsening-upward trends. Log examples from the Cardium (left) and Lea Park (center) formations, Western Canada Sedimentary Basin, and modified from Vail and Wornardt (1990) and Kolla (1993) (right). Abbreviations: GR—gamma ray log; LST—lowstand systems tract; TST—transgressive systems tract; HST—highstand systems tract; FSST—falling-stage systems tract.

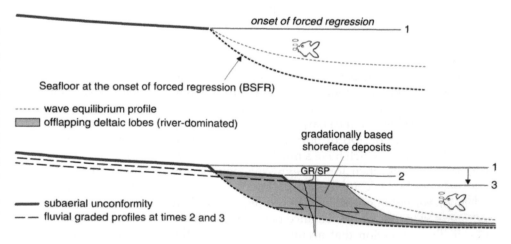

FIGURE 4.26 Shallow-marine deposits of the falling stage, in a *river-dominated* deltaic setting. Since the angle of repose of the prograding clinoforms is steeper than the wave equilibrium profile, no wave scouring affects the seafloor during forced regression. As a result, the basal surface of forced regression (BSFR) is preserved along the entire shallow-marine profile, and the forced regressive shoreface deposits are gradationally based—for a field example, see Fig. 3.30. GR/SP—synthetic gamma ray/spontaneous potential log.

FIGURE 4.27 Outcrop examples of the 'basal surface of forced regression,' showing the base of the submarine fan complex in discrete locations within the deep-water setting. A—contact between pelagic sediments and the overlying gravity-flow facies: the base of the submarine fan complex (contact between the Whitehill and Collingham formations, Early Permian, Ecca Pass, Karoo Basin); B—contact between pelagic sediments and the overlying gravity-flow facies: the base of the submarine fan complex (Miette Group, Precambrian, Jasper National Park, Alberta). The turbidites comprise the divisions A to C of the Bouma sequence, and belong to a proximal frontal splay; C—contact between pelagic sediments and the overlying gravity-flow facies: the base of the submarine fan complex (detail from B). A potential pitfall of this method of mapping the basal surface of forced regression is that, due to autocyclic shifts in the *locus* of deposition of the different lobes of the submarine fan complex, the base of a particular lobe may not correspond to the *earliest* manifestation of gravity flows associated with the forced regression of the shoreline. Hence, some of these surfaces are just facies contacts, *younger* than the basal surface of forced regression (see Chapters 5 and 6 for more details).

no physical expression in a conformable succession of shallow-water deposits (Figs. 4.25 and 4.29). It is most probable, however, that inner shelf deposits with a thickness greater than a couple of meters are normal regressive in nature (highstand), whereas outer shelf mudstones directly underlying the regressive surface of marine erosion are forced regressive. The presence of isolated gutter casts filled with hummocky cross-stratified sands within the dominantly fine-grained succession underlying a regressive surface of marine erosion suggests base-level fall accompanied by

seafloor scouring and reduced accommodation, and hence a forced regressive origin (Plint, 1991; Fig. 4.30).

Above the regressive surface of marine erosion, the prograding upper to middle shoreface deposits are swaley cross-stratified (Fig. 3.29), and sharp-based (Figs. 3.28 and 4.28). Most of these sharp-based shoreface deposits are forced regressive, with the exception of the earliest normal regressive (lowstand) lobe which accumulates on top of the seawardmost portion of the regressive surface of marine erosion (e.g., vertical profile E in Fig. 4.24; Fig. 4.29). This means that, as the

FIGURE 4.28 Regressive surface of marine erosion at the contact between forced regressive shoreface (above) and outer shelf (below) facies (Late Cretaceous Marshybank Formation, Alberta; photo courtesy of A.G. Plint). The sharp-based shoreface deposits have large, shore-normal gutter casts at their base (arrows).

regressive surface of marine erosion expands basinward until the end of base-level fall, the youngest forced regressive shoreface deposits are sharp-based (Figs. 4.20, 4.21, 4.23, and 4.24). Consequently, where preserved from subsequent subaerial or transgressive wave-ravinement erosion, the gradationally based forced regressive shoreface deposits are always placed *landward* relative to their sharp-based counterparts, near the shoreline position at the onset of forced regression (Figs. 4.20, 4.21, and 4.23–4.25). The sharp-based forced regressive deposits are thinner than the depth of the fairweather wave base, commonly with a thickness in a range of meters. This is because they do not include the entire shoreface profile, but only the upper to middle shoreface facies, and also, they are generally truncated at the top by the subaerial unconformity or the transgressive ravinement surface. Basinward relative to the seaward termination of the subaerial unconformity, the thickness of sharp-based shoreface deposits may, however, increase due to the fact that they amalgamate forced regressive and overlying

FIGURE 4.29 Well-log expression of the regressive surface of marine erosion (arrows; modified from Catuneanu, 2002, 2003). See Fig. 4.9 for a summary of diagnostic features of the regressive surface of marine erosion. Note that the sharp-based shoreface deposits are thicker basinward relative to the seaward termination of the subaerial unconformity, as they include forced regressive and lowstand normal regressive strata. Log examples from the Cardium Formation (left) and the Lea Park Formation (right), Western Canada Sedimentary Basin. Abbreviations: GR—gamma ray log; LST—lowstand systems tract; FSST—falling-stage systems tract; HST—highstand systems tract.

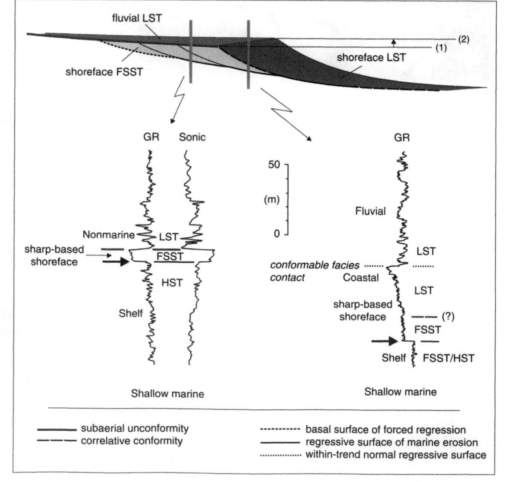

FIGURE 4.30 Isolated gutter casts filled with hummocky cross-stratified sands, indicating the forced regressive origin of the shelf facies underlying a regressive surface of marine erosion (Late Cretaceous Marshybank Formation, Alberta; photos courtesy of A.G. Plint). The scale bar is 20 cm in length.

lowstand normal regressive shoreface facies (Fig. 4.29). The thickness of this expanded sharp-based shoreface package depends on the shoreline trajectory during lowstand normal regression, being inversely proportional to the rates of regression and directly proportional to the rates of sedimentation.

Perhaps the most important feature of the regressive surface of marine erosion is its time-transgressive character, as it continues to form and expand basinward for the entire duration of base-level fall. Consequently, the regressive surface of marine erosion is highly diachronous, with the rate of shoreline forced regression (Fig. 4.9). For this reason, such wave scours, or any portions thereof, are *not part of prograding clinoforms*. Instead, the regressive surface of marine erosion truncates older clinoforms, and is downlapped by the younger clinoforms of the prograding sharp-based shoreface deposits (Figs. 4.9 and 4.23). It is therefore important to note that the regressive surface of marine erosion cuts across the shallow-marine forced regressive succession, merging with the correlative conformity

sensu Posamentier *et al.* (1988) in a landward direction and with the correlative conformity *sensu* Hunt and Tucker (1992) in a basinward direction (Fig. 4.24). As such, the regressive surface of marine erosion is to a large extent the counterpart of the transgressive ravinement surface, which is also highly diachronous merging with the maximum regressive surface basinward and with the maximum flooding surface landward. These two highly diachronous sequence stratigraphic surfaces differ, however, in timing of formation (i.e., during stages of base-level fall and transgression, respectively), *locus* of scouring (i.e., lower shoreface and coastal to upper shoreface, respectively), and the direction of expansion (i.e., seaward and landward, respectively).

The above discussion shows that there are circumstances where the regressive surface of marine erosion may develop *within* the systems tract that includes all shallow-marine forced regressive deposits, with forced regressive shelf deposits below and forced regressive shoreface facies above (e.g., profile D in Fig. 4.24). In such cases, this surface may not be used as a systems tract or sequence boundary. It is also possible that the regressive surface of marine erosion may be found at the base of forced regressive deposits, where it reworks the basal surface of forced regression (e.g., profile C in Fig. 4.24), or even at the top of forced regressive deposits and implicitly at the base of the overlying lowstand normal regressive strata (e.g., profile E in Fig. 4.24). For these reasons, Plint and Nummedal (2000) conclude that the regressive surface of marine erosion 'is neither a logical nor practical surface at which to place the sequence boundary.' Instead, and in a most general scenario, the base of all forced regressive deposits only includes the oldest (stratigraphically lowest) portion of the regressive surface of marine erosion (Posamentier *et al.*, 1992b; Plint and Nummedal, 2000). Where no forced regressive shelf deposits are preserved, the regressive surface of marine erosion attains the status of systems tract boundary (or sequence boundary, depending on the model), and is associated with a stratigraphic hiatus that increases in a basinward direction.

Sharp-based shorefaces, underlain by the regressive surface of marine erosion, are often detached and form shore-parallel sand bodies that mark successive positions of the regressive shoreline (Posamentier and Morris, 2000). These elongated sand bodies are subject to subaerial erosion for the duration of the falling stage, and are left behind the regressive shoreline at progressively lower elevations. Recent examples of such forced regressive shoreface deposits may be observed in areas affected by Holocene post-glacial isostatic rebound (Fig. 4.31), but numerous ancient

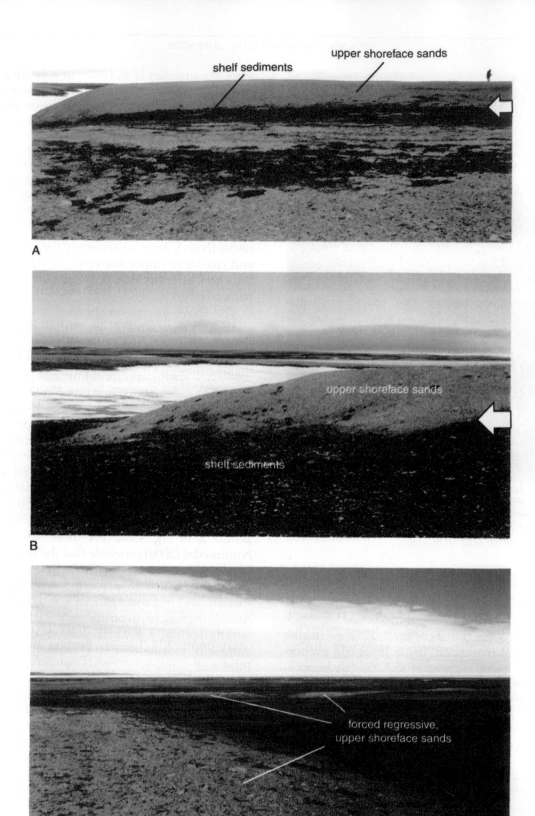

FIGURE 4.31 Forced regressive setting associated with Holocene post-glacial isostatic rebound (Melville Island, Arctic Canada). A—regressive surface of marine erosion (arrow). The photograph shows one offlapping lobe, prograding to the left in the direction of forced regression. Aerial photographs show that these offlapping lobes are detached and parallel to each other, marking successive positions of the paleoshoreline. They are elongated sand bodies, left behind by shoreline regression at progressively lower elevations, and are now subject to subaerial erosion. B—regressive surface of marine erosion (arrow). C—forced regressive shoreface sands, separated from the underlying shelf fines by the regressive surface of marine erosion. The sands are subject to subaerial erosion, and are often preserved as isolated patches generally aligned parallel to the shoreline.

examples have been documented in the rock record as well (Plint, 1988, 1991, 1996; Posamentier *et al.*, 1992b; Ainsworth, 1994; Plint and Nummedal, 2000; Posamentier and Morris, 2000; Fig. 4.28).

The regressive surface of marine erosion is one of the most prominent sequence stratigraphic surfaces, with a strong physical expression in the rock record due to the contrast in facies across the scoured contact, even though both the underlying and overlying deposits are coarsening-upward, as being part of a regressive succession (Figs. 4.9, 4.28, 4.29, and 4.31). The process of wave scouring during forced regression leads to the exhumation of semi-lithified marine sediments, resulting in the formation of firmgrounds colonized by the *Glossifungites* ichnofacies tracemakers (MacEachern *et al.*, 1992; Chaplin, 1996; Buatois *et al.*, 2002). Such firmgrounds separate deposits with contrasting ichnofabrics, largely due to the abrupt shift in environmental conditions that prevailed during the deposition of the juxtaposed facies across the contact. Both MacEachern *et al.* (1992) and Buatois *et al.* (2002) provide case studies where the regressive surface of marine erosion, marked by the *Glossifungites* ichnofacies, separates finer-grained shelf deposits with *Cruziana* ichnofacies from overlying shoreface sands with a *Skolithos* assemblage. The basinward extent of the forced regressive *Glossifungites* firmground is limited to the area affected by fairweather wave erosion, beyond which the stratigraphic hiatus collapses, being replaced by the correlative conformity *sensu* Hunt and Tucker (1992) (Fig. 4.24). Synonymous terms for the regressive surface of marine erosion include the *regressive ravinement surface* (Galloway, 2001) and the *regressive wave ravinement* (Galloway, 2004).

Maximum Regressive Surface

The maximum regressive surface (Catuneanu, 1996; Helland-Hansen and Martinsen, 1996) is defined relative to the transgressive-regressive curve, marking the change from shoreline regression to subsequent transgression (Fig. 4.6). Therefore, this surface separates prograding strata below from retrograding strata above (Fig. 4.32). The change from progradational to retrogradational stacking patterns takes place *during* the base-level rise at the shoreline, when the increasing rates of base-level rise start outpacing the sedimentation rates (Fig. 4.5). As a result, the end-of-regression surface forms within an aggrading succession, sitting on top of lowstand normal regressive strata, and being onlapped by transgressive 'healing phase' deposits (Figs. 4.9 and 4.32). As the youngest clinoform associated with shoreline regression, the maximum regressive surface downlaps the pre-existing seafloor in a basinward direction, and drapes the preceding regressive clinoforms. Hence, the underlying lowstand normal regressive strata do not terminate against the maximum regressive surface (Fig. 4.9).

The maximum regressive surface is generally conformable (Fig. 4.9), although the possibility of seafloor scouring associated with the change in the direction of shoreline shift at the onset of transgression, which triggers a change in the balance between sediment load and the energy of subaqueous currents, is not excluded (Loutit *et al.*, 1988; Galloway, 1989). The maximum regressive surface may also be scoured in the transition zone between coastal and fluvial environments, in relation to the backstepping of the higher energy intertidal swash zone (transgressive beach) over the fluvial overbank deposits of the lowstand (normal regressive) systems tract (Catuneanu *et al.*, in press; Fig. 4.33). Where conformable, the maximum regressive surface is not associated with any substrate-controlled ichnofacies (Fig. 4.9). Where the transgressive marine facies are missing, the marine portion of the maximum regressive surface is replaced by the maximum flooding surface, and this composite unconformity may be preserved as a firmground or even hardground, depending on the amounts of erosion and/or synsedimentary lithification, colonized by the *Glossifungites* and *Trypanites* ichnofacies, respectively (Pemberton and MacEachern, 1995; Savrda, 1995). As this unconformity forms basinward relative to the shoreline position at the end of regression, within a fully marine environment, no xylic substrates (woodgrounds: the *Teredolites* ichnofacies) are expected to be associated with it.

The end of shoreline regression event (Fig. 4.7) marks a change in sedimentation regimes, as reflected by the balance between sediment supply and environmental energy, in all depositional systems within the sedimentary basin, both landward and seaward relative to the shoreline. As a result, the maximum regressive surface may develop as a discrete stratigraphic contact across much of the sedimentary basin, from marine to coastal and fluvial environments (Figs. 4.9, 4.32, and 4.34). The preservation potential of the end-of-regression surface is highest in the deep- to shallow-marine environments, where it tends to be onlapped by aggrading transgressive strata, and is lower in coastal to fluvial settings, where it may be subject to wave scouring during subsequent shoreline transgression (Fig. 3.21). Landward from the end-of-regression shoreline, the preservation of the maximum regressive surface depends on the balance between the rates of aggradation in the transgressive coastal to fluvial environments and the rates of subsequent transgressive wave-ravinement erosion in the upper shoreface. There are cases where this transgressive wave scouring may remove not only the

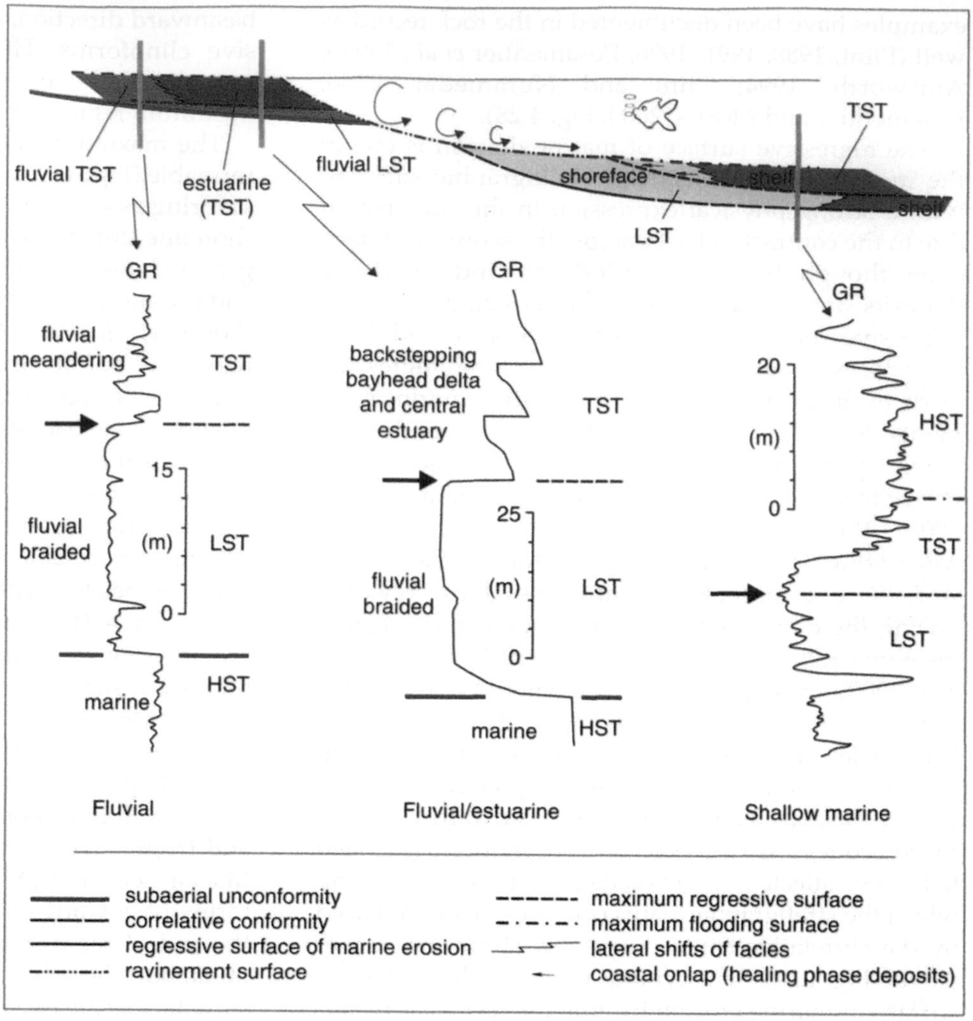

FIGURE 4.32 Well-log expression of the maximum regressive surface (arrows; modified from Catuneanu, 2002, 2003). See Fig. 4.9 for a summary of diagnostic features of the maximum regressive surface. Log examples from Kerr *et al.* (1999) (left and center) and Embry and Catuneanu (2001) (right). Abbreviations: GR—gamma ray; LST—lowstand systems tract; TST—transgressive systems tract; HST—highstand systems tract.

transgressive coastal to fluvial deposits, but also all underlying coastal to fluvial lowstand normal regressive deposits as well. In such cases, the transgressive wave scour, the maximum regressive surface and the subaerial unconformity are all amalgamated in one unconformable contact (Embry, 1995). In a more general scenario, however, the preservation of coastal to fluvial lowstand normal regressive deposits in the rock record depends on the duration of normal regression and the rates of sediment aggradation in coastal to fluvial environments prior to the transgressive wave scouring. Prolonged stages of lowstand normal regression may result in the formation of relatively thick topsets of aggrading and prograding coastal to fluvial strata, which drape the subaerial unconformity and are preserved from subsequent transgressive wave-ravinement erosion (Fig. 4.34). In such cases, the maximum regressive surface has the potential of being mappable across much of the sedimentary basin, within both marine and fluvial successions (Fig. 4.34).

In deep-marine deposits, the maximum regressive surface is most difficult to identify within the facies succession of the submarine fan complex on the basin floor, because the end-of-regression event occurs *during* a stage of waning down in the amount of terrigenous sediment that is delivered to the deep-water environment. For this reason, no physical criteria for outcrop, core, or well-log analysis have been developed to map the maximum regressive surface within the gravity-flow deposits that accumulate on the basin floor. More detailed discussions on the nature of gravity-flow deposits that accumulate in the deep-water environment during the various stages of the base-level cycle are provided in Chapters 5 and 6 of this book. On continental slopes, the maximum regressive surface is the *youngest prograding clinoform* which is onlapped by the overlying transgressive 'healing phase' deposits (Fig. 4.34). Where afforded by high resolution seismic data, the extension of this youngest prograding slope clinoform into the deeper portions of the basin may

FIGURE 4.33 Outcrop photograph of a maximum regressive surface (yellow arrow) at the contact between fluvial normal regressive strata (facies A) and the overlying backstepping beach deposits (facies B) (Bahariya Formation, Lower Cenomanian, Bahariya Oasis, Western Desert, Egypt). The maximum regressive surface is scoured by high-energy swash currents during the earliest stage of shoreline transgression. Underlying the maximum regressive surface, the fluvial strata correlate with a prograding and aggrading delta, and are part of the lowstand systems tract. The backstepping beach is the only preserved portion of the transgressive systems tract. The beach deposits are truncated at the top by a subaerial unconformity (red arrow, base of incised valley), and are overlain by coarse fluvial channel fills (facies C, part of a younger lowstand systems tract; Catuneanu *et al.*, in press). Note the landward shift of facies recorded across the maximum regressive surface, in contrast with the basinward shift of facies associated with the subaerial unconformity.

provide a clue of where to trace the maximum regressive surface within the basin-floor succession.

In shallow-marine systems, the maximum regressive surface is relatively easy to recognize at the top of coarsening-upward (prograding) deposits (Figs. 4.35–4.37). Depending on the rates of subsequent transgression, as well as on the location within the basin, the maximum regressive surface may or may not be associated with a sand/shale lithological contrast. Cases A, B, C, and D in Fig. 4.35 provide examples of maximum regressive surfaces that correspond to a *sand/shale contact*, suggesting rapid transgression and/or an abrupt cut-off of sediment supply as the transgression was initiated. Under these conditions, the sediment is trapped within the retrograding shoreline systems at the onset of transgression, leading to sediment starvation offshore and hence an abrupt facies change at the maximum regressive surface (Loutit *et al.*, 1988). Where the transgression is

slower and/or the sediment supply is high and continues to be delivered offshore, the peak of coarsest sediment may occur *within the sand*, and the sand/shale contact is above the maximum regressive surface, within the overlying transgressive succession (Fig. 4.35E). Farther offshore relative to the paleoshoreline, into lower shoreface and shelf systems, the maximum regressive surface occurs within silty–shaly successions, marking the peak of coarsest sediment (end of progradation; Figs. 4.36 and 4.37). In such settings, the position of the maximum regressive surface is often evident from the breaks in slope gradients that can be observed in outcrops (Figs. 4.36 and 4.37). The end-of-progradation event (top of coarsening-upward trend) does not necessarily correspond to the peak of shallowest water depth, especially in offshore areas. The peak of shallowest water is usually recorded within the underlying regressive (lowstand) deposits, while the maximum regressive surface

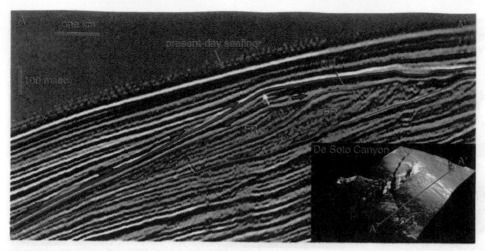

FIGURE 4.34 Maximum regressive surface (red line) on a dip-oriented, 2D seismic transect (location shown on the 3D illuminated surface) (De Soto Canyon area, Gulf of Mexico; image courtesy of H.W. Posamentier). This surface tops all fluvial to deep-marine strata that accumulate during lowstand normal regression. The maximum regressive surface may onlap the subaerial unconformity in a landward direction (fluvial onlap), and is onlapped by transgressive facies in the deep-water environment (marine onlap; blue arrows). The white arrow indicates the shoreline trajectory during lowstand normal regression. It is inferred that the normal regressive facies are marine seaward from the white arrow (downlapping the underlying forced regressive deposits; red arrow), and nonmarine in the opposite direction (onlapping the subaerial unconformity; green arrow—fluvial onlap). In a marine environment, the maximum regressive surface is the *youngest clinoform associated with shoreline regression*. For scale, the channel on the 3D illuminated surface is approximately 1.8 km wide, and 275 m deep at shelf edge. The illuminated surface is taken at the base of forced regressive deposits. Abbreviations: FR—forced regressive deposits; NR—normal regressive deposits; T—transgressive deposits.

forms within deepening water—see discussion in Chapter 7. For this reason it is preferable to describe the trends in terms of *observed* grading (coarsening- *vs.* fining-upward) as opposed to *inferred* bathymetric changes (shallowing- *vs.* deepening-upward).

In coastal settings, the maximum regressive surface underlies the earliest estuarine deposits (Fig. 4.6). The contact between estuarine and underlying fluvial facies diverges from the maximum regressive surface beyond the initial length of the estuary at the onset of

FIGURE 4.35 Outcrop examples of maximum regressive surfaces in proximal shallow-water settings. A—maximum regressive surface (arrow) in a conformable marine succession. The top of the prograding (coarsening-upward) shoreface is marked by a concretionary layer of siderite-cemented sandstone, indicating the preferential fluid migration pathway during diagenesis. In this example, the onset of transgression is accompanied by an abrupt cut-off of sediment supply to the marine environment. Sediment trapping within the retrograding shoreline systems results in sediment starvation on the shelf (Loutit et al., 1988) (contact between Demaine and Beechy members, Bearpaw Formation, Late Campanian, Saskatchewan, Western Canada Sedimentary Basin); B—high-frequency maximum regressive surfaces (arrows) in a conformable deltaic succession. Maximum regressive surfaces are marked by concretionary layers (coarsest sand, prone to preferential precipitation of diagenetic cements), and are overlain by thin transgressive shales (Late Permian Waterford Formation, Ecca Group, southern Karoo Basin); C—maximum regressive surface (arrow) in a conformable marine succession, at the top of coarsening-upward prograding shoreface sands. The sharp lithological contrast across this surface indicates rapid transgression and/or a cut-off of sediment supply as the transgression is initiated (contact between Ardkenneth and Snakebite members, Bearpaw Formation, Late Campanian, Saskatchewan, Western Canada Sedimentary Basin); D—maximum regressive surface (top of coarsening-upward prograding shoreface sands) exposed by the subaerial erosion of the overlying (and more recessive) transgressive shales (top of the Kipp Member, Bearpaw Formation, Late Campanian, Oldman River, Alberta, Western Canada Sedimentary Basin); E—maximum regressive surface (white arrow) in a conformable marine succession, at the top of coarsening-upward prograding shoreface sands. Note that in this case the transition to the overlying transgressive facies is more subtle, and the facies contact between sand and shale (*flooding surface*, grey arrow) is above the maximum regressive surface (top of the Ryegrass Member, Bearpaw Formation, Late Campanian, Alberta, Western Canada Sedimentary Basin).

A   B

FIGURE 4.36 Outcrop examples of maximum regressive surfaces in distal shallow-water settings (arrows). A—maximum regressive surface in a conformable lower shoreface to shelf succession (top of the Magrath Member, Bearpaw Formation, Late Campanian, St. Mary River, Alberta, Western Canada Sedimentary Basin); B—maximum regressive surface in a conformable shelf succession (Beechy Member, Bearpaw Formation, Late Campanian, Saskatchewan, Western Canada Sedimentary Basin). In both cases, the slope breaks indicate textural changes across the maximum regressive surfaces, from coarsening-upward (below) to fining-upward (above).

transgression, becoming progressively younger in an upstream direction (a within-trend facies contact that forms *during* shoreline transgression; Figs. 4.6 and 4.38). Therefore, where dealing with fluvial to estuarine successions it is important to differentiate between the *stratigraphically lowest* surface that defines the base of estuarine facies, which is the low-diachroneity maximum regressive surface, and the highly diachronous facies contact that becomes younger landward with the rate of shoreline transgression. The distinction between these two types of contacts may be made on the basis of juxtaposed facies: the maximum regressive surface separates fluvial from overlying central estuary facies, whereas the within-trend (transgressive) facies contact separates fluvial from overlying bayhead deltas (in a wave-dominated estuarine

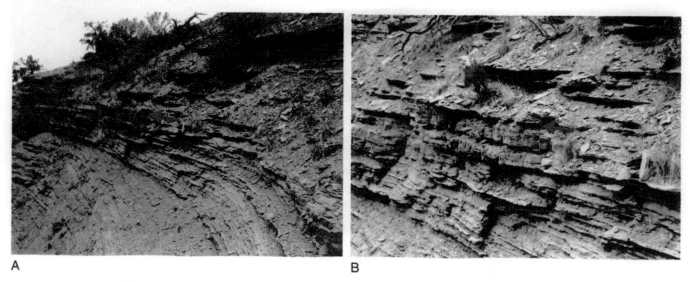

A   B

FIGURE 4.37 Maximum regressive surface (arrows) in a conformable succession of prodelta facies (Campanian Panther Tongue Formation, Utah). The break in slope gradients indicates textural changes across the surface, from coarsening-upward (below) to fining-upward (above). Photograph B: detail from A.

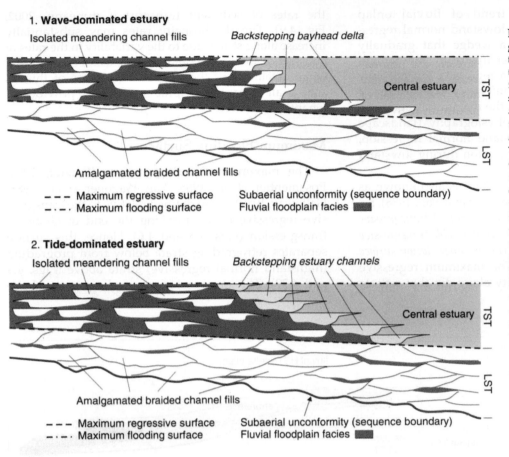

1. Wave-dominated estuary

Isolated meandering channel fills   *Backstepping bayhead delta*

Central estuary   TST

LST

Amalgamated braided channel fills

– – – Maximum regressive surface
– · – · Maximum flooding surface

Subaerial unconformity (sequence boundary)
Fluvial floodplain facies ▮

2. Tide-dominated estuary

Isolated meandering channel fills   *Backstepping estuary channels*

Central estuary   TST

LST

Amalgamated braided channel fills

– – – Maximum regressive surface
– · – · Maximum flooding surface

Subaerial unconformity (sequence boundary)
Fluvial floodplain facies ▮

FIGURE 4.38 Dip-oriented stratigraphic cross-sections through fluvial to estuarine successions in wave- and tide-dominated settings (modified from Kerr *et al.*, 1999). The lowstand systems tract (LST) is composed of amalgamated braided channel-fill facies resting on a sequence boundary with substantial erosional relief. The transgressive systems tract (TST) is composed of meandering fluvial deposits (isolated ribbons encased in well-developed floodplain facies) and correlative estuarine facies towards the coastline. The maximum regressive surface may be traced at the base of central estuary facies, and at the contact between braided and meandering systems farther inland. Beyond the landward limit of the estuary at the onset of transgression, the facies contact between estuarine and fluvial facies becomes highly diachronous (a within-trend facies contact, within the TST), and may be traced at the base of backstepping bayhead deltas (in wave-dominated settings) or at the base of backstepping estuary channels (in tide-dominated settings).

setting) or estuary channels (in a tide-dominated estuarine setting) (Fig. 4.38).

The extension of the maximum regressive surface into the fluvial part of the basin is much more difficult to pinpoint, but at a regional scale it is argued to correspond with an abrupt decrease in fluvial energy, i.e., a change from amalgamated braided channel fills to overlying meandering systems (Kerr *et al.*, 1999; Ye and Kerr, 2000; Fig. 4.38). This shift in fluvial styles across the maximum regressive surface is suggested by the grain size threshold in Fig. 4.6, and is attributed to the formation of the low energy estuarine system at the beginning of transgression, which would induce a lowering in fluvial energy upstream. The link between the formation of estuaries and the coeval lowering in fluvial energy upstream is provided by the increased rates of coastal aggradation at the onset of transgression, which result in a decrease in the slope gradient of the fluvial graded profile and a corresponding change in fluvial energy levels, fluvial styles, and sediment load. Notwithstanding these general principles, much work is still needed to properly document the physical attributes of the nonmarine portion of

maximum regressive surfaces. There is increasing evidence that the commonly inferred 'braided' nature of the lowstand fluvial systems (Kerr *et al.*, 1999; Ye and Kerr, 2000; Figs. 4.32 and 4.38), even though valid in many cases, may not be representative as a generalization. Lowstand fluvial systems of meandering type have also been documented (e.g., Miall, 2000; Posamentier, 2001; see also the discussion in Chapter 5 regarding the nature of lowstand fluvial deposits), especially within incised valleys, and in such cases the identification of the nonmarine portion of the maximum regressive surface may require more in-depth studies than the simple observation of fluvial styles. Where the maximum regressive surface develops within a succession of meandering stream deposits (lowstand normal regressive below and transgressive above), the stratigraphically lowest sedimentary structures, fossils and trace fossils associated with tidal influences may provide the evidence for the onset of transgression. In this case, well-log and seismic data are not sufficient for unequivocal interpretations, and core or outcrop studies need to be performed for detailed facies analyses.

Following the general trend of fluvial onlap recorded by the underlying lowstand normal regressive deposits, which form a wedge that gradually expands and becomes thinner upstream, the nonmarine portion of the maximum regressive surface may also onlap the subaerial unconformity. The location of the landward termination of the maximum regressive surface depends on basin physiography (landscape gradients), duration of lowstand normal regression, and the rates of fluvial aggradation during lowstand normal regression.

The maximum regressive surface is also known as the *transgressive surface* (Posamentier and Vail, 1988), *top of lowstand surface* (Vail et al., 1991), *initial transgressive surface* (Nummedal et al., 1993), *conformable transgressive surface* (Embry, 1995), and *maximum progradation surface* (Emery and Myers, 1996). The maximum regressive surface has a low diachroneity along dip that reflects

the rates of sediment transport (Catuneanu, 2002; Fig. 4.9). The diachroneity rates may substantially increase along strike, due to the variability in the rates of subsidence and sedimentation (Catuneanu et al., 1998b). More details about the temporal attributes of this, as well as all other stratigraphic surfaces, are provided in Chapter 7.

Maximum Flooding Surface

The maximum flooding surface (Frazier, 1974; Posamentier et al., 1988; Van Wagoner et al., 1988; Galloway, 1989) is also defined relative to the transgressive–regressive curve, marking the end of shoreline transgression (Figs. 4.5 and 4.6). Hence, this surface separates retrograding strata below from prograding (highstand normal regressive) strata above (Figs. 4.9 and 4.39). The presence of prograding strata above

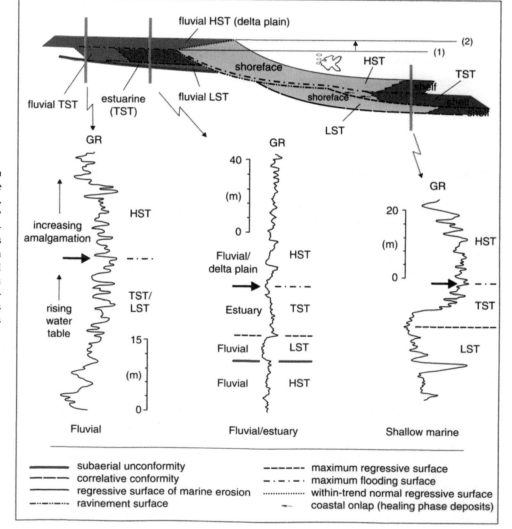

FIGURE 4.39 Well-log expression of the maximum flooding surface (arrows; modified from Catuneanu, 2002, 2003). See Fig. 4.9 for a summary of diagnostic features of the maximum flooding surface. Log examples from the Wapiti Formation, Western Canada Sedimentary Basin (left and center), and Embry and Catuneanu (2001) (right). Abbreviations: GR—gamma ray; LST—lowstand systems tract; TST—transgressive systems tract; HST—highstand systems tract.

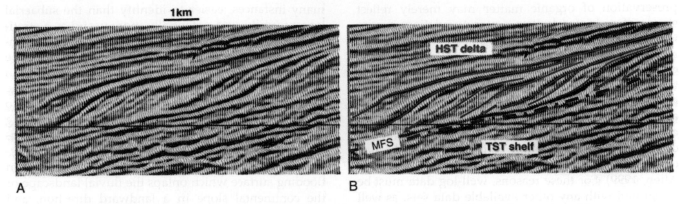

A   B

FIGURE 4.40 Seismic expression of a maximum flooding surface in a coastal to shallow-marine setting (A—uninterpreted seismic line; B—interpreted seismic line; modified from Brown et al., 1995). The maximum flooding surface overlies transgressive shelf facies, and is downlapped by a highstand (normal regressive) delta. For this reason, the maximum flooding surface is also known as a 'downlap surface'.

identifies the maximum flooding surface as a *downlap surface* on seismic data (Fig. 4.40). The change from retrogradational to overlying progradational stacking patterns takes place *during* base-level rise at the shoreline, when sedimentation rates start to outpace the rates of base-level rise (Fig. 4.5). The maximum flooding surface is generally conformable, excepting for the outer shelf and upper slope regions where the lack of sediment supply coupled with instability caused by rapid increase in water depth may leave the seafloor exposed to erosional processes (Galloway, 1989; Fig. 4.41). The maximum flooding surface is also known as the *maximum transgressive surface* (Helland-Hansen and Martinsen, 1996) or *final transgressive surface* (Nummedal et al., 1993). The maximum flooding surface has a low diachroneity along dip that reflects the rates of sediment transport (Catuneanu, 2002; Fig. 4.9). As in the case of the maximum regressive surface, the diachroneity rates may substantially increase along strike due to the variability in subsidence and sedimentation rates (Catuneanu et al., 1998b).

Maximum flooding surfaces are arguably the easiest stratigraphic markers to use for the subdivision of stratigraphic successions, especially in marine to coastal plain settings, because they lie at the heart of areally extensive condensed sections which form when the shoreline reaches maximum landward positions (Galloway, 1989; Posamentier and Allen, 1999). Such condensed sections are relatively easy to identify and correlate on any type of data, as they consist dominantly of fine-grained, hemipelagic to pelagic deposits accumulated during times when minimal terrigenous sediment is delivered to the shelf and deeper-water environments. Condensed sections are typically marked by relatively transparent zones on seismic lines, due to their lithological homogeneity. They also tend to exhibit a high gamma-ray response caused by their common association with increased concentrations of organic matter and radioactive elements. One must note, however, that the generally inferred correlation between condensed sections and organic-rich sediments is subject to exceptions, as the deposition and

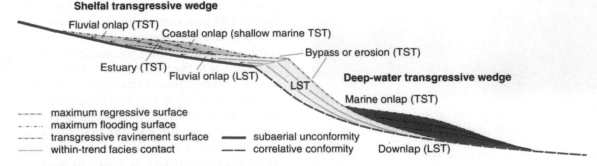

FIGURE 4.41 Stratigraphic expression of transgressive strata. Note that the transgressive systems tract may consist of two distinct wedges, one on the continental shelf and one in the deep-water environment, separated by an area of sediment bypass or erosion around the shelf edge.

preservation of organic matter may merely reflect stages of restricted bottom-water circulation, diminished terrigenous sediment supply, and/or the accumulation of carbonaceous mudstones in paralic environments, which may not necessarily correspond to times of maximum shoreline transgression (Posamentier and Allen, 1999). At the same time, condensed sections associated with stages of maximum flooding may contain glauconite and/or siderite, or other carbonates or biochemical precipitates (Fig. 4.42) which may exhibit a wide range of log motifs (Posamentier and Allen, 1999). For these reasons, well-log data must be integrated with any other available data sets, as well as with the observation of the regional stratal stacking patterns, for more reliable interpretations. As a general principle, '... the identification of a condensed section and a maximum flooding surface should be based on the identification of a convergence of time horizons rather than degree of radioactivity. Converging well-log correlation markers, converging seismic reflections, or converging strata can indicate convergence of time horizons.' (Posamentier and Allen, 1999).

Maximum flooding surfaces have a high preservation potential, being overlain by aggrading and prograding highstand normal regressive deposits, and can be identified in all depositional environments of a sedimentary basin, seaward and landward from the shoreline, on the basis of stratal stacking patterns (Fig. 4.9). The broad areal extent, as well as its consistent association with fine-grained, low energy systems across the basin, makes the 'maximum flooding' a surface that is, in

many instances, easier to identify than the subaerial unconformity, and potentially more useful as a stratigraphic marker for basin-wide correlations. The basin-wide extent of the transgressive tract may, however, be hampered by the absence of transgressive deposits in the area around the shelf edge. For this reason, the transgressive systems tract usually comprises two distinct wedges, one on the continental shelf consisting of fluvial to shallow-marine facies, and one in the deep-water environment (Fig. 4.41). Each of these transgressive wedges is topped by a conformable maximum flooding surface which onlaps the fluvial landscape or the continental slope in a landward direction, and downlaps the shallow or deep-marine seafloor in a basinward direction (Figs. 4.9 and 4.41). The downlap type of stratal terminations may, however, be only apparent in a transgressive context, as potentially marking the base of a sedimentary unit at its erosional rather than depositional limit (Fig. 4.2), which is why depositional downlap is commonly restricted to regressive deposits (Fig. 4.3). Where transgression is accompanied by sediment aggradation, the transgressive strata do not terminate against the maximum flooding surface, which rather drapes the underlying deposits. This principle is valid for all conformable stratigraphic surfaces listed in Fig. 4.9, meaning that sedimentary strata do not terminate against a younger conformable surface. Where the transgressive facies are absent, the maximum flooding surface truncates the underlying regressive deposits (Fig. 4.9).

In a marine succession, the maximum flooding surface is placed at the top of fining-upward (transgressive) deposits. This trend is generally valid in both deep-water settings, where the maximum flooding surface marks the top of waning-down gravity-flow deposits (base of highstand pelagics—see the following chapters for more details), as well as in shallow-water environments. Seaward from the shoreline, on the shelf, the transgressive deposits may be reduced to a condensed section, or may even be missing. In the latter situation, the maximum flooding surface is superimposed on and reworks the maximum regressive surface. Figure 4.43 provides an example where the transgressive deposits are present, and hence the succession is conformable. In this case, the maximum flooding surface corresponds to the peak of finest sediment, marking the top of a fining-upward (transgressive) succession. This surface is not easy to pinpoint in outcrop or core, as it is not associated with a lithological contrast, and it requires thin section textural analysis for unequivocal identification. However, such a conformable maximum flooding surface is easier to recognize on well logs, which are more sensitive in recording changes in grain size. Under restricted detrital supply conditions,

FIGURE 4.42 Coal seam (1 m thick) in a coastal setting, overlain by a 50 cm thick limestone bed (photograph courtesy of M.R. Gibling; Pennsylvanian Sydney Mines Formation, Sydney Basin, Nova Scotia). The coal lies within the transgressive systems tract. The limestone bed (arrow) marks a maximum flooding level with restricted detrital supply, and it is overlain by the highstand systems tract.

FIGURE 4.43 Maximum flooding surface in a conformable shallow-marine succession, at the base of the transition zone between shelf and overlying shoreface facies. The succession is younging to the left. The vertical dashed line marks the peak of finest sediment (top of retrograding succession). This conformity is difficult to pinpoint in the field because of the lack of lithological contrast, and requires thin section textural analysis for accurate identification. Transition between the Sherrard and Demaine members (Bearpaw Formation, Late Campanian), Saskatchewan, Western Canada Sedimentary Basin.

the maximum flooding level may also be marked by condensed sections of carbonate facies (Fig. 4.42). Where the transgressive deposits are missing, the maximum flooding surface is scoured and replaces the maximum regressive surface. In this case, the maximum flooding surface is associated with a lithological contrast and separates two coarsening-upward successions (Fig. 4.44).

Where transgressive deposits are present and the succession is conformable, the top of fining-upward retrograding marine facies does not necessarily correspond to the peak of deepest water, especially in offshore areas. The peak of deepest water is usually recorded within the overlying regressive (highstand) deposits (see Chapter 7 for more details). This is why, as in the case of the maximum regressive surface, grading terms that reflect *observations* (coarsening- *vs.* fining-upward) are preferred over bathymetric terms that reflect *inferred* changes in water depth (shallowing- *vs.* deepening-upward).

The ichnological signature of maximum flooding surfaces in a marine succession is highly variable, depending on the dominant synsedimentary process (i.e., sediment aggradation *vs.* bypass, erosion and/or

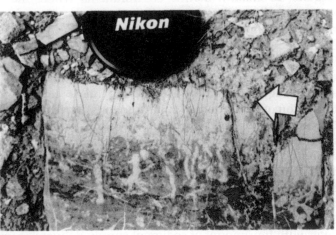

FIGURE 4.44 Outcrop examples of maximum flooding surfaces (scoured) that rework the underlying maximum regressive surfaces. The transgressive facies are missing. A—Young Creek Member (Bearpaw Formation, Early Maastrichtian), Castor area, Alberta, Western Canada Sedimentary Basin; B—firmground associated with the *Glossifungites* ichnofacies, formed as a result of prolonged sediment starvation (Mississippian Shunda Formation, Talbot Lake area, Jasper National Park).

lithification) that affects the seafloor during the maximum transgression of the shoreline. Due to the decrease in sediment supply to the marine environment during shoreline transgression, the maximum flooding surface is often associated with firmgrounds or hardgrounds, as a function of degree of seafloor cementation (Fig. 4.44B), although softgrounds may also form where sedimentation rates are high enough to maintain an unconsolidated seafloor (Fig. 4.43) (Pemberton and MacEachern, 1995; Savrda, 1995; Ghibaudo et al., 1996). Ghibaudo et al. (1996) provide a case study where the maximum flooding surface is represented by a firmground with burrows infilled with glauconitic sandstone. This stratigraphic contact ('omission' surface) is interpreted to correspond to a period of very low sedimentation rates or nondeposition, where the lack of clastic input allowed for glauconite formation and concentration, intense seafloor burrowing and increased cohesiveness of the substrate (Ghibaudo et al., 1996). In this example, the formation of the firmground was accompanied by a decrease in the water's oxygen levels at the seafloor, as evidenced by the preservation of plant debris as well as by the abundance of *Phycosiphon incertum* and *Planolites* traces (Ghibaudo et al., 1996). The landward shift of facies during transgression is also confirmed by the change in softground ichnofacies across the firmground, from *Cruziana* below to *Zoophycos* above (Ghibaudo et al., 1996). The latter ichnofacies is consistent with an oxygen-deprived setting (Pemberton and MacEachern, 1995; Ghibaudo et al., 1996), although the association between maximum flooding surfaces and oxygen-deficient ichnocoenoses is not necessarily a valid generalization, especially in the proximal regions of shallow-marine environments where the water may be well oxygenated during times of maximum shoreline transgression (Savrda, 1995). At the opposite end of the spectrum, Siggerud and Steel (1999) provide a case study where the maximum flooding surface formed during a time of continuous seafloor aggradation, which did not allow for the formation of firmgrounds or hardgrounds. In this case, the position of the maximum flooding surface is inferred on the basis of changes in ichnofabrics, corresponding to the point of highest bioturbation index. The increased level of bioturbation at the maximum flooding surface softground, which is not necessarily accompanied by any abrupt changes in ichnofacies across the conformable stratigraphic contact, correlates with the amount of sediment supply delivered to the marine environment (and the corresponding rates of seafloor aggradation), which is lowest during the time of maximum shoreline transgression. This example is relevant to all conformable shallow-marine successions, where *sediment supply* (as opposed to

inferred changes in water depth) is the main switch that controls the observed grading patterns, sedimentation rates, and associated levels of bioturbation. Besides softgrounds, firmgrounds and hardgrounds, maximum flooding surfaces may also be represented by woodgrounds especially in coastal regions where marine flooding results in the inundation of forested coastal plains (Savrda, 1995). Such woodgrounds are common at all flooding surfaces that form during shoreline transgression, and are preserved *within* the transgressive systems tract, so it is only the *youngest* woodground of any transgressive succession that indicates the position of the maximum flooding surface. It can be concluded that all substrate-controlled ichnofacies may, under different circumstances, be associated with maximum flooding surfaces (Fig. 4.9), although softgrounds characterized by increased bioturbation indexes and changes in ichnofabrics in conformable marine successions should not be ruled out (Savrda, 1995; Siggerud and Steel, 1999).

In coastal settings, the maximum flooding surface is placed at the top of the youngest estuarine facies, marking the turnaround point to subsequent delta plain sedimentation (Figs. 4.6, 4.38, and 4.39). Landward from the coastline, criteria for the recognition of the maximum flooding surface in the fluvial portion of the basin have been provided by Shanley et al. (1992), mainly based on the presence of tidal influences in fluvial sandstones. Sedimentary and biogenic structures that may suggest a tidal influence in fluvial strata include sigmoidal bedding, paired mud/silt drapes, wavy and lenticular bedding, shrinkage cracks, multiple reactivation surfaces, inclined heterolithic strata, complex compound cross-beds, bidirectional cross-beds, and trace fossils including *Teredolites*, *Arenicolites*, and *Skolithos* (Shanley et al., 1992). Tidal influences in fluvial strata generally extend for tens of kilometers inland from the coeval shoreline (Shanley et al., 1992), although, depending on river discharge and tidal range, such influences, including tidal-current reversals, may occur as far as 130 km (Allen and Posamentier, 1993) or even over 200 km inland from the river mouth (Miall, 1997). Farther upstream, the maximum flooding surface corresponds to the highest level of the water table relative to the land surface (Fig. 4.39), which, given a low sediment input and the right climatic conditions, may offer good conditions for peat accumulation at the basin scale. As a result, the position of the maximum flooding surface may be indicated by regionally extensive coal seams (Hamilton and Tadros, 1994; Tibert and Gibling, 1999). Given its association with high water table conditions, the maximum flooding surface is likely included within floodplain and/or lacustrine sediments, and it

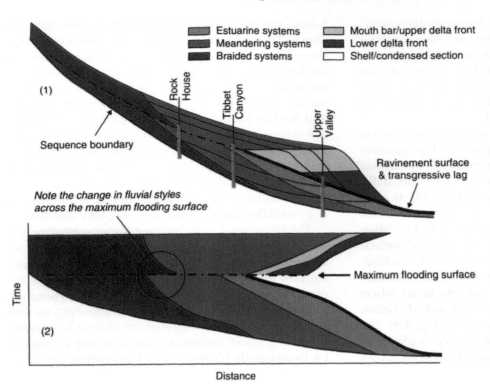

Estuarine systems
Meandering systems
Braided systems
Mouth bar/upper delta front
Lower delta front
Shelf/condensed section

(1)

Sequence boundary

Note the change in fluvial styles
across the maximum flooding surface

Rock House

Tibbet Canyon

Upper Valley

Ravinement surface
& transgressive lag

Maximum flooding surface

Time

(2)

Distance

FIGURE 4.45 Dip-oriented stratigraphic cross-section (1) and chronostratigraphic (Wheeler) diagram (2) of a fluvial to shallow-marine depositional sequence (modified from Shanley et al., 1992). This case study, based on the Upper Cretaceous succession in southern Utah, suggests that the end of the estuary life time (end of transgression) is accompanied by an abrupt shift in fluvial styles upstream, which provides a criterion for the recognition of the nonmarine portion of the maximum flooding surface. The landward shift through time of the boundary between braided and meandering stream facies explains the fining-upward trend within the fluvial part of each systems tract. This trend is also suggested in Fig. 4.6 (including the threshold of facies shift across the maximum flooding surface), and has been observed in other case studies as well (e.g., Catuneanu and Elango, 2001).

may be prograded by crevasse deltas and lake deltas as the balance between accommodation and sedimentation shifts again in the favor of the latter.

The position of the maximum flooding surface in fully fluvial successions may also be indicated by an abrupt increase in fluvial energy, from meandering to overlying braided fluvial systems, as the end of the estuary life time triggers a rapid seaward shift of the river mouth (Shanley et al., 1992; Fig. 4.45). This change in fluvial styles across the maximum flooding surface is suggested by a grain size threshold in Fig. 4.6. It should be noted, however, that this scenario only reflects the particular circumstances of a case study, and it may not be adequate as a generalization. Depending on the patterns of differential subsidence and sediment supply of each basin, other changes in fluvial styles may also be envisaged across maximum flooding surfaces. As a general principle, continuous coastal aggradation during transgression and subsequent highstand normal regression contributes towards a gradual *decrease* in the gradient of fluvial graded profiles. As a result, a lowering with time in fluvial energy should be expected, unless the effects of tectonism and differential subsidence overprint this trend. Irrespective of the actual change in fluvial energy levels and corresponding fluvial styles across the maximum flooding surface, which should therefore be studied on a case-by-case basis, the highstand

fluvial deposits overlying a maximum flooding surface record an abrupt decline in tidal structures, as well as a gradual increase in the degree of channel amalgamation as the amount of available accommodation decreases towards the end of base-level rise (Fig. 4.39; Wright and Marriott, 1993; Shanley and McCabe, 1993; Emery and Myers, 1996). Most of the current models of fluvial sequence stratigraphy acknowledge these changes in sedimentary structures and the ratio between fluvial architectural elements, without accounting for a shift in fluvial styles across the maximum flooding surface.

Transgressive Ravinement Surfaces

Transgressive ravinement surfaces are scours cut by tides and/or waves during the landward shift of the shoreline. In the majority of cases, the two types of transgressive ravinement surfaces (i.e., tide- and wave-generated) are superimposed and onlapped by the transgressive shoreface (i.e., coastal onlap). Such amalgamated transgressive scours form commonly in open shoreline settings, and, where all retrograding facies are preserved, separate backstepping (transgressive) beach deposits below from transgressive shoreface strata above. Depending on the amount of ravinement scouring during transgression, the beach and underlying fluvial transgressive facies may not be preserved, and in this case the transgressive ravinement surface may truncate older, normal regressive (lowstand or even highstand) strata. For this reason, the facies that may be found below a transgressive ravinement surface are variable, from fluvial to coastal or shallow-marine, whereas the facies above are always shallow-marine (Fig. 4.9).

In transgressive river-mouth settings, either wave- or tide-dominated, the two types of transgressive ravinement surfaces may be preserved as distinct scoured contacts separated by the sandy deposits of the estuary-mouth complex (Figs. 4.46 and 4.47). In such cases, the tidal and wave scouring during shoreline transgression take place at the same time but in different areas, within the estuary and the upper shoreface, respectively (Figs. 4.46 and 4.47). As a result, the tidal-ravinement surface is placed at the contact between central estuary muds (or older variable facies where central estuary sediments are not preserved) below, and the estuary-mouth complex above (Fig. 4.9). The age-equivalent wave-ravinement surface is placed at the contact between the estuary-mouth complex

below, and the transgressive shallow-marine deposits above. This scenario is based on the assumption that the rates of aggradation of the estuary-mouth complex are higher than the rates of subsequent wave-ravinement erosion, because otherwise the wave-ravinement surface would rework the tidal-ravinement surface, and the two contacts would be superimposed. Where the estuary-mouth complex is preserved in the rock record, the wave-ravinement surface is always intercepted in vertical profiles at a higher stratigraphic level than the tidal-ravinement surface, due to the retrogradational shift of facies during transgression (e.g., see Allen and Posamentier, 1993, for a case study).

The transgressive ravinement surfaces provide the most favorable conditions for the formation of substrate-controlled ichnofacies, as they are omission surfaces that are always scoured and overlain by marginal-marine to shallow-marine facies. Depending on the amount of tidal and/or wave scouring, as well as on the nature of facies that are subject to erosion, the transgressive ravinement surfaces may be marked by firmgrounds (*Glossifungites* ichnofacies; Figs. 2.25

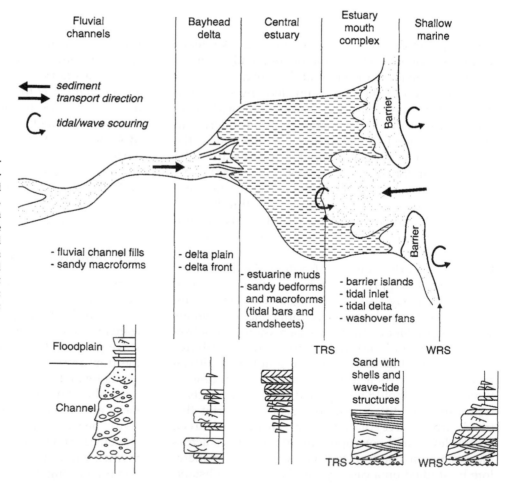

FIGURE 4.46 Tidal- and wave-ravinement surfaces in a wave-dominated estuarine setting (modified from Dalrymple *et al.*, 1992; Reinson, 1992; Zaitlin *et al.*, 1994; Shanmugam *et al.*, 2000). As facies retrograde during transgression, both tidal- and wave-ravinement surfaces expand in a landward direction, truncating central estuary and estuary-mouth complex facies, respectively. Abbreviations: TRS—tidal-ravinement surface; WRS—wave-ravinement surface.

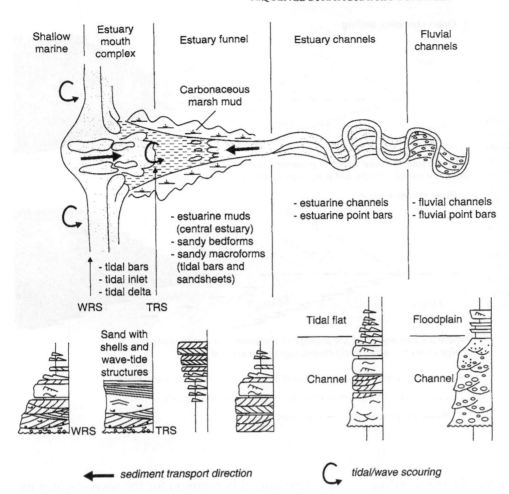

FIGURE 4.47 Tidal- and wave-ravinement surfaces in a tide-dominated estuarine setting (modified from Allen, 1991; Dalrymple *et al.*, 1992; Allen and Posamentier, 1993; Shanmugam *et al.*, 2000). As facies retrograde during transgression, both tidal- and wave-ravinement surfaces expand in a landward direction, truncating central estuary and estuary-mouth complex facies, respectively. Abbreviations: TRS—tidal-ravinement surface; WRS—wave-ravinement surface.

and 2.26), hardgrounds (*Trypanites* ichnofacies; Fig. 2.27), or woodgrounds (*Teredolites* ichnofacies; Fig. 2.28). The colonization of these substrates takes place within a relatively short interval of time, depending on the rates of shoreline transgression, either during or immediately after the ravinement surface is cut (MacEachern *et al.*, 1992). Following the formation of substrate-controlled ichnofacies, transgressive ravinement surfaces are gradually onlapped by the landward-shifting marginal-marine to shallow-marine facies (i.e., coastal onlap: Figs. 4.2 and 4.9). Numerous case studies documenting the ichnology of transgressive ravinement surfaces have been published from both modern settings and ancient successions (e.g., MacEachern *et al.*, 1992, 1999; Taylor and Gawthorpe, 1993; Pemberton and MacEachern, 1995; Ghibaudo *et al.*, 1996; Krawinkel and Seyfried, 1996; Pemberton *et al.*, 2001; Gingras *et al.*, 2004).

Wave-Ravinement Surface

The wave-ravinement surface is a scour cut by waves in the upper shoreface during shoreline transgression,

in an attempt to maintain the shoreface profile that is in balance with the wave energy (Bruun, 1962; Swift *et al.*, 1972; Swift, 1975; Dominguez and Wanless, 1991; the 'wave scour' in Fig. 3.20). This erosion may remove as much as 10–20 m of substrate (Demarest and Kraft, 1987; Abbott, 1998), as a function of the wind regime and related wave energy in each particular coastal region. Under exceptional circumstances, in coastal settings characterized by extreme wave energy, the thickness of material being removed by ravinement scouring may reach 40 m, as documented along the Canterbury Plains of New Zealand (Leckie, 1994). At the opposite end of the spectrum, the amount of erosion associated with transgressive wave scouring may be negligible where the transgressed surface is indurated by various pedogenic processes (Fig. 4.14). The wave-ravinement surface is onlapped during the retrogradational shift of facies by transgressive (fining-upward) shoreface deposits (coastal onlap), and it may overlie any type of depositional system (fluvial, coastal, or marine). The wave-ravinement surface is highly diachronous, with the rate of shoreline transgression (Fig. 4.9).

1. Open shoreline setting

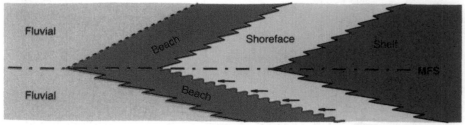

Where a transgressive beach is preserved, the wave ravinement surface does not merge with the within-trend normal regressive surface.

2. River mouth setting

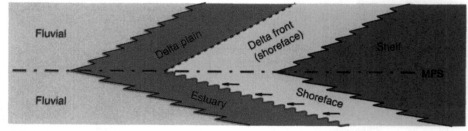

Where the estuary facies are preserved, the wave ravinement surface merges with the within-trend normal regressive surface.

— · — maximum flooding surface (MFS)   〰 ravinement surface

· · · · · · within-trend normal regressive surface (facies contact)

‾‾⌐‾ facies changes   ◄— coastal onlap

FIGURE 4.48 Architecture of facies and stratigraphic surfaces at the point of maximum shoreline transgression (from Catuneanu, 2002). The position of the within-trend normal regressive surface varies with the type of coastline, between open shoreline and river-mouth settings. The wave-ravinement surface always sits at the base of transgressive shoreface facies. The maximum flooding surface separates retrograding from overlying prograding geometries. Within the transgressive systems tract, the facies contact between shoreface sands and the overlying shelf shales defines the within-trend flooding surface.

In a vertical profile that preserves the entire succession of facies, the wave-ravinement surface separates coastal strata below (backstepping foreshore and backshore facies in an open shoreline setting, or estuarine facies in a river-mouth setting) from shoreface and shelf deposits above (Figs. 4.6, 4.48, and 4.49). Where the transgressive coastal and fluvial deposits are not preserved, the wave-ravinement surface may rework the underlying lowstand normal regressive strata and even the subaerial unconformity (Embry, 1995; Fig. 4.49). In the latter case, the wave-ravinement surface becomes part of the sequence boundary. The chances for a wave-ravinement surface to replace the underlying subaerial unconformity depend on the balance between the thickness of the lowstand normal regressive strata and the amount of subsequent wave-ravinement erosion, and are highest in the case of short stages of lowstand normal regression and/or low rates of aggradation during the lowstand normal regression. Where stages of lowstand normal regression result in the deposition of thick (> 20 m) fluvial to coastal deposits, the subaerial unconformity is preserved as such in the rock record (Fig. 4.34).

In stratigraphic sections located immediately landward from the shoreline position at the onset of transgression, it is common for the wave-ravinement surface to rework the maximum regressive surface and the underlying lowstand beach, coastal plain or delta plain strata, and therefore to be found within a fully shallow-marine succession (Fig. 4.49). In such cases, the distinction between a wave-ravinement surface and the marine portion of a maximum regressive surface (scoured and conformable contacts, respectively, both separating coarsening-upward strata below from fining-upward strata above; Figs. 4.9, 4.32, and 4.49), solely from the study of well logs, may be difficult (compare the well logs in Figs. 4.32 and 4.49). Under these circumstances, additional information (e.g., core material) is required for the unequivocal identification of the wave-ravinement surface, in order to document the scoured nature of this stratigraphic contact (Fig. 2.26). Owing to their mode of formation, wave ravinement surfaces are commonly marked by the concentration of transgressive lag deposits, which can be best observed in outcrop or core (Fig. 4.50). Where developed within fully marine successions, wave-ravinement surfaces are commonly demarcated by firmgrounds (*Glossifungites* ichnofacies; MacEachern *et al.*, 1992) or hardgrounds (*Trypanites* ichnofacies; e.g., case study by Krawinkel and Seyfried, 1996,

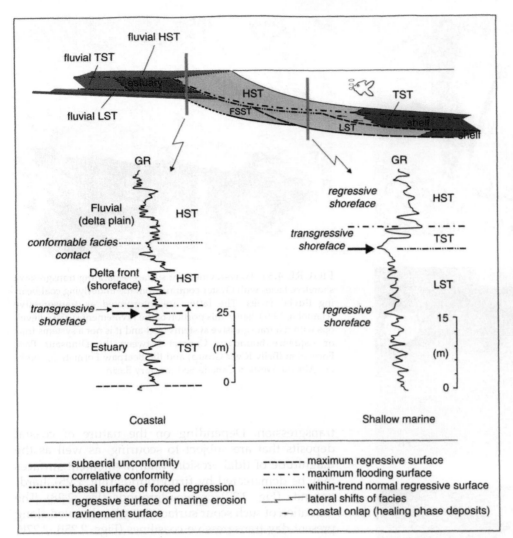

FIGURE 4.49 Well-log expression of the transgressive wave-ravinement surface (arrows; modified from Catuneanu, 2003). See Fig. 4.9 for a summary of diagnostic features of the transgressive wave-ravinement surface. Note that in fully shallow-marine successions, the transgressive wave-ravinement surface may replace the maximum regressive surface, if the log is located landward relative to the shoreline position at the onset of transgression. Log examples from the Bearpaw Formation (left) and Embry and Catuneanu (2001) (right). Abbreviations: GR—gamma ray; LST—lowstand systems tract; TST—transgressive systems tract; FSST—falling-stage systems tract; HST—highstand systems tract.

where the wave-ravinement surface is a wave-cut platform with *Gastrochaenolites* borings and a thin veneer of transgressive lag, cut into regressive shoreface deposits and overlain by transgressive shoreface facies). In stratigraphic sections located farther inland relative to the shoreline position at the onset of transgression, the chances of preservation of nonmarine deposits beneath a wave-ravinement surface are higher, and as a result such transgressive scours are commonly cut into rooted nonmarine facies capped by firmgrounds (*Glossifungites* ichnofacies) or woodgrounds (*Teredolites* ichnofacies) (MacEachern *et al.*, 1992; Pemberton *et al.*, 2001). The presence of coal beds within the nonmarine succession that is subject to transgressive wave scouring may limit the amount of downcutting, due to the more resilient nature of coal, and as a result many wave-ravinement surfaces are found directly on top of xylic substrates (Fig. 4.51).

The term 'wave-ravinement surface' was introduced by Swift (1975); synonymous terms include the *transgressive surface of erosion* (Posamentier and Vail, 1988), *shoreface ravinement* (Embry, 1995) and *transgressive ravinement surface* (Galloway, 2001). Figure 4.51 provides a field example of a ravinement surface that separates coal-bearing fluvial floodplain strata from the overlying transgressive shoreface facies. In this example, no coastal deposits are preserved following the wave-ravinement erosion, and the fluvial deposits are transgressive (fluvial transgressive facies in Fig. 4.6). As a result, this particular wave-ravinement surface develops *within* a transgressive systems tract, and it is not part of a systems tract or sequence boundary.

Tidal-Ravinement Surface

The tidal-ravinement surface is a scour cut by tidal currents in coastal environments during shoreline

FIGURE 4.51 Wave-ravinement surface separating transgressive shoreface facies with Oyster coquina from the underlying coal-bearing fluvial facies. The latter are interpreted as transgressive (Hamblin, 1997), hence this portion of the ravinement surface develops within a transgressive systems tract and it is not a systems tract or sequence boundary. Contact between the Dinosaur Park Formation (Belly River Group) and the Bearpaw Formation, southern Alberta, Western Canada Sedimentary Basin.

FIGURE 4.50 Outcrop examples of transgressive lag deposits associated with wave-ravinement surfaces. A—plan view of a wave-ravinement surface, showing the presence of transgressive lag deposits (plant debris in this case). In this example, the wave-ravinement surface is at the top of coarsening-upward shoreface deposits (a 'parasequence'), and it is overlain by transgressive shales (not shown). This type of sharp lithological contact, from sands below to shales above, qualifies the wave-ravinement surface as a 'flooding surface' (Late Cretaceous Blackhawk Formation, Utah); B, C—wave-ravinement surface associated with transgressive lag deposits ('TL'—coarse sandstone with shell fragments). In this example, the wave-ravinement surface is at the top of forced regressive delta front deposits ('FR'), is overlain by transgressive marine shales (not shown), and reworks the subaerial unconformity. This wave-ravinement surface also fits the definition of a 'flooding surface,' and, in this case, is part of the sequence boundary (Campanian Panther Tongue Formation, Gentle Wash Canyon, Utah).

transgression. Depending on the nature of coastal deposits that are subject to scouring, as well as the magnitude of tidal erosion, tidal-ravinement surfaces may be demarcated by firmgrounds (Fig. 2.25), hardgrounds (Fig. 2.27) or woodgrounds (Fig. 2.28). The formation of such scour surfaces may be observed along present-day transgressive coastlines (Figs. 2.25B, 2.27B, and 2.28), or in the rock record where the fill of tidal channels is preserved from subsequent transgressive wave-ravinement erosion (Figs. 2.25A and 2.27A). The process of tidal reworking of the underlying transgressive or normal regressive (lowstand or even highstand) deposits is equally important in open shoreline and river-mouth settings, although the type of coastline is a critical factor that controls the preservation of the tidal-ravinement surface as a distinct stratigraphic contact in the stratigraphic record. In open shoreline settings, tidal reworking in the intertidal to coastal plain areas is followed by wave erosion in the upper shoreface, as the shoreline shifts in a landward direction during transgression. For this reason, the tidal-ravinement surface is generally replaced, shortly after formation, by the landward-expanding wave-ravinement surface. This is why the wave-ravinement surface is commonly the only type of transgressive ravinement scour that is referred to in the majority of studies.

The chances of preservation of the tidal-ravinement surface as a distinct stratigraphic contact are enhanced

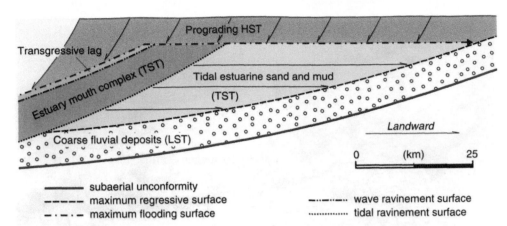

subaerial unconformity
maximum regressive surface
maximum flooding surface
wave ravinement surface
tidal ravinement surface

FIGURE 4.52 Stratigraphic model of an incised-valley fill, based on the Gironde estuary (modified from Allen and Posamentier, 1993). Note the spatial relationship between sequence stratigraphic surfaces, as well as their relation with the various facies of the incised-valley fill. This case study provides a most complete scenario, where all systems tracts that form during base-level rise are represented in the rock record of the valley fill.

in transgressive river-mouth settings, where the rates of aggradation of the estuary-mouth complex outpace the rates of subsequent wave-ravinement erosion (Figs. 4.46 and 4.47). In such settings, the two transgressive ravinement surfaces are separated by the sandy deposits of the estuary-mouth complex (Fig. 4.52). In a most complete scenario, where most estuarine facies are preserved, the tidal-ravinement surface occurs at the contact between central estuary muds below and estuary-mouth sands above (Allen and Posamentier, 1993; Fig. 4.52). The preservation of the underlying central estuary muds depends on the balance between the rates of aggradation in the central estuary and the rates of subsequent tidal erosion, as all subenvironments shift in a landward direction. In turn, these two opposing forces, of sedimentation *vs.* erosion, are a function of several variables, including sediment supply, available accommodation, and tidal range. Higher sediment supply contributes towards increased rates of aggradation, whereas a higher tidal range increases the magnitude of tidal scouring, counteracting the effect of sedimentation.

The documentation of tidal-ravinement surfaces is most common in case studies involving incised-valley fills, as the bulk of such deposits is generally tidally influenced and estuarine in origin. The Gironde estuary in France provides a classic example of a mixed tide- and wave-influenced coastal setting, where the fill of the incised valley preserves a full succession of lowstand fluvial, transgressive estuarine, and highstand deltaic sedimentary facies (Fig. 4.52; for core photographs, see Fig. 6 of Allen and Posamentier, 1993). This case study provides a good example of a tidal-ravinement surface at the contact between central estuary and overlying estuary-mouth facies (Allen and Posamentier, 1993). In coastal settings characterized by rapid transgression following the onset of base-level rise, high tidal range, and/or reduced accommodation, the lowstand fluvial deposits, as well

as the low energy central estuarine facies may not be preserved in the rock record. In such cases the tidal-ravinement surface reworks the subaerial unconformity, and the underlying highstand facies may range from fluvial to shallow-marine (Figs. 4.9, 4.53, and 4.54). Irrespective of the nature of underlying facies, the preservation of a tidal-ravinement surface as such requires the presence of estuary-mouth complex deposits on top (Fig. 4.9). As with the wave-ravinement surface, the tidal-ravinement surface is highly diachronous, with a rate that matches the rate of shoreline transgression.

WITHIN-TREND FACIES CONTACTS

In addition to the seven sequence stratigraphic surfaces described above, facies contacts associated with a strong physical expression may also be recognized *within* the various systems tracts. Such lithological discontinuities may be caused by shifts in depositional environments accompanied by corresponding changes in environmental energy and sediment supply during transgressions or regressions, and are surfaces of lithostratigraphy or allostratigraphy. They are not proper sequence stratigraphic surfaces as they do not serve as systems tract boundaries. In a sequence stratigraphic approach, within-trend facies contacts need to be dealt with only after the framework of sequence stratigraphic surfaces has been constructed. A discussion of the most prominent types of within-trend facies contacts follows below.

Within-trend Normal Regressive Surface

The within-trend normal regressive surface is a conformable facies contact that develops *during* normal regressions at the top of prominent shoreline

FIGURE 4.53 Incised-valley fill within the Muddy Formation (Ft. Collins, Colorado), showing a *tidal-ravinement surface* (arrow) at the contact between highstand shelf deposits below (Ft. Collins Member) and a transgressive estuary-mouth complex above (Horsetooth Member) (photograph courtesy of H.W. Posamentier). The tidal-ravinement surface reworks the subaerial unconformity, thus becoming part of the sequence boundary. In this example, neither lowstand nor central estuary (transgressive) deposits are preserved following the tidal-ravinement scouring. The transgressive *wave-ravinement surface* is expected to rework the top of the estuary-mouth complex.

sands (Figs. 4.9 and 4.55). The formation of this facies contact therefore requires coeval progradation and aggradation, which bring lower energy supratidal sediments on top of higher energy subtidal to intertidal facies. The underlying prominent coarser deposits may

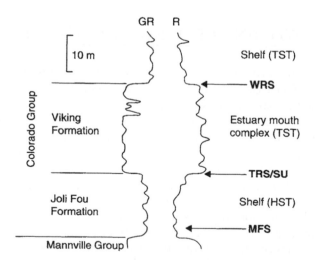

FIGURE 4.54 Well-log expression of a tidal-ravinement surface at the contact between highstand shelf deposits below and transgressive estuary-mouth sands above (Colorado Group, Crystal Field, Alberta). The estuary-mouth complex forms the fill of an incised valley, and is capped by a wave-ravinement surface. The tidal-ravinement surface reworks the subaerial unconformity, thus becoming part of the sequence boundary. Abbreviations: GR – gamma ray log; R – resistivity log; TST – transgressive systems tract; HST – highstand systems tract; WRS – wave ravinement surface; TRS – tidal ravinement surface; SU – subaerial unconformity; MFS – maximum flooding surface.

be represented by beach sands in an open shoreline setting, or by delta front sands in a river-mouth setting (Fig. 4.48), and are usually overlain by alluvial deposits dominated by floodplain fines. Due to its formation during a stage of coastal aggradation, the within-trend normal regressive surface is not demarcated by any substrate-controlled ichnofacies (Fig. 4.9). Instead, this facies contact may be associated with intertidal softground ichnofacies such as *Psilonichnus* or *Skolithos* (Fig. 2.21). This surface has a strong physical expression (i.e., an abrupt facies shift from sand to overlying mud; Fig. 4.55), which makes it easy to identify in outcrop and subsurface, and has the potential to form over large distances, depending on the duration and rates of normal regression. In spite of its prominent physical characteristics and possible regional extent, the within-trend normal regressive surface has little value for chronostratigraphic correlations as it is highly diachronous, with the rate of shoreline normal regression (Fig. 4.9).

It is important to note that the mere contrast in lithologies (mud over sand) is not sufficient for the proper identification of this facies contact as a within-trend normal regressive surface, as other facies contacts, such as some flooding surfaces for example, may also exhibit a similar juxtaposition of facies. Therefore, in addition to the observation of lithologies, other key attributes of the underlying and overlying deposits need to be explored, including depositional trends, bathymetric contrasts, and the direction of syndepositional shoreline shift. For example, even though within-trend normal regressive surfaces and flooding surfaces

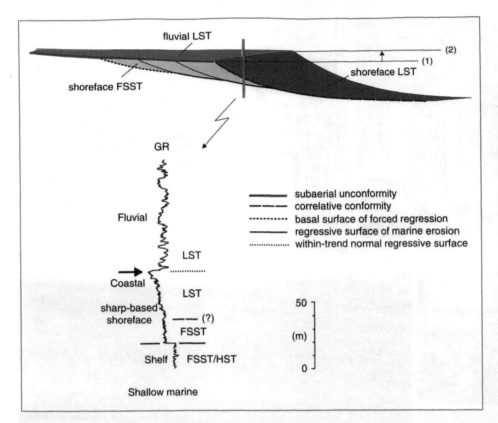

FIGURE 4.55 Well-log expression of the within-trend normal regressive surface (arrow). See Fig. 4.9 for a summary of diagnostic features of the within-trend normal regressive surface. Log example from the Lea Park Formation, Western Canada Sedimentary Basin. Note that such conformable facies contacts may occur within either lowstand or highstand systems tracts, where preserved from subsequent transgressive ravinement or subaerial erosion, respectively. Abbreviations: GR—gamma ray log; LST—lowstand systems tract; HST—highstand systems tract; FSST—falling-stage systems tract.

may display similar lithological signatures, the former are generated during *progradation* and water *shallowing* in the nearshore area, whereas the latter form during shoreline *transgression* and reflect water *deepening* in the coastal region. As explained in Chapter 3, and further detailed in Chapter 7, the association between regression and water shallowing, as well as between transgression and water deepening, is safely valid only for the shallow-water environment in the vicinity of the shoreline.

Even where a seaward shift of facies across a sand-to-overlying mud contact is documented, ruling out the interpretation of the contact as a flooding surface, the identification of a within-trend normal regressive surface solely based on well logs may be difficult, due to the possible confusion with the subaerial unconformity (e.g., compare the well-log expression of the two surfaces in Figs. 4.13, 4.29, and 4.55). For unequivocal identification, additional evidence from core or nearby outcrops is required to document the nature (scoured *vs.* conformable) of the stratigraphic contact under investigation (Fig. 4.9). In contrast to the subaerial unconformity, which truncates the underlying deposits and is also associated with offlap and fluvial onlap, the within-trend normal regressive surface is part of a conformable succession where no stratal terminations are recorded in relation to the adjacent, older and younger strata (Fig. 4.9).

Within-trend normal regressive surfaces may form during both lowstand and highstand normal regressions. In the case of highstand normal regressions, the within-trend normal regressive surface may or may not connect with the landward termination of the transgressive wave-ravinement surface, depending on the type of coastal setting (Fig. 4.48). In the case of lowstand normal regressions, the within-trend normal regressive surface connects with the basinward termination of the subaerial unconformity (Fig. 4.55). Field examples of within-trend normal regressive surfaces are provided in Figs. 3.36 and 4.56. The preservation potential of within-trend normal regressive surfaces may be hampered by subsequent transgressive ravinement erosion, in the case of lowstand systems tracts, or by subaerial erosion in the case of highstand systems tracts. Even where preserved from such larger-scale erosional processes, the within-trend normal regressive surface may be scoured locally by distributary channels in coastal plain or delta plain environments (Fig. 4.56B).

Besides within-trend normal regressive surfaces, as defined above in coastal settings, other, but less prominent facies contacts may be identified as well within normal regressive systems tracts. Notably, within the shallow-water environment, the facies contact between prodelta (deltaic bottomset) and the overlying delta

FIGURE 4.56 Outcrop examples of within-trend normal regressive surfaces. A—within-trend normal regressive surface separating beach sands from the overlying coal-bearing fluvial strata. This facies contact is conformable, mappable over a relatively large area, but is highly diachronous, with the rate of shoreline regression. Contact between the uppermost regressive shoreline sands of the Bearpaw Formation and the overlying Horseshoe Canyon Formation (Early Maastrichtian, Castor area, Western Canada Sedimentary Basin); B—distributary channel that scours locally the within-trend normal regressive surface in photograph A; C—within trend normal regressive surface (top of prograding strandplain) exposed by the erosion of the overlying fluvial floodplain deposits (contact between the Ecca and Beaufort groups, Late Permian, Karoo Basin); D—within-trend normal regressive surface (larger arrow) at the conformable facies contact between delta front (deltaic foreset) and the overlying coal-bearing delta plain deposits (deltaic topset). The photograph shows the river-dominated, normal regressive Ferron delta prograding from right to left (Late Cretaceous, Utah). Abbreviation: NR—normal regressive.

front (deltaic foreset) in river-mouth settings, or between shelf facies and the overlying prograding shoreface in open shoreline settings, may be identified as a mappable surface (sharp contact) in some cases, although in general the transition between these depositional environments tends to be gradational (Fig. 3.36). A possible reason for this gradual transition, as opposed to an abrupt and mappable facies contact, is that normal regressions are generally *slow*, hence there is sufficient time for wave-driven sediment mixing between the subtidal and the deeper-water environments. This makes it difficult, in most cases, to pinpoint a single

surface as the base of delta front or subtidal facies in a normal regressive systems tract. This situation is often in contrast to what is expected in the case of forced regressions, as explained in the following section of this chapter.

The within-trend normal regressive surface is a lithologic discontinuity that may be used in lithostratigraphic and allostratigraphic analyses, but it is not part of a systems tract boundary or of a sequence boundary. For this reason, the within-trend normal regressive surface is not a proper sequence stratigraphic surface (Fig. 4.8). It may, however, be used to fill in the internal facies details of sequences and systems tracts once the main sequence stratigraphic framework is outlined by mapping and correlating the sequence stratigraphic surfaces.

Within-trend Forced Regressive Surface

The within-trend forced regressive surface is a conformable facies contact that develops *during* forced

regressions at the base of prograding delta front facies of *river-dominated deltas* (Figs. 4.9 and 4.57). This type of within-trend facies contact does not develop in wave-dominated settings, either river-mouth or open shorelines, because in such settings the regressive surface of marine erosion forms instead (Fig. 4.23). It is also noteworthy that the within-trend normal regressive surface does not have an equivalent in a forced regressive coastal setting, where delta plain and fluvial

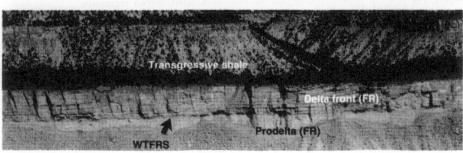

FIGURE 4.57 Outcrop examples of within-trend forced regressive surfaces at the conformable facies contact between prodelta (deltaic bottomset) and the overlying coarser-grained delta front deposits (deltaic foreset) (river-dominated, forced regressive Panther Tongue delta, Late Cretaceous, Utah). For scale, note person in image A. The delta front succession may reach up to 20 m in thickness. The within-trend forced regressive surface forms only in *river-dominated* deltaic settings, and it is highly diachronous, younging basinward with the rate of forced regression. A—steep delta front clinoforms (approximately 27°) associated with grain flow deposits (sand avalanches in a Gilbert-type delta); B—finer-grained delta front deposits (relative to A) associated with lower-angle clinoforms (approximately 10°) and the manifestation of turbidity flows; C—panoramic view showing that the forced regressive deltaic succession is truncated at the top by a composite unconformity that represents a transgressive wave-ravinement surface reworking a subaerial unconformity. Abbreviations: FR—forced regressive; WTFRS—within-trend forced regressive surface; WRS—transgressive wave-ravinement surface; SU—subaerial unconformity.

deposits are missing, being replaced in the rock record by the subaerial unconformity (Figs. 4.20—wave-dominated setting, and 4.26—river-dominated setting). As with any within-trend facies contact, the within-trend forced regressive surface is characterized by high diachroneity, becoming younger in a basinward direction with the rate of shoreline's forced regression (Fig. 4.9).

The conformable facies contact between prodelta and overlying delta front facies of forced regressive river-dominated deltas tends to be sharper than the corresponding facies contact in normal regressive settings, because forced regressions are relatively *fast*, and hence there is less time for mixing between delta front and prodelta sediments. As a result, the within-trend forced regressive surface tends to be prominent (sharp lithological contact), and therefore relatively easy to map in outcrop and subsurface (Figs. 4.57 and 4.58). Although the within-trend forced regressive surface appears, from a distance, to be a unique facies contact between prodelta and delta front facies (Fig. 4.57), detailed analyses from a closer range reveal that the change from prodelta to the overlying delta front facies takes place within a relatively narrow zone of facies transition; as such, no single lithological contact can be picked unequivocally as the within-trend forced regressive surface, which, in reality, amalgamates a few

meters thick transitional interval (e.g., about 4–5 m on the well log in Fig. 4.58). For this reason and in spite of the relatively sharp lithological contrast that defines the within-trend forced regressive surface at a larger scale (Fig. 4.57), the forced regressive delta front deposits in a river-dominated setting are still 'gradationally based', rather than 'sharp-based' (Fig. 3.27), because the succession is conformable (i.e., no regressive surface of marine erosion is present) and the change from prodelta to delta front facies is gradational even though the transition takes places rapidly, within a relatively narrow interval (compare the log in Fig. 4.58, which shows gradationally based delta front deposits in a conformable succession, with the logs in Fig. 4.29, which show a much sharper, and unconformable, facies contact at the base of the forced regressive shoreface or delta front deposits that prograde during forced regression in a wave-dominated setting). Also, in contrast to the sharp-based delta front or shoreface deposits that accumulate in wave-dominated settings, the gradationally based delta front succession that overlies the within-trend forced regressive surface is potentially thicker than the depth of the fairweather wave base (assuming preservation from subsequent subaerial and transgressive ravinement erosion), because the toe of the delta front clinoforms that prograde in a river-dominated

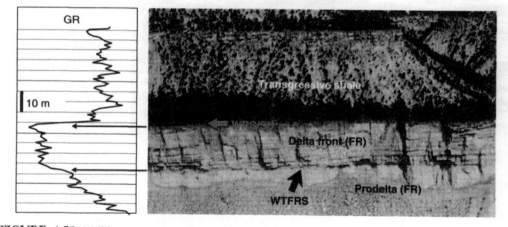

FIGURE 4.58 Well-log expression of a within-trend forced regressive surface (modified from images provided by H.W. Posamentier). The outcrop photograph shows the river-dominated, forced regressive Panther Tongue delta (image C in Fig. 4.57). Note that the deltaic succession, including the transition from prodelta to delta front facies, is conformable. The delta front interval (about 20 m in this example) is likely thicker than the depth of the fairweather wave base, because the toe of the delta front clinoforms that prograde in a river-dominated setting may reach depths greater than the fairweather wave base. Note that, from a distance, the within-trend forced regressive surface looks like a unique and well-defined facies contact (see also additional outcrop examples in Fig. 4.57). From close range, however, no single surface can be picked unequivocally as a unique lithological contact between prodelta and delta front facies. In reality, the within-trend forced regressive surface corresponds to a narrow zone of facies transition that may reach a few meters in thickness. As such, the delta front facies of river-dominated forced-regressive deltas are gradationally based (see also Fig. 3.27, and compare this well log with the logs provided in Fig. 4.29). Abbreviations: GR—gamma ray log; FR—forced regressive; WTFRS—within-trend forced regressive surface; WRS—transgressive wave-ravinement surface; SU—subaerial unconformity.

setting may reach depths greater than the fairweather wave base (Figs. 3.27 and 4.58).

The within-trend forced regressive surface may be used as a proxy for the basal surface of forced regression (seafloor at the onset of base-level fall—the conformable portion of Posamentier and Allen's, 1999, sequence boundary), even though the latter is known to be placed below, within the underlying finer-grained facies (Posamentier and Allen, 1999). This approximation is permitted by (1) the high rates of forced regression, coupled with (2) the low rates of sedimentation on the continental shelf in front of the prograding delta front. These two conditions imply that the within-trend forced regressive surface (above) and the basal surface of forced regression (below) are relatively close spatially (with and without a physical expression, respectively), although, due to the time required by the shoreline to regress, the two surfaces diverge in a basinward direction.

Within-trend Flooding Surface

The flooding surface is defined as 'a surface separating younger from older strata across which there is evidence of an abrupt increase in water depth. This deepening is commonly accompanied by minor submarine erosion or nondeposition' (Van Wagoner, 1995). Even though widely used in sequence stratigraphic work, the term 'flooding surface' is one of the most controversial concepts in sequence stratigraphy, as it allows for multiple meanings. The ambiguous nature of the above definition was discussed by Posamentier and Allen (1999) who emphasized that it is not clear whether the flooding surface forms merely as a result of increasing water depth in a marine (or lacustrine) environment, or actual flooding of a previously emergent landscape. What is clear is that flooding surfaces, commonly marked by abrupt facies shifts from sand to overlying mud in shallow-water settings, form invariably during shoreline transgression, and are topped by marine (or lacustrine) strata. The nature of the underlying deposits is however contentious, as they can vary from fluvial to coastal and shallow-water (Fig. 4.9).

At a semantic level, the usage of the word 'flooding' as a generic term that fits all the above scenarios of facies juxtaposition was challenged by Posamentier and Allen (1999) who proposed that 'flooding' should be restricted to situations where water overflows onto land that is normally dry. This definition is consistent with the common meaning of the word 'flooding', and implies subaerial exposure of the section below, prior to inundation. Following this rationale, and in order to avoid semantic confusions, Posamentier and Allen (1999) suggest replacing the term 'flooding surface' as defined by Van Wagoner (1995) with the more generic term 'drowning surface' to indicate a stratigraphic contact across which an abrupt water deepening is recorded. In this terminology, flooding surfaces become a special case of drowning surfaces, where shallow-water facies overlie nonmarine deposits. A practical problem with this approach is that evidence for subaerial exposure prior to the marine (or lacustrine) flooding is required in order to identify a stratigraphic contact as a 'flooding surface' *sensu* Posamentier and Allen (1999). Such evidence, however, may or may not be preserved in the rock record, depending on the intensity of transgressive ravinement erosion which may remove paleosols, root traces, or any other proof of subaerial exposure prior to flooding. On practical grounds, therefore, the more generic 'drowning surface' (or flooding surface *sensu* Van Wagoner, 1995) is easier to work with in terms of designating facies contacts generated by shoreline transgression, irrespective of the nature of the underlying deposits. In spite of the terminological arguments discussed by Posamentier and Allen (1999), the generic term of 'flooding surface' as defined by Van Wagoner (1995) is still the one that is most commonly used in current sequence stratigraphic work. Part of the reason is that the 'flooding surface' is heavily entrenched in the literature, despite the possible misleading connotation associated with the meaning of the word 'flooding'. In addition to this, the term 'drowning' was already coined as part of the 'drowning unconformity' concept, which is widely used in the context of carbonate sequence stratigraphy (Schlager, 1989).

Flooding surfaces are best observed in coastal to nearshore shallow-marine settings, where evidence of water deepening based on facies relationships is unequivocal (Fig. 4.59). Typical flooding surfaces may cap regressive successions (i.e., deltaic lobes in rivermouth settings or beach/shoreface deposits in open shoreline settings; Figs. 4.35B and 4.59), or transgressive sands (Figs. 4.35E and 4.60). In the former case, the transgressive deposits are typically absent or very thin, and the flooding surface may represent the only evidence of transgression in addition to the occasional transgressive lags (Kamola and Van Wagoner, 1995). Flooding surfaces have correlative surfaces in the coastal plain and shelf environments (Kamola and Van Wagoner, 1995), and possibly beyond, into the alluvial plain and deep-water settings, respectively. However, the identification of such correlative surfaces in nonmarine or deep-water deposits, unless based on uniquely correlatable strata such as volcanic ash beds, serves little purpose and may only be a source of confusion (Posamentier and Allen, 1999).

The definition provided by Van Wagoner (1995) is general enough to allow different types of stratigraphic

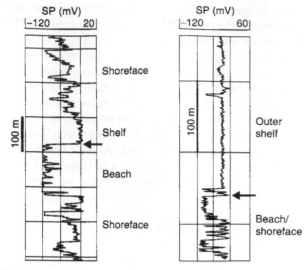

FIGURE 4.59 Flooding surfaces (arrows) at the contact between normal regressive shoreface and beach sands, and the overlying shelf mudstones. In these examples, the flooding surfaces are most likely represented by transgressive wave-ravinement surfaces. Above the flooding surfaces, the transgressive deposits may be very thin.

contacts to be candidates for flooding surfaces. The *transgressive ravinement surface* is often considered a 'flooding surface' (Posamentier and Allen, 1999: 'an overflowing of water onto land that is normally dry') (Fig. 3.30), but other surfaces that form in fully marine successions satisfy the definition of a flooding surface as well: the *maximum regressive surface*, where there is an abrupt cut-off of sediment supply at the onset of transgression (cases A, B, C, and D in Fig. 4.35); the *maximum flooding surface*, where the transgressive strata are missing and the maximum flooding surface reworks the maximum regressive surface (Fig. 4.44); or a *within-trend facies contact*, where the sand/shale contact occurs within the transgressive succession (Figs. 4.35E and 4.61). As the transgressive ravinement, maximum regressive, and maximum flooding surfaces are already defined in an unequivocal manner, the within-trend type of flooding surface is the only new surface left to be considered (Figs. 4.35E, 4.60, and 4.61). This within-trend facies contact, separating transgressive sands from the overlying transgressive

FIGURE 4.60 Well-log expression of the within-trend flooding surface (arrow; modified from Catuneanu, 2003). See Fig. 4.9 for a summary of diagnostic features of the within-trend flooding surface. Log example from Embry and Catuneanu (2001). Abbreviations: GR—gamma ray; LST—lowstand systems tract; TST—transgressive systems tract; HST—highstand systems tract; FSST—falling-stage systems tract.

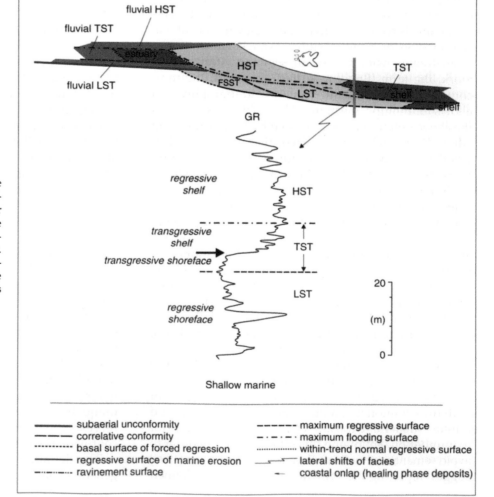

FIGURE 4.61 Within-trend flooding surface (arrow in image A) at the contact between transgressive shoreface deposits (Bad Heart Formation, Coniacian) and the overlying transgressive outer shelf shales (Puskwaskau Formation, Santonian) (photographs courtesy of Andrew Mumpy). This flooding surface corresponds to an episode of abrupt water deepening within the marine basin, which led to sediment starvation and the development of a firmground on the seafloor. The substrate immediately underlying the flooding surface is burrowed, and indurated by subaqueous seafloor cementation. The stage of nondeposition required by the formation of this firmground ('omission' surface) provided a proper environment for the formation of substrate-controlled ichnofacies. No lag deposits, or other evidence of scouring, are associated with this flooding surface. Images A—D show the indurated nature of the firmground (approximately the top 20 cm of the sediment underlying the flooding surface); images E and F show the fabric of the substrate-controlled ichnofacies.

shales, is not in a position to serve as a systems tract or sequence boundary, which is why it is not a surface of sequence stratigraphy. Similar to the within-trend normal regressive and forced regressive surfaces, the within-trend flooding surface may, however, be used to resolve the internal facies architecture of a systems tract (transgressive systems tract in this case) once the sequence stratigraphic framework is established.

As the flooding surface may change its meaning depending on case study, from a within-trend facies contact to an actual sequence stratigraphic surface, its defining features, associated stratal terminations and temporal attributes may vary significantly (Fig. 4.9). For this reason, the diagnostic features listed in Fig. 4.9 for the flooding surface cover a spectrum wide enough to allow for all possible scenarios. For example, where the transgressive facies are missing, the flooding surface may 'borrow' the characteristics of a maximum flooding surface, truncating the strata below, being downlapped by the strata above, and separating two normal regressive successions (Figs. 4.9 and 4.44). Similarly, a transgressive wave-ravinement surface may also qualify as a flooding surface (Fig. 4.57C), displaying, in this case, a high diachroneity, variable underlying facies, and onlapping shallow-marine deposits on top (Fig. 4.9). When possessing the significance of a maximum regressive or maximum flooding surface, the flooding surface itself may onlap and downlap the pre-existing landscape and seascape in a landward and seaward direction, respectively (Fig. 4.9). In a most general sense, therefore, the flooding surface may, in terms of field attributes, fit the profile of several different types of stratigraphic contacts depending on circumstances. The common thread, however, is the fact that *flooding surfaces are always overlain by marine/lacustrine shales*, either transgressive (e.g., Figs. 4.35, 4.57C, 4.60, and 4.61) or regressive (e.g., Fig. 4.44A), accumulated in a deeper-water environment relative to the underlying facies. Some flooding surfaces may be *conformable*, where sedimentation is continuous during their formation. This is likely the case where flooding surfaces are represented by maximum regressive surfaces (cases A, B, C, and D in Fig. 4.35), or by non-omission (i.e., with no substrate-controlled ichnofacies associated with them) within-trend facies contacts (e.g., the conformable surface indicated by the grey arrow in Fig. 4.35E). Often, however, flooding surfaces are represented by 'omission' contacts, associated with a stratigraphic hiatus caused by a lack of sediment supply, sediment bypass or erosion, and as a result they are potentially demarcated by substrate-controlled ichnofacies (Figs. 4.9 and 4.61). The actual type of substrate that marks a flooding surface may vary with the location within the

basin, with firmgrounds and hardgrounds forming in fully marine environments (e.g., Fig. 4.61), and all types of substrate-controlled ichnofacies (firmgrounds, hardgrounds, and woodgrounds) possibly occurring where the underlying facies are coastal or nonmarine. Such *unconformable* flooding surfaces are typically represented by maximum flooding surfaces and transgressive ravinement surfaces (e.g., Figs. 4.44 and 4.57C), but also by within-trend facies contacts that are associated with significant stages of water deepening and sediment starvation of the seafloor during transgression (e.g., Fig. 4.61).

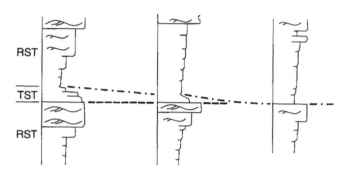

1. Sequence stratigraphic interpretation

- - - - maximum regressive surface
- · - maximum flooding surface

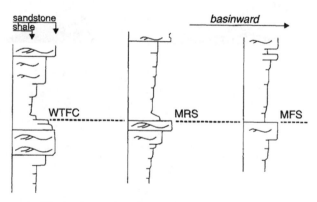

2. Allostratigraphic interpretation

······ flooding surface (lithologic discontinuity)
⌒ storm-related structures (HCS, SCS)

FIGURE 4.62 Shallow-marine (shoreface to shelf) succession of sands and shales interpreted in sequence stratigraphic and allostratigraphic terms. The thickness shown is about 12 m. Note that the transgressive facies thin basinward, to the point where the maximum flooding surface reworks the maximum regressive surface. The flooding surface is placed at the strongest lithological contrast. The example is from the Cardium Formation, Western Canada Sedimentary Basin. Abbreviations: WTFC—within-trend facies contact; MRS—maximum regressive surface; MFS—maximum flooding surface; TST—transgressive systems tract; RST—regressive systems tract; HCS—hummocky cross-stratification; SCS—swaley cross-stratification.

The unconformable flooding surfaces may or may not be associated with erosion of the seafloor. Where scoured, flooding surfaces are commonly overlain by a thin veneer of lag deposits, including coarse sand, granules or rip-up clasts, indicating that variable amounts of erosion have taken place in the process of their formation (Pemberton *et al.*, 2001). The amount of erosion varies with the type of flooding surface, being higher in the case of transgressive ravinement surfaces and maximum flooding surfaces, and minimal (if any) in the case of maximum regressive surfaces and within-trend flooding surfaces. As a rule of thumb, the higher the amount of erosion, the greater the chance for the formation of well developed transgressive lags and substrate-controlled ichnofacies, although the latter may also form in relation to stages of sediment starvation, in the absence of any discernable scouring (Fig. 4.61). Irrespective of the stratigraphic significance of the flooding surface, the shift to deeper-water facies across the contact usually triggers an increase in faunal abundance and ichnodiversity following the flooding event (Pemberton *et al.*, 2001), as well as a sharp increase in the bioturbation index (Siggerud and Steel, 1999). This change in ichnofabric across the flooding surface is accompanied by an increase in water load, which may contribute to further compaction that will enhance the firmness of the substrate, and hence generate substrate-controlled ichnofacies (Snedden, 1991).

Due to its generic nature, the flooding surface is thus too general, or vague, as a concept to pinpoint the exact type of stratigraphic contact under analysis. The usage of more specific terms, or surface types, is therefore preferred whenever sufficient data are available for the unequivocal identification of the actual type of stratigraphic contact. In a generic sense, as a lithological contact with or without sequence stratigraphic significance, the flooding surface is more appropriate for allostratigraphic studies. For sequence stratigraphic work, however, the vague nature of flooding surfaces hampers the communication of precise genetic meanings, and hence the usage of sequence stratigraphic surfaces, whenever possible, is recommended. Figure 4.62 illustrates the conceptual difference between the approaches used for sequence stratigraphic *vs.* allostratigraphic correlations. The main lithological discontinuity (the sand/shale contact, i.e., the flooding surface) is the surface of choice for allostratigraphic correlations. This surface not only transgresses time, but also changes in significance along dip, from a within-trend facies contact, to a maximum regressive surface, and finally to a maximum flooding surface (Fig. 4.62). This allostratigraphic approach is descriptive, as opposed to the sequence stratigraphic interpretation that provides a genetic framework for the rock record under analysis.

5

Systems Tracts

INTRODUCTION

The concept of systems tract was introduced to define a linkage of contemporaneous depositional systems, forming the subdivision of a sequence (Brown and Fisher, 1977; Fig. 1.9). It is fundamental to note that no thickness was implied in the original definition, nor any time connotations (see discussion on the *Concept of scale* in Chapter 1). Systems tracts are interpreted based on stratal stacking patterns, position within the sequence and types of bounding surfaces, and are assigned particular positions along an inferred curve of base-level changes at the shoreline (Fig. 4.6). The definition of systems tracts was gradually refined from the earlier work of Exxon scientists (Vail, 1987; Posamentier *et al.*, 1988; Posamentier and Vail, 1988; Van Wagoner *et al.*, 1988, 1990) with the subsequent contributions of Galloway (1989), Hunt and Tucker (1992), Embry and Johannessen (1992), Embry (1993, 1995), Posamentier and James (1993), Posamentier and Allen (1999), and Plint and Nummedal (2000).

As recently described by Galloway (2004), systems tracts correspond to 'genetic stratigraphic units that incorporate strata deposited within a synchronous sediment dispersal system.' Sediment dispersal systems, describing the way sediments are distributed within a sedimentary basin, are relatively stable during the deposition of each particular systems tract. The significant changes, or reorganizations in sediment dispersal systems, occur at systems tract boundaries, which correspond to the four main events of the base-level cycle (Fig. 4.7). Each systems tract is defined by a specific type of stratal stacking pattern, closely associated with a type of shoreline shift (i.e., forced regression, normal regression, or transgression), and represents 'a specific sedimentary response to the interaction between sediment flux, physiography, environmental energy, and changes in accommodation' (Posamentier and Allen, 1999).

The early Exxon sequence model accounts for the subdivision of depositional sequences into four component systems tracts, as first presented by Vail (1987) and subsequently elaborated by Posamentier and Vail (1988) and Posamentier *et al.* (1988). These are the lowstand, transgressive, highstand, and shelf-margin systems tracts. These systems tracts were first defined relative to a curve of eustatic fluctuations (Posamentier *et al.*, 1988; Posamentier and Vail, 1988), which was subsequently replaced with a curve of relative sea-level (base-level) changes (Hunt and Tucker, 1992; Posamentier and James, 1993).

The lowstand and the shelf-margin systems tracts are similar concepts, as being both related to the same portion of the reference sea-level curve (the stage of fall—early rise), so they were used interchangeably as part of a depositional sequence (Vail, 1987; Posamentier and Vail, 1988; Vail *et al.*, 1991). A sequence composed of lowstand, transgressive and highstand systems tracts was defined as a 'type 1' sequence, whereas a combination of shelf-margin, transgressive and highstand systems tracts was said to have formed a 'type 2' sequence (Posamentier and Vail, 1988). The differentiation between lowstand and shelf-margin systems tracts, and implicitly between types 1 and 2 sequences, therefore relies largely on the recognition of types 1 and 2 bounding unconformities. The definition of types 1 and 2 sequence boundaries was first provided by Vail *et al.* (1984), for the tectonic setting of a divergent continental margin. According to these authors, a type 1 sequence boundary forms during a stage of rapid eustatic sea-level fall, when the rates of fall are greater than the rate of subsidence *at the shelf edge*. By implication, as the rates of subsidence decrease in a landward direction across a continental shelf, the rates of sea-level fall exceed even more the rates of subsidence *at the shoreline*, leading to a fast retreat (forced regression) of the shoreline and significant erosion of the exposed shelf.

In contrast, a type 2 sequence boundary forms during stages of slow eustatic sea-level fall, when the rates of fall are less than the rate of subsidence *at the shelf edge* (Vail *et al.*, 1984). As the rates of subsidence decrease in a landward direction, such type 2 unconformities are inferred to be associated with very slow rates of relative sea-level fall *at the shoreline* (slow eustatic fall > slower subsidence), and as a result with only minor subaerial exposure and erosion of the continental shelf (Vail *et al.*, 1984). In this latter scenario, the relative sea-level fall at the shoreline is coeval with a relative sea-level rise at the shelf edge. It is important to note that both types 1 and 2 sequence boundaries include subaerial unconformities and their correlative conformities, with the main difference consisting in the amount of erosion and areal development of the subaerial unconformities. As such, a type 1 sequence boundary includes a 'major' subaerial unconformity that is characterized

by significant erosion and areal extent across the continental shelf, whereas a type 2 sequence boundary includes a 'minor' subaerial unconformity associated with minimal erosion and a limited areal extent (Fig. 5.1). The definition of types 1 and 2 sequence boundaries was subsequently reworded by Posamentier and Vail (1988), by eliminating reference to the rates of subsidence *at the shelf edge*. According to this latter paper, the occurrence of a type 1 or type 2 unconformity depends on whether the rate of eustatic fall exceeds or is less than the rate of subsidence *at the shoreline*. In this view, a type 2 unconformity would form during relative sea-level rise at the shoreline, which poses more conceptual problems than the original definition of Vail *et al.* (1984) because stages of base-level rise are not expected to result in the formation of subaerial unconformities. The situation described by Posamentier and Vail (1988), with a slow relative sea-level rise at

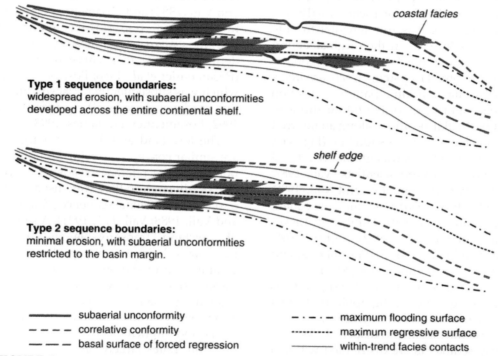

FIGURE 5.1 Definition of 'type 1' and 'type 2' sequence boundaries (modified from Vail *et al.*, 1984, and Galloway, 1989). Both types 1 and 2 sequence boundaries consist of subaerial unconformities and correlative conformities. A type 1 sequence boundary includes a subaerial unconformity that is associated with widespread erosion and development across the entire continental shelf. A type 2 sequence boundary includes a subaerial unconformity that is restricted to the basin margin (minimal erosion and limited areal extent). The formation of subaerial unconformities requires relative sea-level fall *at the shoreline* (eustatic fall > subsidence). *At the shelf edge*, however, the formation of a type 1 sequence boundary assumes relative sea-level fall (eustatic fall > subsidence), while a type 2 sequence boundary assumes relative sea-level rise (eustatic fall < subsidence). The difference in relative sea-level changes between the shoreline and the shelf edge areas, in the case of type 2 sequence boundaries, is made possible by the differential rates of subsidence recorded along the depositional dip (see text for details). The two candidates for the conformable portion of the depositional sequence boundary, marked with red in the diagram, include the 'correlative conformity' *sensu* Hunt and Tucker (1992) and the 'basal surfaces of forced regression' (i.e., the correlative conformity of Posamentier and Allen, 1999).

the shoreline, is rather conducive to the manifestation of normal regressions, when aggradation is favored in all environments across the basin.

The introduction of types 1 and 2 sequences and bounding unconformities into the literature was generally detrimental to the application of the sequence stratigraphic method, due to confusions regarding their definition and identification criteria. From a theoretical standpoint, estimation of the relative rates of eustasy and subsidence at the shelf edge during the formation of unconformities on the shelf is rather difficult and potentially subjective. On practical grounds, the differentiation between a type 1 and a type 2 unconformity was supposed to be based on the amount of associated erosion, widespread *vs.* minimal, respectively (Vail *et al.*, 1984). The estimation of the magnitude and extent of erosion is often difficult, however, especially when dealing with relatively low-resolution multichannel seismic data, but also in outcrops where age data, differential incision or angular relationships are missing. After more than a decade of confusion and controversy, Posamentier and Allen (1999) advocated elimination of types 1 and 2 in favor of a single type of depositional sequence and sequence boundary. With the fall of the type 2 unconformity, the shelf-margin systems tract (part of the type 2 sequence) exited the sequence stratigraphic arena as well. As a result, the Exxon depositional sequence model is now regarded as a tripartite scheme that includes lowstand, transgressive, and highstand systems tracts as the basic subdivisions of a sequence (Posamentier and Allen, 1999).

Perhaps the primary weakness of the early Exxon sequence model, which triggered additional debates that still perpetuate today, was the initially limited recognition of sediments deposited on the shelf during relative sea-level fall. This idea, based on overall seismic lap-out geometries, led to the early postulation of 'instantaneous' base-level fall, as reflected by the 'sawtooth' sea-level curve of Vail *et al.* (1977) (Fig. 5.2). This curve was constructed by mapping seismic reflection terminations onlapping the basin margins, which were generally interpreted as 'coastal' onlap (Mitchum, 1977), even in the absence of facies information on the seismic lines. It is now understood that this original 'coastal' onlap includes in fact a combination of fluvial and coastal onlap (Figs. 4.2 and 4.3), reflecting accumulation during both lowstand and highstand normal regressions, as well as during transgressions, and hence deposition during the entire stage of base-level rise. The apparent absence of forced regressive deposits on the shelf, as inferred in 1977, simplified the issue of the sequence boundary position in a succession of nonmarine to shallow-marine strata, as no choice had to be made with regards to where the

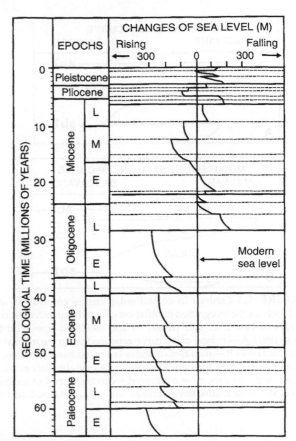

FIGURE 5.2 Global cycle chart of sea-level changes based on the interpretation of coastal onlap on seismic lines (redrafted and modified from Vail *et al.*, 1977).

boundary should be placed if falling-stage shelf deposits were present. In this view, the sequence boundary was simply separating packages of strata (sequences) characterized by continuous landward migration of 'coastal' onlap, thus corresponding to an abrupt seaward shift in 'coastal' onlap (shown as instantaneous on the sea-level charts of Vail *et al.*, 1977; Figs. 5.2 and 5.3). Subsequent work by the Exxon research group scientists led to the recognition of the possibility of shelf deposition during base-level fall, resulting in 'shelf-perched' deposits (Posamentier and Vail, 1988; Van Wagoner *et al.*, 1990). The recognition of forced regressive shelf deposits opened a new line of sequence stratigraphic debate regarding their placement within the sequence and relative to the sequence boundary. Posamentier and Vail (1988) assigned the forced regressive shelf deposits to the lowstand systems tract, thus placing the sequence boundary at their base, whereas Van Wagoner *et al.* (1990) placed the sequence boundary at the subaerial erosion surface on top of falling-stage shallow-marine strata (see depositional sequences II and III in Fig. 1.7). The latter approach is illustrated in

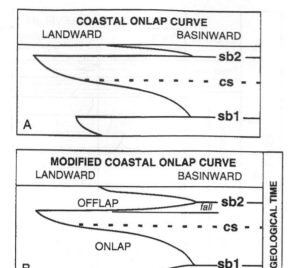

FIGURE 5.3 Contrast in coastal onlap curves constructed with and without the recognition of offlapping forced regressive deposits (stages of base-level fall) (redrafted and modified from Christie-Blick, 1991). A — coastal onlap curve generated using the methods of Vail *et al.* (1977); B — modified coastal onlap curve, based on the recognition of offlapping forced regressive deposits. Abbreviations: sb — sequence boundaries; cs — condensed sections (interval of sediment starvation in the marine environment); fall — stage of base-level fall.

the modified coastal onlap curve of Christie-Blick (1991) (Fig. 5.3).

The lowstand systems tract, as defined by Posamentier *et al.* (1988), includes a 'lowstand fan,' accumulated during falling sea level, and a 'lowstand wedge,' representing deposition during sea-level lowstand and early rise (depositional sequence II in Fig. 1.7). The lowstand fan systems tract consists of autochthonous (shelf-perched deposits, offlapping slope wedges) and allochthonous gravity-flow (slope and basin-floor fans) facies, whereas the lowstand wedge systems tract includes part of the aggradational fill of incised valleys, and a progradational wedge which may downlap onto the basin-floor fan (Posamentier and Vail, 1988). A major source of controversy in the late 1980s and early 1990s was related to the position of the sequence boundary in relation to the falling-stage lowstand fan deposits. While everybody in the Exxon team agreed to place the boundary at the base of the deep-water allochthonous facies (onset of base-level fall), the boundary was traced either at the top (Van Wagoner *et al.*, 1990: end of base-level fall) or at the base (Posamentier *et al.*, 1988, 1992b: onset of base-level fall) of the autochthonous facies. This disagreement resulted in the use of different names for the *in situ* (autochthonous) forced regressive deposits, from lowstand systems tract (Posamentier *et al.*, 1988) to highstand systems tract

(Van Wagoner *et al.*, 1988, 1990; Christie-Blick, 1991). The 'lowstand' interpretation of these deposits accounts for a sequence boundary that is placed at their base, in which case they become the oldest strata of the sequence they belong to. The 'highstand' terminology argues that the sequence boundary is at the top of these deposits, which therefore become the youngest within the sequence (Fig. 1.7). In fact none of these approaches is perfectly satisfying from a terminology viewpoint because the stage of base-level fall starts from highstand and ends at the lowstand position. In this case, neither the 'lowstand' nor the 'highstand' terms would technically apply for the entire suite of forced regressive deposits: the early falling-stage strata are closer to highstand, whereas the late ones accumulate as the base level approaches the lowstand position. Beyond just a nomenclatural issue, this debate also hinged on the temporal significance of depositional sequence boundaries. The approach proposed by Van Wagoner *et al.* (1990) implied that coeval falling-stage shallow- and deep-water deposits were separated by a highly diachronous sequence boundary, or that the deposition of deep-water strata post-dated deposition of the shelf-perched deposits. In contrast, the approach promoted by Posamentier and Vail (1988), further advocated by Posamentier and Allen (1999), preserved the chronostratigraphic significance of the sequence boundary, and the age-equivalence of falling-stage shallow- and deep-water deposits.

The inconsistency of terminology that stemmed from the Exxon research group was highlighted by Hunt and Tucker (1992), who proposed a solution by redefining the lowstand fan deposits as the 'forced regressive wedge systems tract.' In doing so, they placed the sequence boundary at the top of the newly defined systems tract (i.e., at the end of base-level fall), and the base of all falling-stage deposits (i.e., the correlative conformity of Posamentier *et al.*, 1988) became the 'basal surface of forced regression' (depositional sequence model IV in Fig. 1.7; Fig. 4.6). The advantage of this approach is that the highstand and lowstand systems tracts are now restricted to the late and early stages of base-level rise, closely associated with the actual highstand and lowstand positions of the base level, respectively. In Hunt and Tucker's (1992) approach, the correlative conformity meets the seaward termination of the subaerial unconformity (Figs. 4.17, 5.4, and 5.5). Hunt and Tucker (1992) also modified the timing of the various systems tracts relative to a reference curve of base-level changes, using the highstand and lowstand points as the temporal boundaries of the new forced regressive wedge systems tract. This is in contrast to Posamentier and Vail's (1988) approach, where the boundaries of the lowstand fan systems tract were

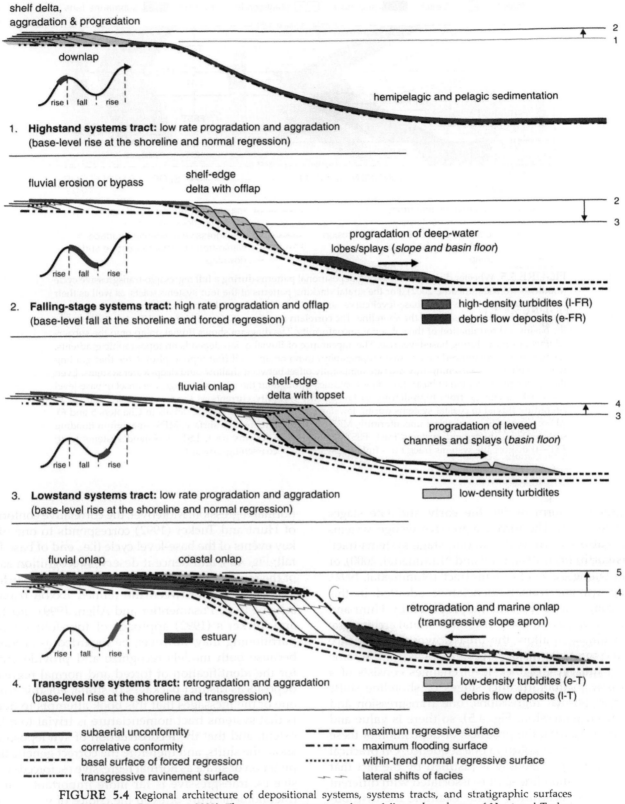

shelf delta,
aggradation & progradation

downlap

rise | fall | rise

hemipelagic and pelagic sedimentation

1. **Highstand systems tract:** low rate progradation and aggradation
   (base-level rise at the shoreline and normal regression)

fluvial erosion or bypass

shelf-edge
delta with offlap

rise | fall | rise

progradation of deep-water
lobes/splays (*slope and basin floor*)

2. **Falling-stage systems tract:** high rate progradation and offlap
   (base-level fall at the shoreline and forced regression)

- high-density turbidites (l-FR)
- debris flow deposits (e-FR)

fluvial onlap

shelf-edge
delta with topset

rise | fall | rise

progradation of leveed
channels and splays (*basin floor*)

3. **Lowstand systems tract:** low rate progradation and aggradation
   (base-level rise at the shoreline and normal regression)

- low-density turbidites

fluvial onlap

coastal onlap

estuary

rise | fall | rise

retrogradation and marine onlap
(transgressive slope apron)

4. **Transgressive systems tract:** retrogradation and aggradation
   (base-level rise at the shoreline and transgression)

- low-density turbidites (e-T)
- debris flow deposits (l-T)

- subaerial unconformity
- correlative conformity
- basal surface of forced regression
- transgressive ravinement surface
- maximum regressive surface
- maximum flooding surface
- within-trend normal regressive surface
- lateral shifts of facies

FIGURE 5.4 Regional architecture of depositional systems, systems tracts, and stratigraphic surfaces (modified from Catuneanu, 2002). The systems tract nomenclature follows the scheme of Hunt and Tucker (1992). Systems tracts are defined by stratal stacking patterns and bounding surfaces, with an inferred timing relative to the base-level curve at the shoreline. The formation of these systems tracts in a time/distance framework is illustrated in Fig. 5.5. Note that on seismic lines, downlapping clinoforms are concave-up, whereas transgressive 'healing phase' strata associated with coastal and marine onlap tend to be convex-up (Fig. 3.22). Abbreviations: e-FR—early forced regression; l-FR—late forced regression; e-T—early transgression; l-T—late transgression.

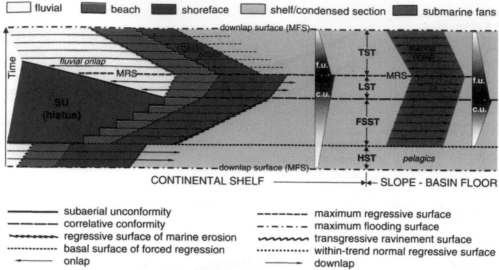

fluvial   beach   shoreface   shelf/condensed section   submarine fans

subaerial unconformity   maximum regressive surface
correlative conformity   maximum flooding surface
regressive surface of marine erosion   transgressive ravinement surface
basal surface of forced regression   within-trend normal regressive surface
onlap   downlap

FIGURE 5.5 Wheeler diagram illustrating depositional patterns during a full regressive-transgressive cycle (modified from Catuneanu, 2002). For the stratal stacking patterns of the four systems tracts, as well as their inferred timing relative to the base-level curve, see Fig. 5.4. The subaerial unconformity extends basinward during the forced regression of the shoreline. The correlative conformity (*sensu* Hunt and Tucker, 1992) meets the basinward termination of the subaerial unconformity. The diagram shows fluvial onlap onto the subaerial unconformity during base-level rise. The appearance of fluvial onlap depends on topographic gradients, ranging from pronounced onlap (steep topography) to no onlap at all (flat topography). Note that grading trends (fining- *vs.* coarsening-upward) are temporally offset between shallow- and deep-water systems. Even though the progradation of basin-floor fans continues throughout the regressive stage, the onset of base-level rise marks a change from high-density to low-density turbidity currents as sand starts to be trapped in aggrading fluvial to coastal systems during lowstand normal regression (more details in Chapters 5 and 6). Abbreviations: SU—subaerial unconformity; MRS—maximum regressive surface; MFS—maximum flooding surface; HST—highstand systems tract; FSST—falling-stage systems tract; LST—lowstand systems tract; TST—transgressive systems tract; f.u.—fining-upward; c.u.—coarsening-upward.

suggested to form *during* the early and late stages of sea-level fall. The forced regressive wedge systems tract is also known as the 'falling-stage systems tract' (Ainsworth, 1992, 1994; Plint and Nummedal, 2000), or as the 'falling sea-level systems tract' (Nummedal, 1992).

The systems tract nomenclature adopted in this book conforms to the scheme proposed by Hunt and Tucker (1992), as sufficient criteria of stratal architecture are available to allow the breakdown of a sequence into the full suite of four systems tracts. At the same time, a full cycle of base-level changes consists of a succession of four distinct stages of shoreline shifts (i.e., two normal regressions, one transgression and one forced regression; Fig. 4.5), so there is value and logic in separating the products of deposition of these four stages in the evolution of a sequence. On practical grounds, this partitioning is justified by the fact that each stage of shoreline shift is associated with different economic opportunities, as for example the distribution of petroleum plays, and hence exploration strategies change markedly between the products of forced regression and the products of subsequent lowstand

normal regression. Moreover, the correlative conformity of Hunt and Tucker (1992) corresponds to one of the key events of the base-level cycle (i.e., end of base-level fall; Fig. 4.7), and hence it deserves recognition as one of the most significant sequence stratigraphic surfaces. The distinction between the refined Exxon tripartite scheme (e.g., Posamentier and Allen, 1999) and Hunt and Tucker's (1992) approach of four-fold sequence partitioning may, however, be regarded as academic, because both models recognize and provide criteria for the identification of forced and normal regressive deposits as distinct packages of strata. For this reason, one of the messages that this book attempts to deliver is that systems tract nomenclature is trivial to a large extent, and that the reconstruction of syndepositional shoreline shifts, and therefore the correct genetic interpretation of strata as normal regressive *vs.* forced regressive *vs.* transgressive is far more important than the tract nomenclature or even the choice of what type of surface should serve as a sequence boundary (Fig. 1.7). This point is further supported by the existence of hybrid models, which use the four systems tracts of

Hunt and Tucker (1992), but follow Posamentier and Vail (1988) in the placement of the sequence boundary at the onset of base-level fall (e.g., Coe, 2003).

All classical sequence stratigraphic models assume the presence of an interior seaway within the basin under analysis, and as a result the systems tract nomenclature makes direct reference to the direction and type of shoreline shifts (Fig. 1.7). In overfilled basins, however, dominated by nonmarine sedimentation, or in basins where only the nonmarine portion is preserved, the definition of systems tracts is based on changes in fluvial accommodation, as inferred from the shifting balance between the various fluvial architectural elements. This chapter reviews the characteristics of all systems tracts, in both underfilled and overfilled basins. Five systems tracts are currently in use in underfilled basins, as defined by the interplay of base-level changes and sedimentation (Fig. 4.6). These are the highstand, falling-stage, lowstand and transgressive systems tracts, as well as a composite 'regressive systems tract' that amalgamates all deposits accumulated during shoreline regression. In addition to these five systems tracts, which assume the presence of a full range of marine to nonmarine depositional systems within the basin separated by a paleoshoreline, two more systems tracts have been defined for fully nonmarine settings. These are the low accommodation and the high accommodation systems tracts. The following sections provide a brief account of all types of systems tracts currently in use, from definition to identification criteria and economic potential. This presentation starts with the suite of three individual regressive systems tracts (i.e., highstand, falling-stage, and lowstand), followed by a discussion of the transgressive, the composite regressive, and the two fluvial systems tracts.

HIGHSTAND SYSTEMS TRACT

Definition and Stacking Patterns

The highstand systems tract, as defined in the context of depositional sequence models II and IV (Fig. 1.7), forms during the late stage of base-level rise, when the rates of rise drop below the sedimentation rates, generating a normal regression of the shoreline (Figs. 4.5 and 4.6). Consequently, depositional trends and stacking patterns are dominated by a combination of aggradation and progradation processes (Figs. 3.35 and 5.4–5.6).

The highstand systems tract is bounded by the maximum flooding surface at the base, and by a composite surface at the top that includes a portion of the subaerial unconformity, the basal surface of forced regression,

and the oldest portion of the regressive surface of marine erosion (Figs. 4.6, 4.23, and 5.4–5.6). As accommodation is made available by the rising, albeit decelerating, base level, the highstand sedimentary wedge is generally expected to include the entire suite of depositional systems, from fluvial to coastal, shallow-marine, and deep-marine. Nevertheless, the bulk of the 'highstand prism' consists of fluvial, coastal, and shoreface deposits, located relatively close to the basin margin (Fig. 5.7). Highstand deltas are generally far from the shelf edge, as they form subsequent to the maximum transgression of the continental shelf, and develop diagnostic topset packages of aggrading and prograding delta plain and alluvial plain strata (Figs. 3.35 and 5.8). Along open shorelines, strandplains are likely to form as a result of beach progradation under highstand conditions of low-rate base-level rise. Shelf edge stability, coupled with the lack of sediment supply to the outer shelf – upper slope area, results in a paucity of gravity flows into the deep-water environment (Fig. 5.7). With a proximal location on the continental shelf, highstand prisms tend to be found stranded relatively close to the basin margins following the rapid forced regression of the shoreline, coupled with the lack of fluvial sedimentation during subsequent base-level fall (Figs. 5.7, 5.9, and 5.10). Also, highstand prisms tend to be subject to preferential fluvial incision during the subsequent stage of base-level fall (Fig. 5.9), as the forefront of the highstand wedge, which inherits the slope gradient of shoreface or delta front environments, is commonly steeper than the fluvial equilibrium profile. Such processes of differential fluvial erosion have been documented by Saucier (1974), Leopold and Bull (1979), Rahmani (1988), Blum (1991), Posamentier et al. (1992b), Allen and Posamentier (1994), Ainsworth and Walker (1994), also consistent with the flume experiments of Wood et al. (1993) and Koss et al. (1994), and are discussed in more detail in the following section that deals with the falling-stage systems tract.

The relative increase in coastal elevation during highstand normal regression, which is the result of aggradation along the shoreline systems, is accompanied by differential fluvial sedimentation, with higher rates in the vicinity of the shoreline. This pattern of sedimentation, which involves progradation and vertical stacking of distributary mouth bars at the shoreline coeval with backfilling of the newly created fluvial accommodation, leads to a decrease in the gradient of the topographic slope and a corresponding lowering with time in fluvial energy (Shanley et al., 1992). This trend, superimposed on continued denudation of the sediment source areas, tends to generate an upward-fining fluvial profile that continues the overall upwards-decrease in grain size recorded by the underlying lowstand and

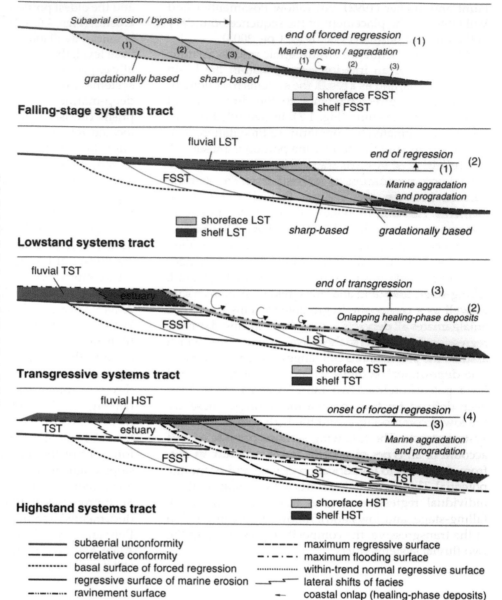

FIGURE 5.6 Detailed architecture of systems tracts and stratigraphic surfaces in the transition zone between fluvial and shallow-marine environments, in a shelf-type setting (modified from Catuneanu, 2002). The falling-stage shallow-marine deposits have a low preservation potential where the shoreline falls below the shelf edge (Fig. 5.4). Note that the earliest falling-stage shoreface deposits are gradationally based, whereas the earliest lowstand shoreface deposits are sharp-based. These are exceptions to the rule, as the falling-stage shoreface strata are generally recognized as sharp-based, in contrast to the lowstand shoreface facies which are generally regarded as gradationally based.

transgressive systems tracts (Fig. 4.6). However, the late highstand may be characterised by laterally interconnected, amalgamated channel and meander belt systems with poorly preserved floodplain deposits, due to the lack of floodplain accommodation once the rate of base-level rise decreases, approaching stillstand (Legaretta *et al.*, 1993; Shanley and McCabe, 1993; Aitken and Flint, 1994). The fluvial portion of the highstand systems tract may therefore be split into a lower part, characterized by isolated channel fills engulfed in finer-grained overbank sediments, and an upper part characterized by a higher degree of channel amalgamation. The early phase of the highstand stage is defined by relatively high rates of base-level rise, albeit lower than the sedimentation rates, which results in a stacking pattern with a strong

aggradational component. Consequently, the ratio between floodplain and channel fill architectural elements also tends to be high. In contrast, the late phase of the highstand stage is defined by much lower rates of base-level rise, which result in a stacking pattern with a stronger *progradational component*, and hence it is prone to an increase in channel clustering and implicitly in the ratio between channel fill and floodplain architectural elements. Progradation therefore accelerates with time during the highstand stage, in parallel with the decrease in the rates of base-level rise and the corresponding decrease in the rates of creation of fluvial and marine accommodation.

The trends recorded by the fluvial portion of the highstand systems tract may be described in two different terms, one referring to energy and related

Highstand normal regression: depositional processes and products

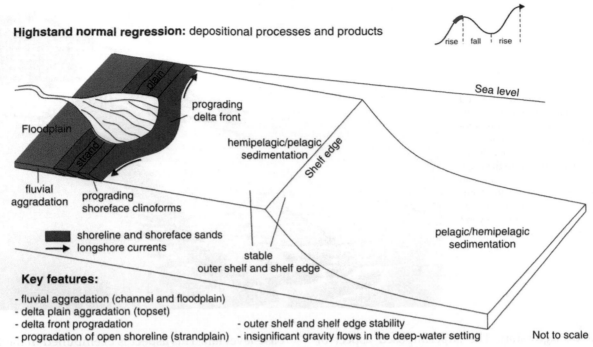

Key features:

- fluvial aggradation (channel and floodplain)
- delta plain aggradation (topset)
- delta front progradation
- progradation of open shoreline (strandplain)
- outer shelf and shelf edge stability
- insignificant gravity flows in the deep-water setting

Not to scale

FIGURE 5.7 Depositional processes and products of the highstand (late rise normal regression) systems tract (modified from Catuneanu, 2003). The deposits of this stage overlie and downlap the maximum flooding surface. The bulk of the 'highstand prism' includes fluvial, coastal, and shoreface deposits. The shelf and deep-marine environments receive mainly fine-grained hemipelagic and pelagic sediments.

FIGURE 5.8 Satellite image of the Indus Delta (Pakistan), showing the aggrading and prograding alluvial and delta plains of a modern highstand prism (image courtesy of H.W. Posamentier). Subaerial accommodation is created by the relative increase in coastal elevation at the shoreline during the highstand normal regression, as defined by the trajectory of the anchoring point of the fluvial graded profile (Fig. 3.35). The delta plain corresponds to the intertidal environment, and it is marked by tidal creeks. Fluvial aggradation is most active along the Indus River, which explains the seaward encroachment of the alluvial plain in the vicinity of the river. For scale, the Indus River is approximately 2.5 km wide.

competence (maximum grain size that can be transported by rivers), and the other referring to the balance between channel sandstones and overbank fines (Fig. 5.11). While the maximum grain size transported by highstand fluvial systems decreases with time, as a result of lowering slope gradients and fluvial energy, the sand/mud ratio increases in response to decelerating base-level rise and the corresponding increase in the degree of channel clustering. The vertical profile of the fluvial highstand deposits may therefore be described as fining-upward, if one plots the maximum grain size observed within channel fills, even though the net amount of sand tends to increase up section. The fining-upward trend is even more evident in most preserved stratigraphic sections, as the amalgamated channels at the top of the highstand systems tract are usually subject to erosion during the subsequent fall in base level. In the interfluve areas of incised-valley systems, which are less affected by erosion during forced regression, the top of the nonmarine highstand systems tract may be preserved, and instead be subject to pedogenic processes (Wright and Marriott, 1993).

An example of low energy, 'sluggish' highstand fluvial systems is presented in the left half of the seismic image in Fig. 5.12. This image captures a system of overlapping, moderate to high sinuosity Pleistocene rivers in the Malay Basin, offshore Malaysia, which were subsequently flooded during the Holocene sea-level

FIGURE 5.9 Oblique aerial photograph of a Pleistocene highstand coastal prism stranded behind and above the forced regressive shoreline of the Great Salt Lake, Utah (photograph courtesy of H.W. Posamentier). The arrow points to localized fluvial incision, which is limited to the highstand prism. The depth of incision decreases downstream, as the landscape gradient becomes in balance with the fluvial graded profile beyond the toe of the highstand prism.

rise and transgression. The superimposed aspect of these highstand rivers is an artefact of the diachronous nature of the seismic time slice (196 ms two-way travel time below the sea level), which thus captures rivers of slightly different ages on the same amplitude extraction map. Note the isolated nature of the channel fills, which are engulfed, and surrounded by extensive floodplain deposits. As discussed above, highstand fluvial systems may have a limited preservation potential due to subsequent subaerial erosion during

base-level fall. This aspect is exemplified in the lower-right area of the seismic image in Fig. 5.12, where the highstand rivers have been removed by processes of valley incision and replaced on the time slice by younger, lowstand fluvial deposits that form the fill of an incised valley (Miall, 2002).

The shallow-marine portion of the highstand systems tract displays a coarsening-upward profile related to the basinward shift of facies (Fig. 5.11), and includes low-rate prograding and aggrading normal

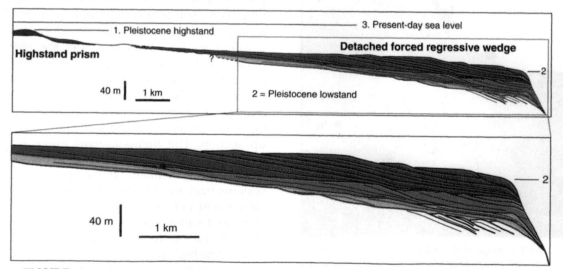

FIGURE 5.10 Cross-section through the uppermost, Pleistocene deposits of the Rhone shelf (offshore southeast France), based on the interpretation of a 2D seismic line (modified from Posamentier *et al.*, 1992b). The profile shows a typical detachment between a highstand coastal prism and the younger shallow-marine forced regressive deposits that accumulated during subsequent base-level fall. The highstand prism has been abandoned on the continental shelf behind the rapidly shifting forced regressive shoreline. The detached forced regressive wedge consists of a succession of offlapping, wave-dominated deltaic and shoreface prograding lobes, and preserves the record of at least three high-frequency episodes of base-level fall. Note that each set of forced regressive lobes pinches out in the landward direction (arrows), being separated from the highstand prism by a zone of sediment bypass.

Systems tract / Environment	HST		FSST		LST		TST	
	maximum grain size	sand-mud ratio	maximum grain size	sand-mud ratio	maximum grain size	sand-mud ratio	maximum grain size	sand-mud ratio
Fluvial			N/A[3]		Upward decrease[4]		Upward decrease[4]	
Shallow water	Upward increase[5]		Upward increase[5]		Upward increase[5]		Upward decrease[6]	
Deep water	N/A[7]		Upward increase[8]		Upward decrease[9]		Upward decrease[10]	

FIGURE 5.11 Grading trends along vertical profiles through the fluvial, shallow- and deep-water portions of the various systems tracts. The trends of change in maximum grain size and sand/mud ratio correlate in general, with the exception of the highstand fluvial systems (shaded area). Notes: (1)—younger channel fills tend to be finer-grained than the older ones due to the decrease with time in slope gradients and associated fluvial competence; (2)—due to increasing degree of channel amalgamation with time; (3)—fluvial degradation and steepening of the slope gradient; formation of sequence boundary; (4)—due to decreasing slope gradients and associated fluvial competence; (5)—due to the progradation of delta front/shoreface facies over finer prodelta/shelf sediments; (6)—due to the retrogradation of facies; (7)—dominant pelagic sedimentation; (8)—transition from mudflow deposits to high-density turbidites; (9)—transition from high-density to low-density turbidites; (10)—transition from high-density turbidites to mudflow deposits.

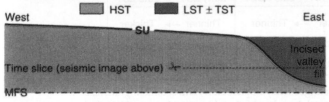

FIGURE 5.12 Amplitude extraction map along a time slice (196 ms two-way travel time below the sea level) in the Malay Basin, offshore Malaysia (modified from Miall, 2002; seismic image courtesy of A.D. Miall). The image shows juxtaposed highstand (left half of map) and lowstand (lower-right side of map) fluvial systems of Pleistocene age, which are physically separated by a subaerial unconformity that formed during an intervening stage of base-level fall—see cross-section for an interpretation. Abbreviations: HST—highstand systems tract; LST—lowstand systems tract; TST—transgressive systems tract; SU—subaerial unconformity; MFS—maximum flooding surface.

regressive strata. Within the overall regressive shallow-marine succession of a sequence, which includes highstand, falling-stage and lowstand deposits, the highstand systems tract occupies the lower part of the coarsening-upward profile (Figs. 4.6 and 5.5). This highstand prism typically includes deltas with topset geometries, in clastics-dominated settings, or carbonate platforms, where the submerged shelf hosts favourable conditions for a 'carbonate factory.'

The internal architecture of a highstand shallow-marine succession depends in part on the pattern of shoreline shift, which can be continuous during the entire duration of the highstand stage or may comprise a succession of higher-frequency transgressive-regressive pulses caused by fluctuations in the rates of sedimentation and/or base-level rise. In the case of a continuous regression, the shallow-marine portion of the highstand systems tract consists of a single upward-coarsening facies succession ('parasequence') that downlaps the maximum flooding surface. In the case of the more complex pattern of highstand regression, the shallow-marine portion of the highstand systems tract includes a succession of stacked prograding lobes ('parasequences'), in which each lobe extends farther seaward relative to the previous one. This shallow-marine architecture is often referred to as a forestepping, or seaward-stepping pattern of basin fill. The degree of vertical overlap of the progressively younger prograding lobes is more pronounced during the early phase of highstand, when the rates of base-level rise are high, and the normal regression has a strong aggradational component. In contrast, the late phase of highstand is characterized by an increased rate of shoreline regression, which is a consequence of the fact that base-level rise decelerates as it approaches still-stand. As a result, the thickness of the topset package, which reflects the degree of vertical overlap between successive prograding lobes, decreases with time, as the balance between aggradation and progradation shifts in favour of the latter. Another consequence of a decelerating base-level rise is the fact that progressively less accommodation is created on the shelf, so the prograding lobes ('parasequences') that fill the available accommodation become thinner with time and in a basinward direction (Fig. 5.13). Nevertheless, as accommodation is limited during late highstand, the youngest coastal to shoreface sandstones of the high-stand systems tract tend to have a wider geographic distribution across the shelf, as autocyclic shifting in the *locus* of lobe deposition is forced upon deltas, and as a result these shallow-marine reservoirs have a better connectivity relative to their early highstand counter-parts (Posamentier and Allen, 1999; Fig. 5.13). At the same time, the gradual lowering in fluvial energy during the highstand stage indicates that the late high-stand deltas are expected to consist of finer-grained sedi-ments relative to the early highstand deltas (Fig. 5.13). In spite of this general trend of grain size decrease from the older to the younger lobes of the highstand deltas, which occupy more proximal *vs.* more distal portions of the shelf, respectively, the vertical profile in any given location still shows an overall coarsening-upward trend due to the progradation of delta front facies over finer prodelta sediments (Fig. 5.11).

The preservation potential of the upper part of the fluvial to shallow-marine highstand prism is hampered by the subaerial and marine erosional processes that are associated with the subsequent fall in base level. It is typical therefore for the highstand systems tract to be truncated at the top by the subaerial unconformity, and to a lesser extent by the regressive surface of marine erosion (e.g., Fig. 4.23).

Economic Potential

Petroleum Plays

The best potential reservoirs of the highstand stage tend to be associated with the shoreline to shoreface depositional systems, which concentrate the largest amounts of sand, with the highest sand/mud ratio (Fig. 5.14). These reservoirs are usually meters to tens of meters thick (Fig. 3.38), and may display very good lateral continuity along the strike of the basin. Both strandplains (open shorelines) and deltas (river-mouth settings) prograde and downlap the maximum flooding surface, which marks the lower boundary of the high-stand normal regressive package (Fig. 4.40). At the top, the highstand reservoirs may be truncated by the subaerial unconformity. Fluvial systems have a moder-ate hydrocarbon potential, with the reservoirs mainly represented by channel fills and crevasse splays inter-bedded with finer-grained floodplain facies (Fig. 5.14). The sand/mud ratio and the reservoir connectivity within the fluvial systems tend to improve upwards, as the decreasing rates of base-level rise during the high-stand normal regression lead to an increase in the degree of channel amalgamation (Fig. 5.11). The distri-bution in plan view of fluvial reservoirs depends of course on their nature (channel fills *vs.* crevasse splays), which needs to be assessed based on sedimentological and geomorphological grounds. No significant reser-voirs are expected to develop during this stage in the shelf and deeper-marine settings (Fig. 5.7).

Systems tracts / Trends	HST deltas		FSST deltas		LST deltas	
	Early HST	Late HST	Early FSST	Late FSST	Early LST	Late LST
Thickness	Thicker → Thinner		Thicker → Thinner		Thinner → Thicker	
Distribution	Localized → Wider		Localized → Wider		Wider → Localized	
Grain size	Coarser → Finer		Finer → Coarser		Coarser → Finer	

FIGURE 5.13 Trends of change in thickness, distribution (geometry in plan view) and sediment grain size of deltas that prograde a shelf-type setting during highstand, base-level fall and lowstand stages. Note that changes in thickness and distribution are linked to each other, as required by the conservation of deltaic lobe volumes associated with similar sediment supply. Thus, given a constant sediment supply, thinner and wider lobes have the same volume as thicker but more localized lobes. The trends of change in sediment grain size are independent of the lobe geometry, and reflect corresponding changes in fluvial energy and competence. Fluvial gradients and energy are lowered during stages of base-level rise, and increase during the falling stage. Also note that even though younger lobes (with a more distal position on the shelf) are finer-grained than the older lobes (with a more proximal position on the shelf) in the lowstand and highstand deltas, verti-cal profiles in any given location still show coarsening-upward grading trends due to the progradation of delta front facies over finer-grained prodelta sediments (Fig. 5.11).

Systems tract	Significance	Fluvial	Coastal	Shallow-water	Deep-water
Highstand Systems Tract	Sediment budget	Good: *aggrading systems*	Good: *deltas and strandplains (coastal prisms)*	Good: *gradationally based shoreface and shelf facies*	Poor
	Reservoir	Fair: *channel fills, crevasse splays*	Good: *shoreline sands*	Good: *shoreface sands*	Poor
	Source and Seal	Poor source, fair seal: *overbank facies*	Poor	Fair: *shelf fines*	Good: *pelagic facies*
Transgressive Systems Tract	Sediment budget	Good: *rapidly aggrading systems, incised and unincised*	Good: *estuaries, deltas, backstepping beaches*	Fair: *onlapping shoreface and shelf facies*	Fair: *low-density turbidity flows and debris flows*
	Reservoir	Fair: *channel fills, crevasse splays*	Good: *estuarine, deltaic, and beach sands*	Fair: *shelf-sand deposits, basal healing-phase wedges*	Fair: *turbidites (basin floor)*
	Source and Seal	Poor source, fair seal: *overbank fines*	Poor source, fair seal: *central estuary facies*	Good: *shelf fines (shelf facies may be missing distally)*	Good: *pelagic facies*
Lowstand Systems Tract	Sediment budget	Good: *amalgamated channel fills, incised and unincised*	Good: *shelf/shelf-edge deltas, strandplains*	Good: *gradationally based shoreface and shelf facies*	Fair: *low-density turbidity flows*
	Reservoir	Good: *channel fills*	Good: *shoreline sands*	Good: *shoreface sands*	Good: *turbidites (basin floor)*
	Source and Seal	Poor	Poor	Fair: *shelf fines*	Fair: *"overbank" pelagics*
Falling-stage Systems Tract	Sediment budget	Poor	Fair: *offlapping deltas, downstepping beaches*	Fair: *sharp-based shoreface, and shelf facies*	Good: *debris flows and high-density turbidity flows*
	Reservoir	Poor	Fair: *detached shoreline sands*	Fair: *shoreface sands*	Good: *turbidites (slope and basin floor)*
	Source and Seal	Poor	Poor	Fair: *shelf fines*	Fair: *"overbank" pelagics*

FIGURE 5.14 Sediment budget and the petroleum play significance of systems tracts. Sediment budget refers to the relative volumes of sediment present in the various portions (fluvial, coastal, shallow-water, and deep-water) of each systems tract. Ranking qualifiers used in this table range from poor to fair and good.

The downside of the increased degree of fluvial to shallow-marine sand amalgamation and connectivity toward the top of the highstand systems tract is the corresponding poorer representation of source and seal rocks (Fig. 5.14). As a result, the interconnected late-highstand sand deposits tend to lack adequate seals. The sealing potential of these reservoir facies is further diminished by the presence of the overlying subaerial unconformity and, where incised valleys are located, by the presence of sand-prone valley-fill deposits above the subaerial unconformity (Posamentier and Allen, 1999). It can be concluded that the petroleum play significance of the highstand systems tract consists in the accumulation of reservoir facies mainly within proximal regions (fluvial to coastal and shoreface environments) and of source and seal facies mainly within the distal areas of the basin (shallow- to deep-water environments).

The primary risk for the exploration of highstand reservoirs is represented by the potential lack of charge due to the insufficient development of seal facies, especially towards the top of the proximal portion of the systems tract. Where present, however, highstand fluvial floodplain shales may provide a seal for the early-highstand isolated channel fills, whereas the overlying lowstand fluvial floodplain shales and/or fluvial or marine transgressive shales may seal the

late-highstand amalgamated reservoirs. The exploration potential of each individual reservoir therefore needs to be assessed on a case-by-case basis.

Coal Resources

Coal exploration is restricted to the nonmarine portion of the basin, where the thickest and most regionally extensive coal seams are generally related to episodes of highest water table relative to the landscape profile. Providing that all favorable conditions required for peat accumulation are met, which involve the interplay of subsidence, vegetation growth and sediment supply, these most significant coal seams tend to be associated with maximum flooding surfaces (Hamilton and Tadros, 1994), hence marking the base of the highstand systems tract (Fig. 5.15).

Following a stage characterized by a high accommodation to sediment supply ratio during the transgression of the shoreline, the time of end of shoreline transgression is arguably the most favorable for peat accumulation and subsequent coal development. During highstand normal regression, the balance between accommodation and sedimentation gradually changes in the favor of the latter. This, coupled with the decelerating rates of base-level rise, diminishes the chance for significant peat accumulations. The lower portion of the highstand systems tract, defined by a predominantly

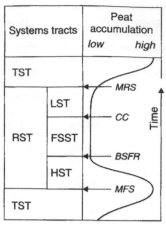

FIGURE 5.15 Generalized trend of peat accumulation during the various stages of a base-level cycle, in response to changes in accommodation. See text for discussion. No temporal scale is implied for the relative duration of systems tracts. Abbreviations: TST—transgressive systems tract; RST—regressive systems tract; HST—highstand systems tract; FSST—falling-stage systems tract; LST—lowstand systems tract; MFS—maximum flooding surface; BSFR—basal surface of forced regression; CC—correlative conformity (*sensu* Hunt and Tucker, 1992); MRS—maximum regressive surface.

aggradational sedimentation pattern, may still include well-developed coal seams interbedded with overbank fluvial facies, above the tidally-influenced transgressive fluvial channel fills. The upper portion of the highstand systems tract commonly lacks coal deposits due to insufficient accommodation and the relatively high sediment input that results in the amalgamation of meander belts. These trends in the likelihood of peat accumulation during highstand normal regressions, as well as all other stages of the base-level cycle, are illustrated in Fig. 5.15.

Placer Deposits

Mineral placers may also be studied within the framework of sequence stratigraphy, as they tend to be associated with specific sequence stratigraphic surfaces. The gold 'reefs' of the Late Archean Witwatersrand Basin, for example, offer a good opportunity to observe the stratigraphic position and significance of placer deposits (Catuneanu and Biddulph, 2001). Regardless of the mechanism of emplacement, detrital or hydrothermal, the gold in the Witwatersrand Basin is always present in the coarser lag deposits that are associated with unconformities. These conglomerates ('reefs') are not alike throughout the basin fill, but may display various textural attributes and relationships to the adjacent facies that argue for different origins. Understanding the origin of each individual placer is the key to the strategy of exploration of that particular deposit, because both the distribution and the changes

in grades along dip are a function of its genesis. Three genetic types of placer deposits may be defined in the context of sequence stratigraphy, and correspond to unconformities that form during the forced regressive and transgressive shifts of the shoreline. These stratigraphic surfaces include the subaerial unconformity, the regressive surface of marine erosion, and the transgressive ravinement surface; all three types of unconformities have the potential of concentrating economic lag deposits (placers) as a result of erosion and sediment reworking.

It can be noted that none of the three types of placers forms *during* the highstand normal regression of the shoreline, but at least portions of the subaerial unconformity and of the regressive surface of marine erosion may be part of the composite boundary at the top of the highstand systems tract (Fig. 4.23). These two placer types are discussed in the following section that deals with the falling-stage systems tract. The placers associated with transgressive scouring in near-shore environments are also described in this chapter, in the section that deals with the transgressive systems tract.

FALLING-STAGE SYSTEMS TRACT

Definition and Stacking Patterns

The falling-stage systems tract corresponds to the 'lowstand fan' of Posamentier *et al.* (1988), and it was separated as a distinct systems tract in the early 1990s, as a result of independent work by Ainsworth (1991, 1992, 1994), Hunt (1992), Hunt and Tucker (1992) and Nummedal (1992). The actual systems tract terminology varied from 'falling-stage' (Ainsworth, 1991, 1992, 1994) to 'forced regressive wedge' (Hunt, 1992; Hunt and Tucker, 1992) and 'falling sea-level' (Nummedal, 1992), with the simplest nomenclature of Ainsworth (1991, 1992, 1994) becoming generally more accepted and subsequently adopted by more recent work (e.g., Plint and Nummedal, 2000).

The falling-stage systems tract includes all strata that accumulate in a sedimentary basin during the forced regression of the shoreline. According to standard sequence stratigraphic models, the forced regressive deposits consist primarily of shallow- and deep-water facies, which accumulate at the same time with the formation of the subaerial unconformity in the nonmarine portion of the basin (Fig. 5.11). The falling-stage systems tract is bounded at the top by a composite surface that includes the subaerial unconformity, its correlative conformity (*sensu* Hunt and Tucker, 1992), and the youngest portion of the regressive surface of marine erosion (Fig. 5.6). At the base, the falling-stage systems tract is bounded by the basal surface of forced regression (= correlative conformity of Posamentier and Allen, 1999), and by the oldest portion of the regressive surface of marine erosion (Fig. 5.6). Departures from the standard sequence stratigraphic models have been pointed out by Blum (1990, 1994) and Blum and Price (1998), who demonstrated that climate shifts may trigger fluvial responses that are opposite relative to what is normally expected from changes in base level. For example, stages of climate cooling (glaciation) result in a decrease in fluvial discharge, which may in turn trigger fluvial aggradation in spite of the sea-level fall. Such 'exceptions' from the predictions of standard models must always be kept in mind in order to avoid dogmatic interpretations of data.

The formation of subaerial unconformities in the nonmarine portion of the basin during base-level fall may involve a combination of processes, including fluvial incision, fluvial bypass, pedogenesis and deflation, as discussed in more detail in Chapter 4. As a general principle, fluvial incision caused by base-level fall occurs only where the base level is lowered below major *topographic breaks* (e.g., depositional or fault scarps, the shelf edge, etc.; Figs. 3.31A, 5.16), thus exposing segments of former seascapes that are *steeper* than the fluvial graded profile (Schumm, 1993; Ethridge *et al.*, 2001; Posamentier, 2001). Under such circumstances, fluvial incision starts from the downstream end of the subaerially exposed steep topographic feature, and gradually propagates landward by the upstream migration of fluvial knickpoints (Figs. 3.31A, 3.32, and 5.16). It is estimated that the migration rates of fluvial knickpoints are generally very high, in a range of tens of meters per year, thus generating nearly instantaneous unconformities over geological time (Posamentier, 2001). The extent of upstream migration of fluvial knickpoints may also be very significant, in excess of 200 km (perhaps even 300 km), as documented in the case of the Java Sea continental shelf that was entirely subaerially exposed during the Late Pleistocene episodes of base-level fall (Posamentier, 2001).

Highstand prisms of fluvial to shallow-marine strata, abandoned on subaerially exposed continental shelves behind forced regressive shorelines, provide classic examples of *depositional* features/scarps that are prone to fluvial incision during base-level fall (Figs. 5.9, 5.16A, and 5.17). The resulting incised valleys are characterized by V-shaped cross-sectional profiles and incised tributaries. Note that, under the scenario presented in Fig. 5.16A, fluvial incision is of limited extent along dip, and is restricted to the highstand prism whose forefront slope is steeper than the fluvial graded profile (Fig. 5.9). Similar processes of fluvial

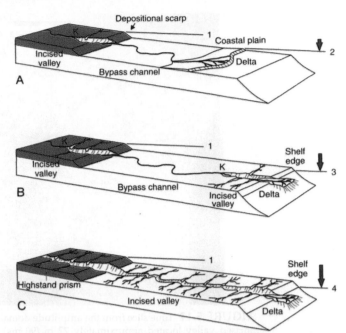

FIGURE 5.16 Incised and unincised (bypass) fluvial systems during base-level fall (modified from Posamentier, 2001). Note that, in contrast to bypass or aggrading systems, incised valleys have characteristic V-shape cross sectional profiles and incised tributaries. Fluvial incision due to base-level fall occurs only where the subaerially exposed seascape is steeper than the fluvial graded profile. Fluvial incision propagates landward *via* the upstream migration of fluvial knickpoints (K). A—early stage of base-level fall, when the forced regressive shoreline is still inboard of the shelf edge. The highstand prism is subject to fluvial incision, but the rest of the subaerially exposed continental shelf may be bypassed only by unincised fluvial systems. B—as the forced regressive shoreline falls below the elevation of the shelf edge, fluvial incision starts affecting the continental shelf. C—late stage of base-level fall, when the entire fluvial system is incised. Time 1 shows the sea level at the onset of base-level fall (end of highstand stage); time 2 shows the sea level at the end of base-level fall.

downcutting on continental shelves may also be triggered by *structural* features such as fault scarps (Fig. 5.18). Irrespective of the nature of the topographic scarp, river incision is caused by abrupt increases in fluvial energy, which in turn are triggered by increases in the slope gradient of the fluvial profile (segments of the profile that are steeper than the fluvial graded profile). Where the exposed seascapes preserve the gradient of the fluvial graded profile, rivers will only *bypass* the continental shelf, as there are no changes in river energy that need to be compensated by erosion or aggradation (Figs. 3.31B, 5.16A, and 5.18). This is commonly the case in situations where the forced regressive shoreline remains inboard of the shelf edge during base-level fall (Figs. 5.16A and 5.18).

The process of fluvial downcutting during the falling stage often leads to the preservation of the

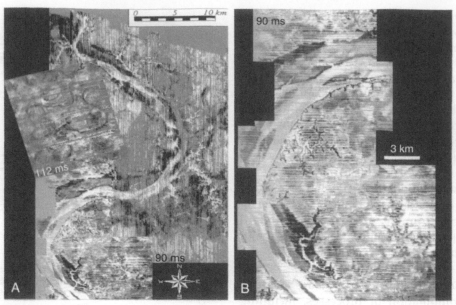

FIGURE 5.17 Time slice from the amplitude domain of a 3D seismic volume, illustrating a Late Pleistocene incised valley located approximately 72 m (90 ms) beneath the modern seafloor, offshore northeast Java (Indonesia) (modified from Posamentier, 2001; images courtesy of H.W. Posamentier). Fluvial incision was caused by a base-level fall in excess of 110 m, which led to the subaerial exposure of the entire continental shelf and upper slope. The main trunk of the valley, shown in image A, is approximately 90 km long. The valley system is characterized by short and incised tributary valleys that display a dendritic pattern in plan view. The paleoflow is inferred to be southward, as the river system appears to widen toward the south (Posamentier, 2001). The inset map of image A shows a time slice at a deeper level (approximately 90 m/ 112 ms subsea) that captures the morphology of older (probably highstand), unincised alluvial systems. Note that the unincised (highstand) and incised (falling-stage) systems have similar meander belt morphologies, excepting that the latter are associated with incised tributaries. The similarity in the morphology of unincised and incised meander belts is due to the fact that the falling-stage valleys are in fact incised, former highstand rivers. In other words, the meander pattern of incised valleys is inherited from the time the rivers were unincised. Image B details the lower portion of image A.

plan-view morphology of highstand rivers, whose channel meander pattern may be inherited by younger systems that are trapped within the confines of incised valleys. As highstand rivers tend to be the lowest energy fluvial systems of a stratigraphic sequence, they are commonly sluggish, of meandering type, being characterized by channels of moderate to high sinuosity. Erosion along such channels leads to the formation of *incised meander belts that may preserve the morphology of highstand rivers beyond the end of base-level rise*, and after changes in slope gradients and fluvial energy occur during subsequent stages of base-level fall, lowstand, or even transgression (Fig. 5.19). This means that the earliest fluvial strata that accumulate on top of a subaerial unconformity are not necessarily braided stream deposits, as commonly inferred by standard sequence stratigraphic models (e.g., Figs. 4.32 and 4.38). Instead, in the case of incised meander belts, the valley fill, which generally consists of lowstand and/or transgressive strata, is more likely to start with meandering stream deposits (Fig. 5.19). The abrupt

shift from meandering to braided systems across a subaerial unconformity (e.g., Fig. 4.16C) is easier to achieve in the case of unincised fluvial systems, as rivers that are not constrained by an erosional landscape can adjust their morphology more rapidly to new energy regimes. Such changes in morphology are more difficult to achieve in the case of incised valleys (Fig. 5.19), where increases in energy levels with time across the sequence boundary are not readily accompanied by corresponding changes in fluvial style. A good example is provided by the Orange River along the South Africa–Namibia border, which retained the meander pattern inherited from its early stage of evolution as a low energy system despite the subsequent differential uplift and steepening of the slope gradient. During its approximately 100 Ma history, the Orange River started as an unincised meandering system flowing on a relatively flat landscape, which gradually evolved as an incised meander belt in response to continental-scale tectonic uplift (J.D. Ward, pers. comm., 1997). Even today, the river preserves

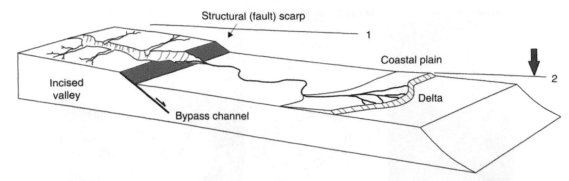

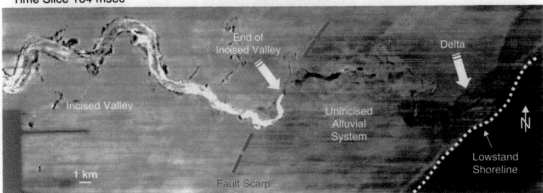

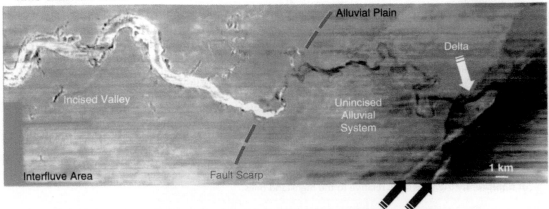

FIGURE 5.18 Incised *vs.* unincised falling-stage alluvial systems on a continental shelf affected by normal faulting (images courtesy of H.W. Posamentier). The block diagram provides an interpretation for the geology observed on the two 3D seismic amplitude time slices from the Late Pleistocene continental shelf of offshore east Java, Indonesia (see also Fig. 2.66 for additional information). Times 1 and 2 in the block diagram show the positions of the sea level at the onset and end of forced regression. The boundary between coeval incised and unincised alluvial systems on the continental shelf is controlled by the topographic break associated with a fault scarp. The process of valley incision affects the more elevated footwall of the normal fault. Since the shoreline at the end of forced regression remained inboard of the shelf edge, the downstream portion of the fluvial system seaward from the fault scarp is unincised. Note that the fault scarp has the same effect on fluvial processes as the depositional/stratigraphic scarp of a highstand prism (compare this case study with Fig. 5.16).

the original meander pattern to a large extent, although a braided style that is more in line with its current energy level is emerging in some areas as a result of meander cut-off, valley widening, and erosion of valley walls (Fig. 5.20).

Unincised fluvial systems of the falling stage are now documented as having a much more common occurrence in the rock record than originally inferred by early, standard sequence stratigraphic models. This is especially the case in shallow-marine basins with

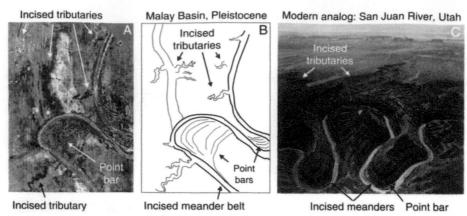

Incised tributaries | Malay Basin, Pleistocene | Modern analog: San Juan River, Utah

Incised tributary | Incised meander belt | Incised meanders   Point bar

FIGURE 5.19 Incised meander belts formed during stages of base-level fall. A—subsurface example depicting a time slice through a 3D seismic volume, placed approximately 77 m subsea in the Malay Basin, offshore Malaysia (detail from Fig. 5.12; image courtesy of A.D. Miall). The Pleistocene meandering stream deposits captured in this image are interpreted as lowstand (normal regressive) strata, and form the lower portion of an incised-valley fill that rests on a subaerial unconformity (Miall, 2002; Fig. 5.12). Note the incised tributaries that feed into the main meander belt. These tributaries are slightly older (formed during base-level fall) than the point bars (younger, lowstand deposits accumulated during early base-level rise). B—interpretation of the features observed in image A. C—modern example of an incised meander belt (modified from Press *et al.*, 2004). Note that these incised meander belts preserve the morphology of pre-existing low energy and high sinuosity highstand rivers which became trapped within their own valleys during subsequent falling-stage incision. As a result, renewed fluvial aggradation at the onset of base-level rise takes place in meandering rather than braided rivers, even though lowstand fluvial systems are commonly expected to be of significantly higher energy relative to the highstand ones.

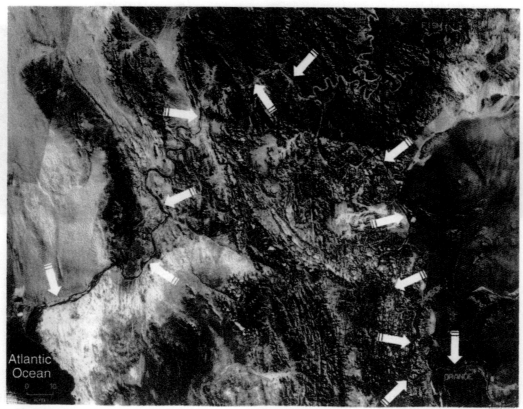

FIGURE 5.20 Satellite photograph of the Orange River (arrows) along the Namibia – South Africa border (image courtesy of J.D. Ward). This incised fluvial system retains the meander pattern inherited from its early stages of evolution, when the river was of low-energy and flowing across a relatively flat landscape, in spite of the subsequent steepening of the slope gradient and the associated increase in fluvial energy. See text for more details.

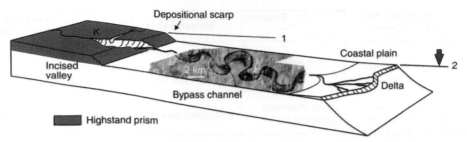

FIGURE 5.21 Unincised (bypass) fluvial system on a continental shelf that is not fully exposed by the fall in base level (offshore Java, Indonesia; modified from Posamentier, 2001; seismic image courtesy of H.W. Posamentier). The seismic amplitude horizon slice shows a high-sinuosity unincised channel flowing to the right (southeast) across the Miocene shelf of the Java Sea.

gently sloping ramp margins, such as intracratonic basins and filled foreland systems, or in continental shelf settings where the forced regressive shoreline does not fall below the elevation of the shelf edge (Posamentier, 2001; Fig. 5.21). The under-representation of falling-stage bypass fluvial systems in the sequence stratigraphic literature may be due to a number of factors, including inadequate well spacing, lack of high-resolution 3D seismic data, overlooked horizon slice imagery, and interpretation approaches unprepared to challenge conventional thinking. Thus, it appears that many published examples of incised valleys may not, in fact, be incised systems (Posamentier, 2001). A look at the continental shelf of the Java Sea during the last 0.5 Ma before present reveals that extensive fluvial valley incision, across the shelf, could only have occurred during three, rather short time intervals when the entire shelf and upper slope were exposed subaerially (Posamentier, 2001; Fig. 5.22). Consequently, incised valleys such as the one captured in Fig. 5.17 are rather the exception than the rule, and for the majority

of Late Pleistocene time, fluvial systems across the Java shelf were mainly characterized by unincised channels (Fig. 5.22).

The separation between incised and unincised falling-stage fluvial systems is fundamental for the design of successful petroleum exploration and production strategies, as the fill of incised valleys and unincised channels are inherently different play types. The differentiation between these play types is important not only for the evaluation of the petroleum potential of fluvial deposits, but it has implications for the evaluation of all deposits of the falling-stage systems tract. For example, the fact that not all episodes of base-level fall result in valley incision and full subaerial exposure of the shelf is highly significant for the development of deep-water reservoirs, and hence for the design of deep-water exploration strategies. Thus, stages of incomplete exposure of the shelf, as inferred from the presence of unincised fluvial systems, are prone to the accumulation of mud-rich sediments in the deep-water environment, as opposed to stages

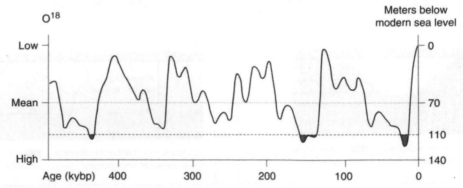

FIGURE 5.22 Late Pleistocene to Holocene sea-level curve based on oxygen isotope data (modified from Bard *et al.*, 1990). The 110 m threshold marks the depth of the Java shelf edge below the modern sea level, and it has been identified as the level below which the Java continental shelf is fully exposed. A fall of less than 110 m below the modern sea level results in the formation of unincised fluvial systems, whereas a sea-level fall in excess of 110 m below the present level results in the formation of incised valleys across the shelf (Posamentier, 2001). Note that the last 0.5 Ma in the evolution of the Java shelf are by far dominated by the presence of unincised fluvial systems.

FIGURE 5.23 Summary of criteria that may be used to differentiate between incised-valley fills and unincised or distributary channel fills. Inadequate data (e.g., a lack of high-resolution 3D seismic data, well logs, or core material) may lead to confusions between these two play types, with negative consequences for the strategy of exploration of the entire falling-stage systems tract – see text for details. The presence of incised tributaries in association with incised valleys is one of the easier and unequivocal features to observe in the preliminary stages of stratigraphic analysis (e.g., Figs. 3.7, 5.17, and 5.19). Note that incised-valley fills tend to form stratigraphic 'anomalies,' which disrupt the continuity of stratigraphic markers and the predictable association of facies that is expected according to Walther's law (compare the two scenarios in Fig. 5.24).

Criteria \ Systems	Incised-valley fills	Unincised fluvial or distributary channel fills
Stratigraphic architecture	Complex, involving depositional systems ranging from fluvial to estuarine and open marine	Simple, commonly including only fluvial deposits
Width:thickness aspect ratio	Low, commonly less than 200:1	High, potentially close to 1000:1
Tributaries	Incised	Unincised
Well log response	Anomalous, showing a lack of correlation with juxtaposed units	Good correlation with juxtaposed units
Well log markers	Commonly truncated by valley incision	Preserved in a relatively conformable succession
Gas/oil production	Potentially very high	Average

of full exposure of the shelf that are likely to result in the formation of deep-water reservoirs – see further discussion below on the effects of low- *vs.* high-magnitude falls in base level on deep-water sedimentation. The definition of criteria for the correct identification of incised *vs.* unincised fluvial systems is therefore highly important. A summary of features that can be used to distinguish between these two play types is presented in Fig. 5.23. A detailed imagery of subsurface fluvial systems, as afforded by 3D seismic, well-log and core data, is necessary to document their stratigraphic

makeup and architecture, aspect ratio, the characteristics of tributary systems, and the position of the fluvial deposits under analysis within the overall stratigraphic context. From the latter perspective, incised-valley fills form stratigraphic 'anomalies,' being genetically unrelated to the adjacent facies, whereas unincised channel fills integrate within the paleogeography of the juxtaposed and underlying depositional environments (Figs. 5.24 and 5.25). Note that all incised valleys regardless of origin (e.g., climate- *vs.* base level-controlled) share similar features (e.g., compare Figs. 3.7 and 5.17),

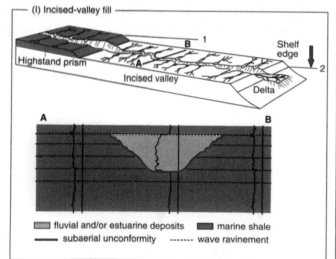
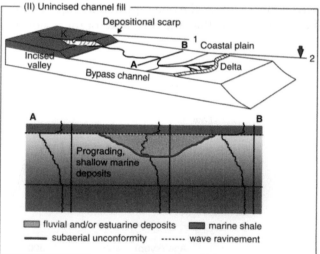

FIGURE 5.24 Synthetic gamma ray logs illustrating the stratigraphic context of (I) incised-valley fills and (II) unincised channel fills (modified from Posamentier and Allen, 1999). Not to scale. Incised-valley fills occupy an anomalous position in the stratigraphic context, being genetically unrelated to the juxtaposed facies—in this example, the valley fill is fully engulfed within outer shelf shales. Unincised valley fills are genetically related to the juxtaposed and underlying facies—in this example, the channel may be part of an alluvial or delta plain environment that progrades over delta front facies. In both block diagrams, 1 represents the sea level at the onset of forced regression, and 2 the sea level at the end of forced regression. K—upstream-migrating fluvial knickpoint.

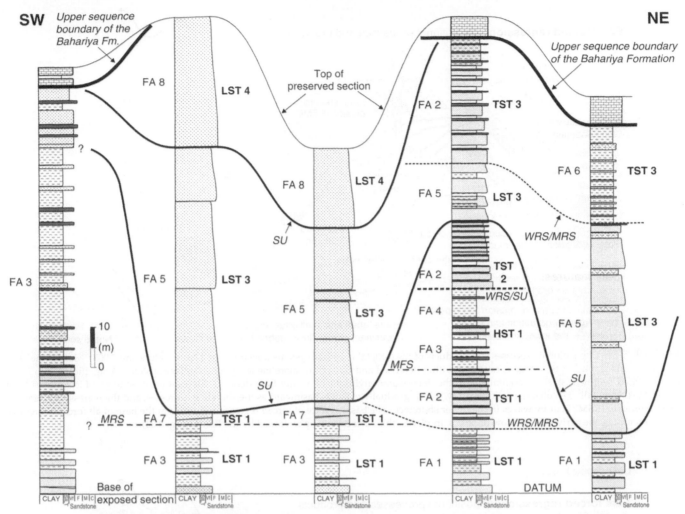

FIGURE 5.25 Sequence stratigraphic framework, and incised-valley systems, of the Lower Cenomanian Bahariya Formation in the Bahariya Oasis, Western Desert of Egypt (from Catuneanu *et al.*, in press). For scale, the cross section covers a horizontal distance of about 100 km. The erosional relief generated by the process of valley incision explains the abrupt facies shifts that may occur laterally over relatively short distances, as well as the absence of some systems tracts in areas most affected by fluvial erosion. The magnitude of base-level fall associated with the formation of each subaerial unconformity (sequence boundary) may be estimated from the amount of valley incision as inferred from the thickness variation of lowstand deposits across the Bahariya Oasis. Accumulation of lowstand fluvial deposits contributes to the peneplanation of the incised-valley topographic relief. Abbreviations: LST—lowstand systems tract; TST—transgressive systems tract; HST—highstand systems tract; SU—subaerial unconformity; MRS—maximum regressive surface; MFS—maximum flooding surface; WRS/MRS—wave-ravinement surface that replaces the maximum regressive surface; WRS/SU—wave-ravinement surface that replaces the subaerial unconformity. Facies associations (FA): FA 1—aggrading and prograding delta plain; FA 2—backstepping parasequences; FA 3 and 4—low-energy fluvial systems; FA 5 and 8—high-energy fluvial systems; FA 6—outer-shelf glauconitic shales; FA 7—beach deposits (see Catuneanu *et al.*, in press, for full details).

so the mere identification of incised valleys is not sufficient for an immediate interpretation of the allogenic controls responsible for the process of valley incision. More criteria that may help to interpret the nature of the allogenic controls on fluvial processes are discussed further in Chapter 6.

Diagnostic for the falling-stage systems tract are the shallow-marine deposits with rapidly prograding and offlapping stacking patterns (Fig. 5.10), which are age-equivalent with the bulk of the deep-water submarine fans (e.g., Hunt and Tucker, 1992; Plint and Nummedal, 2000; Figs. 5.26 and 5.27). This systems tract was independently described by Hunt and Tucker (1992) who specifically referred to the slope and basin-floor settings, and by Plint and Nummedal (2000) who studied the processes and products of forced regressions

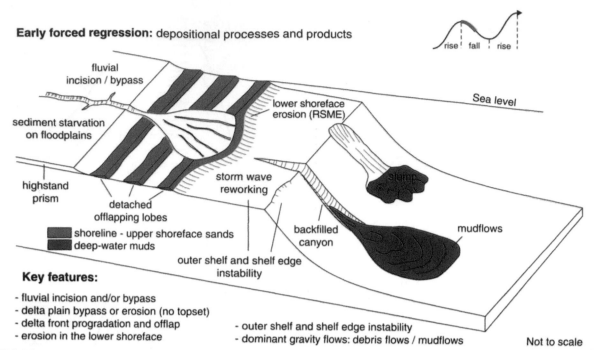

Early forced regression: depositional processes and products

fluvial incision / bypass

lower shoreface erosion (RSME)

Sea level

sediment starvation on floodplains

slump

highstand prism

storm wave reworking

detached offlapping lobes

backfilled canyon

mudflows

■ shoreline - upper shoreface sands
■ deep-water muds

outer shelf and shelf edge instability

Key features:

- fluvial incision and/or bypass
- delta plain bypass or erosion (no topset)
- delta front progradation and offlap
- erosion in the lower shoreface

- outer shelf and shelf edge instability
- dominant gravity flows: debris flows / mudflows

Not to scale

rise | fall | rise

FIGURE 5.26 Depositional processes and products of the *early* falling-stage systems tract (modified from Catuneanu, 2003). Most of the sand that accumulates during this stage is captured within detached and offlapping shoreline to upper shoreface systems. A significant amount of finer-grained sediment starts to accumulate in the deep-water environment as mudflow deposits. Two sequence stratigraphic surfaces form *during* base-level fall: the subaerial unconformity, which gradually expands basinward as the shoreline regresses; and the regressive surface of marine erosion (RSME) cut by waves in the lower shoreface. The basal surface of forced regression is taken at the base of all forced regressive strata, including the early fall mudflow deposits. In places, this surface may be reworked by the RSME (Fig. 4.23).

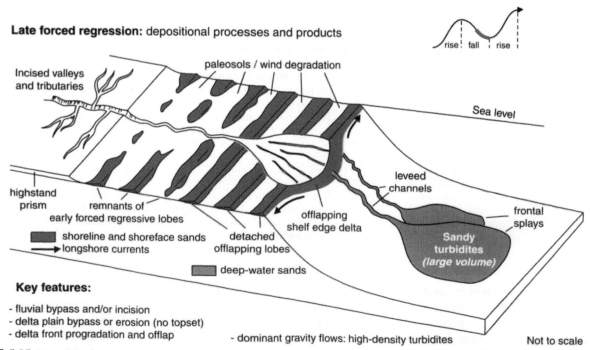

Late forced regression: depositional processes and products

paleosols / wind degradation

Incised valleys and tributaries

Sea level

highstand prism

leveed channels

remnants of early forced regressive lobes

frontal splays

offlapping shelf edge delta

detached offlapping lobes

Sandy turbidites *(large volume)*

■ shoreline and shoreface sands
→ longshore currents

■ deep-water sands

rise | fall | rise

Key features:

- fluvial bypass and/or incision
- delta plain bypass or erosion (no topset)
- delta front progradation and offlap

- dominant gravity flows: high-density turbidites

Not to scale

FIGURE 5.27 Depositional processes and products of the *late* falling-stage systems tract (modified from Catuneanu, 2003). The sediment mass balance changes in the favor of the deep-sea submarine fans, which capture most of the sand. The subaerial unconformity keeps forming and expanding basinward until the end of base-level fall. Once the shoreline falls below the shelf edge, the regressive surface of marine erosion stops forming as the seafloor gradient of the continental slope is steeper than what is required by the shoreface profile to be in equilibrium with the wave energy. Note that fluvial systems are likely to incise into the highstand prism but may only bypass the rest of the subaerially exposed shelf, unless the base level falls below the elevation of the shelf edge. The turbidity currents of the deep basin are dominantly of high-density type, due to the massive amount of sediment supply, and hence they tend to be overloaded and aggradational (sediment load > energy of the flow) along their entire course.

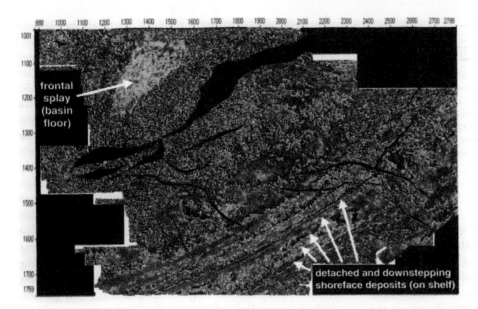

FIGURE 5.28 Amplitude extraction map along a seismic horizon, showing detached and downstepping forced regressive shoreface deposits on the continental shelf (from Catuneanu *et al.*, 2003a; image courtesy of PEMEX). The color code uses blue for sand and orange for shale.

in a shelf-type setting. Posamentier and Morris (2000) also provide a comprehensive synthesis of the diagnostic features of shallow-marine forced regressive deposits, and Posamentier and Kolla (2003) discuss the depositional elements that are most likely to accumulate in the deep-water environment during base-level fall. As pointed out by Plint and Nummedal (2000), the pattern of stratal offlap that is characteristic of forced regressions (Fig. 3.22) may be obliterated by subsequent subaerial or transgressive ravinement erosion. In such cases, the most practical feature that helps to

identify a falling-stage systems tract is the presence of sharp-based shoreface sandbodies in wave-dominated nearshore areas (Plint and Nummedal, 2000; Figs. 4.21 and 4.28). Additional criteria for the recognition of shallow-marine forced regressive deposits include: the presence of zones of separation (detachment) between successive shoreface deposits (Figs. 5.26–5.30); the occurrence of long-distance regression (Fig. 5.27); the absence of alluvial plain, coastal plain, or delta plain deposits at the top of shoreface deposits (e.g., contrast Figs. 3.27 and 3.35); the presence of a seaward-dipping subaerial unconformity on top (Fig. 4.15); the presence of progressively lower-relief clinoforms in a basinward direction (Fig. 5.31); a potential increase in average grain size in a seaward direction, as the gradient

FIGURE 5.29 Detached paleoshorelines (arrows) left behind by the forced regression associated with the fall in base level caused by Holocene glacio-isostatic rebound. The horizontal distance shown in the photograph is about 6 km. Dundas Peninsula, Melville Island, Canadian Arctics.

FIGURE 5.30 Detached and downstepping beaches (arrows) associated with base-level fall and the forced regression of the shoreline caused by Holocene glacio-isostatic rebound. The horizontal distance shown in the photograph is about 6 km. Dundas Peninsula, Melville Island, Canadian Arctics.

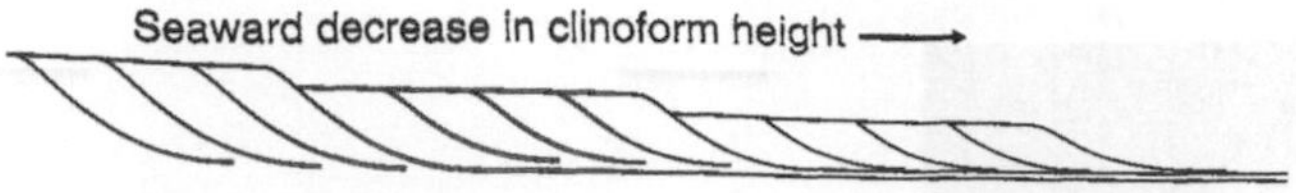

Seaward decrease in clinoform height ⟶

FIGURE 5.31 Decreasing shallow-water clinoform height in a seaward direction in a shelf-type setting, in response to base-level fall (modified from Posamentier and Morris, 2000). This trend is particularly representative of river-dominated deltaic settings, where clinoforms are steeper than the wave equilibrium profile, and as a result no wave scouring affects the lower shoreface to inner shelf environments. This is the reason why no regressive surfaces of marine erosion are shown in the diagram.

of fluvial systems may steepen during base-level fall (Fig. 5.13); and the presence of 'foreshortened' stratigraphic successions (Posamentier and Morris, 2000). The latter criterion describes situations where the decompacted thickness of the regressive succession is significantly less than the paleowater depth at the time of deposition. For example, the Panther Tongue Sandstone (Fig. 3.30) accumulated in 75–100 m deep water, but its decompacted thickness is only 25 m. The difference is primarily accounted for by subaerial erosion during the forced regression of the shoreline (Posamentier and Morris, 2000).

In a most complete scenario, a falling-stage systems tract may include offlapping shoreface lobes on the continental shelf, inner to outer shelf macroforms, shelf-edge deltas that downlap the continental slope, and slope and basin-floor submarine fans (Figs. 4.23, 4.24, and 5.4–5.6). These deposits do not necessarily coexist. The type of falling-stage facies that accumulate at any given time depends largely on the position of the base level relative to the shelf-edge elevation, and implicitly on the location of the shoreline relative to the shelf edge (Figs. 5.26 and 5.27). Furthermore, the type of falling-stage deposits that get preserved in the rock record depends on the magnitude of base-level fall and on the location of the shoreline relative to the shelf edge at the end of fall. Among all portions of the falling-stage systems tract, the shallow-marine facies that accumulate on the continental shelf during forced regression are most susceptible to subsequent subaerial erosion, especially in situations where the base level falls below the elevation of the shelf edge.

For low-magnitude falls in base level, when the base level remains above the elevation of the shelf edge during forced regression (Figs. 4.23 and 5.6), the falling-stage deposits typically include offlapping deltaic and shoreface lobes, shelf macroforms, and deep sea (slope and basin-floor) submarine fans. In such cases, where a portion of the continental shelf is still submerged, no shelf-edge deltas may form and the deep-water fans are dominated by fine-grained sediments (Fig. 5.26). The forced regressive deltas that prograde the continental shelf during the falling stage tend to thin in a basinward direction (Fig. 5.31) and, assuming that equal volumes of sediment are present in each successive deltaic lobe, they also tend to become wider along strike (Fig. 5.13). Reservoir connectivity of forced regressive shelf-delta sands is therefore expected to improve toward the top of the falling-stage systems tract. At the same time, the gradual steepening of the fluvial slope gradient during forced regression results in a coarsening down dip of the sediment that is present in the offlapping delta lobes, which further enhances the reservoir quality of the late falling-stage deltaic sands (Posamentier and Morris, 2000; Fig. 5.13). The shoreface deposits that accumulate in a shelf setting during forced regression are sharp-based, excepting for the earliest falling-stage shoreface strata which are gradationally based (Figs. 4.23 and 5.6). The preservation potential of these shallow-marine falling-stage strata is inversely proportional to the magnitude of base-level fall. As the shoreline approaches the shelf edge during forced regression, shelf-edge deltas will form and supply coarser sediment to the deep-water environment (Fig. 5.27). At the same time, the linear paleoshoreline sandbodies abandoned on the subaerially exposed shelf are now subject to fluvial and wind degradation. For as long as the base level does not fall below the elevation of the shelf edge, fluvial systems may only incise the highstand prism, bypassing the rest of the shelf (Posamentier, 2001; Fig. 5.27).

For higher-magnitude falls in base level, when the base level falls below the elevation of the shelf edge (Fig. 5.4), a shelf-edge delta with offlapping geometries will prograde and downlap onto the continental slope, coeval with the manifestation of significant gravity-flow events in the deep-water environment. These gravity flows consist mainly of high-density turbidity currents, which are potentially rich in sandy riverborne sediment that is supplied by distributary channels directly to the deep-water environment. As the regressive surface of marine erosion is unlikely to form beyond the shelf edge, on the relatively steep continental slope (see discussion in Chapter 4), these falling-stage deposits are bounded at the top by the subaerial unconformity and its correlative conformity, and at the base by the basal surface of forced regression (Fig. 5.4). In this scenario, fluvial systems are likely to incise not only the highstand prism, but also the rest of the shelf that was submerged at the onset of forced regression (e.g., case C in Fig. 5.16). The preservation potential of all shallow-marine falling-stage deposits that accumulated on the shelf during the earlier stages of forced regression is minimal in this case.

Economic Potential

Petroleum Plays

The formation and distribution of reservoir facies may be markedly different between the stratigraphically lower and upper portions of the falling-stage systems tract. For this reason, the discussion below focuses on the distinct processes and products of the early *vs.* late stages of forced regression (Figs. 5.26 and 5.27).

The early forced regression corresponds to the early stage of base-level fall at the shoreline, when a significant portion of the continental shelf is still submerged (Fig. 5.26). The shoreline trajectory is defined by progradation and offlap, accompanied by fluvial erosion or bypass upstream. The lack of aggradation in delta plain or coastal plain environments prevents the formation of regressive coastal topsets. Under these circumstances, the prograding shoreface or delta front deposits are truncated by the subaerial unconformity (Fig. 3.27). Elongated and downstepping beach-upper shoreface sandbodies may be abandoned on the subaerially exposed shelf as the base level falls (Fig. 5.26). These paleoshoreline sands are generally thin (range of meters) and 'detached' (i.e., separated by gaps; Figs. 5.28–5.30). The degree of detachment depends on the interplay of sediment supply and the rates of base-level fall (Fig. 3.33; Posamentier and Morris, 2000). During early fall, the shoreline is still far from the shelf edge, so no river-borne sand is delivered directly to the continental slope. However, lowering of the storm wave base causes instability on the outer shelf, which triggers gravity-flow processes into the deep-water environment. These gravity flows include mainly the fine-grained sediment accumulated on the outer shelf-upper slope area during the previous highstand (late rise) normal regression, as well as during the earliest phases of forced regression.

The reservoirs of the early forced regression stage are mainly represented by the offlapping and downstepping paleoshoreline and shoreface sands abandoned on the shelf during the fall in base level (Figs. 5.14 and 5.26). These sand bodies, even though generally thin and detached, may have very good lateral extent along strike (Fig. 5.26), and may be significant enough to be seen on amplitude extraction maps of 3D seismic data (Fig. 5.28). No fluvial reservoirs are expected to develop during this stage, as, according to standard sequence stratigraphic models, the nonmarine portion of the basin is subject to bypass or downcutting processes (Figs. 5.11, 5.14, and 5.26). The lower (early fall) portion of the deep-water submarine fans displays a poor reservoir quality, due to the low sand/mud ratio, and is usually represented on seismic data by transparent and/or chaotic seismic facies (e.g., transparent facies A in Fig. 5.32). The plastic behavior of these mud-rich cohesive debris flows (mudflows in Fig. 5.26) confers on them additional characteristics that may be observed on 2D and 3D seismic data, including thrust faults and associated compressional ridges, as well as striations and grooves at the base (Posamentier and Kolla, 2003). Such diagnostic features of mud-rich

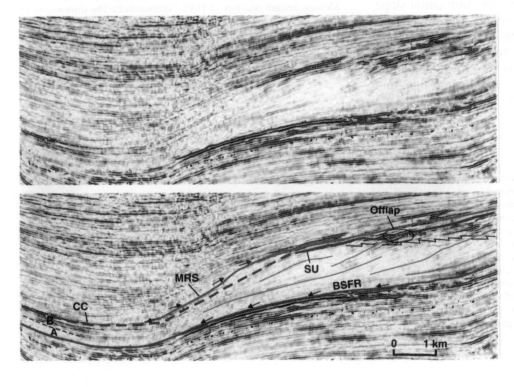

FIGURE 5.32 Uninterpreted and interpreted seismic line showing the contrast in facies between mudflow deposits (facies **A**—early forced regressive) and turbidites (facies **B**—late forced regressive) in a deep-water setting (from Catuneanu *et al.*, 2003a; image courtesy of PEMEX). Note that the top of the coarser-grained facies of the submarine fan is marked by the extension within the basin of the youngest clinoform associated with offlap (i.e., the correlative conformity *sensu* Hunt and Tucker, 1992). In this example, the maximum regressive surface downlaps the correlative conformity, and hence no significant lowstand normal regressive deposits are present above the late forced regressive turbidites. Abbreviations: BSFR—basal surface of forced regression; SU—subaerial unconformity; CC—correlative conformity (*sensu* Hunt and Tucker, 1992), MRS—maximum regressive surface.

deep-water deposits are of course critical to recognize on seismic data prior to drilling. The formation of thrust faults and compressional ridges is caused by the tendency of mudflows to freeze on deceleration, due to the high internal strength (cohesiveness) of the muddy matrix. This 'plastic' rheological behavior, in contrast to the 'fluidal' behavior of turbidity currents, explains why mudflow deposits tend to accumulate in more proximal areas of the deep-water setting relative to turbidites (illustrated by the relative location of frontal lobes/splays in Figs. 5.26 and 5.27). Seismic examples of deep-water cohesive debris flow (mudflow) deposits, which summarize their defining characteristics, are provided in Figs. 5.33–5.36. These diagnostic features include: *high erosional relief* generated by the flow (Fig. 5.33); the presence of *basal grooves* caused by the drag force of the plastic flow over the seafloor (Fig. 5.34); *internal chaotic/contorted seismic facies* that reflects the plastic behavior of the flow, as the flow freezes on deceleration rather than allowing sediments to settle according to weight (Fig. 5.35); and the presence of *internal thrust faults*, and associated *compressional ridges* at the top of mudflow deposits, which once again reflect the plastic rheology of the flow (Fig. 5.36).

The late forced regression corresponds to the late stage of base-level fall at the shoreline, when most, if not all, of the continental shelf becomes subaerially exposed (Fig. 5.27). During this stage, the early forced regressive paleoshoreline sandbodies may loose their original linear geometry due to prolonged fluvial and wind degradation (Figs. 4.31C and 5.27). As the shoreline approaches the shelf edge, the fluvial sediment starts to be delivered straight to the continental slope, causing major gravity-flow events. Additional sediment supply is generated by processes of fluvial incision upstream. The lack of accommodation for the fluvial and shoreline systems explains the large volume of turbidites which accumulates during this time in the deep-water environment (Figs. 5.14 and 5.27).

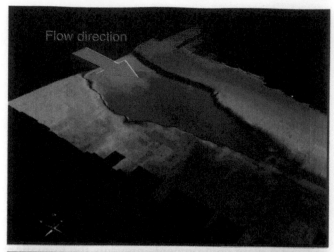

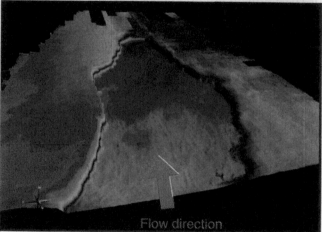

FIGURE 5.33 Erosional relief caused by the motion of a cohesive debris flow (mudflow) over the seafloor (Pleistocene, Gulf of Mexico; images courtesy of H.W. Posamentier). The dimensions of the scoured relief shown in this seismic image are as follows: approximately 30 km in length, approximately 12.5 km maximum width, and approximately 240 m in depth.

FIGURE 5.34 Seismic time slice illustrating grooves at the base of a mudflow deposit (upper-left side of the image; Pleistocene, eastern Gulf of Mexico; modified from Posamentier, 2003; image courtesy of H.W. Posamentier). These striations are caused by the drag force imposed by the plastic flow upon the unconsolidated substrate. The time slice shows reflection amplitudes—see Fig. 2.51 for a visualization of seismic facies along the same surface. This image is time-transgressive, as the lower-right side of the slice shows a slightly older, channelized turbidity system.

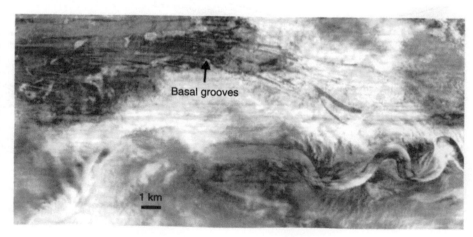

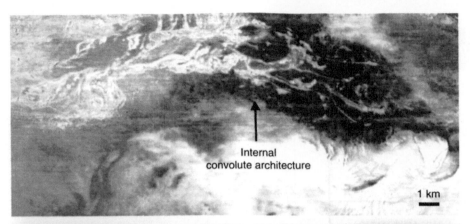

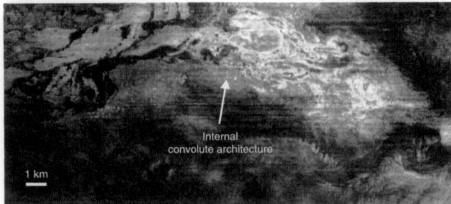

FIGURE 5.35 Seismic time slices showing the internal contorted/convolute architecture of a debris flow deposit (upper-left side of images), caused by the tendency of the flow to freeze on deceleration (Pleistocene, eastern Gulf of Mexico; modified from Posamentier, 2003; images courtesy of H.W. Posamentier). The dense and plastic nature of the sediment/water mixture in mudflows does not allow sediments to settle from suspension; instead, the flow stops abruptly as soon as the slope angle decreases below a critical threshold of approximately 1°. This explains the lack of layering of mudflow deposits, as well as other structural features captured in Fig. 5.36. The two seismic time slices in this Figure are slightly above the image shown in Fig. 5.34.

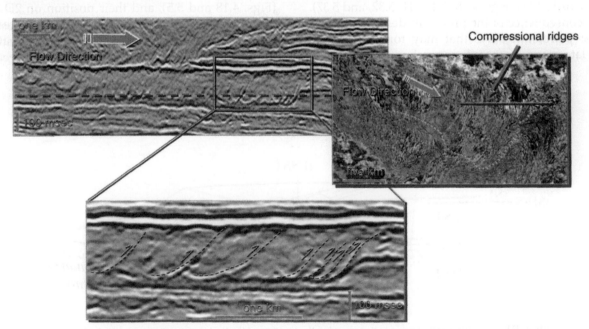

FIGURE 5.36 Compressional structures of mudflow deposits, generated by the tendency of the flow to freeze on deceleration (Pleistocene, Gulf of Mexico; images courtesy of H.W. Posamentier). Owing to its plastic rheology, the flow 'freezes' as soon as the shear stress becomes less than the internal shear strength of the sediment/water mixture. The frontal part of the flow freezes first, when reaching areas of lower slope gradient, acting as a barrier against which the rest of the sediment/water mixture crashes. This causes the formation of internal thrust faults (see cross sectional views), whose expression at the top of mudflows deposits is represented by compressional ridges (see plan view).

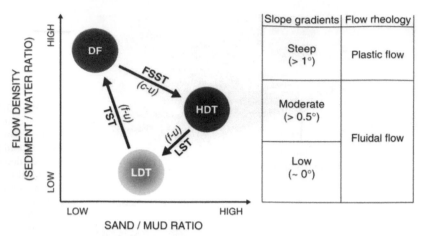

FIGURE 5.37 Trends of change in the main types of gravity flows that operate in the deep-water environment during the formation of the falling-stage, lowstand and transgressive systems tracts. Abbreviations: DF—cohesive debris flows (mudflows); HDT—high-density turbidites; LDT—low-density turbidites; FSST—falling-stage systems tract; LST—lowstand systems tract; TST—transgressive systems tract; c-u—coarsening-upward; f-u—fining-upward. The table indicates the minimum slope gradients required by the three main types of gravity flows to move, as well as the rheology of each flow type. These flow characteristics explain why mudflows tend to travel shorter distances relative to turbidity currents, and also why low-density turbidites may reach farther into the basin than the high-density turbidites. The latter is also facilitated by the lower sand/mud ratio of the low-density turbidity flows, which sustains the formation of levees over larger distances. See text for details (Chapters 5 and 6).

As the sediment entry points get gradually closer to the shelf edge during the falling stage, the overall vertical profile of the forced regressive submarine fans is coarsening-upward, with a transition from mudflows to sandy turbidites (Figs. 4.18, 5.4, 5.11, 5.32, and 5.37).

As a consequence of the processes described above, the best petroleum plays that may form in relation to the late stage of forced regression are the sandy turbidites associated with the deep-water submarine fans (Figs. 5.14 and 5.27). These reservoirs form the coarsest part of the basin-floor fans, are topped by the correlative conformity *sensu* Hunt and Tucker (1992) (Figs. 4.18 and 5.5), and their position on 2D seismic lines may be inferred by mapping the youngest clinoform with an offlap type of stratal termination up dip (Figs. 4.17, 5.32, and 5.38). The turbidity currents

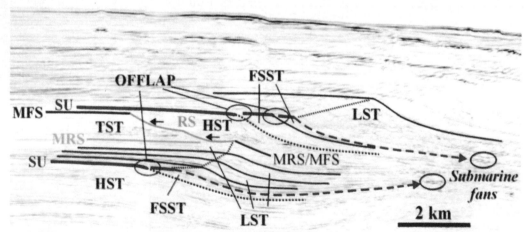

FIGURE 5.38 Interpreted seismic line showing the location of the best deep-water reservoirs in a sequence stratigraphic framework (from Catuneanu *et al.*, 2003a; image courtesy of PEMEX). See Fig. 2.65 for the uninterpreted seismic line. The offlap type of stratal termination is highly significant for deep-water exploration because the youngest clinoform associated with offlap (i.e., the correlative conformity *sensu* Hunt and Tucker, 1992—dashed line in the figure) leads to the top of the coarsest deep-water facies. Abbreviations: FSST—falling-stage systems tract; LST—lowstand systems tract; TST—transgressive systems tract; HST—highstand systems tract; SU—subaerial unconformity; RS—transgressive ravinement surface; MRS—maximum regressive surface; MFS—maximum flooding surface.

triggered during the late stage of forced regression tend to have high sand/mud and sediment/water ratios because of the large amounts of terrigenous sediment supplied by fluvial systems at the shelf edge (Fig. 5.37). Due to their high-density nature (high sediment/water ratio), these turbidity currents tend to be overloaded, which favours channel aggradation and the construction of levees on the continental slope as opposed to canyon incision (Figs. 5.27, 5.39, and 5.40). As levees are built by the finer-grained sediment fraction of the flow, they tend to be relatively short for turbidity currents with a high sand/mud ratio of the initial sediment mixture (i.e., insufficient amount of mud to sustain the construction of levees over large distances). As a result, levees' height decreases rapidly down flow and as the turbidity currents become unconfined the frontal splays accumulate relatively close to the toe of the slope, albeit more distally relative to the mudflow lobes of the previous stage (Figs. 5.4, 5.41, and 5.42; also, compare Figs. 5.26 and 5.27). Both the channel fills on the slope and the frontal splay on the basin floor are sand-prone and hence potentially good reservoirs (Fig. 5.42). The leveed channels on the slope may be confined within older submarine canyons, or may form in areas between slope canyons. These issues concerning the processes and products of the deep-water environment during the various stages of the base-level cycle are discussed further in Chapter 6. Besides the submarine fans and the associated feeding leveed channels, additional exploration potential is offered by the offlapping shelf-edge deltaic lobes that prograde the upper part of the continental slope, as well as by the strike-oriented elongated sandbodies abandoned on the shelf during the forced regression of the shoreline (Fig. 5.27). As in the case of the early forced regression, no fluvial reservoirs are expected to form during this stage (Fig. 5.14).

The primary risk for the exploration of coastal to shallow-marine falling-stage reservoirs accumulated on the continental shelf, is represented by the potential lack of charge due to the insufficient development of seal facies on top. Where present, however, the overlying lowstand fluvial floodplain shales and/or fluvial or marine transgressive shales may seal the detached shoreline to shoreface sands of the falling-stage systems tract. Therefore, as with all systems tracts, the exploration potential of each individual reservoir needs to be assessed on a case-by-case basis. The deep-water reservoirs of the falling-stage systems tract, which arguably form some of the best petroleum plays of an entire stratigraphic sequence, involve fewer risks relative to their coastal to shallow-marine counterparts, as the deep portion of the basin is generally most conducive toward the accumulation of source and seal facies. All underlying and overlying systems tracts (lowstand, transgressive and highstand) may provide source rocks and seal facies, respectively, for the falling-stage turbidite reservoirs.

Coal Resources

The stage of base-level fall is unfavorable for peat accumulation and subsequent coal development because accommodation is negative, and the nonmarine environment is generally subject to fluvial valley incision and/or paleosol development in interfluve areas (Fig. 5.15). Therefore, according to the standard sequence stratigraphic models (Fig. 4.6), the only remnant in the rock record of the surface processes that take place in the nonmarine realm during base-level fall is the subaerial unconformity. This 'standard' view is based on the underlying assumption that fluvial processes are mainly controlled by base-level changes. As discussed in Chapters 3 and 4, this process/response relationship is particularly true for the downstream

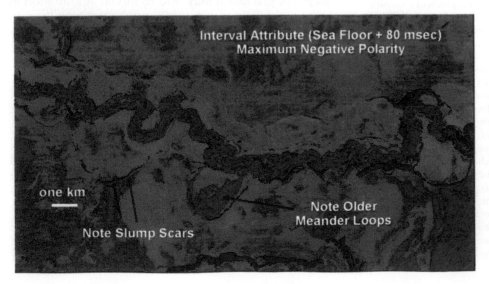

FIGURE 5.39 High-sinuosity channelized turbidity system in a continental slope setting (Late Pleistocene, offshore Nigeria; modified from Posamentier and Kolla, 2003; image courtesy of H.W. Posamentier). Flow direction from right to left. The formation of such leveed channels on a continental slope is most probable in the case of high-density turbidity flows, which may be overloaded (sediment load > energy), and hence aggradational, in high-gradient settings. In contrast, low-density turbidity currents tend to be entrenched (underloaded) on continental slopes. The manifestation of high-density turbidity flows is most likely during late stages of forced regression, when terrigenous sediment supply to the deep-water environment is highest.

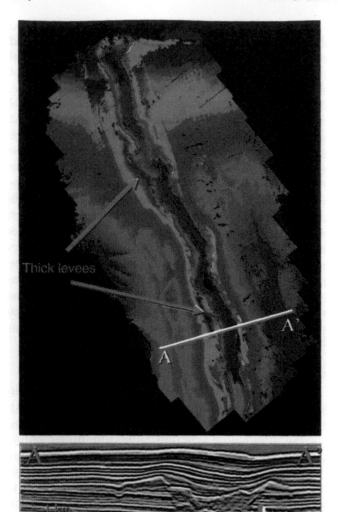

FIGURE 5.40 Turbidite leveed channel in an upper slope continental setting (Late Pleistocene, De Soto Canyon area, Gulf of
Mexico; images courtesy of H.W. Posamentier). As seen on the seismic isochron map, channel sinuosity is significantly higher than
levee sinuosity. The 2D seismic line (A – A') reveals the wedge-
shaped geometry of levee deposits. The presence of levees indicates
a high-density turbidity flow, whose sediment/water ratio is high
enough to allow aggradation even on the steep-gradient continental
slope. Such leveed channels are most likely to form during late stages
of forced regression, when terrigenous sediment supply to the deep-
water environment is highest. This massive sediment influx overcomes the high energy of the flow on the steep-gradient continental
slope, and leads to the aggradation of the channelized turbidity
system. Levees are important depositional elements that keep the
flow confined during the motion of the sediment/water mixture
within an aggrading turbidity system. As levees are built by the finer-
grained sediment fractions of the sediment/water mixture, the flow
collapses when the system runs out of mud, leading to the formation
of frontal splays at the end of leveed channels (Fig. 5.41). Due to the
high sand/mud ratio that commonly characterizes high-density
turbidity flows, the frontal splays of such flows tend to be found in
proximal locations, close to the toe of the continental slope (i.e., insufficient amount of mud to sustain the formation of levees over large
distances). The location of frontal splays may thus provide an additional criterion to evaluate the type of turbidity flows (high- vs. low-
density) at the time of sedimentation. For scale, the leveed channel on
the seismic isochron map is approximately 1.8 km wide.

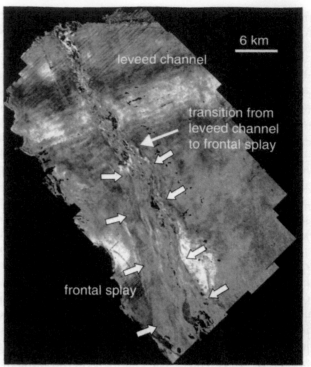

FIGURE 5.41 Leveed channel to frontal splay transition within an
aggrading turbidity system (Late Pleistocene, De Soto Canyon area,
Gulf of Mexico; image courtesy of H.W. Posamentier). The presence
of aggrading leveed channels on the continental slope indicates an
overloaded (sediment load > flow energy) high-density turbidity
flow. The transition from leveed channel to frontal splay takes place
where the levees' height decreases below a critical threshold that is
required to maintain the flow confined. This transition is placed
more proximally in the case of high-density (relative to low-density)
turbidity flows, due to the lower mud content of the former (see also
Fig. 5.40). For scale, the leveed channel in the upper part of the
image is 1.8 km wide.

reaches of fluvial systems, but it may be significantly
distorted due to the interference of the climatic control.
For example, the decrease in fluvial discharge during
stages of glaciation may lead to fluvial aggradation in
spite of the fall in sea level (Blum, 1990, 1994). Even in
such cases, however, the cold climate is not favorable
for the accumulation of any significant peat deposits.

Placer Deposits

Surface processes during the falling leg of the base-
level cycle are generally considered to be most conducive to the formation of placer deposits. According to
standard sequence stratigraphic models, erosional
processes are expected to affect both nonmarine and
wave-dominated shallow-marine environments during
forced regression, resulting in the formation of subaerial unconformities and regressive surfaces of marine
erosion, respectively (Fig. 3.27). These two types of
unconformities have a different geographic distribution within the basin, even though they may partly

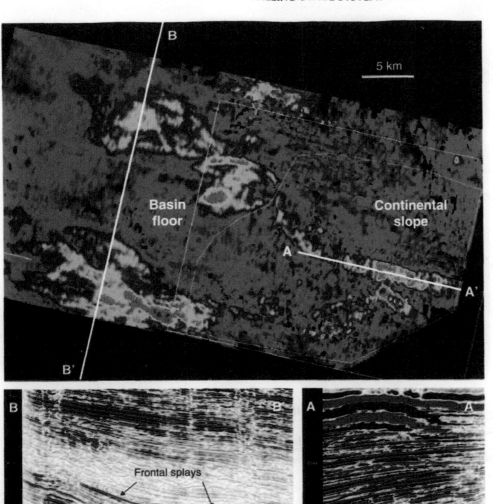

FIGURE 5.42 Deep-water turbidity systems, comprising sand-dominated frontal splays on the basin floor and their associated sand-filled submarine channels on the continental slope (from Catuneanu *et al.*, 2003a; image courtesy of PEMEX). Channel aggradation on the continental slope and the location of frontal splays close to the toe of the slope are indicative of high-density (overloaded, and with a high sand/mud ratio) turbidity currents, which characterize late stages of forced regression. These features provide important criteria to differentiate the high-density turbidity flows of the falling-stage systems tract from the low-density turbidity flows (entrenched channels on slope, and frontal splays located deeper into the basin) of the lowstand and early transgressive systems tracts.

overlap in the region of the paleoshoreline (Fig. 3.27), so their differentiation is important for the design of exploration and production strategies.

Subaerial unconformities are the most important placer-forming stratigraphic surfaces, because of the extended period of time that is available for the erosion and reworking of the underlying highstand deposits. In this case, erosion is linked to fluvial and wind degradation processes, and takes place during the entire stage of base-level fall at the shoreline. The reworking of late highstand fluvial deposits is particularly prone to the development of significant lag deposits, because of the amalgamated nature of fluvial

channels that accumulate under very low accommodation conditions. The concentration of ten 10-cm-thick individual channel lags, for example, may lead to the formation of a 1 m thick lag (placer) deposit. The Zandpan conglomerate that caps the Vaal Reef in the Witwatersrand Basin is one of the many examples of placers associated with subaerial unconformities (Catuneanu and Biddulph, 2001; Fig. 5.43).

Wave scouring within shoreface to inner shelf environments during the same stage of forced regression may lead to the formation of other lag deposits, this time associated with shallow-marine facies. The Upper Vaal reef, up to 0.8 m thick, which is one of the

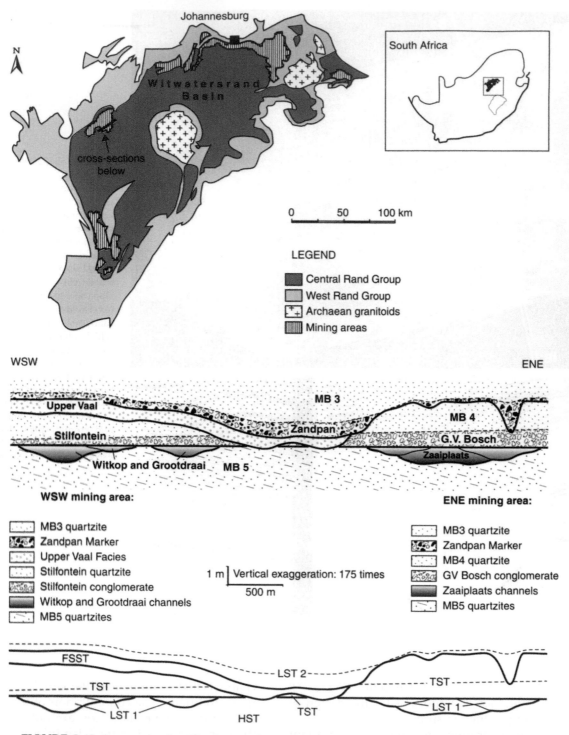

FIGURE 5.43 Cross sectional profile showing the position of gold placers ('reefs': Zandpan, Upper Vaal, B.V. Bosch and basal Stilfontein) in a sequence stratigraphic framework (modified from Catuneanu and Biddulph, 2001). See map for the location of the cross-section. Depositional environments: deltaic (MB 5, Upper Vaal); fluvial (Witkop, Grootdraai, Zaaiplaats, Zandpan, MB 3); and transgressive shallow-marine (Stilfontein, G.V. Bosch, MB 4). Abbreviations: LST—lowstand systems tract; TST—transgressive systems tract; HST—highstand systems tract; FSST—falling-stage systems tract.

important gold placers in the Witwatersrand Basin, has been identified as such, and is associated with a regressive surface of marine erosion (Fig. 5.43).

Facies relationships across these unconformable surfaces are critical to establish their nature and sequence stratigraphic significance (Fig. 4.9), which in turn are important to evaluate the geographic distribution of the associated placers, as well as to predict changes in grades and placer quality along dip. In the case of subaerial unconformities, the level of reworking, and implicitly the textural maturity of the lag deposit, is proportional to the amount of base-level fall and downcutting. During this process, the finer sediment fractions are removed, allowing for a concentration of coarser clasts. As the amount of fluvial incision in response to base-level fall changes along dip, commonly decreasing in an upstream direction, the best reef quality tends to be found adjacent to the paleoshoreline. Such a reef not only loses quality upstream, but it also thins until it eventually disappears beyond the area of influence of base-level changes. Depending on the distance between paleoshoreline and the proximal rim of the basin, such placers may not have a physical expression along the basin margins, and may be missed if exploration is solely based on mapping basin margin unconformities. Similarly, shallow-marine forced regressive placers that overlie regressive surfaces of marine erosion, develop only offshore relative to the paleoshoreline at the onset of base-level fall, and may be missed where exploration is based solely on mapping basin margin unconformities.

The shoreline is therefore a central element in the exploration for placer deposits, because it limits the lateral development of all placer types. The subaerial unconformity-related placers may be found only landward relative to the end-of-fall paleoshoreline, whereas the regressive surface of marine erosion can form only seaward from the onset-of-fall paleoshoreline. Consequently, a successful exploration program must include paleogeographic reconstructions for successive time steps, upon which reliable sequence stratigraphic models may be built.

LOWSTAND SYSTEMS TRACT

Definition and Stacking Patterns

The lowstand systems tract, when defined as restricted to all sedimentary deposits accumulated during the stage of early-rise normal regression (*sensu* Hunt and Tucker, 1992), is bounded by the subaerial unconformity and its marine correlative conformity at the base, and by the maximum regressive surface at the top (Figs. 4.6, 5.4, and 5.5). Where the continental shelf is still partly submerged at the onset of base-level rise, following forced regression, the basal composite boundary of the lowstand systems tract may also include the youngest portion of the regressive surface of marine erosion (Fig. 5.6; also see Fig. 4.23, and the discussion in Chapter 4). The lowstand systems tract forms during the early stage of base-level rise when the rate of rise is outpaced by the sedimentation rate (case of normal regression; Figs. 4.5 and 4.6). Consequently, depositional processes and stacking patterns are dominated by low-rate aggradation and progradation across the entire sedimentary basin. As accommodation is made available by the rising base level, this 'lowstand wedge' is generally expected to include the entire suite of depositional systems, from fluvial to coastal, shallow-marine and deep-marine (Fig. 5.44).

Lowstand deposits typically consist of the coarsest sediment fraction of both nonmarine and shallow-marine sections, i.e., the lower part of a fining-upward profile in nonmarine strata, and the upper part of an upward-coarsening profile in a shallow-marine succession (Fig. 4.6). Sediment mass balance calculations indicate, however, that the grading trends observed within shallow-marine successions do not correlate with the grading trends that characterize the age-equivalent deep-water deposits (Fig. 5.11). Thus, preferential trapping of the coarser sediment fractions within aggrading fluvial and coastal to shallow-marine systems starting at the onset of base-level rise, reduces not only the net amount of sand supplied to the deep-water environment, but also the sand/mud ratio of the sediment load transported by turbidity currents. As a result, the lowstand sediments of the basin-floor submarine fan complex are overall finer-grained relative to the underlying late forced regressive deposits (Fig. 5.5). The maximum grain size of the sediment transported by gravity flows during the lowstand normal regression is also expected to decrease with time, due to the gradual lowering in fluvial slope gradients and related competence following the onset of base-level rise (Fig. 5.11). Consequently, in contrast to the high-density turbidity currents of the late stage of forced regression (Fig. 5.27), the deep-water portion of the lowstand systems tract is dominated by low-density turbidites (Fig. 5.44). The transition from high-density to low-density turbidites at the onset of base-level rise is illustrated in Fig. 5.37. Due to their lower sediment/water ratio, the low-density turbidity currents tend to be underloaded on the continental slope (high energy relative to sediment load), where channel entrenchment rather than aggradation is often recorded (Figs. 5.44 and 5.45). Beyond the toe of the

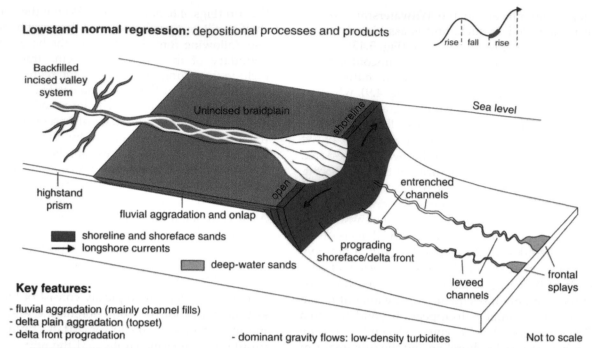

Lowstand normal regression: depositional processes and products

Key features:
- fluvial aggradation (mainly channel fills)
- delta plain aggradation (topset)
- delta front progradation
    - dominant gravity flows: low-density turbidites

FIGURE 5.44 Depositional processes and products of the lowstand systems tract (modified from Catuneanu, 2003). In contrast to the falling-stage systems tract (Figs. 5.26 and 5.27), the sediment of this stage of early-rise normal regression is more evenly distributed between the fluvial, coastal, and deep-water systems. Sand is present in amalgamated fluvial channel fills, beach, and delta front systems, as well as in submarine fans. The 'lowstand prism' gradually expands landward *via* fluvial aggradation and onlap. Aggradation on the continental shelf in fluvial to shallow-marine environments reduces the amount of sediment supply to the deep basin, and hence the turbidity currents of this stage are dominantly of low-density type, being underloaded (entrenched) on the continental slope and aggradational only on the low-gradient basin floor where the energy of the flow drops below the threshold of balance with the sediment load. The top of all early-rise normal regressive deposits is marked by the maximum regressive surface.

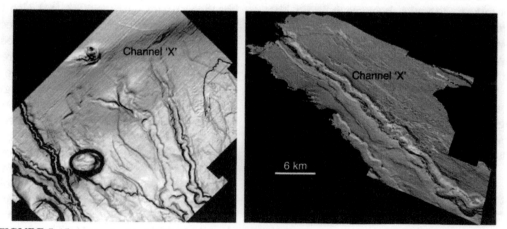

FIGURE 5.45 Entrenched turbidite channels on a continental slope (Late Pleistocene, De Soto Canyon area, Gulf of Mexico; images courtesy of H.W. Posamentier). Channel entrenchment indicates flow energy in excess of sediment load, which is most likely to occur in the case of low-density turbidity currents. Such currents commonly characterize the early stages of base-level rise (lowstand normal regression and early transgression). For scale, channel 'X' is approximately 1.8 km wide. Note that channel entrenchment on the continental slope (lowstand normal regression—early transgression) occurs after the aggradation of late forced regressive leveed channels. Consequently, relict levees may be preserved adjacent to the entrenched channels. These entrenched channels on the continental slope tend to become aggrading leveed channels on the basin floor (see text for details).

continental slope, on the basin floor, the lowstand turbidity currents may become overloaded as energy drops in response to decreasing seafloor gradients. As a result, the basin-floor setting is likely to record aggradation of leveed channels during the stage of lowstand normal regression (Posamentier and Kolla, 2003; Figs. 5.44 and 5.46–5.48). Such leveed channels are expected to develop further into the basin relative to the underlying leveed channels of the falling-stage systems tract because the muddier sediment of the low-density turbidity currents sustains the formation of levees over larger distances (compare the location of frontal splays in Figs. 5.27 and 5.44). More details on the depositional trends recorded in the deep-water environment follow below, as well as in Chapter 6.

Coastal aggradation during lowstand normal regression (i.e., relative increase in coastal elevation; Fig. 5.44) triggers a decrease in slope gradient in the downstream portion of fluvial systems (Fig. 5.6), which induces a lowering with time in fluvial energy and an overall upward decrease in grain size (Figs. 4.6 and 5.11). These decreases in fluvial energy and the grain size of the riverborne sediment that is made available at the shelf edge, explain the decrease in the maximum grain size of the sediment transported by gravity flows to the deep-water environment during the lowstand stage, as discussed above. The increase with time in

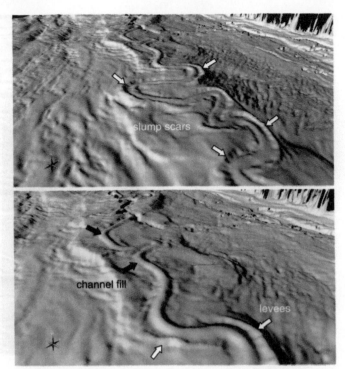

FIGURE 5.47 Depositional elements of a low-density turbidite leveed-channel system on a basin floor (details from Fig. 5.46; images courtesy of H.W. Posamentier). The raised appearance, with a convex-up top, of the channel fill is caused by postdepositional differential compaction. Levees are better developed along the outer channel bends, and their inner margins are characterized by the presence of scoop-shaped slump scars. The relief of the channel belt above the adjacent basin plain is approximately 65 m. The channel fill is approximately 625 m wide.

the rate of base-level rise also contributes to the overall fining-upward fluvial profile, as it creates more accommodation for floodplain deposition and increases the ratio between floodplain and channel sedimentation (Fig. 5.11). Lowstand fluvial deposits typically accumulate on an uneven, immature topography, and contribute to the peneplanation of a nonmarine landscape sculptured by differential erosion during base-level fall (Fig. 5.12). Due to topographic irregularities at the stratigraphic level of the subaerial unconformity, the nonmarine portion of the lowstand systems tract may display a discontinuous geometry, with significant changes in thickness along dip and strike.

Typical examples of lowstand fluvial deposits include amalgamated channel fills (low accommodation systems) overlying subaerial unconformities (Fig. 5.49), which may accumulate within incised valleys (forming the entire valley fill or only the lower portion thereof; Figs. 5.12 and 5.25) or across areas formerly occupied by unincised, bypass falling-stage rivers (Fig. 5.44). The accumulation of lowstand fluvial sediments

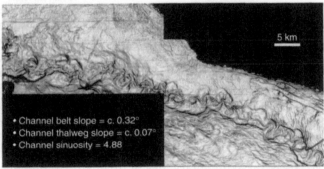

- Channel belt slope = c. 0.32°
- Channel thalweg slope = c. 0.07°
- Channel sinuosity = 4.88

FIGURE 5.46 Aggrading turbidite leveed channel on a basin floor (Late Pleistocene, De Soto Canyon area, Gulf of Mexico; modified from Posamentier, 2003; image courtesy of H.W. Posamentier). Well developed basin-floor leveed channels are typical of low-density turbidity flows, whose mud content is high enough to sustain the formation of levees over large distances. Such basin-floor leveed channels are commonly age-equivalent with entrenched channels on the continental slope (Fig. 5.45). The change in the character of syndepositional processes from erosional on the slope to aggradational on the basin floor relates to changes in slope gradients and associated energy of the flow. Low-density turbidity currents tend to be underloaded on the continental slope (insufficient sediment load relative to energy), but they become overloaded on the basin floor where the energy of the flow decreases significantly. This is in contrast with the high-density turbidity currents, whose larger sediment load allows them to aggrade even on the steeper-gradient continental slope (Figs. 5.39–5.42).

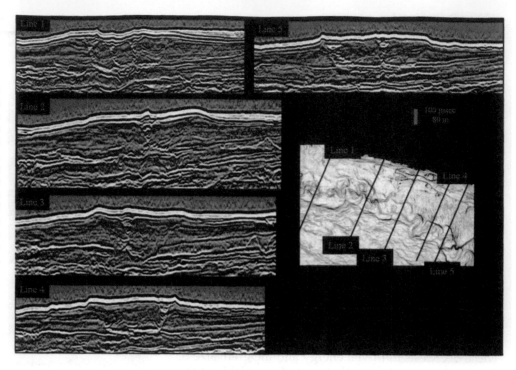

FIGURE 5.48 Transverse sections through the leveed channel in Fig. 5.46 (images courtesy of H.W. Posamentier). The 2D seismic lines indicate channel aggradation, as well as lateral migration with time. Note that the sandy channel fill is characterized by higher amplitude seismic reflections relative to the surrounding finer-grained facies of the overbank environment. Levees are also built by finer-grained material relative to the channel fill. For scale, the channel fill is approximately 625 m wide.

begins within topographic lows, and it is commonly assumed that incised valleys are at least in part filled with such deposits (e.g., models developed by Shanley and McCabe, 1991, 1993, 1994; Wright and Marriott, 1993; Gibling and Bird, 1994). There are also cases, however, where the lowstand fluvial deposits are missing from the stratigraphic architecture of incised-valley fills, due to either nondeposition or erosion during subsequent transgression. In such cases, the fluvially-cut surface at the base of the incised valley is modified into a transgressive surface of erosion, and the incised valley may be entirely filled by transgressive systems tract deposits (e.g., Dalrymple *et al.*, 1992; Ainsworth and Walker, 1994).

Within the lowstand successions of amalgamated channel fills, paleosols may be present, reflecting syndepositional conditions of limited accommodation on floodplains (Fig. 2.17). Such paleosols are often 'wet' and immature, as being formed during stages of base-level rise, and may be associated with poorly developed, if any, coal seams (Fig. 2.17). The poor development of coal seams within the lowstand systems tract is explained by the low rates of creation of accommodation coupled with the generally high influx of clastic sediment.

The inclusion of fluvial deposits under analysis within any conventional systems tract (e.g., 'lowstand' in this case; Fig. 5.49) implies a particular type of *shoreline shift* during the accumulation of fluvial facies

(e.g., lowstand normal regression in this case). The documentation of a process/response relationship between contemporaneous fluvial and marine environments is therefore important to justify the usage of the standard (lowstand – transgressive – highstand) systems tract nomenclature. Figure 5.50 provides examples of sedimentary structures that document the manifestation of marine influences on fluvial processes at the time of sedimentation. In the absence of such evidence, where no relationship can be established between fluvial processes and any shifts of a coeval shoreline, the concepts of low- and high-accommodation systems tracts may provide a more realistic approach to describing fluvial deposits in a sequence stratigraphic framework. This is generally the case in overfilled basins or in areas of sedimentary basins that are beyond the influence of marine base-level changes (e.g., zone 3 in Fig. 3.3). The low- and high-accommodation systems tracts are discussed in more detailed in a subsequent section of this chapter.

Lowstand fluvial strata are commonly depicted as the product of sedimentation within high-energy rivers of braided type, due to the steepening of the landscape gradient that is generally expected during stages of base-level fall. While this may often be the case (e.g., Figs. 4.32 and 4.38), one should not exclude the possibility of meandering lowstand systems, which are particularly prone to develop within incised meander belts (Figs. 5.12 and 5.19). Lowstand rivers of

FIGURE 5.49 Outcrop expression of lowstand fluvial systems (Castlegate Formation, Utah). A—amalgamated channel fills (upper part of the photograph, lighter color: Castlegate Formation) separated from the underlying forced regressive shoreface deposits (lower part of the photograph, reddish color: Blackhawk Formation) by the subaerial unconformity; B, C, D, E—amalgamated channel fills of braided fluvial systems; F—climbing dunes indicating high sediment supply in the high-energy braided streams. Abbreviations: FSST—falling-stage systems tract; LST—lowstand systems tract; SU—subaerial unconformity.

A

B

FIGURE 5.50 Tidal influences in lowstand fluvial systems, indicating proximity to the coeval shoreline (Castlegate Formation, Utah). A—mud drapes associated with cross-bedding; B—sigmoidal bedding, mud drapes, and worm burrows. Evidence of marine influences suggests that fluvial processes are at least in part controlled by marine base-level changes; therefore, the fluvial deposits under analysis accumulated in zone 2 in Fig. 3.3. The process/response relationship between contemporaneous fluvial and marine environments justifies the usage of the standard systems tract nomenclature ('lowstand' in this case), which makes direct reference to syndepositional shoreline shifts. If fluvial systems accumulate beyond the reach of marine influences (e.g., in overfilled basins, or within zone 3 in Fig. 3.3), the usage of the lowstand—transgressive—highstand systems tract nomenclature becomes redundant, and the concepts of low- vs. high-accommodation systems tracts are more appropriate (see discussion below on low- and high-accommodation systems tracts).

braided type are most likely to establish themselves in areas formerly occupied by bypass falling-stage systems, where the channel pattern is not constrained by incised valleys that tend to preserve the plan-view morphology of low-energy highstand rivers. In such cases, where the flow is not constrained by an erosional landscape, rivers may easily adjust their channel pattern (e.g., from meandering to braided) to reflect the change in energy levels from highstand to lowstand conditions. In contrast, incised valleys tend to retain the meander pattern inherited from highstand rivers, and impose that pattern on the younger and higher energy lowstand systems (Fig. 5.20; see also discussion in the section dealing with the falling-stage systems tract).

Following the stage of base-level fall, when most of the shelf becomes subaerially exposed, the lowstand systems tract may include shelf-edge deltas with diagnostic topset geometries (Fig. 5.4). Triggered by the rise in base level at the shoreline, the aggradation of lowstand fluvial strata starts from the delta plain area and gradually extends upstream by onlapping the subaerial unconformity (Figs. 5.4 and 5.5). This trend of fluvial onlap widens the stratigraphic hiatus that is associated with the subaerial unconformity in a landward direction, as the age of the overlying fluvial strata is increasingly younger upstream (Fig. 5.5). This scenario is valid under the assumption that rivers flow along graded profiles upstream of the point of fluvial onlap, which allows for a simple modelling that only takes into account the downstream controls (i.e., base-level changes) on fluvial processes. Under this assumption, the lowstand fluvial deposits that accumulate due to base-level rise form a wedge-shaped prism that thins upstream towards the youngest point of fluvial onlap (Fig. 5.4). The distance along dip that is subject to fluvial onlap depends on several factors, including the duration of the lowstand stage, the amount of sediment supply and the rates of coastal aggradation, and the topographic gradients of the land surface. A flat topography (e.g., in a low-gradient shelf-type setting) coupled with high sediment supply are conducive to fluvial aggradation over a large area, whereas a steep topography (e.g., in a high-gradient ramp setting, such as a continental slope or a fault-bounded basin margin) coupled with low sediment supply restrict the size of the area that is subject to fluvial aggradation (Blum and Tornqvist, 2000). In the latter case, the subaerial unconformity may be overlain directly by transgressive fluvial strata over much of its extent (Embry, 1995; Dalrymple, 1999). Data from the Gulf Coastal Plain of the U.S.A. indicate that the landward limit of fluvial onlap correlates to the amount of sediment supply, and is inversely proportional to the gradient of the onlapped floodplain surface. This distance may vary significantly, from approximately 40 km in the case of the steep-gradient, low-sediment-supply Nueces River to at least 300–400 km for the low-gradient, high-sediment-supply Mississippi River (Blum and Tornqvist, 2000).

The preservation potential of coastal and adjacent lowstand fluvial strata may be low due to subsequent transgressive ravinement erosion (Fig. 5.6). Figure 4.53 illustrates an example where no lowstand fluvial strata are preserved on top of a former subaerial unconformity, which was subsequently reworked by a transgressive tidal-ravinement surface. Where central estuary facies are preserved, they may act as a buffer that protects the underlying fluvial strata from subsequent transgressive scouring. In such cases, fluvial lowstand deposits are likely to be preserved between the subaerial unconformity and the earliest estuarine strata whose timing marks the onset of transgression (Fig. 4.32). The contact between lowstand fluvial and the overlying estuarine facies is the maximum regressive surface (Fig. 5.6). This stratigraphic contact tends to be sharp, because of the rapid development of the estuarine system as soon as the shoreline starts its landward shift (Figs. 4.32 and 4.39). This sequence stratigraphic surface should not be confused with the facies contact between transgressive fluvial facies and the overlying estuarine strata, which develops *within* the transgressive systems tract (Fig. 5.6). The latter facies shift is gradational, with significant interfingering between tidally-influenced fluvial and estuarine deposits, and is highly diachronous with the rate of shoreline transgression. The differentiation between lowstand fluvial and transgressive fluvial facies based solely on well logs may be difficult, unless a more regional understanding of the study area is achieved. Such a distinction is greatly facilitated where core and/or outcrop data are available to allow the observation of sedimentary structures associated with tidal influences, which are much more abundant within the downstream reaches of transgressive fluvial systems.

Seaward from the coastline, the shoreface deposits of the lowstand systems tract are generally gradationally based, excepting for the earliest lowstand shoreface strata which may be sharp-based, as they overlie the regressive surface of marine erosion (Fig. 5.6). Beyond the fairweather wave base, the extent of shelf facies may be limited due to the potential proximity of the shoreline to the shelf edge at the end of forced regression (Fig. 5.4). In this case, the subtidal facies may pass directly into deep-water slope facies, which consist primarily of gravity-flow deposits (Figs. 4.15 and 5.4).

In contrast to the trends discussed in the case of highstand normal regression, the seaward shift of the lowstand shoreline *decelerates* during the lowstand stage because the rates of base-level rise increase with time, from zero until the turnaround from regression to transgression is achieved (Fig. 3.19). As a result, increasingly more sediment is required to fill the newly created accommodation at the shoreline, and so the lowstand normal regressive deltaic lobes ('parasequences') become thicker with time and in an offshore direction (Fig. 5.13). If a portion of the continental shelf is still submerged at the end of forced regression, and as accommodation is limited during early lowstand, the oldest coastal to subtidal sandstones of the lowstand systems tract tend to have a wider geographic distribution across the shelf due to rapid autocyclic shifting imposed upon deltaic lobes (Fig. 5.13). Such shallow-marine reservoirs are expected to have a better connectivity relative to their late lowstand counterparts, which display a more pronounced aggradational component (i.e., vertical rather than lateral stacking). In parallel to these trends of change in the thickness and stacking patterns of the deltaic lobes that accumulate during lowstand normal regression, the average grain size of successive delta lobes is also expected to decrease in a seaward direction in response to the lowering with time in fluvial energy during base-level rise (Fig. 5.13). The latter trend of change in average grain size along dip parallels the one observed in the case of highstand deltas, and is the opposite of what characterizes forced regressive deltas (Fig. 5.13). Irrespective of these trends of change in grain size from older to younger deltaic lobes along dip, vertical profiles in any particular location show invariably a reverse grading (coarsening-upward) due to the progradation of delta front facies over finer prodelta sediments (Fig. 5.11).

Economic Potential

Petroleum Plays

Rising base level during the lowstand normal regression provides accommodation across the entire basin, from fluvial to marine environments. Sediment budget observations indicate a concentration of the coarsest riverborne sediment within fluvial and coastal depositional systems, which arguably form the best reservoirs, with the highest sand/mud ratio, of the lowstand systems tract. The trapping of sand within aggrading fluvial to shallow-marine systems following the onset of base-level rise results in a net decrease in the volume of sediment available for deep-water gravity flows, and also in a lowering of the sand/mud ratio in submarine fans (Figs. 4.18 and 5.11). Shelf-edge deltas and correlative strandplains continue to prograde the upper slope, with the development of a topset in response to coastal aggradation (Figs. 3.22 and 5.4). Increased elevation at the shoreline triggers fluvial aggradation, starting from the shoreline and gradually expanding upstream (fluvial onlap), which explains the wedging out of lowstand fluvial reservoirs toward the basin margins (Figs. 5.4 and 5.44).

The petroleum plays of the lowstand systems tract are therefore diverse in terms of origin and syndepositional processes, ranging from fluvial to coastal, shallow- and deep-marine systems. In this way, the entire dip profile of the basin offers exploration opportunities within this systems tract. A key for recognizing the 'lowstand wedge' on seismic lines is the presence of a topset (rather than offlap and truncation, typical of forced regressions) associated with the shelf-edge deltas and correlative strandplains (Figs. 3.22 and 5.44). Landward from the shelf edge, fluvial reservoirs are mainly represented by amalgamated channel fills, due to the low amounts of accommodation available during this stage. These are the best fluvial reservoirs of the entire base-level cycle, with the highest sand/mud ratio. The shelf-edge deltas that prograde the upper continental slope, and their open shoreline beach and shoreface correlative systems, also trap a significant amount of sand and may form good reservoirs that are laterally extensive along strike (Fig. 5.44). These normal regressive shelf-edge reservoirs are often topped by high amplitude reflectors on seismic lines, marking the change in acoustic impedance from the transgressive and highstand shales above to the underlying lowstand sand-rich reservoirs (Fig. 3.22). Beyond the shelf-edge deltas and their correlative strandplains, riverborne sediment continues to be delivered to the deeper basin but in decreasing amounts and with a decreasing sand/mud ratio in response to the increasing rates of base-level rise at the shoreline (Fig. 5.11). As more and more sand is trapped in aggrading fluvial and coastal systems, the submarine fan receives less and less sand relative to the pelagic fallout, which generates the overall fining-upward trend noted for the slope fans in Fig. 4.18 above the correlative conformity. This fining-upward trend is also shown in Figs. 5.5 and 5.11.

The significance of the lowstand systems tract, and the various portions thereof, in terms of sediment budget, potential petroleum reservoirs, and petroleum source and seal rocks is summarized in Fig. 5.14. As suggested in this table, the lowstand systems tract tends to be the most balanced among all systems tracts in terms of the relatively even distribution of reservoirs across the basin. The lowstand fluvial deposits (amalgamated channel fills), where preserved from subsequent transgressive scouring, form the best reservoirs of the entire fluvial portion of a stratigraphic sequence. Equally good reservoirs may form in coastal, shallow-water and deep-water environments during the lowstand normal regression of the shoreline (Fig. 5.14). The change in the type of dominant gravity flows that manifest during the lowstand time, from high-density turbidity currents at the end of base-level fall/onset of base-level rise to low-density turbidity currents, has important consequences for the lithology, morphology and location of deep-water reservoirs within the basin, as discussed above (Figs. 5.11,

5.27, 5.37, and 5.44). These issues are explored further in Chapter 6.

The main risks for the exploration of lowstand reservoirs relate to charge, seals, and source rocks, especially toward the basin margins. Even within a shelf setting, however, fluvial, coastal, and shallow-water lowstand reservoirs may be sealed by overlying transgressive shales, which may be fluvial, estuarine or shallow-water in nature (Fig. 5.14). The risks of exploration of lowstand reservoirs decrease toward the deep portion of the basin, as lowstand turbidites, which travel farther into the basin relative to the falling-stage gravity flows, stand a good chance of being in direct contact with transgressive/highstand source and seal facies both below and above.

Coal Resources

The lowstand systems tract is defined by high sediment supply in an overall low accommodation setting, and therefore environmental conditions are generally unfavourable for peat accumulation (Fig. 5.15). The architecture of lowstand fluvial systems is commonly described by amalgamated channel fills, partly because of the lack of sufficient accommodation, and partly due to the tendency of lowstand rivers to be of higher energy following the steepening of the topographic profile as a result of tilt and/or differential erosion during the stage of base-level fall (e.g., Catuneanu, 2004a; Figs. 4.32 and 4.38). The limited amount of fluvial accommodation affects particularly the overbank environment, which may also be subject to scouring by laterally shifting fluvial channels, and therefore no significant coal deposits are generally associated with the lowstand systems tract. As the rates of base-level rise increase with time during the lowstand stage, gradually more accommodation becomes available to the overbank environment, and so chances of peat accumulation and subsequent coal development tend to improve toward the top of the lowstand systems tract (Fig. 5.15).

Placer Deposits

No unconformities form *during* the lowstand normal regression, but the lowstand systems tract is closely associated with all three types of unconformity-related placer deposits. The subaerial unconformity and the youngest portion of the regressive surface of marine erosion are found at the base of the lowstand systems tract, whereas the oldest portion of the transgressive ravinement surface commonly truncates the top of lowstand deposits (Fig. 5.6). The first two unconformity-related placer types are described in more detail in the section dealing with the falling-stage systems tract; the placers associated with transgressive ravinement surfaces are discussed in the following section.

Following the end of base-level fall, the accumulation of coarse sediments may continue in amalgamated channels during the early stage of lowstand normal regression. These lowstand deposits are particularly prone to 'reef' facies development in the case of gravel-bed fluvial systems. Such 'depositional' reefs (as opposed to unconformity-related reefs) involve only limited reworking of the underlying sediments, and so they may be of economic significance especially where the mineralization is the result of precipitation from hydrothermal fluids. Where conditions are favorable for the aggradation of coarse-grained clasts following the onset of base-level rise, the depositional lowstand reefs add to the thickness of the underlying placers represented by lag deposits associated with subaerial unconformities or regressive surfaces of marine erosion.

TRANSGRESSIVE SYSTEMS TRACT

Definition and Stacking Patterns

The transgressive systems tract is bounded by the maximum regressive surface at the base, and by the maximum flooding surface at the top. This systems tract forms during the stage of base-level rise when the rates of rise outpace the sedimentation rates at the shoreline (Figs. 3.19, 4.5, and 4.6). It can be recognized from the diagnostic retrogradational stacking patterns, which result in overall fining-upward profiles within both marine and nonmarine successions (Figs. 4.6, 5.5, and 5.11). As the rates of creation of accommodation are highest during shoreline transgression (Fig. 4.5), the transgressive systems tract is commonly expected to include the entire range of depositional systems along the dip of a sedimentary basin, from fluvial to coastal, shallow-marine and deep-marine (Fig. 5.4).

The transgressive fluvial and coastal deposits may potentially be thick, due to the high sedimentation rates stimulated by the available accommodation, although exceptions do occur under particular circumstances (Fig. 3.20; see also discussion below). The trapping of large amounts of terrigenous sediment within aggrading fluvial and coastal systems during transgression results in a cut-off of sediment supply to the marine environment (Loutit *et al.*, 1988). As a consequence, transgressive shallow-marine deposits accumulate primarily in areas adjacent to the shoreline, with correlative condensed sections or even unconformities in the more distal portions of the shelf (Galloway, 1989). Triggered by the lack of sediment supply and a regime of hydraulic instability during

rapid base-level rise, the shelf edge region is generally subject to non-deposition and/or sediment reworking during transgression (Fig. 5.4). As a result, the transgressive systems tract tends to be composed of two distinct wedges separated by an area of non-deposition around the shelf edge, one on the continental shelf consisting of fluvial to shallow-marine deposits, and one in the deep-water setting consisting of gravity-flow deposits and pelagic sediments (Figs. 4.41 and 5.4). Both these wedges shift toward the basin margin during transgression, following the general retrogradational trend, by onlapping the landscape and the seascape, respectively, in a landward direction (Fig. 5.5). The gradual expansion of the transgressive depozone in a continental shelf-type setting is associated with *fluvial onlap* (leading edge of the transgressive wedge). Within the deep-water portion of the basin, the transgressive deposits are often seen onlapping the continental slope, forming a transgressive slope apron associated with *marine onlap* (Galloway, 1989; Figs. 4.2, 4.3, 4.41, 5.4, and 5.5). In addition to the fluvial and marine onlaps that characterize the leading edges of the two transgressive wedges, *coastal onlap* is also an important type of stratal termination, diagnostic for transgression, forming within the continental shelf-based transgressive wedge by the shift of shoreface facies on top of the landward-expanding wave-ravinement surface (Figs. 4.2, 4.3, 5.4, and 5.5).

The fluvial portion of the transgressive systems tract commonly shows evidence of tidal influences (Shanley *et al.*, 1992; Shanley and McCabe, 1993), and is characterized by an overall fining-upward vertical profile (Fig. 4.6). This overall grading trend reflects both an upward decrease in maximum grain size caused by a decline with time in the competence of the rivers, and also a lowering of the sand/mud ratio (channel *vs.* overbank sedimentation) in response to accelerating base-level rise following the lowstand normal regression (Fig. 5.11). The latter feature of the vertical profile also translates into an upward decrease in the degree of amalgamation of transgressive channel-fill sandstones, which are often described as isolated ribbons engulfed within floodplain fines (Shanley and McCabe, 1993; Wright and Marriott, 1993). The decline with time in the energy of transgressive fluvial systems parallels a corresponding decrease in topographic gradients, which in turn is triggered by coastal aggradation coupled with the general pattern of fluvial sedimentation during base-level rise. As in the case of lowstand and highstand normal regressions, the sedimentation of fluvial deposits during transgression *in response to base-level rise* starts from the downstream reaches of rivers, where the fluvial succession is thickest, gradually expanding upstream (Fig. 5.5). This pattern of fluvial onlap explains the wedge-shaped geometry of the fluvial transgressive package, which thins landward from the shoreline, leading to the observed decrease in topographic gradients and fluvial energy during transgression (Fig. 5.4). Following this style of fluvial sedimentation established at the onset of base-level rise, the transgressive fluvial deposits often extend farther toward the basin margins relative to the underlying lowstand fluvial strata, by onlapping the subaerial unconformity (see fluvial onlap in Figs. 5.4 and 5.5). Such predictable trends could, however, be altered if fluvial processes are influenced by controls other than base-level changes, notably by climate and/or source area tectonism. As accommodation is generated rapidly during transgression, and the water table rises in parallel with the base level, the fluvial portion of the transgressive systems tract often includes well developed coal seams (Fig. 4.42).

The transgressive fluvial deposits may form a significant portion of incised-valley fills, or may aggrade in the interfluve areas of former incised valleys. Where incised valleys are inherited from previous stages of base-level fall and are not entirely filled by lowstand deposits, their downstream portions are commonly converted into estuaries at the onset of transgression (Dalrymple et al., 1994). In such cases, the lowstand fluvial deposits that overlie the subaerial unconformity may be scoured, or partly reworked, by estuarine channels and tidal-ravinement surfaces (Rahmani, 1988; Allen and Posamentier, 1993; Ainsworth and Walker, 1994; Breyer, 1995; Rossetti, 1998; Cotter and Driese, 1998). Where not reworked by the tidal-ravinement surface, the contact between lowstand fluvial and the earliest (stratigraphically lowest) overlying estuarine facies is represented by the maximum regressive surface. In this setting, the maximum regressive surface is relatively easy to map in outcrop or core, at the abrupt change from coarse fluvial sand and gravel (lowstand deposits) to the overlying estuarine facies comprising finer-grained and more varied lithologies with abundant tidal structures such as clay drapes and flasers (see Allen and Posamentier, 1993, for the case study of the Holocene Gironde incised valley in southwestern France; Fig. 4.52). This contrast between lowstand fluvial and overlying transgressive estuarine facies may also be strong enough to be seen in well logs, at the contact between 'clean' and blocky sand and the younger, more interbedded and finer-grained lithologies (Fig. 4.32).

In coastal settings, the transgressive systems tract may include backstepping foreshore (beach) deposits, diagnostic estuarine facies (particularly in the case of smaller rivers), and even proper deltas in the case of large rivers (Figs. 5.51 and 5.52). The formation and preservation of transgressive coastal deposits depends on the

rates of base-level rise, sediment supply, the wind regime and the amount of associated wave-ravinement erosion, and the topographic gradients at the shoreline. Coastal aggradation is favoured by high rates of base-level rise, weak transgressive ravinement erosion, and shallow topographic gradients (e.g., in low-gradient shelf-type settings; Fig. 5.6). Steeper topographic gradients (e.g., in high-gradient ramp settings) tend to induce coastal erosion in relation to a combination of factors including higher fluvial energy, wave ravinement, and slope instability (Fig. 3.20). This may explain the common lack of estuarine facies in fault-bounded basins, but also in areas characterized by extreme wind energy and associated strong wave-ravinement erosion (Leckie, 1994).

In the case of erosional coastlines, where transgressive coastal facies are not preserved in the rock record, transgressive fluvial deposits are likely to be missing as well (Fig. 3.20). In this case, the coastal to nonmarine portion of the transgressive systems tract is replaced by a subaerial unconformity with an associated hiatus that is age-equivalent with the marine transgressive deposits. A modern analog is represented by the incised estuaries and fluvial systems of the Canterbury Plains, New Zealand, where the transgressive coastline is dominated by erosional processes (Figs. 3.24–3.26).

In the case of aggrading coastlines (Figs. 3.20 and 5.6), both coastal and fluvial deposits have a high preservation potential. The character of the coastline may change along strike from transgressive to normal regressive as a function of the shifting balance between the rates of base-level rise and the rates of sedimentation in *open shoreline* settings (Fig. 5.52). As such, prograding strandplains are typical of normal regressive coastlines, whereas backstepping beaches define transgressive coastlines (Fig. 5.52). The boundary between coeval transgressive and normal regressive coastlines in Fig. 5.52 may either be constrained by spatial variations in sedimentation rates or by strike variability in subsidence rates, or both. The mechanisms controlling the change in depositional trends along a coastline have been investigated by Wehr (1993), who noted that 'spatial variations in sedimentation rates … might locally shift the onset of progradation to an earlier time and delay the onset of retrogradation.' These issues were further tackled by Martinsen and Helland-Hansen (1995), Helland-Hansen and Martinsen (1996) and Catuneanu et al. (1998b), who summarized the various types of shoreline trajectories that may develop in response to the strike variability in subsidence and sedimentation.

As depicted in Fig. 5.52, the defining element that is common among all types of transgressive coastlines is

River mouth environments		Conditions	
		River mouth	Open shoreline
Deltas	Prograding deltas *(large rivers, high sediment supply)*	Sedimentation > Base-level rise	Base-level rise > Sedimentation
Estuaries	Retrograding ("bayhead") deltas/ Incomplete (drowned) estuaries *(smaller rivers, unincised channels)*	Base-level rise > Sedimentation Drowning of river	Base-level rise > Sedimentation < Transgression of open shoreline
	Complete estuaries *(smaller rivers, incised valleys)*	Base-level rise > Sedimentation Drowning of river	Base-level rise > Sedimentation > Transgression of open shoreline

FIGURE 5.51 River-mouth environments of the transgressive systems tract. Estuaries are commonly regarded as river-mouth environments diagnostic for transgression, as being associated with a retrogradational shift of facies and forming only during the landward shift of the shoreline. While this is true, a wider array of river-mouth environments, ranging from estuaries to proper deltas, may be part of a larger-scale transgressive systems tract as a function of the balance between the rates of sedimentation and the rates of base-level rise at the river mouth. The two end members of this array include the estuaries, usually in the case of smaller rivers (the rates of base-level rise outpace the sedimentation rates at the river mouth), and the prograding (proper) deltas in the case of large rivers (the rates of sedimentation outpace the rates of base-level rise at the river mouth). Note that in the latter case, deltas may only be considered as part of a transgressive systems tract where the adjacent open shoreline is transgressive; otherwise, we deal with normal regressive (lowstand or highstand) deltas (Fig. 5.52). Between proper (prograding) deltas and fully developed estuaries, retrograding ('bayhead') deltas may also develop where estuaries are partly drowned by the rapid transgression of the adjacent open shorelines. This situation is prone to occur in the case of unincised channels, where the transgression of the open shoreline tends to be faster than the rate of drowning of the river (the river's sediment supply is higher than the amounts of sediment available for the construction of backstepping beaches). Such bayhead deltas represent only one of the several sub-environments of a typical (complete, or fully developed) estuary, and hence are marked in the diagram as 'incomplete' estuaries. Complete estuaries tend to form at the mouth of incised valleys, which facilitate the drowning of rivers at rates that are higher relative to the rates of transgression of the adjacent open shorelines.

the retrogradational character of open shorelines. Within this overall transgressive setting, river-mouth environments may show a range of depositional trends, from prograding deltas to retrograding and fully developed estuaries, as a function of the balance between accommodation and riverborne sediment supply. At one end of the spectrum, large rivers with high sediment load may prograde into the basin in spite of the transgression recorded by the adjacent open shorelines (case B in Fig. 5.52; Figs. 5.53 and 5.54). In such cases, the rates of base-level rise are higher than the rates of aggradation of backstepping beaches, but lower than the sedimentation rates at the river mouth. At the other end of the spectrum, smaller rivers that do not supply enough sediment to fill the entire accommodation created by base-level rise are converted into estuaries characterized by a retrogradational shift of facies. In such cases, the rates of base-level rise

outpace the rates of aggradation both within the estuaries and along the adjacent open shorelines. Within this context of retrogradational river-mouth environments, two situations may be envisaged (Figs. 5.51 and 5.52). Firstly, where the transgression of the open shoreline lags (is slower than) the transgression of the river, a fully developed estuary will form, which is commonly the case with incised valleys (Dalrymple et al., 1994; case D in Fig. 5.52). Secondly, where the transgression of the open shoreline is faster than the transgression of the river, which is generally the case with unincised channels, the estuary is drowned (i.e., incompletely developed) and represented only by its retrograding bayhead delta subenvironment (Fig. 5.51; case C in Fig. 5.52). The formation of bayhead deltas is favoured in wave-dominated estuaries, and it is unlikely in tide-dominated settings (Figs. 4.46 and 4.47; Allen, 1991; Reinson, 1992; Dalrymple et al., 1992; Allen

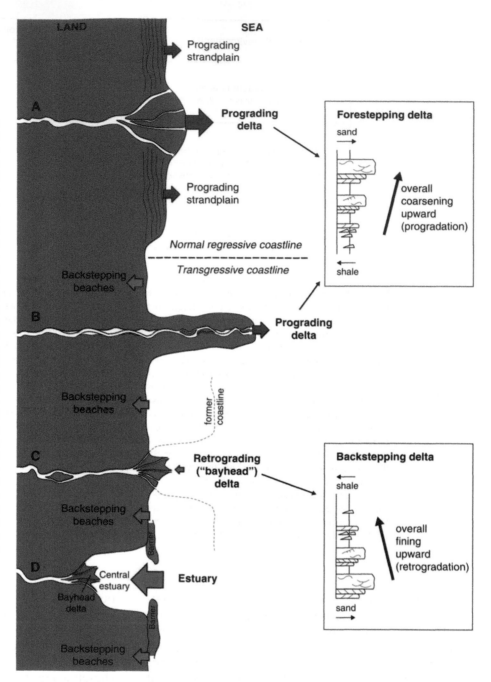

FIGURE 5.52 Types of coastlines that may develop during base-level rise. Not to scale. The progradational or retrogradational character of the shoreline in both river-mouth and open shoreline settings is dictated by the balance between sedimentation rates and the rates of base-level rise. Normal regressive coastlines (lowstand, highstand) are defined by progradation of both river-mouth and adjacent open shoreline settings. Transgressive coastlines are defined by retrogradation of the open shoreline, while the river mouth could be progradational ('proper' deltas) or retrogradational (estuaries, or portions thereof/bayhead deltas) (Fig. 5.51). Note that the definition of deltas and estuaries is based on stratigraphic criteria (progradational *vs.* retrogradational depositional trends, respectively), irrespective of the mechanisms of sediment redistribution adjacent to the coastline (i.e., river-, wave- or tide-dominated settings). A—normal regressive delta (for an example of a prograding and aggrading strandplain in an open shoreline setting, see Fig. 3.22); B— 'proper' (progradational/forestepping) delta in a transgressive setting (see Fig. 5.53 for a modern analogue); C—retrogradational ('bayhead') delta in a transgressive setting (see Fig. 5.51 for the conditions that may lead to the formation of retrograding/backstepping deltas); D—fully developed estuary.

and Posamentier, 1993; Zaitlin *et al.*, 1994; Shanmugam *et al.*, 2000).

The overall vertical profiles of prograding (forestepping) and retrograding (backstepping) deltas of the transgressive systems tract are presented in Fig. 5.52. Note that in both cases, the overall coarsening- and fining-upward profiles, respectively, consist of higher-frequency coarsening-upward successions reflecting short-term progradation of riverborne sediments in each of the two types of transgressive river-mouth settings.

In the longer term, however, the overall trend reflects the dominant direction of facies shift. This overall vertical profile is punctuated by high-frequency flooding events which terminate the deposition of each individual short-term coarsening-upward succession.

The accumulation of shallow-marine facies in a transgressive setting is governed by a set of first principles, including: sediment supply to the shallow-marine environment is limited during shoreline transgression, as most riverborne sediment is trapped

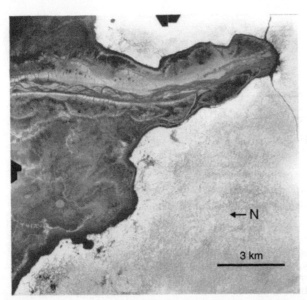

FIGURE 5.53 Aerial photograph showing a river-dominated, prograding delta in an overall transgressive setting (case B in Fig. 5.52; Chads Point, western Sabine Peninsula, Melville Island, Canadian Archipelago; photograph courtesy of J. England). The high sediment supply of the river causes the delta to prograde in spite of the transgression recorded by the adjacent open shorelines. The rate of post-glacial base-level rise is higher than the rate of aggradation of backstepping beaches along the open shoreline, but it is lower than the sedimentation rates at the river mouth.

within rapidly aggrading fluvial and coastal systems (*caveat*: see the case of coastal erosion in Fig. 3.20); an additional source of sediment for the shallow-marine environment is provided by processes of wave erosion in the upper shoreface during transgression; these sediments are transported landward during fairweather to form backstepping beaches or estuary-mouth complexes, and are dispersed seaward on the shelf by storm surges and tidal currents to form sheet-, ridge-, or wedge-shaped deposits; and sedimentation in a transgressive marine environment tends to 'heal' the seascape profile by smoothing out the breaks in seafloor gradients. The latter process leads to the formation of '*healing-phase wedges*' in the low areas of the seafloor in an attempt to re-establish a graded seafloor profile (Posamentier and Allen, 1993). Healing-phase wedges may form in various settings of the transgressive marine environment, each involving similar depositional processes but different scales of observation, from shoreface (wedge thickness in a range of meters; Fig. 3.20), to shelf (wedge thickness up to tens of meters, filling the low area outboard of the youngest regressive clinoform; Figs. 3.21 and 5.55) and even the deep-water environment (wedge thickness up to hundreds of meters, smoothing out the difference in gradients between the continental slope and the basin floor; Figs. 3.22, 5.56, and 5.57). In addition to healing-phase wedges, transgressive shallow-marine

FIGURE 5.54 Satellite image showing a river-dominated, prograding delta in an overall transgressive setting (case B in Fig. 5.52; Mississippi delta, Louisiana; image released by the U.S. Geological Survey National Wetlands Research Center). The high sediment supply of the river causes the delta to prograde in spite of the transgression recorded by the adjacent open shorelines.

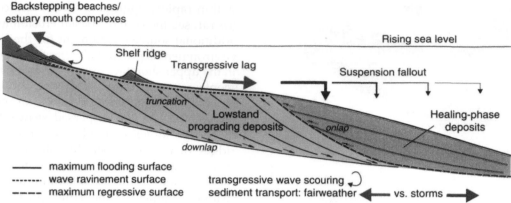

FIGURE 5.55 Coastal to shallow-marine deposits of the transgressive systems tract (not to scale; modified from Posamentier and Allen, 1993, 1999). At the scale of the continental shelf, discernable transgressive deposits include backstepping beaches (open shoreline settings) and estuary-mouth complexes (retrograding river-mouth settings), transgressive lag deposits overlying the wave-ravinement surface that forms in the upper shoreface, sand sheets and/or sand ridges that form within inner shelf environments in relation to storm surges and tidal currents, and more distal healing-phase wedges that fill the low areas of the seafloor outboard of the last regressive clinoform. At a smaller scale, healing-phase wedges may also form in the transgressive lower shoreface environment, smoothing out the slope break created by wave scouring in the upper shoreface (not shown in this diagram; Fig. 3.20). Note that the wave-ravinement surface invariably removes the seaward termination of the nonmarine portion of the maximum regressive surface. Assuming that the rates of fluvial and coastal aggradation during transgression are higher than the amount of transgressive wave scouring in the upper shoreface, the maximum regressive surface and the wave-ravinement surface diverge in a landward direction. The healing-phase wedge shown in this diagram is primarily composed of relatively fine-grained sediment accumulated from suspension. As the fallout rate decreases with distance in a seaward direction, the overall geometry of bedding surfaces changes from concave-up (shape that is inherited from the youngest regressive clinoform) to flat and eventually convex-up.

deposits may also include *transgressive lags* (Figs. 3.30B and 4.50), and *shelf-sand deposits* with a sheet-like or ridge-like geometry (Fig. 5.55). Under restricted detrital supply conditions, the shallow-marine portion of the transgressive systems tract may also be represented by *carbonate condensed sections* (Fig. 4.42). The overall thickness of the shallow-water portion of the transgressive systems tract decreases toward the shelf edge, where transgressive deposits are commonly missing (Galloway, 1989; Figs. 4.41 and 5.4).

The process of wave scouring in the upper shoreface in response to the landward translation of the shoreface profile during shoreline transgression is a key to understand the stratigraphy and the sediment budget of the shallow-marine portion of the transgressive systems tract. The balance between the processes of erosion and sedimentation that are caused by this shift of the shoreface profile (see the lever point between shoreface erosion and shoreface sedimentation in Fig. 3.20) was first pointed out by Bruun (1962), and subsequently incorporated into the sequence stratigraphic theory by Dominguez and Wanless (1991), Posamentier and Chamberlain (1993), and others. The seaward transport of the sediments generated by wave

erosion in the upper shoreface onto the shelf is primarily attributed to storm surges (dispersive sediment transport, resulting in the formation of *shelf sand sheets*) and tidal currents (more channelized style of sediment transport, resulting in the formation of *shelf sand ridges*) (Curray, 1964; Swift, 1968, 1976; Swift and Field, 1981; Belknap and Kraft, 1981; Demarest and Kraft, 1987; Kraft *et al.*, 1987; Rine *et al.*, 1991; Snedden *et al.*, 1994; Snedden and Kreisa, 1995; Abbott, 1998; Snedden and Dalrymple, 1999; Posamentier, 2002). During the process of sediment transport, the coarsest clasts produced by wave scouring or otherwise available within the shoreface environment are left behind as a *transgressive lag* on top of the wave-ravinement surface (Swift, 1976; Figs. 3.30B, 4.50, and 4.51). The finer sediment, which includes most of the sand-size fraction of clasts, is transported farther onto the shelf to form sheet-like and ridge-like *shelf-sand deposits*. *Shelf sand sheets* have been documented in numerous studies (e.g., Swift, 1968; Swift and Field, 1981; Belknap and Kraft, 1981; Demarest and Kraft, 1987; Kraft *et al.*, 1987; Masterson and Paris, 1987; Masterson and Eggert, 1992; Helland-Hansen *et al.*, 1992; Eschard *et al.*, 1993; Abbott, 1998), and are known to form relatively thin (1–3 m thick) but

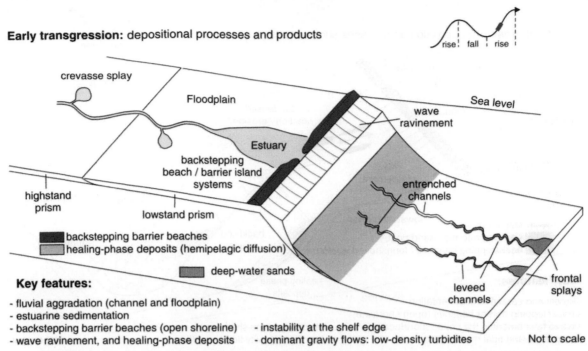

Early transgression: depositional processes and products

rise | fall | rise

crevasse splay

Floodplain

Sea level

wave ravinement

Estuary

backstepping beach / barrier island systems

entrenched channels

highstand prism

lowstand prism

backstepping barrier beaches

healing-phase deposits (hemipelagic diffusion)

deep-water sands

leveed channels

frontal splays

Key features:

- fluvial aggradation (channel and floodplain)
- estuarine sedimentation
- backstepping barrier beaches (open shoreline)
- wave ravinement, and healing-phase deposits
- instability at the shelf edge
- dominant gravity flows: low-density turbidites

Not to scale

FIGURE 5.56 Depositional processes and products of the *early* transgressive systems tract (modified from Catuneanu, 2003). Rapid rates of base-level rise trigger a retrogradational shift of facies on the continental shelf, where most of the riverborn sediment is now trapped in fluvial, coastal and shallow-marine systems. Wave-ravinement processes erode the underlying normal regressive shelf-edge deltas and open shoreline systems, continuing to supply sand for the deep-water turbidity flows. These turbidity flows tend to be of low-density type, similar to the ones of the lowstand systems tract (Fig. 5.44). Such low-density turbidity currents are underloaded on the steep continental slope (flow energy > sediment load, which causes entrenchment), but become overloaded/aggradational on the low-gradient basin floor (sediment load > flow energy). The lowstand and early transgressive low-density turbidity flows travel farther into the basin relative to the high-density late falling-stage flows because the higher proportion of mud sustains the construction of levees over larger distances. Healing-phase wedges are typical for the transgressive systems tract, and infill, or heal over, the bathymetric profile established at the end of regression. Estuaries are diagnostic for transgression, but retrograding or even prograding deltas may also form in river-mouth settings during the transgression of the open shoreline, primarily as a function of degree of channel incision and sediment supply (Figs. 5.51 and 5.52; see text for details).

potentially continuous (over several hundred square kilometres) blankets of upward-fining sandy deposits. *Shelf ridges* are also sand prone, usually consisting of 5–10 m thick and regionally extensive upward-fining successions of well-sorted, cross-bedded to bioturbated fine- to coarse-grained sediment (Snedden *et al.*, 1994). A case study from the Miocene section of offshore northwest Java shelf provides high-resolution seismic images calibrated with well logs and core that reveal some of the geomorphological characteristics of these self ridge deposits (Posamentier, 2002). The features are described as large-scale, up to 17 m thick elongated bodies ranging from 0.3 to 2.0 km wide and more than 20 km long. They are asymmetric, thicker along the sharp leading edge, gradually thinning toward a more irregular trailing edge (Figs. 5.58–5.62). Smaller-scale

sand waves are generally observed on top of shelf ridges, oriented oblique to the long axes of the ridges and also to the direction of ridge migration (Posamentier, 2002). Shelf sand ridges overlie transgressive wave-ravinement surfaces abandoned on the shelf following the retreat of the shoreline, and tend to be oriented parallel to the axes of structural embayments that may channelize the energy of tidal currents. Shelf ribbons, which are the smaller version of shelf ridges, may also concentrate sand in a transgressive shelf setting at scales of less than 5 m thick and less than 100 m wide (Posamentier, 2002). Both types of transgressive shelf-sand deposits (sheet-like and ridge-like) may form excellent regional reservoirs encased in shelf fine-grained seal facies. The formation of such shelf-sand deposits is favored during transgression, when

Late transgression: depositional processes and products

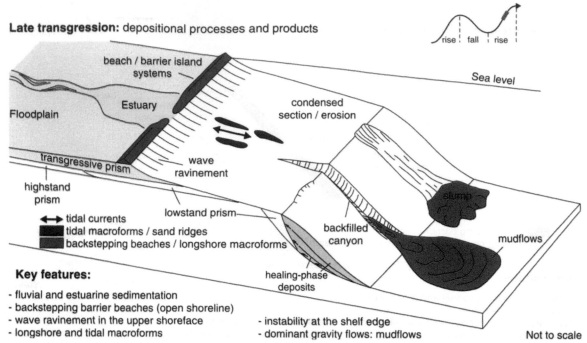

Key features:

- fluvial and estuarine sedimentation
- backstepping barrier beaches (open shoreline)
- wave ravinement in the upper shoreface
- longshore and tidal macroforms

- instability at the shelf edge
- dominant gravity flows: mudflows

Not to scale

FIGURE 5.57 Depositional processes and products of the *late* transgressive systems tract (modified from Catuneanu, 2003). Most of the terrigenous sediment is trapped in the fluvial to shallow-marine transgressive prism, which includes fluvial, estuarine, deltaic, open shoreline, and lower shoreface deposits. Additional sand is incorporated within shelf macroforms (sheets, ridges, ribbons) generated by storm surges and tidal currents. Such shelf-sand deposits are generally associated with the transgressive systems tract, as the best conditions to accumulate and the highest preservation potential are offered to shelf macroforms that form during shoreline transgression (Posamentier, 2002). As base level rises rapidly during transgression, hydraulic instability at the shelf edge generates mudflows in the deep-water environment. The top of all transgressive deposits is marked by the maximum flooding surface. Where the transgressive deposits are missing (e.g., in the outer shelf-upper slope areas subject to nondeposition or erosion), the maximum flooding surface reworks the maximum regressive surface. The river-mouth settings may become estuaries (shown in the diagram) or deltas, depending on the balance between accommodation and sedimentation (Figs. 5.51 and 5.52; see text for details).

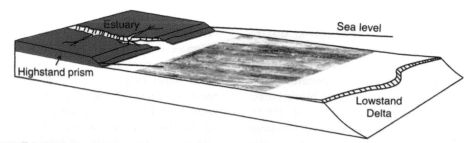

FIGURE 5.58 Reflection amplitude extraction map showing Miocene shelf ridges, offshore northwest Java (not to scale; modified from Posamentier, 2002; seismic image courtesy of H.W. Posamentier). These ridges (white features on the map, corresponding to high negative amplitudes) are several hundred meters wide and several kilometers long, and are observed along a horizon slice approximately 775 m subsea. The formation of such shelf-sand deposits is favored during transgression, when a significant portion of the continental shelf is submerged. Subsequent aggradation during the highstand base-level rise and normal regression confers on these ridges a high preservation potential in the rock record. This is why shelf-sand deposits are now recognized as a significant shallow-water component of the transgressive systems tract (Posamentier, 2002). No other systems tract offers such favorable conditions for the formation and preservation of significant sand-prone shelf macroforms.

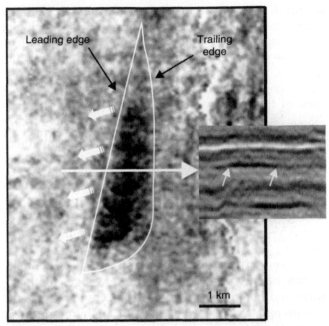

FIGURE 5.59 Reflection amplitude extraction map showing Miocene shelf ridges, offshore northwest Java (modified from Posamentier, 2002; image courtesy of H.W. Posamentier). These ridges (shown as red bands on the map; see arrows) are several hundred meters wide and several kilometers long, and are observed along a horizon slice approximately 810 m subsea. The formation of such shelf-sand deposits is favored during transgression, when a significant portion of the continental shelf is submerged. Subsequent aggradation during the highstand base-level rise and normal regression confers on these ridges a high preservation potential in the rock record. This is why shelf-sand deposits are now recognized as a significant shallow-water component of the transgressive systems tract (Posamentier, 2002). No other systems tract offers such favorable conditions for the formation and preservation of significant sand-prone shelf macroforms.

FIGURE 5.61 Morphology of a Miocene shelf ridge, offshore northwest Java, as seen on an amplitude extraction map from a horizon slice located approximately 775 m subsea (modified from Posamentier, 2002; images courtesy of H.W. Posamentier). Note the cross-sectional expression of the shelf ridge on the 2D seismic line. The shelf ridge has an asymmetrical shape in plan view, with a straight and well-defined leading edge, and a more irregular trailing edge. The direction of ridge migration is indicated by the wide arrows. Due to limitations imposed by vertical seismic resolution, the shape of the shelf ridge in cross sectional view is difficult to assess on the 2D seismic line, although the width of the sandy macroform can be estimated from the amplitude anomaly (the two small arrows indicate the edges of the macroform). The shape of shelf ridges in cross sectional view may be assessed significantly better using well logs (Fig. 5.62).

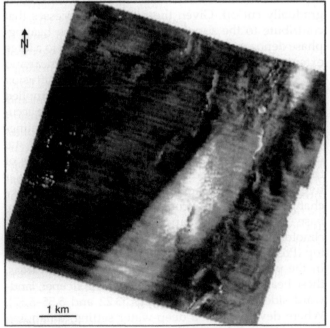

FIGURE 5.60 Reflection amplitude extraction map showing a close-up of a Miocene shelf ridge, offshore northwest Java, from a horizon slice approximately 720 subsea (modified from Posamentier, 2002; image courtesy of H.W. Posamentier). The sharply defined northwestern edge of the ridge (white feature on the map, corresponding to high negative amplitudes) is interpreted as the leading edge of the macroform. See Fig. 5.61 for a contrast between leading and trailing edges.

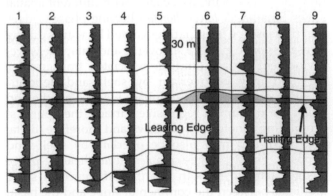

FIGURE 5.62 Well-log cross-section of correlation showing the morphology of a Miocene shelf ridge, and adjacent sand-sheet deposits, located approximately 850 m subsea, offshore northwest Java (modified from Posamentier, 2002; well logs courtesy of H.W. Posamentier). The length of the cross-section is approximately 6 km. Note the asymmetrical shape of the shelf ridge, with a thicker and better defined leading side, and a tapering trailing side. The integration of 3D seismic and well-log data (e.g., Fig. 5.61 and the well logs presented in this Figure) allows for a full 3D reconstruction of the shelf ridge morphology.

a significant portion of the continental shelf is submerged. Subsequent aggradation during the high-stand base-level rise and normal regression confers on these macroforms a high preservation potential in the rock record. This is why shelf-sand deposits are now recognized as a significant shallow-water component of the transgressive systems tract (Posamentier, 2002). No other systems tract offers such favourable conditions for the formation and preservation of significant sand-prone shelf macroforms.

In addition to transgressive lags and shelf-sand macroforms, onlapping healing-phase wedges also form an integral part of, and are diagnostic for the transgressive systems tract (Figs. 3.20, 3.21, 3.22, 5.6, and 5.55–5.57; see also diagrams in Dominguez and Wanless, 1991, and Posamentier and Chamberlain, 1993, based on earlier work by Bruun, 1962). A common feature of all types of healing-phase wedges that may form in different areas of the marine environment and at different scales, is that they fill bathymetric lows in an attempt to re-establish a graded seafloor profile (Posamentier and Allen, 1993; Figs. 3.20–3.22, 5.6, and 5.55–5.57). These healing-phase depozones are invariably asymmetrical, with steeper slope gradients on the landward side, as they inherit the shape of shoreface or delta front profiles in shallow-water settings (Figs. 3.20, 3.21, 5.6, and 5.55), or of the continental slope in deep-water settings (Figs. 3.22, 5.56, and 5.57). The asymmetrical shape of these depozones confers on the healing-phase deposits a wedge-shaped geometry, as they onlap the proximal side of the bathy-metric low and taper gradually in a distal direction (Fig. 5.55). Healing-phase wedges may form in lower shoreface, shelf and deep-water environments, each setting providing different amounts of accommodation and hence being associated with different spatial scales. Small-scale *lower shoreface* healing-phase wedges that fill seascape irregularities carved by waves during transgression (e.g., Fig. 3.20) overlie and onlap the wave-ravinement surface (and its associated transgressive lag) and may be overlain by shelf-sand deposits. Medium-scale *shelf* healing-phase wedges overlie and onlap the maximum regressive surface (the youngest prograding clinoform of the lowstand shoreface/delta; Fig. 5.55), and may also be overlain by shelf-sand deposits. Finally, large-scale *deep-water* healing-phase wedges tend to smooth out the difference in slope gradients between the continental slope and the basin floor, and onlap the maximum regressive surface on the continental slope (Figs. 5.56 and 5.57). Note that only healing-phase wedges that fill bathymetric lows created *during* transgression may overlie and onlap the wave-ravinement surface; healing-phase wedges that fill *existing* bathymetric lows at the onset of

transgression develop basinward relative to the distal termination of the wave-ravinement surface, and hence they overlie and onlap the maximum regressive surface instead.

Irrespective of their location within the basin, on the continental shelf or in the deep-water setting, all healing-phase wedges share common features regarding the processes involved in their formation and the resulting stratal geometry. In the early stage of transgression, when the shoreline is closer to the bathymetric low area that is being infilled, sediment supply is higher and depositional processes are dominated by a combination of gravity flows and suspension sedimentation. The resulting lower portion of the healing-phase wedge is relatively coarse-grained, and may include a significant amount of sand. As transgression proceeds and the shoreline becomes remote relative to the bathymetric low area, sediment supply diminishes and the accumulation of the healing-phase wedge continues primarily from suspension fallout. This upper portion of the healing-phase wedge is relatively fine-grained, being composed mainly of silt and mud. The typical vertical profile of a fully-developed healing-phase wedge is therefore fining-upward, showing an increase in the concentration of sand beds towards the base in relation to the activity of non-channelized hyperpycnal flows. Up section, the balance between hyperpycnal and hypopycnal flow deposits changes in the favour of the latter as the supply of sand is gradually cut off. Given the nature of processes that contribute to the supply of sediment to the healing-phase depozones, which involve wave action to a large extent, sediment sources may be considered linear and the transport of sediment is primarily by diffusion rather than being channelized. As sediment is supplied from the coastline and is moved basinward to the accumulation area, sedimentation rates within the healing-phase depozone decrease accordingly in a distal direction (Fig. 5.55). As a result, the proximal side of the healing-phase wedge grows thicker with time relative to the distal portion, and the clinoform geometry changes accordingly from concave-up towards the base (mimicking the shape of the youngest regressive clinoform) to flat and eventually convex-up towards the top (Posamentier and Allen, 1993, 1999; Figs. 5.55–5.57). In the process of infilling the bathymetric low areas, these healing-phase clinoforms onlap the steeper, landward side of the seascape (Figs. 3.22 and 5.55–5.57). Where developed in a deep-water setting, onlapping the continental slope, such healing-phase wedges correspond to the 'transgressive slope aprons' of Galloway (1989) (Figs. 3.22, 5.56, and 5.57).

In addition to transgressive slope aprons (large-scale healing-phase wedges that onlap the continental slope and form from linear sediment sources), the deep-water portion of the transgressive systems tract may also include submarine fans associated with more localized sediment sources and involving a channelized style of sediment transport. The nature of gravity flows that lead to the formation of such submarine fans is known to change during transgression, from early-transgressive low-density turbidity currents to late-transgressive cohesive debris flows (mudflows) (Posamentier and Kolla, 2003; Figs. 5.11, 5.37, 5.56, and 5.57).

The low-density turbidity currents of the early stage of transgression are similar to the ones of the underlying lowstand systems tract (Figs. 5.45, 5.46, 5.47, and 5.48; also, compare Figs. 5.44 and 5.56), which makes the recognition of the maximum regressive surface in a conformable succession of deep-water turbidites most difficult (Fig. 5.63). The trend of decrease with time in the amount of sand delivered to the deep-water environment, initiated at the onset of base-level rise by the trapping of riverborne sediment in aggrading lowstand normal regressive systems on the continental shelf, continues during transgression. This trend is illustrated by the fining-upward profile of the lowstand – transgressive portion of the basin-floor submarine fan deposits in Fig. 5.63, and is explained primarily by a combination of two different factors. Firstly, the rates of base-level rise increase from the lowstand normal regression to the subsequent early transgression, which means that increasingly more riverborne sediment is trapped within aggrading fluvial to shallow-marine systems. In turn, this leads to a decrease in the amount of riverborne sediment that is made available to the deep-water environment. Secondly, the landward shift recorded by the shoreline during transgression increases the distance between the sediment entry points (river mouths) and the shelf edge, again reducing the chance of the riverborne sediment being delivered to the deep-water environment. In addition to these two factors, the gradual decrease in energy recorded by fluvial systems in relation to the denudation of source areas coupled with coastal aggradation may also explain, although to a lesser extent, the decrease with time in the amount of riverborne sand delivered to the deep-water environment during base-level rise.

During early transgression (Fig. 5.56), the shoreline is still close to the shelf edge and therefore sand can still be delivered to the deep-water environment by low-density turbidity currents. Such low-density turbidity currents are underloaded on the steep continental slope (flow energy > sediment load, which causes entrenchment; Fig. 5.45), but become overloaded/ aggradational on the low-gradient basin floor (sediment load > flow energy; Figs. 5.46–5.48). The lowstand and early transgressive low-density turbidity flows travel further into the basin relative to the high-density late falling-stage flows because the higher proportion of mud sustains the construction of levees over larger distances. During late transgression (Fig. 5.57), the sediment entry points into the marine basin are far from the shelf edge, and hence no riverborne sand is made available to the staging area for the deep-water gravity flows. The vast majority of this riverborne sediment is now trapped in the fluvial to shallow-marine transgressive prism, which includes fluvial, estuarine, deltaic, open shoreline, lower shoreface, and shelf-sand deposits. As base level rises rapidly during transgression, hydraulic instability at the shelf edge results in the erosion of outer shelf – upper slope fine-grained sediments, generating mudflows in the deep-water environment (Figs. 5.57 and 5.63). These mudflow deposits are similar to the ones of the early falling-stage systems tract (Figs. 5.33–5.36), and complete the fining-upward profile illustrated in Fig. 5.63 for the lowstand – transgressive portion of the basin-floor submarine fan complex.

The preservation potential of the transgressive deposits is generally high due to the fact that the subsequent highstand normal regression leads to sediment aggradation across the entire basin (Fig. 5.6). Generally speaking, the transgressive systems tract has the best preservation potential among all systems tracts, from the basin margin to the basin center. By comparison, the falling-stage deposits in continental shelf-type settings are strongly affected by subaerial erosion during base-level fall; the downstream fluvial portion of the lowstand systems tract is commonly affected by wave-ravinement erosion during transgression; and the fluvial to shallow-marine portion of the highstand systems tract is subject to subaerial erosion during the subsequent fall in base level.

Economic Potential

Petroleum Plays

The petroleum plays of the early transgression have a bimodal distribution, some being related to the continental shelf-based transgressive wedge, and others being part of the deep-water wedge (Figs. 4.41 and 5.56). On the continental shelf, likely close to the shelf edge, the best reservoirs are concentrated along the coastline, being represented by backstepping beaches (open shoreline settings), estuary-mouth complexes, retrograding bayhead deltas or even prograding deltas

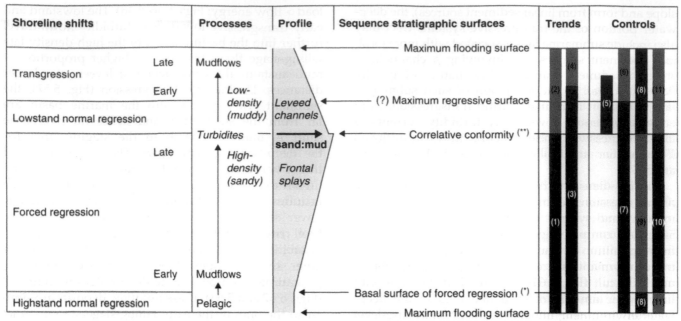

FIGURE 5.63 Composite vertical profile of a basin-floor submarine fan complex that forms during a full cycle of base-level changes, showing overall grading trends and the inferred position of the four low-diachroneity (event-significant) sequence stratigraphic surfaces (modified from Catuneanu, 2003, with additional information from Posamentier and Kolla, 2003). Key: (*) correlative conformity of Posamentier and Allen (1999); (**) *sensu* Hunt and Tucker (1992); [1] coarsening-upward; [2] fining-upward; [3] progradation of the submarine fan complex; [4] retrogradation of the submarine fan complex; [5] accelerating base-level rise (increasingly more terrigenous sediment is trapped within fluvial to shallow-marine systems, and correspondingly less sand is available for the deep-water setting—hence the fining-upward profile); [6] shoreline transgression (retrogradation of sediment entry points into the marine basin—hence the fining-upward profile); [7] shoreline regression (progradation of sediment entry points into the marine basin—hence the coarsening-upward profile); [8] decrease with time in fluvial gradients and energy during base-level rise (caused by denudation of source areas coupled with coastal aggradation—hence the fining-upward profile);[9] increase with time in fluvial gradients and energy during base-level fall (caused by differential fluvial incision or differential tectonism—hence the coarsening-upward profile); [10] no fractionation of the riverborne sediment during base-level fall: all grain-size classes are delivered to the deep-water environment; [11] fractionation of the riverborne sediment: the coarser sediment fractions are preferentially trapped on the continental shelf, leaving only the finer sediment fractions available for deep-water gravity flows (hence the sharp decrease in sand/mud ratio across the correlative conformity). Note that different controls that operate at the same time may tend to generate opposite trends (e.g., controls [5] and [8] promote a fining-upward profile, whereas control [7] promotes a coarsening-upward profile), and it is their interplay that determines the actual trend in the rock record. In this case, the onset of base-level rise, with its subsequent accelerating rates (control [5]), has the most profound influence on the balance of sediment budget across the basin, and hence it triggers the change from high- to low-density turbidites across the correlative conformity. The change from high- to low-density turbidites at the onset of base-level rise means not only a decrease in the volume of terrigenous sediment made available to the deep-water environment, affecting the *sediment/water ratio* of gravity flows, but also a decrease in the *sand/mud ratio* in these flows as the coarser fractions of the riverborne sediment are trapped first in the aggrading fluvial to coastal systems. Also note that autocyclic shifts through time in the *locus* of deposition of the different fan lobes may result in the different portions of this composite profile being found in different locations within the submarine fan complex.

(river-mouth settings). The formation of estuary-mouth complexes depends on the degree of estuary development. For example, a fully established estuary in a wave-dominated setting will have all its subenvironments represented, including the bayhead delta, the central estuary and the estuary-mouth complex (Fig. 5.52).

This is commonly the case with transgressions that flood rivers that flow within incised valleys, where the drowning of the river is faster than the transgression of the adjacent open shoreline (Fig. 5.51). In such settings, the preservation of estuary-mouth complexes (e.g., Fig. 4.53) also indicates that the rates of wave-ravinement erosion are less than the rates of sedimentation within the estuary. The same condition applies for the preservation of backstepping beaches and barrier island systems; otherwise, in coastal settings characterized by extreme wave energy, the preservation of coastal reservoirs is unlikely (Leckie, 1994; Figs. 3.24 and 3.26). An incomplete (drowned) estuary, where the flooding of the river is outpaced by the transgression of the adjacent open shoreline, is only represented by the backstepping bayhead delta, without the establishment of the central estuary and estuary-mouth complex subenvironments (Fig. 5.52). This situation is prone to occur in the case of relatively small rivers (low sediment supply) that flow within unincised channels (Fig. 5.51). This discussion is meant to reveal the complexity that may be encountered in the 'real world,' where a seemingly endless number of possibilities may be envisaged depending on local circumstances. This argues, once again, that sequence stratigraphic modeling needs to be performed on a case-by-case basis, as rigid and 'universal' templates are bound to be misleading when applied indiscriminately to all case studies.

Landward from the shoreline, the potential for petroleum exploration of the transgressive systems tract is generally moderate to poor because of the extensive development of fine-grained floodplain facies in response to the rapid rates of base-level rise. Fluvial reservoirs are represented by isolated channel fills, levees, and crevasse splay deposits engulfed within floodplain fines. Seaward from the shoreline, onlapping healing-phase deposits trap part of the sand supplied by wave-ravinement processes, whereas the surplus of sediment continues to feed the deep-water submarine fans *via* low-density turbidity flows, for as long as the shoreline is still close to the shelf edge (Fig. 5.56). These early transgressive turbidites are commonly expected on the low-gradient basin floor, forming the fill of leveed channels and also building relatively small, distal frontal splays. No such reservoirs are expected on the steeper-gradient continental slope, where channels tend to be entrenched (erosion > sedimentation) due to the underloaded nature (energy flux/transport capacity > sediment load on a steep seascape) of the low-density turbidity flows. As the diluted turbidity currents of the early stage of transgression are increasingly dominated by finer-grained sediment (Fig. 5.63), which sustains the formation of levees, they tend to travel farthest into the basin relative to all other gravity flows that are recorded during a full cycle of base-level changes. These early transgressive turbidity currents are similar in nature to the flows of the previous lowstand stage, but are expected to be of even lower density because of the accelerating base-level rise that allows for more riverborne sediment to be trapped within aggrading fluvial to shallow-marine systems on the continental shelf.

During late transgression, the decreasing rates of base-level rise at the shoreline still outpace sedimentation rates (Fig. 5.57). A significant portion of the shelf is now submerged, and the combined activity of storm surges and tidal currents may lead to the accumulation of shelf-sand deposits, including ridges oriented normal to the shoreline (Posamentier, 2002). Elsewhere, especially in areas closer to the shelf edge, the shelf is generally subject to sediment starvation and condensed sections are likely to form (Loutit *et al.*, 1988). Rapid increases in water depth lead to shelf edge instability, which results in the manifestation of gravity flows (Galloway, 1989). Such flows are mud-rich, involving fine-grained outer shelf sediments that accumulated far from the sediment entry points. As such, the sand/mud ratio of the gravity-flow deposits accumulated in the deep-water environment during rising base level (lowstand normal regression to transgression) records an overall decrease, from turbidites to mudflows (Figs. 5.44, 5.56, 5.57, and 5.63). This fining-upward trend of the rising stage in the deep-water basin is completed by the accumulation of pelagic/hemipelagic sediments of the highstand (late rise) normal regression at the top of the submarine fan complex (Figs. 5.7 and 5.63).

The petroleum plays of the late transgression are concentrated in the fluvial to shallow-marine depositional systems (the transgressive wedge that develops on the continental shelf; Fig. 4.41). The characterization of fluvial, coastal, and lower shoreface reservoirs is similar to what was described above in the case of early transgression. These reservoirs may include sandy fluvial architectural elements (channel fills, levees, crevasse splays) engulfed within floodplain fines; backstepping beaches, bayhead deltas and estuary-mouth complexes; prograding deltas; and healing-phase deposits in the lower shoreface. The new addition to the types of petroleum plays that characterize late transgression stages is represented by the shelf-sand deposits referred to above (Fig. 5.57; Posamentier, 2002). Transgressive shelf macroforms can be recognized both on modern shelves and in the stratigraphic record. As noted by Posamentier (2002), these macroforms 'are thought to have formed as a result of erosion and subsequent reworking of sand-prone deltaic and/or coastal plain deposits by shelf tidal currents... These transgressive systems tract deposits have significant exploration potential because they are commonly sand prone and tend to be encased in shelf mudstone seal facies.' During late transgression, the terrigenous sediment entry points are too far from the shelf edge to make any significant contribution to the deep-water gravity flows, so no more sand is fed to the submarine fans. Mudflows are, however, still active due to the general instability around the shelf edge, both in the outer shelf and upper slope areas (Fig. 5.57).

The main risks associated with the exploration of the transgressive systems tract rest with the identification of reservoirs, even though such facies may be found within all depositional systems that accumulate during the shoreline transgression (Fig. 5.14). The best transgressive reservoirs are commonly related to coastal settings (estuarine, deltaic, and beach sands), although their preservation in the rock record requires a number of conditions to be fulfilled, including a relatively weak wave-ravinement erosion during transgression. Such conditions need to be assessed on a case-by-case basis in the process of sequence stratigraphic analysis. In addition to coastal facies, shelf-sand deposits and deep-water turbidites may also make good prospects for petroleum exploration. The main contribution, however, of the transgressive systems tract to the development of petroleum systems within a sedimentary basin is the accumulation of source rocks and seal facies, within most transgressive depositional environments (Fig. 5.14). Transgressive shallow-marine shales, for example, usually form regionally extensive

covers across continental shelves, which may serve as reference units for stratigraphic correlation that can be easily identified on 2D seismic lines, based on their 'transparent' seismic facies (Fig. 5.64).

Coal Resources

The transgressive systems tract is arguably the best portion of a stratigraphic sequence for coal exploration. The time of end-of-shoreline transgression marks the peak for peat accumulation and subsequent coal development because the water table is at its highest level relative to the landscape profile, following a time characterized by a high accommodation to sediment supply ratio during the transgression of the shoreline (Fig. 5.15). This balance between accommodation and sedimentation, tipped in favor of the former during transgression, represents a fundamental prerequisite that optimizes environmental conditions for significant accumulations of peat deposits. However, the condition that accommodation > sedimentation is necessary but not sufficient, as vegetation growth also depends on climatic constraints. Assuming that all favorable conditions are fulfilled, the best developed coal seams are expected to overlap with the maximum flooding surface (Hamilton and Tadros, 1994; Fig. 5.15). The timing of the maximum flooding surface is also relatively late in the stage of base-level rise (Fig. 4.7), which means that denuded source areas now supply

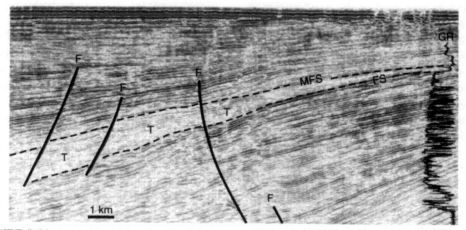

FIGURE 5.64 Seismic line showing a Pliocene to recent succession accumulated within the tectonic setting of a continental shelf (image courtesy of PEMEX). The seismic facies are calibrated with a gamma ray log. Note the regionally extensive transgressive shale that can be mapped on the seismic line as a 'transparent' facies. This transgressive shale forms a stratigraphic marker that can be used for regional correlation, and it is bounded by a flooding surface at the base and by a maximum flooding surface at the top. The transgressive shale accumulated within an outer shelf environment (below the storm wave base), following an episode of abrupt flooding that can be recognized across the basin. The underlying facies (below the flooding surface) accumulated mainly above the storm wave base, in inner shelf to beach environments. The maximum flooding surface is overlain by regressive (highstand) deposits. Abbreviations: T—transgressive shale; F—faults; FS—flooding surface; MFS—maximum flooding surface.

less sediment than in the earlier stages of base-level rise (such as during the lowstand normal regression). The scenario described in this section fits the view of standard sequence stratigraphic models, which predict coastal and fluvial aggradation during stages of shoreline transgression. One has to be aware, however, that exceptions do occur, such as in the situation described in case 2 in Fig. 3.20 (see Chapter 3 for a detailed discussion). In such cases, where coastal erosion prevails in spite of the rising base level, the nonmarine environment may also be dominated by erosional processes or sediment bypass, leading to the formation of subaerial unconformities (Leckie, 1994).

Placer Deposits

Transgressive ravinement surfaces, which are the product of wave or tidal scouring and reworking in near-shore environments during shoreline transgression, may be associated with lag deposits that have the potential of forming economically-significant placers. The G.V. Bosch and Stilfontein reefs of the Witwatersrand Basin are examples of such transgressive placers (Catuneanu and Biddulph, 2001; Fig. 5.43). The geographic distribution of transgressive placers is strictly controlled by the location of paleoshorelines, and, along dip-oriented transects, it is restricted to the area that is limited by the shoreline trajectories at the onset and end of transgressive stages. Once again, as in case of the other two unconformity-related placer types (subaerial unconformities and regressive surfaces of marine erosion – see section on the falling-stage systems tract), the paleoshoreline is a central element in the exploration for placer deposits because it limits the lateral extent of the transgressive reefs. Depending on where the maximum transgressive shoreline is located in relation to the basin margins, transgressive placers may be missed if exploration is solely based on the mapping of basin-margin unconformities.

REGRESSIVE SYSTEMS TRACT

Definition and Stacking Patterns

The regressive systems tract includes all strata that accumulate during shoreline regression, i.e., the entire succession of undifferentiated highstand, falling-stage, and lowstand deposits (Fig. 5.65). As such, this systems tract is defined by progradational stacking patterns across the basin. The concept of regressive systems tract was introduced in the sequence stratigraphic literature by Embry and Johannessen (1992), as part of their transgressive-regressive sequence model (Figs. 1.6

and 1.7), and it was subsequently refined in follow-up publications by Embry (1993, 1995).

The amalgamation of all regressive deposits into one undifferentiated systems tract is particularly feasible where the available data base is insufficient to observe stratal terminations (e.g., offlap) and stacking patterns, and thus to separate between the different genetic types of regressive deposits. In such instances, the use of the regressive systems tract over individual lowstand, falling-stage, and highstand systems tracts is preferable, due to the difficulty in the recognition of some of the surfaces that separate the lowstand, falling-stage, and highstand facies (notably, the correlative conformity and the conformable portions of the basal surface of forced regression; Embry, 1995). The identification of conformable sequence stratigraphic surfaces that serve as systems tract boundaries is virtually impossible in individual boreholes, where only well-log and core data are available. For example, if we only had well logs (2) and (5) in Fig. 5.65, it would be impossible to estimate where the basal surface of forced regression and the correlative conformity, respectively, are placed within the conformable and coarsening-upward succession of prograding shallow-marine strata. Knowledge, however, of the regional architecture and stacking patterns of this succession, as afforded by seismic data for instance, helps to infer where these conformable surfaces are placed along the cross sectional profile. Such additional insights into the stratigraphic architecture of the studied succession allow one to map the basal surface of forced regression as the oldest clinoform associated with offlap, and the correlative conformity as the youngest clinoform associated with offlap (Fig. 5.65). The application of these criteria may, however, be limited by a number of factors, including the degree of preservation of offlapping stacking patterns in the rock record, as discussed in Chapter 4.

The regressive systems tract, as defined by Embry (1995), is bounded at the base by the maximum flooding surface within both marine and nonmarine portions of the basin. At the top, the regressive systems tract is bounded by the maximum regressive surface in a marine succession, and by the subaerial unconformity in nonmarine strata. The latter portion of the systems tract boundary is taken by definition (Embry, 1995), even though there is a possibility that lowstand fluvial strata (still regressive) may be present above the subaerial unconformity. In this practice, all fluvial strata directly overlying the subaerial unconformity are assigned to the transgressive systems tract (Embry, 1995). A drawback of this approach in delineating the upper boundary of the regressive systems tract, which coincides with the boundary of the T–R (transgressive–regressive) sequence, consists in the fact that the

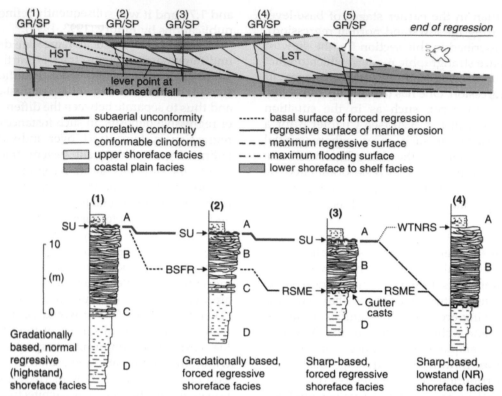

FIGURE 5.65 Anatomy of a regressive systems tract in a wave-dominated shallow-marine setting (modified from Plint, 1988; Posamentier *et al.*, 1992b; Walker and Plint, 1992; Posamentier and Allen, 1999). The five synthetic well logs capture different stratigraphic aspects along the dip-oriented cross sectional profile. Log (1) shows a gradationally based shallow-marine succession truncated at the top by the subaerial unconformity. This succession accumulated during the highstand normal regression, and includes a relatively thick package of shoreface facies that indicates sedimentation during base-level rise. Log (2) also intercepts a gradationally based shallow-marine succession truncated at the top by the subaerial unconformity, but the shoreface deposits are thinner (< depth of the fairweather wave base) and early forced regressive in nature. Log (3) captures a sharp-based, and relatively thin (< fairweather wave base), forced regressive shoreface succession directly overlying outer shelf highstand facies. The shoreface deposits may be topped either by the subaerial unconformity (in the diagram) or by its correlative conformity. Log (4) intercepts a relatively thick succession of lowstand shoreface deposits (sedimentation during base-level rise), which is sharp-based as it overlies the youngest portion of the regressive surface of marine erosion. The top of this shoreface succession is conformable (within-trend normal regressive surface), unless subsequently reworked by a transgressive ravinement surface. Log (5) shows a relatively thick succession of lowstand shoreface deposits (sedimentation during base-level rise), which is gradationally based as it is located seaward relative to the distal termination of the regressive surface of marine erosion. If log (5) is located seaward from the maximum regressive shoreline (as shown in the diagram), the succession of lowstand shoreface facies is topped by a conformable maximum regressive surface unless reworked subsequently by a transgressive ravinement surface. Sedimentary facies: A—coastal plain; B—shoreface (with swaley cross-stratification); C—inner shelf (with hummocky cross-stratification); D—outer shelf fines. Abbreviations: GR/SP—gamma ray/spontaneous potential; HST—highstand systems tract; LST—lowstand systems tract; SU—subaerial unconformity; BSFR—basal surface of forced regression; RSME—regressive surface of marine erosion; WTNRS—within-trend normal regressive surface; NR—normal regressive. For the significance of the lever point at the onset of fall, see Fig. 4.20.

subaerial unconformity and the maximum regressive surface are temporally offset, forming in relation to different stages or events of the base-level cycle (Figs. 4.6 and 4.7). On the other hand, the motivation behind this approach is that the subaerial unconformity is arguably the most significant surface within a nonmarine succession, while the maximum regressive surface is easier to recognize than the basal surface of forced regression and the correlative conformity within the shallow-marine portion of the basin. Other limitations may, however, hamper the practical applicability of this approach, especially in downstream-fluvial and deep-water settings. Within the downstream region of fluvial systems, where the fluvial portion of the lowstand systems tract is commonly thickest (Figs. 5.4–5.6 and 5.65), the physical connection between the subaerial unconformity and the marine portion of the maximum regressive surface may only be achieved where the thickness of the lowstand shore to coastal plain strata is less than the amount of erosion caused by subsequent transgressive ravinement processes (see Fig. 2.5 for a possible geometry of the lowstand wedge on a continental shelf). Otherwise, the upper boundary of the regressive systems tract may be represented by two discrete surfaces separated both temporally and spatially by lowstand shore to coastal plain deposits (Fig. 5.65). Within the deep-water setting, the identification of the maximum regressive surface is as difficult as the recognition of correlative conformities in a shallow-water succession. These issues are discussed in more detail below.

Within the nonmarine portion of the basin, the regressive package may incorporate the subaerial unconformity and its associated stratigraphic hiatus, where lowstand shore, coastal plain or alluvial plain deposits are preserved in the rock record (Figs. 4.6, 5.5, and 5.65). In such cases, the regressive succession includes deposits that are genetically unrelated (i.e., highstand and lowstand strata in contact across the subaerial unconformity), formed in relation to two different cycles of base-level changes. Landward from the edge of the lowstand fluvial wedge, defined by the point where the maximum regressive surface onlaps the subaerial unconformity, the subaerial unconformity is directly overlain by transgressive fluvial strata (Figs. 5.4 and 5.5). In this case, the subaerial unconformity becomes the true boundary between regressive and overlying transgressive deposits (e.g., log (1) in Fig. 5.65). Even within the area of accumulation of lowstand fluvial strata, strong subsequent transgressive ravinement erosion may result in the subaerial unconformity being reworked by the transgressive ravinement surface, in which case this composite unconformity becomes again the true boundary between regressive and overlying transgressive deposits (Embry, 1995; Dalrymple, 1999).

Within the shallow-marine portion of the basin, the regressive package displays a coarsening-upward grading trend which relates to the basinward shoreline shift (Figs. 4.6 and 5.5). This coarsening-upward profile should strictly be regarded as a *progradational* trend, which is not necessarily the same as a *shallowing-upward* trend (Catuneanu *et al.*, 1998b). It is documented that the earliest, as well as the latest deposits of a marine coarsening-upward succession are likely to accumulate in deepening water, especially in areas that are not immediately adjacent to the shoreline (Naish and Kamp, 1997; T. Naish, pers. comm., 1998; Catuneanu

et al., 1998b; Vecsei and Duringer, 2003; more details regarding this topic, as well as examples of numerical modeling, are presented in Chapter 7). The characteristics of the subtidal facies of the regressive systems tract vary with their genetic type, i.e., highstand normal regressive, forced regressive, or lowstand normal regressive (Figs. 5.65 and 5.66). The highstand shoreface deposits are always gradationally based, and tend to be relatively thick (more than the depth of the fairweather wave base) reflecting the tendency of aggradation during base-level rise (e.g., log (1) in Fig. 5.65). The falling-stage shoreface deposits are generally sharp-based in a wave-dominated setting (e.g., log (3) in Fig. 5.65), excepting for the earliest lobe that overlies the conformable basal surface of forced regression (e.g., log (2) in Fig. 5.65). In a river-dominated setting, where the regressive surface of marine erosion does not form, the falling-stage shoreface facies are gradationally based (Fig. 3.27). In either case, the thickness of the falling-stage shoreface sands tends to be less than the fairweather wave base due to the restriction in available accommodation imposed by base-level fall (Figs. 5.65 and 5.66). The lowstand shoreface deposits are generally gradationally based (e.g., log (5) in Fig. 5.65), excepting for the earliest lobe that accumulates on top of the distal termination of the regressive surface of marine erosion (e.g., log (4) in Fig. 5.65). The lowstand shoreface facies also tend to be thicker than the depth of the fairweather wave base, similar to the highstand deposits, due to the fact that they accumulate and aggrade during rising base level (Figs. 5.65 and 5.66).

The regressive systems tract in a deep-water setting records a change with time in the character of gravity flows, from mudflows (early forced regression) to high-density turbidity flows (late forced regression) and finally to low-density turbidity flows (lowstand normal regression). The depositional products of these gravity flows gradually prograde into the basin during shoreline regression, on top of the underlying highstand pelagic sediments (Figs. 5.7, 5.26, 5.27, and 5.44). The composite vertical profile of the deep-water portion of the regressive systems tract therefore includes a lower coarsening-upward succession, which consists of pelagic facies grading upward into mudflow deposits and high-density turbidites, overlain by a fining-upward succession of low-density turbidites accumulated during accelerating base-level rise (Figs. 5.5 and 5.63). The maximum flooding surface (base of regressive systems tract) may be mapped with relative ease at the top of late transgressive mudflow deposits, but the maximum regressive surface (top of regressive systems tract) is much more difficult to identify within a conformable succession of low-density

Systems tract / Shoreface deposits	RST		
	HST	FSST	LST
Thickness	Thick (> FWB)	Thin (< FWB)	Thick (> FWB)
Base (stratigraphic surface)	Gradational (WTFC)	Sharp / gradational (RSME / BSFR)	Gradational / sharp (CC / RSME)
Top (stratigraphic surface)	Truncated (SU)	Truncated / conformable (SU / CC)	Conformable / truncated (WTNRS, MRS / TRS)

FIGURE 5.66 Stratigraphic characteristics of the shoreface deposits of the regressive systems tract. The highstand and lowstand shoreface deposits are commonly thicker than the depth to the fairweather wave base because of aggradation that accompanies base-level rise. Forced regressive shoreface deposits are thinner than the fairweather wave base, as only a portion of the shoreface (commonly the upper shoreface) may receive sediments during base-level fall. The forced regressive shoreface deposits are generally sharp-based, excepting for the earliest lobe that accumulates on top of the conformable basal surface of forced regression. The lowstand shoreface deposits are generally gradationally based, excepting for the earliest lobe that accumulates on top of the youngest portion of the regressive surface of marine erosion. See also Fig. 5.65 for a graphic representation of these types of shoreface facies, and for additional explanations. Abbreviations: RST—regressive systems tract; HST—highstand systems tract; FSST—falling-stage systems tract; LST—lowstand systems tract; FWB—fairweather wave base; WTFC—within-trend facies contact; RSME—regressive surface of marine erosion; BSFR—basal surface of forced regression; CC—correlative conformity (*sensu* Hunt and Tucker, 1992); SU—subaerial unconformity; WTNRS—within-trend normal regressive surface; MRS—maximum regressive surface; TRS—transgressive ravinement surface.

turbidity flow deposits (Fig. 5.63). This limits the applicability of the regressive systems tract in deep-water settings. It is interesting to note that the sequence stratigraphic analysis of deep-water successions poses an entirely different set of challenges relative to what is encountered in the case of shallow-water deposits. Conformable surfaces that are more difficult to identify in shallow-water successions, such as the basal surface of forced regression (correlative conformity *sensu* Posamentier and Allen, 1999) and the correlative conformity *sensu* Hunt and Tucker (1992), have a better physical expression within deep-water strata relative to the maximum regressive surface (Fig. 5.63). This is the opposite of the situation described for shallow-water settings, where the maximum regressive surface has a stronger lithological signature than the more cryptic correlative conformities.

Economic Potential

The regressive systems tract combines all exploration opportunities of the highstand, falling-stage and lowstand systems tracts (Fig. 5.14). The reader is therefore referred to the previous sections in this chapter that deal with the individual systems tracts associated with specific types of shoreline shifts.

LOW- AND HIGH-ACCOMMODATION SYSTEMS TRACTS

Definition and Stacking Patterns

The identification of all regressive (highstand, falling-stage, and lowstand) and transgressive systems tracts, discussed above, is directly linked to, and dependent on the reconstruction of *syndepositional shoreline shifts* (i.e., highstand normal regression, forced regression, lowstand normal regression or transgression, respectively). Therefore, the application of these 'traditional' systems tract concepts requires a good control of both marine and nonmarine portions of a basin, and, most importantly, the preservation of paleocoastline and near-shore deposits that can reveal the type of shoreline shift during sedimentation. The patterns of progradation or retrogradation of facies and sediment entry points into the marine basin are thus critical for the identification of any of the systems tracts presented above. There are situations, however, where sedimentary basins are dominated by nonmarine surface processes (e.g., overfilled basins; Fig. 2.64), or where only the nonmarine facies are preserved or available for analysis. In such cases, any reference to syndepositional shoreline shifts becomes superfluous, and therefore the usage of the traditional systems tract nomenclature lacks the fundamental justification provided by the evidence of shoreline transgressions or regressions. The solution to this problem was the introduction of low- and high-accommodation systems tracts, designed specifically to describe fluvial deposits that accumulated in isolation from marine/lacustrine influences, or for which the relationship with coeval shorelines is impossible to establish because of preservation or data availability issues (Dahle *et al.*, 1997). These systems tracts are defined primarily on the basis of *fluvial architectural elements*, including the relative contribution of channel fills and overbank deposits to the fluvial rock record, which in turn allows inference of the amounts of fluvial accommodation (low *vs.* high) available at the time of sedimentation. The low- and high-accommodation 'systems tracts' have also been referred to as low- and high-accommodation 'successions' (e.g., Olsen *et al.*, 1995; Arnott *et al.*, 2002).

The application of sequence stratigraphy to the fluvial rock record is a relatively recent endeavor, which started in the early 1990s with works such as those by Shanley *et al.* (1992) and Wright and Marriott (1993), whose models were subsequently refined with increasing detail (e.g., Shanley and McCabe, 1993, 1994, 1998). Generally, however, these models of fluvial sequence stratigraphy are still tied to a coeval marine record, describing changes in fluvial facies and architecture within the context of marine base-level changes and using the traditional lowstand – transgressive – highstand systems tract nomenclature. In this context, the fluvial (low- and high-accommodation) systems tracts of Dahle *et al.* (1997) represent a conceptual breakthrough in the sense that they define nonmarine stratigraphic units independently of marine base-level changes and associated shoreline shifts. The differentiation between low- and high-accommodation systems tracts involves an observation of the distribution of fluvial architectural elements in the rock record, which then can be interpreted within a sequence stratigraphic context of *changing fluvial accommodation conditions through time*. The low- and high-accommodation systems tracts replace the tripartite lowstand – transgressive – highstand sequence stratigraphic model, although a correlation between these concepts may be attempted based on general stratal stacking patterns (e.g., Boyd *et al.*, 1999; Ramaekers and Catuneanu, 2004; Eriksson and Catuneanu, 2004a).

When referring to models of nonmarine sequence stratigraphy, it is important to make the distinction between low- and high-accommodation *systems tracts* and low- and high-accommodation *settings*. Even though these concepts use a similar terminology ('low-accommodation,' 'high-accommodation'), they are fundamentally different in the way unconformity-bounded

fluvial depositional sequences are subdivided into component systems tracts. The low- and high-accommodation *systems tracts* are the building blocks of a fluvial depositional sequence that is studied in isolation from any correlative marine deposits, and they succeed each other in a vertical succession as being formed during a stage of varying rates of positive accommodation. It is thus implied that, following a stage of negative fluvial accommodation when the sequence boundary forms, sedimentation resumes as fluvial accommodation becomes available again, starting with lower and continuing with higher rates. In contrast, low- *vs.* high-accommodation *settings* indicate particular areas in a sedimentary basin that are generally characterized by certain amounts of accommodation, such as high or low in the proximal and distal sides of a foreland system, respectively. The definition of low- and high-accommodation settings is therefore based on the subsidence patterns of a tectonic setting, and is independent of the presence or absence of marine influences on fluvial sedimentation. Consequently, both zones 2 and 3 in Fig. 3.3 may develop within low- or high-accommodation settings. As such, the low- and high-accommodation settings may host fluvial depositional sequences that conform to the standard sequence stratigraphic models, consisting of the entire succession of traditional lowstand – transgressive – highstand systems tracts (e.g., Leckie and Boyd, 2003), or they may host fully fluvial successions accumulated independently of marine base-level changes (e.g., Boyd *et al.*, 2000; Zaitlin *et al.*, 2000, 2002; Arnott *et al.*, 2002; Wadsworth *et al.*, 2002, 2003; Leckie *et al.*, 2004). The criteria that separate low- from high-accommodation *settings*, based on a series of papers by Boyd *et al.*, 1999, 2000; Zaitlin *et al.*, 2000, 2002; Arnott *et al.*, 2002; Wadsworth *et al.*, 2002, 2003; Leckie and Boyd, 2003; Leckie *et al.*, 2004, are presented in Chapter 6. The discussion below focuses on low- *vs.* high-accommodation *systems tracts*.

Low-Accommodation Systems Tract

Within fluvial successions, low accommodation conditions result in an incised-valley-fill type of stratigraphic architecture dominated by multi-storey channel fills and a general lack of floodplain deposits. The depositional style is progradational, accompanied by low rates of aggradation, often influenced by the underlying incised-valley topography, similar to what is expected from a lowstand systems tract (Boyd *et al.*, 1999; Fig. 5.67). The low-accommodation systems tract generally includes the coarsest sediment fraction of a fluvial depositional sequence, which may in part be related to rejuvenated sediment source areas and also to the higher energy fluvial systems that commonly build up the lower portion of a sequence. These features

Systems tract / Features	Low-accommodation systems tract	High-accommodation systems tract
Depositional trend	early progradational[1]	aggradational
Depositional energy	early increase, then decline	decline through time
Grading	coarsening-upward at base[1]	fining-upward
Grain size	coarser	finer
Geometry	irregular, discontinuous[2]	tabular or wedge-shaped[3]
Sand:mud ratio	high	low[4]
Reservoir architecture	amalgamated channel fills	isolated ribbon sandstones[4]
Floodplain facies	sparse	abundant[4]
Thickness	tends to be thinner[5]	tends to be thicker[5]
Coal seams	minor or absent[6]	well developed[7]
Paleosols	well developed[8]	poorly developed[9]

FIGURE 5.67 Defining features of the low- and high-accommodation systems tracts (modified from Catuneanu, 2003, with additional information from Leckie and Boyd, 2003). Notes: [1]—the progradational and associated coarsening-upward trend at the base of a fluvial sequence are attributed to the gradual spill over of coarse terrigenous sediment into the basin, on top of finer-grained floodplain or lacustrine facies. Once fluvial sedimentation is re-established across the basin, the rest of the overall profile is fining-upward. The basal coarsening-upward portion of the sequence thickens in a distal direction, and its facies contact with the rest of the sequence is diachronous with the rate of coarse sediment progradation; [2]—this depends on the landscape morphology at the onset of creation of fluvial accommodation, which is a function of the magnitude of fluvial incision processes during the previous stage of negative fluvial accommodation. Irregular and discontinuous geometries form where fluvial deposits prograde and infill an immature landscape; [3]—this depends on the mechanism that generates accommodation, i.e., sea-level rise or differential subsidence, respectively; [4]—this is valid for Phanerozoic successions, where vegetation is well established and helps to confine the fluvial systems. The fluvial systems of the vegetationless Precambrian are dominated by unconfined braided and sheetwash facies, which tend to replace the vegetated overbank deposits of Phanerozoic meandering systems; [5]—this depends on the rates of creation of fluvial accommodation, and the relative duration of systems tracts; [6]—where present, they are commonly compound coals; [7]—simpler (fewer hiatuses), more numerous, and thicker; [8]—commonly multiple and compound; [9]—thinner, widely spaced, and organic-rich.

give the low-accommodation systems tract some equivalence with the lowstand systems tract, reflecting early and slow base-level rise conditions (or low rates of creation of fluvial accommodation, in the absence of marine influences) that lead to a restriction of accommodation for floodplain deposition. The dominant sedimentological features of the low-accommodation systems tract are illustrated in Fig. 5.68.

Low-accommodation systems tracts typically form on top of subaerial unconformities, reflecting early

A   B

FIGURE 5.68 Low-accommodation systems tract—outcrop examples of fluvial facies that are common towards the base of fluvial depositional sequences. A—amalgamated braided channel fills (Katberg Formation, Early Triassic, Karoo Basin).

(Continued)

FIGURE 5.68 Cont'd B, C—massive sandstone channel fills and downstream accretion macroforms, products of high-energy braided streams (Balfour Formation, late Permian-earliest Triassic, Karoo Basin); D—amalgamated braided channel fills. Note the base of a channel scouring the top of an underlying channel fill. Very small amounts of floodplain sediment may be preserved in this succession (left of the geological hammer) (Molteno Formation, Late Triassic, Karoo Basin); E, F—amalgamated braided channel fills and downstream accretion macroforms (E—Molteno Formation, Late Triassic, Karoo Basin; F—Frenchman Formation, Maastrichtian, Western Canada Sedimentary Basin); G, H—mudstone rip-up clasts at the base of amalgamated channel fills, eroded from the floodplains during the lateral shift of the unconfined braided channels. The low accommodation, coupled with channel erosion, explain the lack of floodplain facies within the low-accommodation systems tract (G—Katberg Formation, Early Triassic, Karoo Basin; H—Frenchman Formation, Maastrichtian, Western Canada Sedimentary Basin).

stages of renewed sediment accumulation within a nonmarine depozone, while the amount of available fluvial accommodation is still limited ('low'). Depending on the location within the basin, and the distance relative to the sediment source areas, the base of the low-accommodation systems tract may display a coarsening-upward profile, referred to above as a 'progradational' depositional trend (Fig. 5.67). Such progradational trends have been recognized in different sedimentary basins, ranging in age from Precambrian (e.g., Ramaekers and Catuneanu, 2004) to Phanerozoic (e.g., Heller *et al.*, 1988; Sweet *et al.*, 2003, 2005; Catuneanu and Sweet, 2005), and reflect the gradual spill over of coarse terrigenous sediments from source areas into the developing basin, on top of finer-grained floodplain or lacustrine facies. As it takes time for the coarser facies to reach the distal parts of the basin, it is expected that the basal progradational (coarsening-upward) portion of the low-accommodation systems tract will be wedge-shaped, thickening in a distal direction and with a diachronous top facies contact that youngs away from the source areas. Consequently, the most proximal portion of a fluvial sequence may not include a coarsening-upward profile at the base, as the lag time between the onset of sedimentation and the arrival of the coarsest sediments adjacent to the source

areas is insignificant, whereas such profiles are predictably better developed, in a range of several meters thick, towards the distal side of the basin (Sweet *et al.*, 2003, 2005; Ramaekers and Catuneanu, 2004). Figure 5.69 provides an example of such a facies transition within the basal portion of a fluvial sequence, illustrating the progradation of gravel-bed fluvial systems on top of finer-grained deposits that belong to the same depositional cycle of positive accommodation. Notwithstanding the scours at the base of channel fills, this facies transition may be regarded as 'conformable,' as being formed during a stage of continuous aggradation. The actual sequence boundary (base of the low-accommodation systems tract) is in a stratigraphically lower position, occurring within the underlying finer-grained facies (Sweet *et al.*, 2003, 2005; Catuneanu and Sweet, 2005). The more distal portion of this sequence boundary, as well as the conformable facies contact between the earliest fine-grained facies and the overlying coarser-grained fluvial systems of the low-accommodation systems tract, are shown in Fig. 5.70. In this example, the accumulation of relatively thick lacustrine facies of the Battle Formation corresponds to the lag time required by the coarse terrigenous sediments to reach the distal side of the foredeep depozone. Details of the internal architecture of the

FIGURE 5.69 Low-accommodation systems tract facies, showing the progradation of gravel-bed fluvial systems over finer-grained deposits. This lithostratigraphic facies contact between the Brazeau Formation and the overlying Entrance Conglomerate of the basal Coalspur Formation (Maastrichtian, Alberta Basin) is diachronous, younging in a basinward direction (i.e., the direction of progradation/coarse sediment spill over). The actual subaerial unconformity (sequence boundary) is in a stratigraphically lower position, and demonstrated palynologically to occur within the fine clastics of the Brazeau Formation (Sweet *et al.*, 2005).

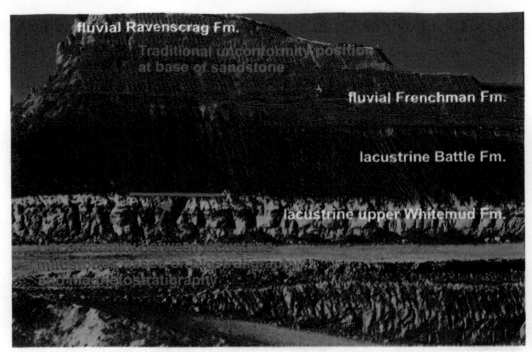

FIGURE 5.70 Unconformable contact (yellow line) between a high-accommodation systems tract (the lacustrine deposits of the upper Whitemud Formation) and the overlying low-accommodation systems tract (photo courtesy of A.R. Sweet). The low-accommodation systems tract consists of a lower fine-grained portion (the lacustrine deposits of the Battle Formation) overlain by the prograding coarser-grained facies (amalgamated channel fills) of the Frenchman Formation. The relatively thick fine-grained basal portion of the low-accommodation systems tract is characteristic of distal settings of sedimentary basins, and incorporates the time required by the influx of coarse clastics to reach these distal areas. The facies contact between the lacustrine and fluvial facies of the low-accommodation systems tract (red line in photo) is conformable and diachronous, younging in a basinward direction. The facies contact shown in this photograph is in the physical continuation of, but younger than, the facies contact in Fig. 5.69.

amalgamated fluvial channel fills of the Frenchman Formation, which prograded on top of the earliest lacustrine facies of the depositional sequence and are characteristic of the low-accommodation systems tract, are presented in Fig. 5.68. Additional core photographs of low-accommodation sedimentary facies that accumulated immediately above subaerial unconformities, and typify the lower portion of fully nonmarine depositional sequences, are shown in Fig. 5.71. These case studies question the validity of the commonly accepted axiom that major subaerial unconformities always occur at the base of regionally extensive coarse-grained units, and demonstrate the value of biostratigraphic documentation of stratigraphic hiatuses (Sweet *et al.*, 2003, 2005; Catuneanu and Sweet, 2005).

The basal progradational portion of the low-accommodation systems tract also indicates an increase in depositional energy, from initial low-energy floodplain and/or lacustrine environments to higher-energy bedload-dominated fluvial systems (Sweet *et al.*, 2003, 2005; Catuneanu and Sweet, 2005; Figs. 5.69 and 5.70). These bedload rivers generally represent the highest

energy fluvial systems of the entire depositional sequence; once they expand across the entire overfilled basin, depositional energy tends to decline gradually through time until the end of the positive accommodation cycle in response to the denudation of source areas and the progressive shallowing of the fluvial landscape profile. The relatively coarse sediments of the low-accommodation systems tract usually fill an erosional relief carved during the previous stage of negative accommodation (e.g., driven by tectonic uplift or climate-induced increase in fluvial discharge), and therefore this systems tract is commonly discontinuous, with an irregular geometry. The low amount of available accommodation also controls additional defining features of this systems tract, including a high channel fill-to-overbank deposit ratio, the absence or poor development of coal seams, and the presence of well-developed paleosols (Fig. 5.67).

High-Accommodation Systems Tract

High accommodation conditions (attributed to higher rates of creation of fluvial accommodation) result in a simpler fluvial stratigraphic architecture that

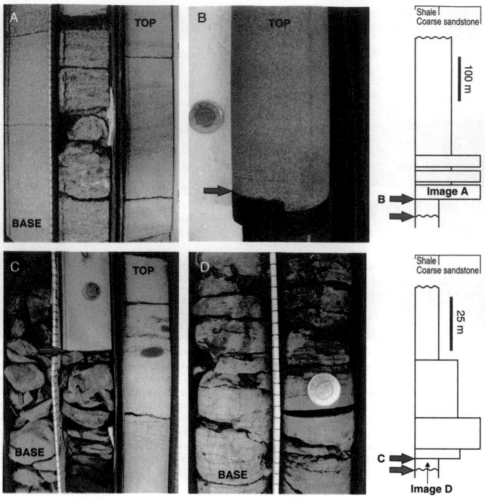

FIGURE 5.71 Core examples of facies associations of low-accommodation systems tracts (Maastrichtian-Paleocene, central Alberta). Subaerial unconformities (sequence boundaries; not shown in the photographs, marked with blue arrows on the vertical profiles) are cryptic from a lithological standpoint, and occur within fine-grained (low depositional energy) successions that underlie the coarser-grained portions of each depositional sequence. Photographs A and B illustrate facies that overlie a Paleocene-age sequence boundary; photographs C and D show facies that overlie a Maastrichtian-age sequence boundary. Each facies association starts with fine-grained deposits, which grade upward to coarser facies (increase with time in depositional energy). These two main components of the low-accommodation systems tract are separated by 'conformable' facies contacts (red arrows). Paleocene low-accommodation systems tract: A—amalgamated channel fills (Lower Paskapoo Formation); B—conformable facies contact between overbank mudstones (Upper Scollard Formation) and the overlying fluvial channel sandstones (Lower Paskapoo Formation). Maastrichtian low-accommodation systems tract: C—conformable facies contact between lacustrine mudstones (Battle Formation) and the overlying fluvial channel sandstones (Lower Scollard Formation, which is age-equivalent with the Frenchman Formation in Figs. 5.68 and 5.70); D—lacustrine mudstones that overlie directly the subaerial unconformity (Battle Formation—see also Fig. 5.70).

includes a higher percentage of finer-grained overbank deposits, similar in style to the transgressive and highstand systems tracts. The depositional style is aggradational, with less influence from the underlying topography or structure (Boyd *et al.*, 1999). The high-accommodation systems tract is characterized by a higher water table relative to the topographic profile, a lower energy regime, and the overall deposition of finer-grained sediments. Channel fills are still present in the succession, but this time isolated within floodplain facies (Fig. 5.67). The dominant sedimentological features of the high-accommodation systems tract are illustrated in Fig. 5.72.

The deposition of the high-accommodation systems tract generally follows the leveling of the sequence boundary erosional relief, which is attributed to the

FIGURE 5.72 High-accommodation systems tract—outcrop examples of fluvial facies that are common towards the top of fluvial depositional sequences (Burgersdorp Formation, Early-Middle Triassic, Karoo Basin). A—isolated channel fill (massive to fining-upward) within overbank facies. Note the erosional relief at the base of the channel; B—lateral accretion macroform (point bar) in meandering stream deposits; C—proximal crevasse splay (approximately 4 m thick, massive to coarsening-upward) within overbank facies. Note the sharp but conformable facies contact (no evidence of erosion) at the base of the crevasse splay; D—floodplain-dominated meandering stream deposits, with isolated channel fills and distal crevasse splays. All sandstone bodies of the high-accommodation systems tract may form petroleum reservoirs engulfed within fine-grained floodplain facies. These potential reservoirs lack the connectivity that characterizes the reservoirs of the low-accommodation systems tract (Fig. 5.68).

early fluvial deposits infilling lows and prograding into the developing basin, and so this systems tract has a much more uniform geometry relative to the underlying low-accommodation systems tract. The accumulation of fluvial facies under high accommodation conditions continues during a regime of declining depositional energy through time, which results in an overall fining-upward profile. These fining-upward successions form the bulk of each fluvial depositional sequence, as documented in numerous case studies from different sedimentary basins (e.g., Catuneanu and Sweet, 1999, 2005; Catuneanu and Elango, 2001; Sweet *et al.*, 2003,

2005; Ramaekers and Catuneanu, 2004). Additional criteria for the definition of the high-accommodation systems tract include the potential presence of well-developed coal seams (e.g., high water table in an actively subsiding basin, coupled with decreased sediment supply; Fig. 5.73) and the poor development of paleosols (Fig. 5.67).

Discussion

The usage of the low- and high-accommodation systems tracts is most appropriate in overfilled basins, or in portions of sedimentary basins that are beyond

FIGURE 5.73 Well-developed coal seams within a high-accommodation systems tract (Early Paleocene, Coalspur Formation, Western Canada Sedimentary Basin). In contrast with the low-accommodation systems tracts, high-accommodation systems tracts are more likely to host economic coal seams due to environmental factors (higher water table, less sediment influx) that are conducive to peat accumulation under high-accommodation conditions.

the influence of marine base-level changes (i.e., zone 3 in Fig. 3.3). Within such depozones, sedimentation is controlled primarily by tectonism in the sediment source areas and within the basin itself, and also by climate-induced changes in the efficiency of weathering, erosion, and sediment transport processes.

The underlying assumption behind the low- *vs.* high-accommodation systems tract terminology is that following the stages of negative accommodation that result in the formation of sequence boundaries (subaerial unconformities), the rates of creation of fluvial accommodation gradually increase from low to high during each depositional cycle. This allows for more and more floodplain and associated low energy facies to be deposited as the sequence thickens. Besides accommodation, changes in sediment supply through time also contribute to the observed upwards increase in the abundance of finer-grained sediment fractions. Over time, the gradual denudation of source areas during the deposition of each sequence, coupled with a decrease in slope gradients of the fluvial landscape, contribute to the lowering of the amount of coarse terrigenous sediment delivered to the basin, and implicitly to the frequently observed fining-upward trends (e.g., Catuneanu and Elango, 2001; Ramaekers and Catuneanu, 2004). Each such depositional cycle is terminated by an episode of source area rejuvenation, commonly of tectonic nature, during which time subaerial unconformities form in parallel with the

steepening of the fluvial landscape profile (e.g., the overfilled phase in Fig. 2.64; see also discussion in Catuneanu and Elango, 2001).

The general correlation between low-accommodation and lowstand systems tracts, and also between high-accommodation and transgressive to highstand systems tracts is only tentative, based on similarities in fluvial architecture. These terms should not be used interchangeably unless a good control on the patterns of the age-equivalent shoreline shifts is also available. In the absence of such control, the 'maximum regressive surface' should not be used as the boundary between the low- and high-accommodation systems tracts, as there is no evidence that this contact corresponds to a turnaround point between regressive and transgressive conditions. In fact it is common that the change from the low- to the overlying high-accommodation systems tract is gradational rather than abrupt, as seen in a number of case studies of overfilled foredeeps (Fig. 5.74).

Examples of fluvial depositional sequences that display a change through time from low- to high-accommodation conditions are found in numerous basins around the world, including the Ainsa Basin of Spain (Dahle *et al.*, 1997), the Karoo Basin of South Africa (e.g., Catuneanu and Bowker, 2001; Catuneanu and Elango, 2001), the Western Canada Sedimentary Basin (Catuneanu and Sweet, 1999; Arnott *et al.*, 2002; Zaitlin *et al.*, 2002; Wadsworth *et al.*, 2002, 2003; Leckie

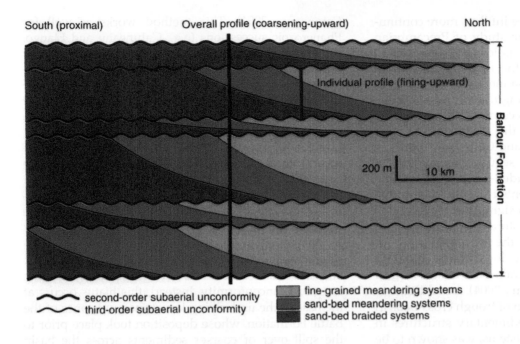

South (proximal)   Overall profile (coarsening-upward)   North

Individual profile (fining-upward)

200 m   10 km

Balfour Formation

~~~ second-order subaerial unconformity
~ third-order subaerial unconformity

fine-grained meandering systems
sand-bed meandering systems
sand-bed braided systems

FIGURE 5.74 Fluvial depositional sequences of the Balfour Formation, Karoo Basin (modified from Catuneanu and Elango, 2001). Note that each sequence displays a fining-upward profile, due to the change with time in fluvial styles from higher- to lower-energy systems. At the same time, the overall vertical profile of the Formation is coarsening-upward in response to the progradation of the orogenic front. The change from low- to high-accommodation conditions during the deposition of each sequence is gradational.

et al., 2004), the Athabasca Basin of Canada (Ramaekers and Catuneanu, 2004), and the Transvaal Basin of South Africa (Eriksson and Catuneanu, 2004a). The Late Permian to Middle Triassic Beaufort Group of the Karoo Basin is a classic example of a succession of fluvial depositional sequences that display fining-upward trends related to changes through time in fluvial styles, from higher- to lower-energy systems. The early high-energy systems of each sequence resulted in the accumulation of amalgamated channel fills, interpreted to reflect deposition under low-accommodation conditions (i.e., low-accommodation systems tracts). The overlying low-energy systems of each fluvial sequence are preserved as ribbon-like channel-fill sandstones engulfed within overbank fines, and are interpreted to reflect sedimentation under high-accommodation conditions (i.e., high-accommodation systems tracts). The upwards change from low- to high-accommodation systems tracts within each sequence is gradational, and so any attempt to place a systems tract boundary between them may only be regarded as tentative (e.g., no such separation is attempted in Fig. 5.74). In this case study, the change from low- to high-accommodation conditions during each depositional cycle correlates to a gradual decrease in topographic gradients during stages of orogenic loading and differential subsidence (Catuneanu and Elango, 2001). Sequence boundaries correspond to periods of time of differential isostatic rebound (Fig. 2.64), and are associated with stratigraphic hiatuses that

mark stages of basin reorganization, as suggested by changes in paleocurrent directions across the subaerial unconformities (Fig. 2.11).

The concepts of low- and high-accommodation systems tracts were initially developed for Phanerozoic sequences, where vegetation favors the preservation of thick overbank fines and isolated channel fills under high-accommodation conditions. More recently, these concepts have been applied to the Precambrian stratigraphic record as well (Ramaekers and Catuneanu, 2004; Eriksson and Catuneanu, 2004a). As noted in these studies, the less confined fluvial systems of the vegetationless Precambrian require new or additional criteria that are more applicable to such conditions. The general lack of overbank fines within Precambrian fluvial sequences may be attributed to the dominance of unconfined fluvial systems, where sheetwash facies tend to replace the vegetated overbank deposits of Phanerozoic meandering systems. The lack of fines in a sand-rich vegetationless environment may also be related to a greater eolian influence, as dust storms may remove mud more efficiently from barren surfaces (Ramaekers and Catuneanu, 2004). The ratio between sand and mud, and the associated fluvial architectural elements, seem therefore to be of less importance when trying to distinguish between low- and high-accommodation systems tracts in Precambrian deposits. Among the criteria defined for Phanerozoic fluvial sequences (Fig. 5.67), changes in the overall grading trends, as well as the geometry of fluvial deposits (irregular, immature-landscape infill *vs.* more continuous) are still applicable to the study of Precambrian deposits. The gradual progradation of coarser facies from outside the basin and the mixing with locally eroded muds, sands, and channel bank debris may also generate crudely coarsening-upward trends at the base of Precambrian low-accommodation systems tracts, as documented in the Early Proterozoic Athabasca Basin (Ramaekers and Catuneanu, 2004).

Since Precambrian fluvial sequences may consist entirely of unconfined, braided-style systems, the change in architectural elements from the base to the top of each sequence may be insignificant. This confers upon the succession a monotonous character, and, under these circumstances, the documentation of grading trends may require logarithmic plots to enhance the differences in grain size along vertical profiles (D. Long, pers. comm., 2004). In addition to this, the degree of preservation of trough cross-beds in cosets, which are common sedimentary structures in higher-energy braided-type systems, was shown to be particularly useful in the interpretation of low- or high-accommodation environments (Ramaekers and Catuneanu, 2004). As documented in the Athabasca Basin, under low-accommodation conditions only the toes of the troughs are generally preserved and the sections show apparent horizontal bedding to low-angle cross-bedding. The correct interpretation of these sedimentary structures is difficult in core, but easier where outcrop exposures are available. In contrast, the preservation of cosets of thicker and therefore readily recognizable trough cross-beds is more likely under high-accommodation conditions.

The time-transgressive progradation of coarse sediment into the basin at the onset of each depositional cycle, in both Precambrian and Phanerozoic settings, may allow for sequence boundaries to develop within fine clastics, separating sediments deposited during the waning phase of a prior sequence from similar lithologies deposited during the next cycle of positive accommodation but before the coarse sediments spill over the basin (Sweet *et al.*, 2003, 2005; Catuneanu and Sweet, 2005). This challenges conventional thinking that sequence boundaries are always expected at the base of coarse clastics. Additional methods or criteria need to be applied in order to locate the major hiatuses in the stratigraphic succession, and hence the position of sequence boundaries. In the case of Precambrian deposits, where high-resolution time control is difficult to achieve, major hiatuses usually correspond to stages of basin reorganization, and hence they may be evidenced by shifts in paleocurrent directions across sequence-bounding unconformities (Ramaekers and

Catuneanu, 2004). This method works as well for Phanerozoic successions (e.g., Catuneanu and Elango, 2001), but additional constraints are also afforded by biostratigraphy, magnetostratigraphy and high-resolution radiochronology. An example is offered by the correlative Scollard and Coalspur formations (Late Maastrichtian – Early Paleocene, Alberta foredeep), which form the bulk of an unconformity-bounded fluvial depositional sequence. The conventional placement of the lower sequence boundary has been at the base of the Coalspur 'Entrance' conglomerate (Fig. 5.69), based on lithological criteria. However, as demonstrated by palynology, the hiatus occurs within the fine clastics of the underlying Brazeau Formation (Sweet *et al.*, 2005). More distally, along the dip of the same depositional sequence, the base of the amalgamated Scollard Formation sandstones has been considered as overlying a regional unconformity. Instead, this hiatus occurs at the base of the underlying lacustrine mudstones of the Battle Formation, whose deposition took place prior to the spill over of coarser sediments across the basin (Catuneanu and Sweet, 1999, 2005; Sweet *et al.*, 2005). A similar situation has been documented in the case of the coal-bearing Santonian to Campanian Bonnet Plume Formation in east-central Yukon Territory (Sweet *et al.*, 2003). This 300 m thick coal-bearing interval consists of eight depositional sequences, each including basal coarsening-upward (coal-mudstone to conglomerate) and overlying fining-upward (conglomerate to mudstone) portions. With few exceptions, palynological zones start near or at the base of a coal seam and terminate in the mudstones overlying coarse clastic units. The magnitude of the inferred hiatuses within the fine-grained component of each cycle is of sufficient duration to allow the recognition of discrete zones within an overall continuum of change (Sweet *et al.*, 2003). These case studies shed new light on the value of time-control in stratigraphic analysis, and afford a better understanding of the depositional processes that take place during the accumulation of low- and high-accommodation systems tracts.

Economic Potential

The low- and high-accommodation systems tracts combine all natural resources that are commonly expected within the nonmarine portions of the lowstand, transgressive and highstand systems tracts, and which have been discussed in more detail in the previous sections of this chapter. This statement does not imply a direct correlation between fluvial systems tracts (low- and high-accommodation) and the conventional lowstand – transgressive – highstand systems tracts, but it merely indicates that changes in accommodation are somewhat predictable within any depositional sequence, and therefore depositional patterns follow similar trends.

Petroleum Plays

Figures 5.68 and 5.72 provide field examples of low-accommodation (amalgamated channel fills) and high-accommodation (floodplain-dominated fluvial successions) systems tracts. Within each fluvial sequence, the best petroleum reservoirs are related to the low-accommodation systems tract, where the channel fills tend to be amalgamated and hence there is a good connectivity between individual sandstone bodies (Fig. 5.68). Reservoirs may, however, be found in high-accommodation systems tracts as well, as isolated point bars, channel fills, or crevasse splays (all with different morphologies in plan view), encased within finer-grained floodplain facies (Fig. 5.72).

Coal Resources

Coal seams are best developed within high-accommodation systems tracts (Figs. 5.67 and 5.73) due to a combination of factors conducive to peat accumulation, including high rates of creation of fluvial accommodation, high water table relative to the topography, and the associated low depositional energy that results in the accumulation of finer-grained sediment fractions. Assuming that climate is favorable as well, and vegetation is available, these are the best conditions for peat accumulation during an entire depositional cycle of positive accommodation. The best developed coal seams of the high-accommodation systems tract are expected to form when the rates of creation of fluvial accommodation are at a maximum, which happens before the latest stage of the depositional cycle when the formation of accommodation decelerates to zero, before becoming negative. These coals are the equivalent of nonmarine maximum flooding surfaces in the conventional (lowstand – transgression – highstand) sequence stratigraphic models, considering that the highest water table (maximum 'flooding') in an overfilled basin occurs during times of highest rates of creation of fluvial accommodation (e.g., peaks of most active subsidence).

Low-accommodation systems tracts are unlikely to host any significant amounts of coal, due to a lack of sufficient accommodation, and when they do the coal seams tend to be thin and closely spaced (compound coals; Leckie and Boyd, 2003; Fig. 5.67). It can be noted therefore that the occurrence of interconnected petroleum reservoirs (amalgamated sand bodies) and of coal seams of economic importance is out of phase, as

their genesis requires mutually exclusive conditions. The former are characteristic of the low-accommodation systems tract, as being favored by limited amounts of accommodation, whereas the latter tend to be associated with the high-accommodation systems tract, as requiring high rates of creation of fluvial accommodation, which in turn translate into a high water table relative to the topographic profile.

Placer Deposits

The most significant placers that may be associated with fluvial depositional sequences are represented by the lag deposits that accumulate on top of subaerial unconformities (sequence boundaries). The quality of a placer is commonly proportional to its thickness and textural maturity. Both these parameters may change within the area of occurrence of the placer deposit, particularly along dip, in response to changes in the magnitude of erosional processes during the formation of the associated unconformity. Thus, the amount of reworking (which controls the textural maturity of the placer deposit), as well as the placer's thickness, are proportional to the amount of negative accommodation during the formation of the subaerial unconformity. In overfilled foredeeps, for example, the amount of isostatic rebound during stages of orogenic unloading (negative accommodation) is highest adjacent to the orogen, and decreases with distance in a distal direction (Fig. 2.64). In such settings, the best placer deposits (thickest, and most mature texturally) tend to develop along the basin margins, and their quality decreases towards the basin. This is the opposite of what is expected from a placer associated with a subaerial unconformity that forms in response to a fall in marine base level (zone 2 in Fig. 3.3), whose quality improves towards the coastline due to the fact that the amount of erosion and reworking increase in that direction (case A in Fig. 3.31). Such placers wedge out away from the coastline, and may be missed if exploration is carried out solely along the basin margins. Therefore, a careful analysis of the nature and genesis of the unconformity that the placer deposit is associated with is of fundamental importance for the design of a successful exploration program. Examples of placer deposits associated with subaerial unconformities, as well as other types of unconformities, may be observed in the gold-bearing Witwatersrand Basin of South Africa (Catuneanu, 2001; Catuneanu and Biddulph, 2001). The upper portion of the Neoarchean Witwatersrand Basin fill accumulated in a fluvial-dominated overfilled foredeep where the best placers ('reefs') associated with subaerial unconformities develop along the basin margins, at the base of low-accommodation systems tracts.

This discussion on the quality of placers associated with different genetic types of subaerial unconformities emphasizes that the difference between fully fluvial depositional sequences (composed of low- and high-accommodation systems tracts) and the fluvial portions of standard depositional sequences (composed of the traditional lowstand – transgressive – highstand systems tracts) is far more significant than just at a semantic level. Subaerial unconformities that separate high- and low-accommodation systems tracts are commonly associated with stratigraphic hiatuses that increase towards the basin margins, as being primarily related to 'upstream' controls (e.g., source area tectonism, or climate). In contrast, subaerial unconformities that separate highstand and lowstand systems tracts tend to be increasingly significant towards the basin, up to the coeval coastline, as being primarily related to 'downstream' controls (e.g., marine base-level fall). Therefore, the systems tract terminology carries important genetic connotations, and should be used carefully and appropriately in the context of each individual case study. For these reasons, fluvial systems tracts (low- and high-accommodation) and standard systems tracts (shoreline-related: lowstand, transgressive, highstand) should not be used interchangeably, even though broad similarities (e.g., between the low-accommodation systems tract and the lowstand systems tract, and between the high-accommodation systems tract and the transgressive-highstand systems tracts) may exist in terms of stacking patterns of fluvial architectural elements.

6

Sequence Models

INTRODUCTION

The 'sequence' is the fundamental stratal unit of sequence stratigraphy, and it corresponds to the depositional product of a full cycle of base-level changes or shoreline shifts depending on the sequence model that is being employed (Fig. 4.6). The concept of sequence is independent of scale, either spatial or temporal, and the general definition is provided in Fig. 1.9. The definition proposed by Mitchum (1977) in the early days of seismic and sequence stratigraphy (Fig. 1.9) builds on the original definition of a 'sequence' by Sloss (1963), but expands the mappability of sequences across entire basins by making reference to 'correlative conformities' beyond the areas of development of bounding unconformities (Figs. 1.4 and 1.5). The unconformity-bounded sequence of Sloss (1963) still remains at the core of modern stratigraphy, as unconformities delineate the *relatively conformable successions of genetically related strata* that are referred to in all more recent and revised definitions of a 'sequence.' Also, unconformities define the position of correlative conformities in the rock record, and hence they represent the fundamental element for the definition of sequences. Nevertheless, the introduction of correlative conformities as part of the definition of a 'sequence' may be regarded as a conceptual breakthrough, as they allow sequences to be delineated beyond the area where one or both bounding unconformities die out in a basinward direction (Fig. 1.5). To make the distinction between the unconformity-bounded 'sequence' of Sloss (1963) and the stratigraphic unit bounded by unconformities *or their correlative conformities* (Mitchum, 1977), the latter is referred to as a *depositional sequence*.

Following the introduction of the 'depositional sequence' concept by Mitchum (1977), subsequent refinements have been proposed to the original definition. Notably, Posamentier *et al.* (1988) elaborated that a

depositional sequence is 'composed of a succession of systems tracts and is interpreted to be deposited between eustatic-fall inflection points,' thus implying a genetic relationship between sequence development and eustatic sea-level changes. This inference has been discarded in subsequent publications, based on the recognition that stratal stacking patterns form in response to *relative* sea-level changes (subsidence and eustasy, often difficult to separate in the rock record), rather than to just eustasy (Posamentier and Allen, 1999). Posamentier and Allen (1999) further suggest a return to the more general term of 'sequence' rather than 'depositional sequence,' by eliminating direct reference to correlative conformities in the definition of sequences. Instead, they propose that reference to correlative conformities be included as *an addendum to the definition*. In this light, a 'sequence' is defined as a 'stratigraphic unit composed of a relatively conformable succession of genetically related strata bounded at its top and base by unconformities. Where the hiatal breaks associated with the bounding unconformities narrow below resolution of available geochronologic tools, the time surfaces (i.e., chronohorizons) that are correlative with the 'collapsed' unconformities constitute the sequence boundaries; these surfaces form the correlative conformities of Mitchum's (1997) definition.' Whether correlative conformities are included in the main definition of a 'sequence,' or only in an addendum to the definition, becomes somewhat irrelevant because the identification of both unconformable and conformable portions of the sequence boundary is equally important in sequence stratigraphic analysis.

Correlative conformities are thus an integral part of modern stratigraphy, and their inclusion into the stratigraphic literature coincides with the birth of seismic and sequence stratigraphy in the 1970s. While correlative conformities mark a significant advance in the development of the method of stratigraphic correlation, they have also been a source of confusion and disagreements with respect to their timing and physical attributes in the rock record. The various opinions regarding the timing of formation of correlative conformities relative to the main events of a reference base-level cycle (i.e., onset of base-level fall, end of base-level fall, end of regression and end of transgression; Fig. 1.7) resulted in the publication of several sequence stratigraphic models which differ primarily in the position of sequence boundaries, and particularly in the position of correlative conformities. Figure 6.1 illustrates the position (timing of formation) of correlative conformities in six different sequence stratigraphic models. Interesting to note is the fact that, with the exception of Galloway (1989), five out of these six models use the *subaerial unconformity* as the unconformable portion of the sequence boundary. Therefore, the difference between these models tends to be more evident within the marine portions of sedimentary basins, where sequence boundaries are picked within conformable packages of strata. The various candidates for the status of sequence boundary are also illustrated in Fig. 6.2.

As model 'F' (Posamentier and Allen, 1999) represents an evolution of model 'A' (Posamentier *et al.*, 1988)

in Fig. 6.1, we are left with five sequence stratigraphic models currently in use (models B – F in Fig. 6.1), all stemming from the original depositional sequence of seismic stratigraphy (Fig. 1.6). These models may be grouped into two main categories: one group defines correlative conformities relative to the base-level curve (*timing of sequence boundaries independent of sedimentation rates*: depositional sequences II, III, and IV in Fig. 1.6), whereas the other group defines correlative conformities relative to the transgressive-regressive curve (*timing of sequence boundaries dependent on sedimentation rates*: genetic and transgressive-regressive sequences in Fig. 1.6). The timing of formation of the conformable portion of the sequence boundary for each of these models is presented in Figs. 1.7 and 6.1. These correlative conformities correspond to different types of stratigraphic surfaces presented in Chapter 4, including the 'correlative conformity' *sensu* Hunt and Tucker (1992) (same age as the basinward termination of the subaerial unconformity), the 'basal surface of forced regression' (= 'correlative conformity' *sensu* Posamentier and Allen, 1999; older than the basinward termination of the subaerial unconformity), the 'maximum regressive surface' (= 'conformable transgressive surface' of

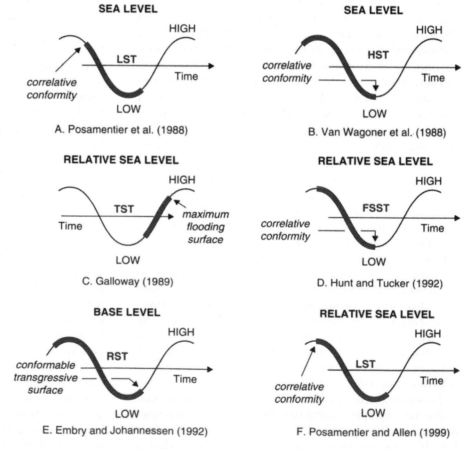

FIGURE 6.1 Correlative conformities as defined in various sequence stratigraphic models. The timing of formation of correlative conformities may be *independent of sedimentation rates* (models A, B, D, and F), or *dependent of sedimentation rates* (models C and E). With the exception of model C (Galloway, 1989), all other models take the subaerial unconformity as the unconformable portion of the sequence boundary. Each correlative conformity shown in this diagram corresponds to a particular type of stratigraphic surface described in Chapter 4: the 'basal surface of forced regression' (models A and F), the correlative conformity *sensu* Hunt and Tucker (1992) (models B and D), the 'maximum flooding surface' (model C) and the 'maximum regressive surface' (model E).

A. Posamentier et al. (1988)

B. Van Wagoner et al. (1988)

C. Galloway (1989)

D. Hunt and Tucker (1992)

E. Embry and Johannessen (1992)

F. Posamentier and Allen (1999)

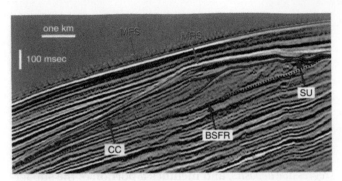

FIGURE 6.2 Stratigraphic surfaces that may serve as sequence boundaries according to different sequence stratigraphic models (image courtesy of H.W. Posamentier). This seismic line shows a Pleistocene to Recent succession in the Gulf of Mexico. Abbreviations: SU—subaerial unconformity (it truncates offlapping lobes); CC—correlative conformity *sensu* Hunt and Tucker, 1992 (the youngest clinoform associated with offlap, at the top of forced regressive deposits); BSFR—basal surface of forced regression (= correlative conformity *sensu* Posamentier and Allen, 1999; the oldest clinoform associated with offlap, at the base of forced regressive deposits); MRS—maximum regressive surface; MFS—maximum flooding surface. Surfaces represented in red are defined relative to the base-level curve (independent of sedimentation rates). Surfaces represented in yellow are defined relative to the transgressive-regressive curve (dependent on sedimentation rates, and hence potentially diachronous along strike). Note that the MFS is approximated with the modern seafloor in this image, but as the transgression still continues today, the actual MFS is yet to be formed. Also, see Fig. 3.22 for additional interpretations of this seismic line.

Embry and Johannessen, 1992; younger than the basinward termination of the subaerial unconformity), and the 'maximum flooding surface' (within a genetically related package of strata bounded by subaerial unconformities).

The concept of 'sequence' may be applied to any portion of a sedimentary basin fill, from underfilled to filled and overfilled (e.g., Fig. 2.64). In the early days of sequence stratigraphy, sequences were always described as including the entire array of depositional systems, from fluvial to deep-marine (Posamentier *et al.*, 1988; Van Wagoner *et al.*, 1990; Vail *et al.*, 1991). At the same time, the underlying assumption was that depositional processes leading to the formation of sequences were primarily controlled by sea-level changes, or by a combination of sea-level changes and tectonism. As part of unconformity-bounded sequences, fluvial deposits were thus inferred to have accumulated under the influence of marine base-level changes, and hence in direct relationship with particular stages of shoreline shift.

It is now known that unconformity-bounded sequences may also be found in fully nonmarine environments, formed independently of marine base-level changes and shoreline shifts, with an origin exclusively

related to tectonic and/or climatic controls (e.g., Mutti *et al.*, 1988; Blum, 1990, 1994; Legaretta *et al.*, 1993; Dam and Surlyk, 1993; Allen *et al.*, 1996; Catuneanu and Elango, 2001; Gibling *et al.*, 2005). Under such circumstances, only depositional sequences may be used for sequence stratigraphic analysis, since subaerial unconformities are the only available candidates for sequence boundaries (no maximum flooding or maximum regressive surfaces may be defined in the absence of a coeval shoreline; e.g., see Catuneanu and Sweet, 1999, and Catuneanu and Elango, 2001, for case studies of fluvial sequences developed in overfilled foredeeps). At the opposite end of the spectrum, there are cases in marine basins where stratigraphic cyclicity forms during continuous base-level rise, in relation to variations in the rates of base-level rise and sedimentation (Catuneanu *et al.*, 1999). In this case, the depositional sequence model may not work, since subaerial unconformities may not form, but the sequence stratigraphic framework of regressive and transgressive systems tracts may be resolved by using the genetic stratigraphic or the transgressive-regressive sequence model. It is thus clear that all models have merits and limitations; each model may work best under particular circumstances, and no one model is applicable to the entire range of case studies. Flexibility is therefore recommended when choosing the sequence model that is most appropriate for a particular case study.

The following is a brief discussion of the main sequence stratigraphic models currently in use. The intention of this discussion is not only to explain the philosophy behind each model, but also to provide a common platform between these different approaches. Ultimately, all models describe the same rocks using different terminology and a different style of conceptual packaging, and it is the purpose of this book to 'translate' this language and show how these approaches 'correlate' to each other. This should facilitate communication among practitioners embracing alternative approaches to stratigraphic analysis. Even more so, such a discussion should help the 'uninitiated' understand the meaning of apparently conflicting stratigraphic information released by members of the various schools of thought.

TYPES OF STRATIGRAPHIC SEQUENCES

Depositional Sequence

The depositional sequence uses the subaerial unconformity and its marine correlative conformity as a composite sequence boundary. The timing of the subaerial unconformity is equated with the stage of base-level fall at the shoreline (Figs. 4.6 and 4.7). The correlative conformity is either picked as the seafloor at the onset of forced regression (depositional sequence II in Figs. 1.6 and 1.7; model 'F' in Fig. 6.1; Fig. 6.2), or as the seafloor at the end of forced regression (depositional sequences III and IV in Figs. 1.6 and 1.7; models 'B' and 'D' in Fig. 6.1; Fig. 6.2). Depositional sequences III and IV are similar, with the exception that a fourth, falling-stage systems tract, is recognized in the latter. The depositional sequence illustrated in Fig. 4.6 is the depositional sequence IV. In overfilled basins, or on the continental side of basins where fluvial processes are independent of marine base-level changes (i.e., zone 3 in Fig. 3.3), depositional sequences are unconformity-bounded, and thus they become equivalent to Sloss' (1963) 'sequences.' In this case, the timing of sequence-bounding unconformities is not defined relative to a reference curve of base-level shifts, but it is controlled by tectonism and/or climate shifts.

The main area of debate within the 'depositional sequence' school is with respect to the position of the sequence boundary relative to the shallow-marine forced regressive deposits (Figs. 1.7, 6.1, and 6.2). Figure 6.3 illustrates the architecture of 'lowstand' fluvial to shoreface strata, as the lowstand systems tract is defined conceptually by Posamentier et al. (1988) and Posamentier and Allen (1999). This succession includes forced regressive ('early lowstand') and normal regressive ('late lowstand') deposits (Figs. 1.7 and 6.3).

According to the depositional sequence II model, the sequence boundary is taken *at the base of forced regressive deposits*, and includes a portion of the subaerial unconformity, the correlative conformity (*sensu* Posamentier et al., 1988, which is the basal surface of forced regression of Hunt and Tucker, 1992), and the proximal (older) portion of the regressive surface of marine erosion that may rework the correlative conformity (Fig. 6.3). The criticism that this model is faced with is that the subaerial unconformity is a continuous physical surface that, with a decreasing stratigraphic hiatus, develops in a basinward direction up to the point that defines the position of the shoreline at the end of forced regression (Fig. 5.5); and yet, only part of it is used as a sequence boundary (Fig. 6.3). In addition, practical limitations related to data availability

FIGURE 6.3 Methods of delineation of the depositional sequence boundary within the region of fluvial to shoreface facies transition. Cross-section A shows the architecture of forced and normal regressive deposits, and the nature of their associated bounding surfaces. Cross-sections B and C indicate the sequence boundary position in the view of the different depositional sequence models. Note that the regressive surface of marine erosion may be part of the sequence boundary in either model, where it replaces the correlative conformity. Abbreviations: HST—highstand systems tract; SU—subaerial unconformity; RSME—regressive surface of marine erosion; c.c.[(1)]—correlative conformity, *sensu* Posamentier et al. (1988); c.c.[(2)]—correlative conformity, *sensu* Hunt and Tucker (1992).

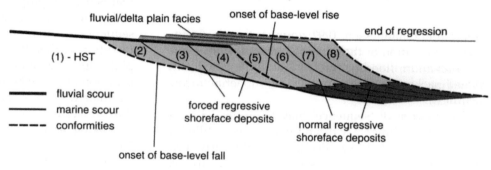
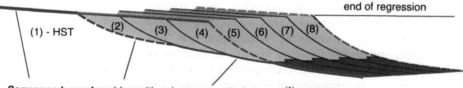
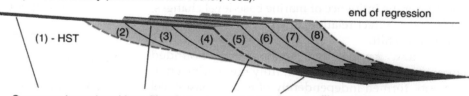

A. Lowstand systems tract (*sensu* Posamentier et al., 1988) - fluvial to shoreface:

B. Sequence boundary (*sensu* Posamentier et al., 1988):

Sequence boundary (depositional sequence II): SU + c.c.[(1)] + RSME

C. Sequence boundary (*sensu* Hunt and Tucker, 1992):

Sequence boundary (depositional sequences III & IV): SU + c.c.[(2)] + RSME

and resolution, as well as facies preservation, may make subjective the choice of where the subaerial unconformity loses its attribute of sequence boundary. For example, subaerial erosion during base-level fall may obliterate the offlapping pattern of stratal terminations, which may make it impossible to recognize which surface ('correlative conformity') corresponds to the oldest clinoform associated with offlap (i.e., the paleo-seafloor at the onset of forced regression).

According to the depositional sequence III and IV models, the sequence boundary is taken *at the top of forced regressive deposits*, and includes the entire subaerial unconformity, the correlative conformity (*sensu* Hunt and Tucker, 1992, which is the youngest clinoform associated with offlap), and the distal (younger) portion of the regressive surface of marine erosion that is overlain by lowstand normal regressive strata (Fig. 6.3). With identical sequence boundaries, depositional sequences III and IV differ in terms of their partitioning into systems tracts, as illustrated in Fig. 1.7. Although this difference is merely of semantic nature, it may create considerable confusion for a reader who is not aware of the nomenclatural preferences of various groups involved in stratigraphic research.

It is important to emphasize that regardless of what depositional sequence model is employed, the definition of the sequence boundary as the subaerial unconformity and its correlative conformity is oversimplified. In reality, there is a good probability that at least part of the correlative conformity that forms in the shallow-marine environment is reworked and replaced by the regressive surface of marine erosion (Figs. 4.23, 4.24, and 6.3). In this case, the regressive surface of marine erosion becomes part of the composite depositional sequence boundary, whether the correlative conformity is taken at the base or at the top of forced regressive deposits (Figs. 4.23, 4.24, and 6.3).

Irrespective of the depositional sequence model of choice, the key to a valid interpretation is the proper identification of facies relationships and syndepositional shoreline shifts. Such analyses allow identification of the key surfaces that can be used to build the sequence stratigraphic framework, which in turn can be used for genetic interpretations and predictive exploration. The choice of sequence type, based on what surfaces should mark the beginning and the end of full cycles of changes in depositional trends, becomes of secondary importance.

A conceptual merit of the depositional sequence models is that sequence boundaries are defined relative to the base-level curve (as opposed to the transgressive–regressive curve; Fig. 4.6), and hence they are independent of sedimentation rates. Variations in sedimentation rates along strike may result in the formation of highly

diachronous maximum flooding and maximum regressive surfaces, as demonstrated by numerical computations (e.g., Martinsen and Helland-Hansen, 1995; Catuneanu *et al.*, 1998b; more details on this topic are discussed in Chapter 7). This problem is bypassed by the depositional sequence boundaries, as correlative conformities (both *sensu* Posamentier *et al.*, 1988, and *sensu* Hunt and Tucker, 1992) can be equated more reliably with chronostratigraphic markers. Also, subaerial unconformities form arguably the most important type of stratigraphic surface, being associated with the largest hiatuses and separating genetically related packages of strata. A practical pitfall of these models is that the shallow-water portion of the correlative conformities is typically invisible in small- to average-size outcrops, in cores, or on wireline logs, although its approximate position may be inferred from larger-scale outcrops and seismic data within 10^0–10^1 m intervals. In deep-water settings, however, correlative conformities may be easier to pinpoint relative to other types of stratigraphic surfaces, based on changes in depositional elements that are likely triggered by the events marking the onset and end of base-level fall at the shoreline (e.g., Fig. 5.63).

Soon after the initial definition of depositional sequences by Mitchum (1977), an attempt was made to differentiate between sequences bounded by unconformities associated with widespread erosion ('type 1') and sequences bounded by surfaces associated with minimal erosion ('type 2') (Vail *et al.*, 1984). This distinction was conceptualized in terms of relative magnitudes of sea-level fall and subsidence at the shelf edge (Vail *et al.*, 1984) or at the shoreline (Posamentier and Vail, 1988), with the dominance of the former resulting in the formation of type 1 sequence boundaries and the dominance of the latter resulting in the formation of type 2 sequence boundaries. However, since the effects of sea-level change and subsidence are often difficult to separate in the rock record, the introduction of types 1 and 2 depositional sequences has proved to cause more confusion than benefit. Consequently, Posamentier and Allen (1999) advocate elimination of type 1 and 2 designations in favor of a single type of depositional sequence. More details regarding the definition of type 1 and type 2 sequences and sequence boundaries are provided in Chapter 5.

Lastly, additional confusion related to the concept of depositional sequence was caused by the temporal connotations introduced by a sequence hierarchy system based on the cyclicity of eustatic fluctuations (Vail *et al.*, 1991). Even though the original definitions of depositional sequences did not imply any spatial or temporal scales (Mitchum, 1977; Posamentier *et al.*, 1988), Vail *et al.* (1991) proposed to restrict the concept of depositional sequence to cyclothems formed by so-called 'third-order' eustatic cycles of 0.5–3.0 Ma duration. This approach poses significant practical problems, as it is the norm rather than the exception to lack the required time control for a precise measurement of the time involved in the formation of any particular sequence. In addition to this, depositional sequences may form over a broad range of temporal and spatial scales, all sharing similar characteristics and requiring the same type of analyses for the recognition of bounding surfaces and internal systems tracts (Posamentier *et al.*, 1992a; Wood *et al.*, 1994; Catuneanu, 2002). It has been proposed, therefore, that the term 'depositional sequence' be kept independent of scale, and the reference to scale, in either an absolute or a relative sense, be resolved by using modifiers such as first-order, second-order, etc. (Posamentier and Allen, 1999; Catuneanu *et al.*, 2004).

Genetic Stratigraphic Sequence

The genetic stratigraphic sequence (Galloway, 1989; Fig. 1.6) uses maximum flooding surfaces as sequence boundaries, both in the marine and in the continental portions of a sedimentary basin (Figs. 6.1 and 6.2). One of the main arguments for this choice of bounding surface is that the 'principal changes in the paleogeographic distribution of depositional systems and depocenters' occur during times of maximum shoreline transgression (Galloway, 1989). In turn, such changes in the distribution of depositional systems and depocenters mark significant shifts in sediment dispersal patterns across the maximum flooding surface, which is commonly identified as a 'downlap surface' in geometric terms (Schlager, 1991; Galloway, 2004).

The genetic stratigraphic sequence is subdivided into highstand (late rise), lowstand (fall and early rise), and transgressive systems tracts, using the same systems tract terminology as the depositional sequence II (Fig. 1.7). This model overcomes the problems related to the recognition of depositional sequence-bounding correlative conformities (surfaces formed at the onset and end of base-level fall) in shallow-marine successions, and has the merit that maximum flooding surfaces are relatively easy to map across a basin. In fact, due to their common association with regionally extensive shale units, maximum flooding surfaces are often easier to map on well logs and seismic lines than the subaerial unconformities. This practical aspect adds a significant bonus to the genetic sequence stratigraphic approach, and it is the reason why many geologists, regardless of their sequence stratigraphic 'affinity' (i.e., model of choice), prefer to start their stratigraphic analysis with mapping maximum flooding surfaces.

The criticism that this model has received is two-fold. Firstly, the genetic stratigraphic sequence includes the subaerial unconformity *within* the sequence (Fig. 4.6), which contravenes the generally accepted notion that sequences consist of genetically related packages of strata. Thus, the presence of subaerial unconformities within genetic stratigraphic sequences allows for the possibility that strata unrelated genetically may be put together into the same 'genetic' package. Secondly, the timing of maximum flooding surfaces depends on the interplay of base-level changes and sedimentation, and hence these surfaces may be diachronous (Posamentier and Allen, 1999). The rate of diachroneity of maximum flooding surfaces defined on stratal stacking patterns is, however, considered to be very low along dip, but it may become significant along strike depending on the fluctuations in terrigenous sediment influx to the various sediment entry points into the marine basin (Catuneanu *et al.*, 1998b). Thus, the onset of highstand normal regression may be delayed in areas of low sediment supply, where maximum flooding surfaces are younger than in areas of high sediment supply. More details about the temporal significance of the maximum flooding surface are provided in Chapter 7. In spite of these limitations, the genetic stratigraphic sequence retains the advantage of being bounded by a single and easily identifiable sequence stratigraphic surface. The basin-wide extent of the maximum flooding surface means that the genetic stratigraphic sequence is bounded by the same surface within both continental and marine portions of the sedimentary basin, which is a feature that is unique to this type of 'sequence.' In contrast, both the depositional sequence and the transgressive–regressive (T–R) sequence are bounded by composite surfaces, making the task of their delineation somewhat more difficult.

The genetic stratigraphic sequence model is linked to the manifestation of shoreline regressions and transgressions, and so it requires evidence of the type of syndepositional shoreline shifts for the proper identification of 'transgressive' deposits, maximum flooding surfaces, etc. Therefore, this model may not be applied to overfilled basins, or to the fluvial portions of basins where fluvial processes are independent of marine base-level changes (e.g., zone 3 in Fig. 3.3). On the other hand, because the genetic stratigraphic sequence does not use subaerial unconformities as sequence boundaries, the model can be applied to marine basins characterized by continuous base-level rise, where, in the absence of subaerial unconformities, stratigraphic cyclicity is controlled entirely by the interplay between the rates of base-level rise and sedimentation.

Transgressive–Regressive (T–R) Sequence

The transgressive–regressive (T–R) sequence (Embry and Johannessen, 1992) is bounded by composite surfaces that include subaerial unconformities on the basin margin and the marine portion of maximum regressive surfaces farther seaward. This model offers an alternative way of packaging strata into sequences, in an attempt to bypass some of the pitfalls of the depositional sequence and the genetic stratigraphic sequence. The proponents of the T–R sequence model recognized the value of subaerial unconformities as sequence boundaries, following the approach that was pioneered by the depositional sequence school, but eliminated the 'correlative conformities' (onset or end of base-level fall surfaces) as part of the sequence boundary due to the recognition problems they may pose in shallow-marine successions, particularly when seismic data are not available for analysis. At the same time, the T–R model avoids the problem of having subaerial unconformities within the sequence by using them as part of the sequence boundaries.

The 'correlative conformity' of the T–R sequence model is represented by the marine portion of the maximum regressive surface (Fig. 6.1). This stratigraphic surface has the advantage of being recognizable in *shallow-water settings* on virtually any type of outcrop or subsurface data, but it may pose recognition problems in *deep-water settings* where it is likely to develop within a conformable succession of low-density turbidite facies (Figs. 5.5 and 5.63). The recognition problems posed by the correlative conformities of the depositional sequence and the T–R sequence models are thus reciprocated between the shallow-water and deep-water settings. Another potential problem with using the maximum regressive surface as a sequence boundary is that the timing of its formation depends on sedimentation rates, and hence this surface may record a significant diachroneity along strike. As the influx of terrigenous sediment to the various sediment entry points into the marine basin may change considerably along strike, the onset of transgression may be delayed in areas of high sediment supply. In such areas, the maximum regressive surface is younger than in other areas characterized by a lower sediment supply, and such age differences may become significant enough to be resolved by biostratigraphy (e.g., case study by Gill and Cobban, 1973). More details on the numerical modeling of the temporal significance of maximum regressive surfaces, and of other stratigraphic surfaces, are provided in Chapter 7. For the nonmarine portion of the basin, the subaerial unconformity is used as the sequence boundary because it corresponds to the most important break in sedimentation, and therefore it should not be included within the

sequence. Maximum flooding surfaces are used to subdivide the T–R sequence into transgressive and regressive systems tracts (Figs. 1.7 and 4.6).

The amalgamation of different genetic types of deposits (highstand normal regressive, forced regressive and lowstand normal regressive) into one single unit, i.e., the 'regressive systems tract' (Fig. 1.7), provides a simple way of subdividing the rock record into systems tracts, and may be the only option in particular case studies (e.g., where stratigraphic cyclicity developed during continuous base-level rise, due to a shifting balance between the rates of subsidence and sedimentation), or where data are insufficient to afford the separation between the different genetic types of regressive deposits. However, this approach is not practical from an exploration perspective, because amalgamation of forced and normal regressive facies leads to a loss of critical 'resolution' in terms of the genetic aspect of stratigraphic analysis, which is the primary function of sequence stratigraphy. As discussed in Chapter 5, the sediment budget and the distribution of reservoirs across a basin change significantly between the four main stages of the base-level cycle (Figs. 5.7, 5.14, 5.26, 5.27, 5.44, 5.56, and 5.57). Figure 5.38 provides an example where the separation between forced regressive and overlying lowstand normal regressive deposits (systems tracts), and implicitly the mapping of their bounding 'correlative conformity' (*sensu* Hunt and Tucker, 1992) is the key for the identification of deep-water reservoirs. The amalgamation of forced and normal regressive deposits into one undifferentiated 'regressive systems tract' fails to provide useful criteria for finding the best deep-water reservoirs in this region, which formed during the late stages of forced regression and are now preserved *within* the 'regressive systems tract.'

A pitfall of the T–R sequence model is that the nonmarine and marine portions of the sequence boundary (the subaerial unconformity and the maximum regressive surface, respectively) are *temporally offset with the duration of the lowstand normal regression* (Figs. 4.6 and 5.5). The physical connection between these two surfaces may be made by the transgressive ravinement surface, assuming that the wave erosion in the upper shoreface during transgression removes all lowstand fluvial strata that accumulated in the vicinity of the shoreline (e.g., Fig. 5.6). This may happen only, however, where the thickness of the nearshore lowstand fluvial strata is less than 20 m, which is the maximum amount of scouring that is normally attributed to wave-ravinement processes (Demarest and Kraft, 1987). For lowstand fluvial wedges in excess of 20 m thick, a lowstand topset above the subaerial unconformity is preserved from subsequent transgressive wave scouring, and the maximum regressive surface may be mapped within both marine and fluvial portions of the basin (Fig. 6.2). In this case, the maximum regressive surface is separated from the subaerial unconformity by the fluvial portion of the lowstand systems tract, and hence the two surfaces do not form one single bounding surface as required by the definition of the T–R sequence. The possibility that the transgressive wave-ravinement surface may not remove all the lowstand nonmarine strata has been recognized and discussed by Embry (1995).

It may be concluded that the T–R sequence model 'works' only where the original (pre-transgression) thickness of the coastal to fluvial lowstand normal regressive deposits (i.e., lowstand 'topset' that accumulates on top of the subaerial unconformity) is limited to a range of meters (or, as a general rule, a thickness that is less than the amount of scouring associated with subsequent transgression). Where this condition is not fulfilled, the maximum regressive surface extends across the continental shelf, above the subaerial unconformity (Fig. 6.2). Such situations represent the norm particularly in sedimentary basins that are characterized by low-gradient depositional surfaces (e.g., continental shelves, 'filled' foreland basins dominated by shallow-water environments) and high sediment supply. Classic examples are provided by the Gulf of Mexico, where a significant portion of the basin fill is formed by lowstand normal regressive wedges generated as a result of high sediment supply derived from tectonically uplifted source areas located to the south. Thus, a borehole drilled on the continental shelf may intercept both the (younger) maximum regressive surface and the (older) subaerial unconformity along the same vertical profile, separated by lowstand normal regressive deposits (Fig. 6.2). This model flaw may be resolved only if the lowstand normal regressive deposits are assigned to the transgressive systems tract, i.e., considered to be the initial progradational 'pulse' of transgression (A.F. Embry, pers. comm., 2005). While providing the means for the 'maximum regressive surface' (base of transgressive systems tract) to become the true correlative conformity of the subaerial unconformity, this approach is invalidated by the overall grading trends: a borehole drilled in the shallow-water succession would show a coarsening-upward trend up to the last (youngest) clinoform of regression (i.e., the real maximum regressive surface), which is stratigraphically higher than the surface that connects to the basinward termination of the subaerial unconformity (i.e., the correlative conformity *sensu* Hunt and Tucker, 1992; Figs. 6.2 and 6.4). This is because regression continues following the end of base-level fall, resulting in the progradation of the shoreline farther seaward relative to its position at the end of forced regression. In this context, the inclusion of lowstand normal regressive deposits within the transgressive systems tract would result in inconsistencies in mapping the same stratigraphic surface based on different data sets: a maximum

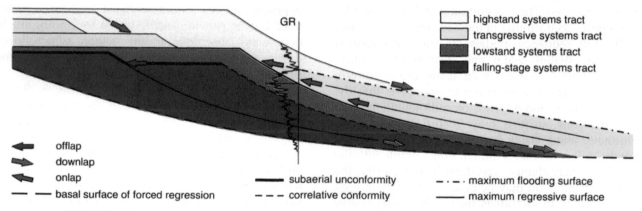

FIGURE 6.4 Conceptual and practical limitations of the transgressive–regressive (T–R) sequence model: where lowstand normal regressive deposits are preserved, the marine portion of the maximum regressive surface does not connect with the basinward termination of the subaerial unconformity. This dip-oriented cross-section illustrates a most general scenario for the architecture of forced regressive, normal regressive, and transgressive deposits in a fluvial to shallow-water setting (not to scale; modified from Schlager, 1994, 2002, 2005; Duval *et al.*, 1998; Posamentier and Allen, 1999; Catuneanu, 2002). The gamma ray (GR) log shows the overall grading trends in the shallow-water succession within the sequence stratigraphic framework. Note that the maximum regressive surface (top of the lowstand systems tract) marks the top of a coarsening-upward trend, as the progradation of the shoreline continues during the lowstand normal regression. This denies the option of assigning the lowstand normal regressive deposits to the transgressive systems tract, which would ensure the universal applicability of the T–R sequence model. In the latter approach, the maximum regressive surface would become effectively the correlative conformity *sensu* Hunt and Tucker (1992). This approach, however, would be misleading because well-log and seismic data would not place the maximum regressive surface at the same stratigraphic level (see text for more details).

regressive surface defined on well-log criteria (top of coarsening-upward trend; Fig. 4.32) would be stratigraphically higher/younger than a maximum regressive surface picked as the youngest forced regressive clinoform on seismic data (Fig. 6.4). In reality, the latter surface is the correlative conformity of Hunt and Tucker (1992).

Shoreline trajectories that are typical of lowstand normal regressions (i.e., progradation and aggradation following the end of base-level fall and preceding the onset of transgression; Figs. 3.35, 5.44, and 6.4) are documented in both siliciclastic and carbonate successions. The separation of their depositional products as distinct 'lowstand' systems tracts is warranted by stratal stacking patterns and depositional trends (e.g., continued progradation following the end of base-level fall and associated coarsening-upward trends in shallow-water clastic successions), and it has been reported from a variety of depositional settings ranging from fluvial (e.g., Kerr *et al.*, 1999; Leckie and Boyd, 2003), to clastic coastal and shallow-water (e.g., Plint, 1988; Plint and Nummedal, 2000; Hampson and Storms, 2003; Ainsworth, 2005), clastic deep-water (e.g., Posamentier and Kolla, 2003), and carbonate platforms (e.g., Cathro, 2003; Schlager, 2005). As shown above, the separation between lowstand and transgressive systems tracts is important not only for assessing the applicability of the T–R sequence model, but more significantly, for the correct identification of different types of sequence stratigraphic surfaces (e.g., maximum regressive surfaces *vs.* correlative conformities).

This discussion makes of course reference to the architecture of *same-order* stratigraphic surfaces, i.e., surfaces that belong to the same order of stratigraphic cyclicity. Higher-frequency (i.e., lower-order/rank) surfaces complicate the internal makeup of each systems tract, and should not be used to change the stratigraphic significance of higher-order/rank surfaces. The stratigraphic significance of third-order surfaces, for example, is not altered by the fourth-order surfaces that may be present in the same stratigraphic section, as the latter are less important and represent only 'details' within the third-order stratigraphic framework (e.g., a fourth-order maximum regressive surface that happens to be superimposed on a third-order correlative conformity does not change the significance of the third-order surface in the bigger-picture framework; in other words, one should wear the same 'glasses' for a stratigraphic analysis that focuses on a certain level/order of cyclicity—more discussions on the issue of sequence stratigraphic hierarchy are provided in Chapter 8).

As with the genetic stratigraphic sequence model, the T–R model is intrinsically related to shoreline shifts, and therefore it cannot be applied in overfilled basins or in fluvial successions that accumulated independently of marine base-level changes. In those cases, the depositional sequence subdivided into low- and high-accommodation systems tracts remains the only viable alternative of sequence stratigraphic analysis.

Parasequences

The 'parasequence' is a stratigraphic unit defined as 'a relatively conformable succession of genetically related beds or bedsets bounded by flooding surfaces' (Van Wagoner, 1995). Parasequences are commonly identified with the coarsening-upward prograding lobes in coastal to shallow-marine settings (Figs. 6.5 and 6.6). The deposition of each prograding lobe is terminated by events of abrupt water deepening, which lead to the formation of marine flooding surfaces. Such parasequences are usually the higher-frequency building blocks of successions associated with overall trends of coastal progradation or retrogradation (Figs. 5.52 and 6.6), so they may be part of larger-scale systems tracts. Depending on the scale of observation, parasequences could be placed within the context of larger-scale systems tracts, or they could be studied in relation to discrete cycles of changing depositional trends. Overall, there has been more confusion than advantage associated with the usage of the parasequence concept, as discussed below.

The main problem with the concept of parasequence rests with its bounding surfaces, i.e., the flooding surfaces. As explained in Chapter 5, the flooding surface is a poorly defined term that allows for multiple meanings, such as transgressive ravinement surface, maximum regressive surface, maximum flooding surface, or within-trend facies contact (e.g., Fig. 4.62). Depending on what type of stratigraphic surface the flooding surface actually is, parasequences may be anything from T–R sequences (where there is an abrupt cut-off of sediment supply at the onset of transgression, so the flooding surface is a maximum regressive surface; cases A, B, C, and D in Fig. 4.35), to genetic stratigraphic sequences (where the transgressive strata are missing altogether and the maximum flooding surface reworks the maximum regressive surface; Fig. 4.44) and allostratigraphic units (where the flooding surface is a facies contact within a transgressive package; Fig. 4.35E). The usefulness of such equivocal terminology is thus questionable, especially since each of the parasequence types is already covered by clearly defined terms.

Another topic of debate rests with the relationship between 'sequences' and 'parasequences.' Parasequences may or may not correspond to full cycles of change in depositional trends, depending on what type of

FIGURE 6.5 Outcrop examples of stacked parasequences (Woodside Canyon, Utah). Parasequences are prograding, coarsening-upward successions bounded by flooding surfaces. Parasequence boundaries (i.e., flooding surfaces) mark events leading to abrupt increases in water depth (arrows).

stratigraphic contact the flooding surface is in each case study. Where flooding surfaces are maximum regressive or maximum flooding surfaces, parasequences do correspond to full cycles of changing depositional trends (T–R and genetic stratigraphic sequences, respectively). Where flooding surfaces are transgressive ravinement surfaces or within-trend facies contacts, parasequences may be either incomplete sequences or units that have a longer duration than a single stratigraphic cycle. It is thus clear that parasequences are not just 'smaller-scale sequences,' as often implied, but a different type of stratigraphic unit altogether. Further discussions on the usage and misuse of the parasequence concept are provided by Posamentier and James (1993), Arnott (1995), Kamola and Van Wagoner (1995), Posamentier and Allen (1999), and Embry (2005). As pointed out by Posamentier and Allen (1999), 'contrary to sequences, parasequences are not defined as being unconformity-bounded (or correlative-conformity-bounded) stratigraphic units, and therefore do not constitute sequences in the sense of Mitchum (1977), small scale or otherwise.'

It is also important to clarify that the concept of parasequence, as originally intended by Van Wagoner *et al.* (1990), was devoid of scale connotations. The original definition was, however, modified in subsequent publications, generally restricting the usage of the term to smaller-scale sequences formed during specific time intervals. For example, in the sequence stratigraphic scheme of Krapez (1996), parasequences are equated with 'fourth-order' sequences that form during 90 000 – 400 000 year time intervals. This approach was criticized by Posamentier and Allen (1999) who proposed a return to the scale-free originally intended meaning of the concept of parasequence.

One of the original reasons for introducing the concept of parasequence was to use it as a tool for regional correlation, based on the fact that they are widespread and possibly associated with changes in sea level (Van Wagoner *et al.*, 1990). It is now known that many deltaic lobes (parasequences) accumulate as a result of autocyclic shifting in the location of depocenters, having a limited lateral extent and thus being only of local significance. Such lobes are indistinguishable in the field from other, more regionally extensive parasequences, which, in the absence of rigorous time control or physical correlation markers, poses a real problem to the process of stratigraphic correlation. Another attempt to expand the usefulness of parasequences for regional correlation was represented by the extrapolation of the term to describe cyclothems in a range of depositional systems that is broader than the coastal to shallow-marine settings for which the parasequence concept was originally defined. Thus, the use of parasequences in fully fluvial or in deep-water settings, where evidence of episodes of abrupt water deepening based on the relationship between juxtaposed depositional elements cannot be produced, is inadequate.

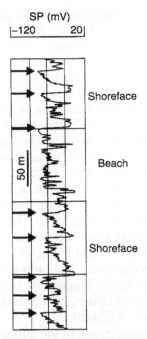

FIGURE 6.6 Well-log example of a succession of parasequences (Pliocene, Gulf of Mexico). Parasequence boundaries are marked by arrows (flooding surfaces). Each parasequence corresponds to a stage of shoreline regression, terminated by an episode of abrupt water deepening when flooding surfaces form. Parasequences may be associated with overall aggradational, progradational or retrogradational coastlines, representing higher-frequency stages of coastal regression within the overall trend of shoreline shift. Thus, they may be part of either standard (progradational) deltaic systems, or retrogradational bayhead deltas (Fig. 5.52). In this well-log example, the lower succession of six parasequences, which is topped by the beach sands, displays an overall progradational trend since a transition is made from shoreface (more distal) to beach (more proximal) depositional systems.

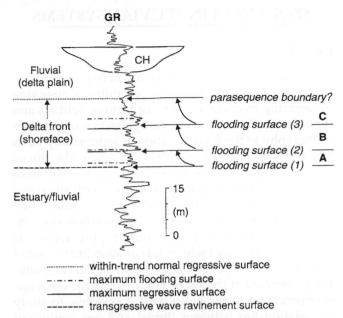

FIGURE 6.7 Concepts of 'parasequence' and 'flooding surface,' as exemplified by a shoreface/delta front succession of prograding facies. Flooding surface (1) is a transgressive wave-ravinement surface; flooding surface (2) is a maximum regressive surface; flooding surface (3) is a within-trend facies contact. A, B, and C are parasequences, i.e., dominantly coarsening-upward shallow-marine successions. The gamma ray (GR) log shown captures the transition between Belly River, Bearpaw, and Horseshoe Canyon lithostratigraphic units in the Western Canada Basin, in the region of the Bearpaw maximum transgressive shoreline. CH—channel fill.

The above discussion, as well as the discussion of flooding surfaces in Chapter 4, point out a number of inconsistencies in the usage of the parasequence concept, which stem in part from the looseness of the original definitions and also from subsequent modifications of the original meaning. It is therefore recommended that, if the term 'parasequence' must be used, it be restricted to prograding successions in coastal to shallow-water settings, where evidence of abrupt episodes of water deepening (flooding surfaces) can be produced. Such prograding successions should also be devoid of scale connotations and sea-level implications (Posamentier and Allen, 1999). However, even where the depositional setting affords the recognition of 'flooding surfaces' and 'parasequences,' alternative terminology is preferred for conveying stratigraphic information in unequivocal terms. Figure 6.7 provides an example of three parasequences (coarsening-upward units in a shoreface/delta front setting), bounded by different sets of stratigraphic surfaces. Hence, no two parasequences in this succession have the same sequence stratigraphic significance. Flooding surface #1 is a transgressive wave-ravinement surface, flooding surface #2 is a maximum regressive surface, and flooding surface #3 is a within-trend facies contact. Parasequence A is an incomplete T–R sequence, since part of the transgressive systems tract (estuarine facies) is preserved below the wave-ravinement surface. Parasequence B is larger than a T–R sequence, as it includes part of the transgressive deposits of the overlying T–R sequence. Parasequence C is bounded by two within-trend facies contacts, and it is even more problematic because its upper boundary, which marks the top of a coarsening-upward prograding lobe, is not even a flooding surface according to the definition. It is thus recommended that more unequivocal terminology be used instead of 'parasequences' and 'flooding surfaces,' where sufficient data are available to identify the nature of stratigraphic contacts observed in outcrop or subsurface. This alternative terminology is already available, and facilitates a better communication between the practitioners of sequence stratigraphy.

SEQUENCES IN FLUVIAL SYSTEMS

Introduction

Fluvial deposits are among the better understood depositional systems, due to the possibility of observing present day rivers in a variety of tectonic settings and climatic conditions. And yet, the application of sequence stratigraphy to the fluvial portion of sedimentary basin fills is most challenging, especially where the fluvial deposits under analysis are isolated or far away from coeval shorelines and marine influences. Such situations are often encountered in overfilled basins, dominated by nonmarine sedimentation, in basins where only the nonmarine portion of the stratigraphic record is preserved, or in cases where data availability is limited to the nonmarine portion of the basin. In the latter situation, every effort should be made to enlarge the scope of observation to the basin scale if possible, to study the relationship between fluvial and age-equivalent coastal and marine systems. Ultimately, however, modern stratigraphy is sufficiently versatile to provide a genetic approach to the analysis of any part of the stratigraphic record, from the scale of a single depositional system to the scale of entire basins.

Fluvial systems respond to a number of allogenic controls on sedimentation, which include eustasy, climate, source area tectonism, and basin subsidence (Shanley and McCabe, 1994, 1998; Fig. 6.8). The separation between their relative roles in fluvial successions is most challenging, although criteria started to

emerge from field studies (e.g., Isbell and Cuneo, 1996; Holbrook and Schumm, 1998; Shanley and McCabe, 1998; Catuneanu and Sweet, 1999; Fielding and Alexander, 2001; Posamentier, 2001; Catuneanu *et al.*, 2003b) and experimental work (e.g., Wood *et al.*, 1993; Koss *et al.*, 1994; Paola, 2000; Paola *et al.*, 2001). Tectonic influences may be interpreted from changes through time in syndepositional tilt directions and landscape gradients, as well as variations in burial depths, as inferred from analyses of paleocurrent directions, architectural elements, fluvial styles, and late diagenetic clay minerals (e.g., Holbrook and Schumm, 1998; Holbrook and White, 1998; Catuneanu *et al.*, 2003b; Ramaekers and Catuneanu, 2004). The relative roles of sea-level and climatic controls have also been evaluated by means of lithofacies and biofacies analyses, coupled with isotope geochemistry and petrographic studies of framework and early diagenetic constituents (e.g., Blum, 1994; Blum and Price, 1998; Sylvia and Galloway, 2001; Ketzer *et al.*, 2003a, b).

The relative importance of the allogenic controls on fluvial accommodation and sedimentation varies between the source areas and the shoreline, as suggested by Shanley and McCabe (1994) (Fig. 3.3). As a result, the application of sequence stratigraphic concepts to fluvial systems also changes with the location within the basin, and with the dominant controls on fluvial processes. Emphasis on eustasy and tectonism has led to the development of sequence models that predict variations in depositional trends and fluvial styles in relation to changes in marine accommodation

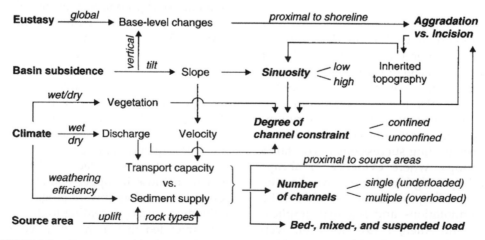

FIGURE 6.8 Allogenic controls on fluvial sedimentation. Fluvial processes of aggradation or incision may be influenced by *downstream controls* ('proximal to shoreline': mainly sea-level change and basin subsidence) and *upstream controls* ('proximal to source areas': particularly climate, basin subsidence and source area uplift). Note that climate may also be a downstream control by modifying the energy of waves and currents in the shallow-marine environment, and hence the position of the base level during fairweather *vs.* storms (Figs. 3.3, 3.4, and 3.15). The controls on the three key parameters that are used in the morphological classification of fluvial systems (sinuosity, degree of channel constraint and number of channels) are also indicated in this diagram.

(e.g., Wright and Marriott, 1993; Shanley and McCabe, 1993, 1994; Marriott, 1999). Emphasis on climate changes has produced models that explain shifts in depositional trends and fluvial styles based on the effects of climate on sediment supply, river discharge, and vegetation cover (e.g., Blum, 1994; Blum and Price, 1998). The essence of these models needs to be fully understood before tackling any case study of fluvial sequence stratigraphy. A set of first principles of fluvial sequence stratigraphy has been summarized by Miall (2002) (Fig. 6.9).

It is important to note that marine base-level changes, and the associated shoreline shifts, may only influence fluvial processes within a limited distance upstream from the coeval shoreline (i.e., zone 2 in Fig. 3.3). This distance is generally in a range of tens of kilometers (e.g., in the case of the Colorado River of Texas the influence of base-level fluctuations extends for about 90 km upstream, beyond which point the river has been affected primarily by climatic changes; Blum, 1994), up to more than 200 km in the case of rivers flowing on low-gradient landscapes (e.g., the Pleistocene fluvial systems of the Java continental shelf; Posamentier, 2001). Beyond the landward limit of the base-level control, the river responds primarily to a combination of climatic and tectonic mechanisms (Figs. 3.3 and 6.8). The course of a river, and implicitly the fluvial portion of a sedimentary basin, may therefore be split into a distal area under the influence of *downstream controls* (i.e., zone 2 in Fig. 3.3), and a proximal area under the influence of *upstream controls* (i.e., zone 3 in Fig. 3.3) (Fig. 6.8).

Fluvial systems that are under the influence of *downstream controls* (zone 2 in Fig. 3.3) respond to the interplay of sea-level changes, basin subsidence and the fluctuations in environmental energy imposed by climate shifts (Fig. 6.8). In such settings, fluvial processes of aggradation or incision correlate in a predictable manner with the pattern of shoreline shifts and the associated depositional processes in coastal and marine environments (e.g., Holbrook and Wright Dunbar, 1992; Tandon and Gibling, 1994, 1997; Feldman *et al.*, 1995; Holbrook, 1996; Heckel *et al.*, 1998; Miller and Eriksson, 2000). Therefore, these fluvial systems may be integrated within the standard sequence stratigraphic models that use the lowstand, transgressive, and highstand systems tract nomenclature. As a function of the tectonic setting, the downstream-controlled fluvial systems may develop within low- or high-accommodation *settings*, depending on the subsidence patterns recorded by that particular region. For example, the foredeep portion of a foreland system qualifies as a *high-accommodation setting*, due to the high rates of subsidence recorded adjacent to the orogen, whereas the distal portion ('backbulge') of the foreland system, adjacent to the craton, is referred to as a *low-accommodation setting* (Leckie and Boyd, 2003). Within each of these settings, however, the interaction between fluvial and marine processes allows the partitioning of fluvial deposits between lowstand, transgressive, and highstand systems tracts (Leckie and Boyd, 2003).

Fluvial systems that are under the influence of *upstream controls* (zone 3 in Fig. 3.3) record fluctuations in discharge regimes and sediment supply in a manner that is independent of base-level changes and dependent on the interplay of climate, source area tectonism and basin subsidence (Fig. 6.8). In such settings, the lowstand—transgressive—highstand systems tract terminology does not apply anymore, since fluvial processes of aggradation or incision are independent of any coeval shoreline shifts (e.g., Catuneanu and Elango, 2001). Instead, unconformity-bounded fluvial sequences may be subdivided into low- and high-accommodation *systems tracts*, based on the relative abundance of fluvial architectural elements (see Chapter 5 for more details on fluvial systems tracts). The usage of these

---

1. Fluvial incision may occur during periods of base-level fall, increased discharge, or reduced sediment load.
2. Fluvial aggradation may occur during periods of base-level rise, increased sediment load, or reduced discharge.
3. Fluvial responses to base-level shifts are mainly related to tectonism and eustatic fluctuations. Fluvial responses to changes in sediment load and discharge are primarily climate related.
4. Low-sinuosity fluvial systems, such as braided rivers or rivers with alternate bars, are most likely to occur during times of low accommodation.
5. Anastomosed fluvial systems are commonly associated with high rates of base-level rise, such as during transgression.
6. High-sinuosity (meandering) fluvial systems commonly characterize periods of low to moderate rates of base-level rise.
7. Straight fluvial systems that show little evidence of lateral migration are typical of areas of very low slope and low accommodation.
8. Incised valleys may be filled by fluvial systems of all types.
9. Evidence of marine influence within fluvial systems, such as tidal features, indicates flooding episodes (accommodation in excess of sedimentation).

FIGURE 6.9 First principles of fluvial sequence stratigraphy (modified from Miall, 2002).

systems tracts for the stratigraphic study of fluvial deposits represents a departure from the first-generation sequence stratigraphic models, whose systems tracts and predicted stratal architectures were intrinsically linked to changes in sea level or relative sea level (e.g., Vail et al., 1977; Jervey, 1988; Posamentier and Vail, 1988; Posamentier et al., 1988). The applicability of these early models to fully fluvial, proximal successions has been questioned by Blum (1990, 1994), Miall (1991), Schumm (1993), Wright and Marriot (1993) and Shanley and McCabe (1994), the debate culminating with the definition of low- vs. high-accommodation systems tracts (or 'successions') in the mid 1990s (e.g., Olsen et al., 1995; Dahle et al., 1997).

As with the downstream-controlled fluvial systems, the upstream-controlled fluvial systems may also develop in various tectonic settings characterized by different amounts of available accommodation. Even though the usage in conjunction of low- and high-accommodation settings and systems tracts seems cumbersome to some extent, fluctuations with time in the amounts of available fluvial accommodation in any tectonic setting permit the recognition of low- and high-accommodation systems tracts in both low- and high-accommodation settings. For example, a fluvial sequence developed within a high-accommodation 'setting' may include a succession of low- and high-accommodation 'systems tracts,' reflecting changes in subsidence rates during a full cycle of positive accommodation (e.g., Olsen et al., 1995; Arnott et al., 2002). The same may be said about fluvial sequences developed within low-accommodation settings, although in such cases fluvial sequences tend to consist almost exclusively of low-accommodation systems tracts (e.g., Olsen et al., 1995; Arnott et al., 2002).

From the above discussion it may be inferred that fluvial sequence stratigraphic models may be classified from two different viewpoints, one that emphasizes the presence or absence of marine influences during the accumulation of fluvial deposits, and one that lays emphasis on the amount of fluvial accommodation that is available during sedimentation. The first group of models makes the distinction between zones 2 and 3 in Fig. 3.3, in which fluvial systems relate to downstream and upstream controls, respectively, requiring the usage of different systems tract terminology. In this classification, downstream-controlled fluvial systems are part of standard lowstand – transgressive – highstand systems tracts, whereas the upstream-controlled fluvial sequences, formed independently of base-level fluctuations, are partitioned into low- and high-accommodation systems tracts. The second group of models focuses on the differences between the fluvial stratigraphy developed within low- vs. high-accommodation settings, irrespective of the presence or absence of marine influences on fluvial processes. The following is a brief discussion of the existing models of fluvial sequence stratigraphy.

Fluvial Cyclicity Controlled by Base-level Changes

The base-level control on fluvial cyclicity represents the essence of the first-generation sequence stratigraphic models, which assume a direct correlation between rising and falling base level, on the one hand, and fluvial aggradation and downcutting on the other, respectively (e.g., Jervey, 1988; Posamentier and Vail, 1988; Posamentier et al., 1988). The predictable relationship between fluvial processes and base-level changes reflects a most likely scenario, but exceptions do occur as discussed in Chapter 3 (e.g., Figs. 3.20 and 3.31). This relationship is valid for the downstream reaches of fluvial systems (zone 2 in Fig. 3.3), where rivers respond to 'downstream controls' (i.e., interplay of sea-level changes, basin subsidence, and climate-induced fluctuations in environmental energy flux). In such settings, which may be characterized by either low or high accommodation in Leckie and Boyd's (2003) scheme of fluvial stratigraphy, fluvial deposits may be integrated within the standard lowstand, transgressive, and highstand systems tracts.

Interesting to note is that in the case of fluvial processes controlled by base-level changes, both areas of fluvial aggradation and incision expand through time from the shoreline in an upstream direction, via the landward shift of depositional or erosional knickpoints. Thus, the gradual expansion of the depositional area during base-level rise results in a pattern of fluvial onlap (Figs. 5.4 and 5.5), whereas the landward expansion of incised valleys during base-level fall is linked to the upstream migration of erosional knickpoints (case A in Fig. 3.31; Figs. 3.32 and 5.16). Within this context, both fluvial aggradation and incision are explained by changes in fluvial-energy flux in response to corresponding changes in the slope gradient of the downstream portion of the fluvial landscape. As such, coastal aggradation during base-level rise leads to a shallowing of the fluvial profile in the vicinity of the shoreline, which in turn triggers fluvial aggradation. During base-level fall, a subaerially exposed seafloor that is steeper than the fluvial graded profile at the onset of forced regression initiates fluvial erosion, which starts from the shoreline, expanding gradually upstream. As depositional and erosional knickpoints migrate upstream during stages of base-level rise and fall, respectively, the difference in slope gradients between the new and the old fluvial profiles diminishes with increasing distance from the coeval shoreline, up to the point where rivers do not respond to base-level shifts anymore. This point represents the threshold between zones 2 and 3 in Fig. 3.3.

The ratio between channel and floodplain architectural elements during stages of positive accommodation depends on the rates of base-level rise. Rapid base-level rise leads to increased floodplain aggradation, which results in overall finer-grained successions. This often describes the architecture of transgressive and early highstand systems tracts. Slower base-level rise results in amalgamated channel fills, as very little (if any) accommodation is available for the overbank areas. At the same time, channel stacking during times of reduced accommodation may be accompanied by frequent avulsion, which helps to spread excess sediments laterally (Holbrook, 1996). Channel amalgamation under conditions of low accommodation is usually the case with the lowstand and late highstand systems tracts. As the late highstand amalgamated channel fills have a low preservation potential due to the subsequent erosion associated with the subaerial unconformity, the fluvial portion of the depositional sequence commonly displays a fining-upward profile (Fig. 4.6). These general principles of fluvial stratigraphy, which relate the stacking patterns of fluvial architectural elements to changes in base level and available accommodation, have also been documented in the case of fan-delta systems, which are governed by similar process/response relationships between fluvial processes within alluvial fans and the base-level fluctuations of the standing bodies of water into which they prograde (e.g., Burns *et al.*, 1997).

These theoretical considerations are incorporated into the fluvial sequence stratigraphic model of Shanley and McCabe (1991, 1993, 1994), which illustrates valley incision during base-level fall, aggradation of amalgamated channels during lowstand, tidally influenced transgressive systems, and highstand successions that end with laterally amalgamated meander belts (Fig. 6.10). The latter channel fills are usually not preserved, as explained above, which is why they are not illustrated in other similar diagrams (e.g., Fig. 15.9 of Miall, 1997). In parallel with the commonly observed fining-upward profiles, fluvial styles also change through the sequence, from higher energy, unconfined and amalgamated channel fills at the base (high width/height and sand/mud ratios; Fig. 5.49) to lower energy, confined and isolated ribbons towards the top (low width/height and sand/mud ratios). Tidal influences are particularly common within the transgressive systems tract, and their presence helps to identify the position of the maximum flooding surface in the nonmarine portion of the basin (Shanley *et al.*, 1992).

One aspect that is not captured in the model in Fig. 6.10 is the effect of the estuary on the fluvial styles established upstream. The estuary forms at the onset of transgression, and is replaced by a deltaic system at the onset of regression. The study of Late Cretaceous sequences in southern Utah (Shanley *et al.*, 1992) suggests that at the end of transgression, the termination of the low energy estuarine environment triggers an abrupt downstream shift of the threshold between age-equivalent braided and meandering fluvial systems (Fig. 4.45), providing an additional criterion for the recognition of maximum flooding surfaces within fluvial deposits. This abrupt change in fluvial styles across the maximum flooding surface may be attributed to the rapid downstream shift of the river mouth (anchoring point of the fluvial graded profile) at the onset of highstand normal regression, as the damming effect of the estuary is removed. One other important aspect to note in Fig. 4.45 is the trend of landward shift with time of the boundary between coeval braided and meandering systems within each systems tract, which explains the overall fining-upward profiles observed in most fluvial sequences. The trajectory through time of the boundary between braided and meandering fluvial systems in Fig. 4.45, including the abrupt shift in fluvial styles across the maximum flooding surface, has been used to generalize the vertical profile of fluvial deposits in Fig. 4.6.

A similar analysis is pertinent to the maximum regressive surface, in terms of changes in fluvial styles induced upstream by the formation of the estuary at the onset of transgression (Fig. 4.38). The work initiated by Kerr *et al.* (1999), and followed up by Ye and Kerr (2000), suggests that the formation of an estuary induces a lowering in fluvial energy upstream that causes an abrupt change in style from braided to meandering across the maximum regressive surface (Fig. 4.38). This may be attributed to the high rates of coastal aggradation that may accompany transgression, which in turn provide increased accommodation for floodplain sedimentation and result in the rapid shallowing of the fluvial landscape gradients. The transition from the high-energy lowstand rivers to the sluggish transgressive systems may therefore be quite rapid, assuming a significant increase in the rates of coastal aggradation at the onset of transgression. The observed abrupt change in fluvial styles across the nonmarine portion of the maximum regressive surface is also incorporated in the generalized vertical profile of fluvial deposits in Fig. 4.6.

The two abrupt lateral shifts of the threshold between coeval braided and meandering systems corresponding to the onset and end of the estuary lifespan, as well as the generalized fluvial vertical profiles of each systems tract, are illustrated in Fig. 6.11. Note that the overall fining-upward profile recorded by a fluvial sequence is suggested by the shift direction of the threshold between braided and meandering systems.

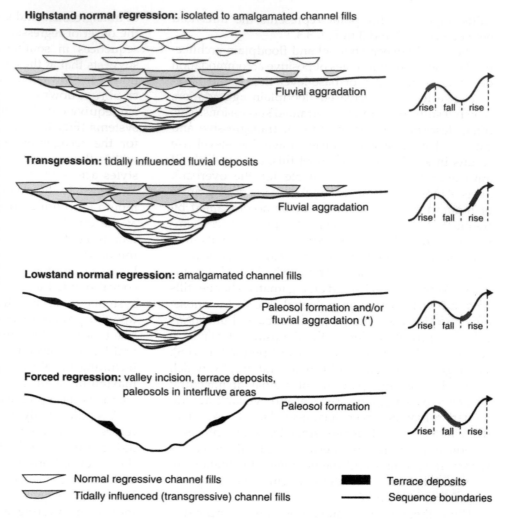

FIGURE 6.10 Stratigraphic architecture of a fluvial depositional sequence (modified from Shanley and McCabe, 1993). Note that the upper sequence boundary usually truncates the highstand deposits, which is why the uppermost amalgamated channel fills of the sequence have a low preservation potential. (*)The formation of the sequence-bounding paleosol may continue during the lowstand normal regression, depending on the magnitude of the previous valley incision. The formation of the sequence-bounding paleosol stops when sufficient accommodation is created to allow a floodplain to be established outside the incised valley. The changes in the height/width ratio of the channel fills suggest that the degree of channel confinement tends to increase upwards, from braided to meandering types of streams, in parallel with the gradual decrease in topographic gradients and fluvial energy.

This trend is the basis for the construction of the generalized vertical profile for fluvial sequences that is represented in Fig. 4.6. The overall fining-upward profile of a fluvial sequence is mainly a consequence of continuous coastal aggradation and the associated *shallowing of fluvial graded profiles* during base-level rise, coupled with the denudation of source areas. In contrast, the formation of the subaerial unconformity (stage of base-level fall) is accompanied by a *steepening of fluvial graded profiles* (case A in Fig. 3.31). The *caveat* for this generalization, as mentioned above, is that the predictable relationship between fluvial processes and base-level changes only reflects a most likely scenario, to which exceptions are known to occur (e.g., Figs. 3.20 and 3.31).

Fluvial Cyclicity Independent of Base-level Changes

Outside the range of influence of base-level fluctuations (zone 3 in Fig. 3.3), rivers respond primarily to a combination of 'upstream controls' which include climate shifts, source area tectonism, and basin subsidence. In such settings, fluvial deposits may not be integrated anymore within the standard lowstand, transgressive, and highstand systems tracts because no relationship may be established between fluvial processes and any coeval shoreline shifts. Instead, unconformity-bounded upstream-controlled fluvial sequences may be partitioned into low- and high-accommodation systems tracts based on the relative abundance of fluvial architectural elements (Figs. 5.68 and 5.72). This is quite different from the distinction between low- and high-accommodation settings, which refer to the subsidence patterns of particular tectonic settings, and may each include low- and high-accommodation systems tracts as discussed above.

Upstream-controlled fluvial systems are located closer to the source areas (Fig. 6.12), and therefore they are more susceptible to source area tectonism (Fig. 6.13), in addition to the effects of climate on discharge and sediment supply, and the overall amounts of fluvial accommodation made available by basin subsidence.

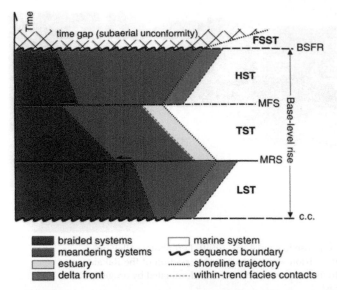

braided systems · marine system
meandering systems · sequence boundary
estuary · shoreline trajectory
delta front · within-trend facies contacts

FIGURE 6.11 Fluvial response to downstream controls during base-level rise (based on case studies by Shanley *et al.*, 1992, and Kerr *et al.*, 1999). Each systems tract includes a fining-upward fluvial succession (see the landward shift with time of the boundary between coeval braided and meandering fluvial systems) that develops in response to coastal aggradation and the corresponding shallowing of fluvial graded profiles. The onset and end of transgression are marked by abrupt shifts in fluvial styles across the maximum regressive and maximum flooding surfaces (see text for details). The shift with time of the boundary between coeval braided and meandering systems provides the basis for the generalized fluvial vertical profile in Fig. 4.6. Abbreviations: LST—lowstand systems tract; TST—transgressive systems tract; HST—highstand systems tract; FSST—falling-stage systems tract; c.c.—correlative conformity *sensu* Hunt and Tucker (1992); MRS—maximum regressive surface; MFS—maximum flooding surface; BSFR—basal surface of forced regression.

The stratigraphic cyclicity of fluvial deposits accumulated independently of base-level changes may be driven by climatic cycles superimposed on a steady tectonic regime, or by tectonic cycles superimposed on a longer-term climatic background.

Climatic Cycles

Fluvial sequences reflecting climate changes are primarily controlled by orbital forcing, i.e., Milankovitch cycles of glaciation and deglaciation at temporal scales of 10^4–10^5 years. Climatic fluctuations have a direct impact on the river discharge (Fig. 6.8), hence altering the balance between the transport capacity of the river and its sediment load. Any change in the loading index modifies the position of the fluvial graded profile, which may shift either above or below the topography. Increasing transport capacity (energy) relative to the sediment load places the new graded profile below the topography (negative fluvial accommodation), which triggers fluvial incision. Increasing sediment load relative to the transport capacity places the new graded profile above the topography (positive fluvial accommodation), which triggers fluvial aggradation. During times of deglaciation, melting of the ice increases the discharge of fluvial systems, which in turn leads to fluvial incision. This is the opposite of what is expected during a time of glacioeustatic rise. During stages of glaciation, the low river discharge modifies the loading index in the favor of sediment load, which results in fluvial aggradation. This is the opposite of what is expected from a stage of glacioeustatic fall.

FIGURE 6.12 Active sediment source area, and colluvial fans, in the Southern Alps of New Zealand (Arthur's Pass, South Island, New Zealand).

FIGURE 6.13 Modern examples of fluvial incision in response to source area uplift. Left photo: valley incision in the Southern Alps of New Zealand. Right photo: incision along the upstream portion of the Rakaia River, South Island, New Zealand. Both rivers are close to the Southern Alps, in areas dominated by orogenic uplift, and drain towards the Canterbury Plains to the East.

This climate-driven model has been documented by Blum (1994, 2001) based on the study of the Late Cenozoic fluvial record of the Gulf Coast rivers. These case studies demonstrate that fluvial cycles controlled by climate changes may be completely out of phase relative to those driven by base-level changes.

The effect of climate change on fluvial processes of aggradation and degradation is particularly evident in tectonically-stable inland regions, such as along the cratonic margins of foreland systems, that are remote from the influence of sea-level fluctuations (Gibling et al., 2005). The unconformity-bounded late Quaternary fluvial sequences in the southern Gangetic Plains provide a case study where fluctuations in monsoonal precipitation triggered by climate shifts generated cycles of floodplain aggradation and degradation over time scales of 10^4 years. In this example, sequences record periods when floodplains were inundated and experienced sustained aggradation, whereas declining flood frequency on parts of the interfluves resulted in low-relief degradation surfaces and badland ravines, as well as local soil development (Gibling et al., 2005).

Tectonic Cycles

Higher-frequency tectonic cycles of subsidence and uplift superimposed on a longer-term background of climatic stability may also lead to the development of cyclicity in fluvial deposits. Models for such unconformity-bounded fluvial sequences, which form in isolation from marine influences, need to be customized for individual tectonic settings, as the mechanisms and patterns of subsidence and uplift vary considerably among the different types of sedimentary basins. As an example, the model established for the foredeep

portion of retroarc foreland systems is based on the cyclicity of thrusting and offloading stages in the adjacent fold-thrust belts (e.g., the overfilled phase in Fig. 2.64; Catuneanu and Sweet, 1999; Catuneanu and Elango, 2001). In the case of the Karoo Basin, the Balfour Formation of the Beaufort Group is composed of a succession of six third-order fluvial depositional sequences bounded by subaerial unconformities (Fig. 5.74; Catuneanu and Elango, 2001). These fluvial sequences formed in isolation from eustatic influences, with a timing controlled by orogenic cycles of thrusting (loading) and quiescence (erosional or extensional offloading). Sediment accumulation took place during stages of flexural subsidence and shallowing of the landscape gradient, whereas bounding unconformities formed during stages of isostatic uplift and steepening of the landscape gradient (Fig. 2.64). The vertical profile of each sequence displays an overall fining-upward trend related to the gradual decrease in topographic slope during orogenic loading that is caused by the pattern of differential subsidence, with higher rates towards the orogen (Catuneanu and Elango, 2001). At the same time, the lowering of the slope gradients during the deposition of each sequence is accompanied by an upward change in fluvial styles, from initial higher to final lower energy systems. The actual fluvial styles in each location depend on paleoslope gradients and the position of the stratigraphic section relative to the orogenic front. Proximal sequences show transitions from braided to meandering systems, whereas more distal sequences show changes from sand-bed to fine-grained meandering systems (Fig. 5.74). The average duration of the Balfour stratigraphic cycles is 0.66 Ma, i.e., six cycles during 4 Ma. No climatic fluctuations are recorded during this time, with the long-term climatic background being represented by temperate to humid conditions. In this example, the creation of fluvial accommodation during the deposition of each sequence is entirely attributed to flexural subsidence. This is in contrast with cases of fluvial accommodation created by marine base-level rise or by decreases in fluvial discharge caused by stages of climate cooling.

Low- vs. High-Accommodation Settings

Irrespective of the presence or absence of marine influences during the accumulation of fluvial deposits, the pattern of syndepositional subsidence has a profound influence on the architecture of unconformities and depositional elements within the fluvial succession. Such variability in the amounts of available fluvial accommodation may affect both zones 2 and 3 in Fig. 3.3, leaving a significant mark on the stratigraphic characteristics of fluvial successions. The differences in fluvial architecture observed between slowly- and rapidly-subsiding basins, or portions thereof, led to the distinction between low- and high-accommodation *settings* (Boyd *et al.*, 1999, 2000; Zaitlin *et al.*, 2000, 2002; Arnott *et al.*, 2002; Wadsworth *et al.*, 2002, 2003; Leckie and Boyd, 2003; Leckie *et al.*, 2004).

The low- and high-accommodation settings may host either standard lowstand – transgressive – highstand systems tracts, if a correlation between fluvial

deposits and the shifts of a coeval shoreline may be established, or low- and high-accommodation systems tracts in the case of tectonically- or climatically-controlled fluvial sequences. This model of fluvial sequence stratigraphy is therefore centered on the characteristics of the tectonic setting rather than the under-filled or overfilled nature of the basin fill. Refinements to the fluvial accommodation-based sequence stratigraphic model led to a diversification of the described tectonic settings, from low-accommodation to intermediate-, high- and very-high accommodation (Leckie and Boyd, 2003). The 'standard' nonmarine sequence stratigraphic model, however, which is also easier to apply, makes the basic distinction between low- and high-accommodation settings. Some of the fundamental criteria that are used to separate between these two settings are presented in Fig. 6.14.

SEQUENCES IN COASTAL TO SHALLOW-WATER CLASTIC SYSTEMS

Introduction

Coastal settings include river-mouth environments, which represent sediment entry points into the marine basin, as well as open shorelines in the areas between river mouths (Figs. 2.3 and 2.4). Coastal environments

Criteria	Setting	
	Low-accommodation	High-accommodation
Sequence boundaries	Multiple, closely spaced	Rare, widely spaced
Incised valleys	Multiple and compound	Rare
Stratigraphic sections	Thin, HST deposits commonly missing	Thick, with deposition in all systems tracts
Channel sandbodies	More amalgamation, potentially isolated near MFS	Amalgamated near SB, isolated near MFS
Floodplain fines	Rare, potentially present near MFS	Common and abundant
Tidal deposits	Most abundant near MFS	Most abundant near MFS
Underlying topography	Enhanced control	Weak control
Coal seams	Commonly absent; compound where present	Abundant, thicker, and simpler (fewer hiatuses)
Paleosols	Well-developed, multiple and compound	Thinner and widely spaced, more organic-rich

FIGURE 6.14 Stratigraphic criteria that are used to differentiate between low- and high-accommodation settings in the context of fluvial sequence stratigraphy (modified from Leckie and Boyd, 2003). The definition of low- and high-accommodation settings is based on the subsidence patterns of the tectonic setting, and is independent of the presence or absence of marine influences on fluvial sedimentation. As such, fluvial sequences formed within low- and high-accommodation settings may consists of either standard lowstand—transgressive—highstand systems tracts (zone 2 in Fig. 3.3), or fully fluvial low- and high-accommodation systems tracts (zone 3 in Fig. 3.3) (see text for details).

continue offshore with shallow-marine environments up to the shelf edge, which marks the 'boundary' between shallow- and deep-water settings.

As discussed in Chapter 5, many petroleum plays are genetically related to coastal and shallow-marine systems, so the understanding of their processes and products represents a critical pre-requisite for successful exploration. Coastlines are also important for the coal and mineral resources industries, as they limit the lateral extent of the stratigraphic units of interest. In addition to this, coastlines are a key element of standard sequence stratigraphic models, representing the link between the nonmarine and marine portions of the basin. Also, coastline processes, and their transgressive and regressive shifts, control the timing of all seven sequence stratigraphic surfaces, as discussed in detail in Chapters 4 and 7. All these aspects provide coastlines with particular relevance to sequence stratigraphy, as the main switch that controls sediment supply to the marine basin and, implicitly, the formation and architecture of systems tracts – the building blocks of stratigraphic sequences.

Physical Processes

Coastal to shallow-water environments are shaped by the interaction between sediment supply and basinal processes of sediment reworking. This section presents the basic mechanisms of sediment transfer between the subenvironments of this key region of a sedimentary basin. Understanding of these processes is not only relevant to Process Sedimentology, but it is also fundamental for Sequence Stratigraphy due to the genetic nature of this approach of strata analysis (Fig. 1.2).

Sediment Supply and Transport Mechanisms

Most of the clastic sediment is terrigenous in origin, transported from source areas to the receiving basin by water (rivers) or wind. Additional sediment supply may derive from coastal (cliff) erosion (Figs. 3.20 and 3.24), as well as from marine erosion in the shoreface or deeper areas (e.g., Figs. 3.20, 3.21, and 3.27).

Transport and reworking of sediments within the coastal – shallow-marine environments may be related to several factors, including tides and fairweather waves, episodically enhanced by storms; hyperpycnal (gravity) flows, denser than the ambient seawater; and hypopycnal (buoyant) plumes, which are less dense than the ambient seawater. **Fairweather waves** give rise to a range of currents which may be directed offshore (*rip currents*), parallel to the shore (*longshore currents*), obliquely (*obliquely-directed currents*) and onshore (*onshore residual motions*). **Tides** primarily

affect shorelines by raising and lowering sea level, thus not only shifting the site of wave action but also generating *tidal currents* as large volumes of water move between the sea and the land across the intertidal (foreshore) area. **Storms** interrupt fairweather processes by increasing their intensity and by giving rise to heightened turbulence and sudden movement of water and sediment both offshore and onshore (Reading and Collinson, 1996). In addition to (fairweather and storm) waves and tides, gravitational reworking is also important in many shallow-marine settings. **Gravity flows** may be triggered on any slope where the gravitational shear exceeds the internal shear strength of the water/sediment mixture. In shallow-marine environments, gravity-driven subaqueous flows are particularly common in the steeper (approximately 0.3° or more) delta front/shoreface areas (Fig. 6.15), helping to disperse the terrigenous sediment into the deeper prodelta and shelf environments. **Hypopycnal plumes** may also provide a mechanism whereby riverborne sediment bypasses the river mouth and is transported out onto the shelf as buoyant suspended load that is less dense than the ambient seawater. Depending on the prevailing winds, waves, and currents, hypopycnal plumes may carry terrigenous sediment far from the river mouth, even beyond the shelf edge, until the plume loses momentum and

FIGURE 6.15 Slump features associated with delta front facies indicating the participation of gravity flows in the process of sediment transport to the deeper lower shoreface—prodelta areas. The effect of gravity, in combination with the seafloor gradients, modifies the commonly assumed linear relationship between depositional energy and water-depth changes (i.e., environmental energy in a deeper-water setting may sometimes be higher than the depositional energy in a shallower-water setting). The photograph shows a detail from the Waterford Formation (Ecca Group, Late Permian), Karoo Basin.

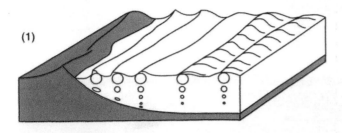

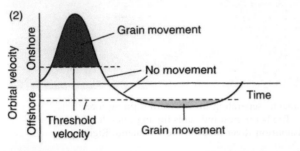

FIGURE 6.16 The littoral energy fence (modified from Swift and Thorne, 1991). (1) Wave transformation as a shoreline is approached. The orbital diameter decreases with depth and becomes asymmetrical as it nears the seafloor and frictional drag increases. Below the fairweather wave base, this wave-driven orbital motion results in a to-and-fro motion of the sediment on the seafloor. (2) The effects on sediment movement during the passage of a shoaling wave. The onshore stroke of the wave as the crest passes carries more sediment than the offshore stroke associated with the passage of the trough.

the suspended load accumulates into the receiving basin. These sediment transport mechanisms may alter the commonly assumed linear relationship between depositional energy, grain size, and water depth, and explain the progradation of coarsening-upward successions into deepening water (see Chapter 7 for more detailed discussions on this topic).

Terrigenous, coastal erosion-derived and shoreface-sourced sediment, other than mud, does not easily pass out on to the shelf. This is because during fair-weather, shoaling and breaking waves have the potential to move more sediment landward than seaward, hence keeping the sand within the beach and shoreface area ('littoral energy fence': Fig. 6.16). In addition to this energy fence, the only offshore-directed fairweather wave-generated currents (rip currents) die out in the lower shoreface. This fence can be broken or 'bypassed' by (1) fluvial processes, especially during the flood stages of rivers, which may generate hyperpycnal (gravity) flows and/or hypopycnal plumes in front of river-dominated deltas; (2) tidal currents, which may transport 'allochthonous' sediment from estuaries out onto the shelf (Swift and Thorne, 1991); (3) storm surges, which may erode any coastal setting (river-mouth or open shoreline) and redistribute the sediment offshore, into the inner shelf; and (4) gravity flows, other than fluvial-related, that may redistribute sediment sourced by coastal or

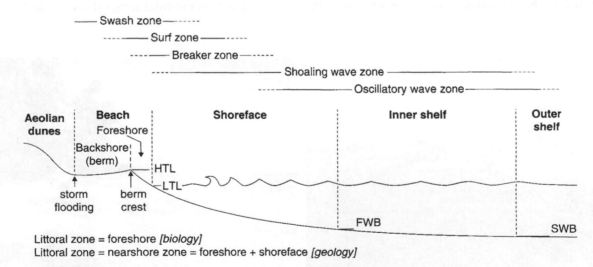

FIGURE 6.17 Dip-oriented coastal to shallow-marine profile, showing the various subenvironments as defined relative to the high tide level (HTL), low tide level (LTL), fairweather wave base (FWB), and storm wave base (SWB) (modified from Walker and Plint, 1992, and Reading and Collinson, 1996). The effect of wave energy on the seafloor increases towards the coastline, which is balanced by a concave up graded (hydraulic equilibrium) profile. The tendency of preservation of this shoreface profile that is in equilibrium with the wave energy during transgressions and forced regressions is the key for the formation of wave-ravinement surfaces and regressive surfaces of marine erosion, respectively (Figs. 3.20 and 3.27).

FIGURE 6.18 Berm crests in open shoreline, beach environments, separating the seaward-dipping fore-shore from the landward-dipping backshore subenvironments. The berm crest indicates the high tide level (Fig. 6.17). Left image: Canterbury Plains (just south of the Ashburton River mouth), New Zealand. Right image: Melville Island, Canadian Arctics.

shoreface erosion beyond the fairweather wave base. Any of these bypass mechanisms may result in the sediment being moved beyond the shoreface, on to the shelf.

Zonation of the Coastal – Shallow-marine Profile

The environments of the coastal to shallow-marine settings are illustrated in Fig. 6.17. The limit between the beach (coastal environment in an open shoreline setting; Figs. 2.3 and 2.4) and the shallow sea is marked by the low tide level. Landward from the shoreline, the limit of storm flooding marks the boundary between the beach and the nonmarine environment (Fig. 2.4).

Within the beach itself, the crest of the berm, which corresponds to the landward limit of marine influences during fairweather (high tide level in Figs. 6.17 and 6.18), splits the open shoreline coastal environment into foreshore and backshore subenvironments (Figs. 2.4, 6.17, and 6.19).

The backshore (Fig. 6.19) only receives marine water during storms, so it is subaerially exposed for most of the time and may be characterized by stagnant water and swampy conditions. In front of the berm crest, the foreshore dips gently towards the sea (Fig. 6.19), is placed within the tidal range (Figs. 2.4 and 6.17), and

FIGURE 6.19 Beach subenvironments in open shoreline settings. See Figs. 2.4 and 6.17 for the definition of backshore and foreshore subenvironments. Left image: backshore (supratidal) subenvironment, with ponds of stagnant water that accumulates during flooding events (storms). Such areas are usually soft wetlands with a high water table (Melville Island, Canadian Arctics). Right image: foreshore (intertidal) subenvironment in a wave-dominated coastal setting. Note the dip direction of the topographic profile, which leads to the formation of low-angle stratification that dominates foreshore deposits (west coast of Barbados).

FIGURE 6.20 Lithified foreshore deposits in a wave-dominated coastal setting (west coast of Barbados). Note the scour surfaces (arrows) that separate distinct foreshore packages consisting of low-angle stratified beds. The deposition of each package corresponds to a period of time of fairweather, whereas the scour surfaces mark storm events.

is characterized by packages of low-angle stratified sands separated by scour surfaces (Fig. 6.20). The scour surfaces mark storm events, when the beach landscape has been reshaped, whereas the intervening packages of low-angle stratified sandstone correspond to periods of time of fairweather when the beach is rebuilt. The contrast in dip angle between the stratified sands that are in contact across the scour surfaces reflects changes in beach morphology as a result of differential storm erosion. In addition to low-angle stratification, both lower and upper flow regime sedimentary structures may form in the intertidal (foreshore) environment, as a function of the energy level of the dominant sediment-reworking process. Tidal currents may generate asymmetrical bedforms that migrate in the direction of the dominant ebb or flood flow (Fig. 6.21). Where both tidal currents are equally strong, herringbone structures may form in the adjacent subtidal environment (Fig. 6.22). Waves may generate even higher-energy landward- and seaward-directed flows in the foreshore area (swash and backwash flows, respectively), at higher frequency than the tidal currents, producing upper flow regime structures such as parting lineation (Figs. 6.23 and 6.24).

The capacity of wave energy to move sediments on the seafloor in coastal to shallow-marine environments increases from the open sea towards the land, with the shallowing of the water. This change in energy

FIGURE 6.21 Foreshore (intertidal) environment in a tide-dominated coastal setting (Christchurch, New Zealand). Unidirectional (asymmetrical) current ripples, related to the dominant flood tidal flow, represent the main sedimentary structures that form in this setting. Left image: note the presence of a tidal channel close to the camera. Right image: detail of tidal current ripples, indicating flow from left to right.

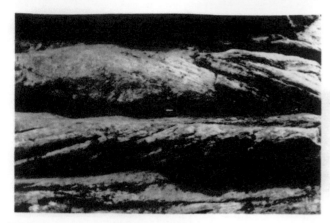

FIGURE 6.22 Herringbone cross-stratification in upper subtidal deposits, indicating equally strong, ebb and flood, tidal currents. The photograph comes from the Gog Group (Early Cambrian), Athabasca Falls, Jasper National Park, Alberta.

regimes requires a corresponding change in seafloor gradients, being balanced by steeper gradients towards the coastline. This explains the concave up profile of the nearshore (foreshore + shoreface) zone (Figs. 6.16 and 6.17), which is in equilibrium with the wave energy. The preservation of this shoreface 'graded' profile is the main reason for the formation of wave-ravinement surfaces and regressive surfaces of marine erosion during transgressions and forced regressions, respectively (Figs. 3.20 and 3.27). Hence, processes of wave scouring are the result of the difference in slope gradients between the steeper upper shoreface and the shallower fluvial and lower shoreface profiles.

Wave energy starts to have an effect on the seafloor as soon as the water depth becomes as shallow as about half of the wave wavelength (Walker and Plint, 1992). This is how the fairweather wave base and the storm wave base are defined (Fig. 6.17), as storm waves have higher magnitudes and corresponding longer wavelengths. Above the fairweather wave base, the effect of wave energy on the seafloor is increasingly important towards the shoreline, and the pattern of orbital motion related to the passage of waves becomes more and more asymmetrical with the shallowing of the water (Fig. 6.16). These changes in energy regime, pattern of orbital motion, and level of frictional drag allow distinguishing of five energy zones according to the dominant wave processes (Fig. 6.17; Reading and Collinson, 1996):

(1) *Oscillatory wave zone.* The passage of each wave results in a symmetrical, straight-line, to-and-fro motion in the direction of wave propagation at the sediment surface.

(2) *Shoaling wave zone.* The waves are extensively modified and change from a symmetrical, sinusoidal form to an asymmetrical, solitary form: wave velocity and wavelength decrease; wave height and steepness increase; only wave period remains constant. Wave motions at the sediment surface involve a brief landward-directed surge, and a rather longer, but weaker, seaward-directed return flow. Thus more sediment is carried landward than seaward (Fig. 6.16).

(3) *Breaker zone.* Progressive steepening of the wave as it approaches the shoreline causes it to oversteepen and break in a landward direction. High energy

FIGURE 6.23 Swash zone of a foreshore environment in a high-energy, wave-dominated coastal setting (Canterbury Plains, New Zealand). Left image: high-energy backwash currents generated by waves, whose powerful scouring action may generate parting lineation. Right image: parting lineation (ridges and grooves) generated by the scouring action of high-energy backwash currents.

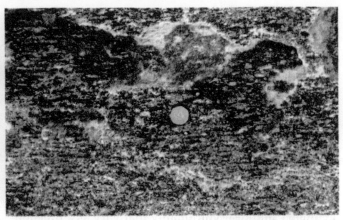

FIGURE 6.24 Parting lineation (upper flow regime) along the bedding planes of low-angle stratified and lithified foreshore deposits (west coast of Barbados).

conditions in the breaker zone cause fine sand to be suspended temporarily while coarser sand is concentrated at the bed.

(4) *Surf zone.* Breaking of waves, particularly as surging breakers, generates the surf zone in which a shallow, high velocity bore is directed up the shoreface on to the foreshore. Coarse sediment is transported landward while finer sand and silt are suspended briefly in bursting clouds.

(5) *Swash zone.* This zone occurs at the landward limit of wave penetration, where each wave produces a

shallow, high velocity, landward-directed swash flow followed almost immediately by an even shallower, seaward-directed backwash which may disappear by infiltration into the bed. Upper plane bed or standing wave/antidune conditions are prevalent in this zone.

The general trend across the shoreface (Fig. 6.17) is therefore one in which oscillatory flow transforms into asymmetrical, landward-directed flow of increasing power. Bedforms reflect this change, showing transitions from lower flow regime conditions in the

FIGURE 6.25 The intertidal to upper subtidal facies accumulate under the highest energy conditions of the coastal to shallow-marine environments, and hence they tend to include the coarsest sediment fractions (i.e., potentially the best petroleum reservoirs of the coastal to shallow-marine systems, with the least amount of mud). Left image: foreshore to upper shoreface sandstones, with a total thickness in excess of 30 m. Both wave- and tide-related sedimentary structures are present in these rocks (paleoflow directions of longshore and tidal currents are perpendicular to each other). Such deposits may form excellent petroleum reservoirs, with very good lateral extent along strike. The outcrop photograph shows the Gog Group (Early Cambrian), Athabasca Falls, Jasper National Park, Alberta. Right image: gravelly beach deposits, suggesting a high energy coastal environment. This is an ancient analogue of the present day gravel beaches of the Canterbury Plains (Figs. 3.24 and 6.18). The photograph shows the Enon Formation (Jurassic), Plettenberg Bay, South Africa.

lower/middle shoreface (oscillatory and shoaling zones dominated by ripples and dunes) to upper flow regime conditions in the upper shoreface and foreshore (breaker, surf and swash zones, dominated by upped flat beds with parting lineation; Figs. 6.20, 6.23, and 6.24).

Overall, the foreshore to upper shoreface deposits represent the product of sedimentation under the highest energy conditions of the coastal to shallow-marine environments, and tend to include the coarsest sediment fraction, with the lowest amounts of mud (Fig. 6.25). These deposits therefore form a prime target for petroleum exploration, with a stratigraphic geometry and reservoir quality that vary as a function of sediment supply, and also with the direction and type of shoreline shift (Figs. 5.7, 5.26, 5.27, 5.44, 5.56, and 5.57, and associated discussions in Chapter 5).

Sediment Budget: Fairweather vs. Storm Conditions

During fairweather, oscillatory and shoaling wave processes operate in the lower part of the shoreface, and breaker/surf zone processes in the upper shoreface. Relatively weak rip currents and longshore currents may operate on the upper shoreface on sandy barred shorelines but otherwise are insignificant. The shoreface tends to have a smooth concave up profile, with no bars (Fig. 6.26). The lower shoreface and offshore zones are not affected by waves and so fine-grained sediment is deposited from suspension and reworked by organisms. On the upper shoreface and foreshore, wave-induced currents, associated with the shoaling, breaker, and surf zones, transport sediment landward. Little sediment is lost *via* seaward-directed rip currents and the beach therefore aggrades (Reading and Collinson, 1996).

During storms, dissipative conditions prevail. The upper shoreface and beach are extensively eroded, and the seafloor gradient is shallower relative to the fairweather conditions. Sediment is both redeposited

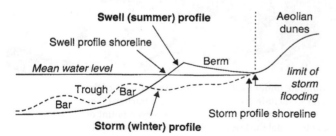

FIGURE 6.26 The beach cycle, of an alternate swell (fairweather) profile, when a pronounced berm is built up and the shoreface profile is smooth, and a succeeding storm profile, when the beach is eroded, the sediment is redistributed to the shoreface and shelf, and offshore bars are formed (modified from Komar, 1976).

landward as washover fans in lagoons, and swept seaward by both enhanced rip currents and wind-driven storm currents generating offshore bars. Therefore beaches aggrade during fairweather and are eroded during storms, a process termed the *beach cycle* (Reading and Collinson, 1996; Fig. 6.26).

Cyclicity of Coastal to Shallow-water Systems in Relation to Shoreline Shifts

Normal Regressive Settings

Normal regressive settings are defined by coeval aggradation and progradation in coastal to shallow-marine environments (Fig. 3.35). As a result, the shoreline shifts seawards and increases its elevation at the same time. Normal regressive coastal environments may either be represented by regressive river mouths (deltas) or by open shorelines. In either case, significant amounts of sand may be trapped in strandplain, shoreface, and delta front systems, some of which may have a potentially good lateral extent along strike (Figs. 5.7 and 5.44). In coastal and near shore shallow-water settings, sedimentation rates (with sediment provided by rivers and longshore currents) exceed the rates of generation of accommodation by base-level rise.

The nature of the depositional environment that is established beyond the shoreface depends on the type of normal regression, and implicitly on the position of the coastline relative to the shelf edge. In the case of highstand normal regressions, inner and outer shelf environments are well-established because the coastline, following the transgressive stage, is far from the shelf edge (Fig. 5.7). Sandy sediment is supplied to the inner shelf mainly by storm currents, generating diagnostic successions of interbedded sand (storm events) and shale (product of deposition during fairweather) with hummocky cross-stratification. Beyond the storm wave base (Fig. 6.17), sedimentation in the outer shelf is dominated by fine-grained pelagic fallout.

In contrast, lowstand normal regressive coastlines are closer to the shelf edge, or even on the upper slope, so the shoreface style of sedimentation passes straight into deep-water conditions along dip, without the intervening shelf transition. In this case, the entire coastal to shallow-marine sedimentation may be restricted to shelf-edge deltas and their laterally correlative open shoreline systems (Fig. 5.44).

Forced Regressive Settings

During base-level fall, fairweather processes favor the transfer of sediment from the lower shoreface to the upper shoreface and beach, in relation to the landward sediment transport associated with the asymmetric shoaling waves (Bruun, 1962; Dominguez and

Wanless, 1991; Fig. 4.20). Sedimentation takes place in the upper shoreface in spite of the high energy conditions, due to the even stronger sediment supply from the lower shoreface, fluvial systems, and longshore currents (i.e., sediment supply > environmental energy flux). Coeval with sedimentation in the upper shoreface, erosion in the lower shoreface generates the regressive surface of marine erosion (Fig. 4.20; energy flux > sediment supply). This scour surface is gradually downlapped by the prograding and downstepping (offlapping) upper shoreface forced regressive lobes. Sedimentation and erosional processes operate in such a way that the concave-up shoreface profile is preserved during the forced regression of the shoreline, which is the reason why the regressive surface of marine erosion forms in the first place (Fig. 4.20). Within the shoreface environment, the change from depositional to erosional regimes takes place at a point of balance between sediment supply and environmental energy flux (lever point in Fig. 4.20).

During storms, the increased wave energy erodes the beach and part of the upper shoreface lobes, contributing to the sediment redistribution to the deeper shelf environment. The regressive surface of marine erosion continues to form, and ultimately becomes part of the depositional sequence boundary (*sensu* Hunt and Tucker, 1992) seaward of the last (youngest) forced regressive shoreface clinoform (Figs. 4.23 and 4.24).

Seawards of the coastal and shoreface systems, the shelf environment shrinks rapidly as a result of the high-rate shoreline regression, and is generally subject to sediment reworking due to the instability induced by the lowering of the storm wave base (Fig. 5.26). In the case of high-magnitude falls in base level, the entire shelf may become subaerially exposed, in which case the entire coastal to shallow marine-environment is reduced to the downstepping shelf-edge deltas and their correlative open shoreline systems (Fig. 5.27). In the case of low-magnitude falls in base level, when the shoreline does not reach the shelf edge, forced regressive shelf sediments may get preserved between the basal surface of forced regression (below) and a composite surface that includes the regressive surface of marine erosion and the correlative conformity (above) (Figs. 4.23 and 4.24). These forced regressive shelf sediments have similar sedimentological characteristics to the underlying highstand shelf facies, including hummocky cross-stratified sand/shale successions in the inner shelf and fine-grained suspension sedimentation deposits in the outer shelf. Therefore, the separation in core and well logs between highstand and forced regressive shelf sediments across the basal surface of forced regression

is most challenging (Fig. 4.25), although at a larger scale the differences in the rates of progradation and the overall stratigraphic geometries may offer useful hints.

Transgressive Settings

A rise in base level accompanied by transgression leads to erosion of the foreshore and upper shoreface and deposition in the lower shoreface (Bruun, 1962; Dominguez and Wanless, 1991; Figs. 3.20 and 5.6). The erosional processes are generated by waves in the upper shoreface as the shoreline transgresses, in an attempt to preserve the concave-up graded profile of the shoreface. The resulting scour (wave-ravinement) surface is highly diachronous, with the rate of shoreline transgression. Part of the sediment derived from the erosion of the upper shoreface is transferred towards the land in relation to the asymmetrical wave motion, contributing to the formation of backstepping beaches or estuary-mouth complexes, whereas some finer sediment is transported seaward to 'heal' the bathymetric profile of the lower shoreface (Figs. 3.20, 3.21, and 5.6).

In addition to deposition in backstepping coastal and onlapping lower shoreface systems, mainly related to wave processes, sandy macroforms may also form on the shelf in relation to tidal currents (Figs. 5.57–5.62; Posamentier, 2002). Such 'shelf ridges' may constitute the reservoirs of ideal stratigraphic traps, being encased in transgressive fine-grained sediments, and form as a result of tidal reworking of the underlying regressive coastline deposits (Posamentier, 2002). Among all systems tracts, the formation and preservation of sandy shelf sheets and ridges is most likely linked with the transgressive systems tract, because of the significant extent of the shallow-marine environment during transgression, coupled with the accumulation of overlying highstand facies that protect the transgressive deposits from subsequent subaerial erosion. Excepting for these tidal macroforms, the rest of the shelf is mainly subject to pelagic sedimentation, as the coarser terrigenous sediment is generally trapped within the rapidly aggrading fluvial and coastal systems. Sediment starvation on the shelf during transgression may also lead to the development of condensed sections, or even scouring of the underlying regressive deposits (Loutit *et al.*, 1988; Galloway, 1989).

Summary

It can be noted that the terrigenous sediment supply to the shelf environment changes significantly between transgressions and regressions. This switch is controlled by the balance between accommodation and sedimentation at the shoreline. During regressions, there is not enough accommodation at the shoreline to capture the entire amount of riverborn sediment, so the surplus of terrigenous detritus is shed into the shoreface and shelf environments *via* various transport mechanisms (see earlier discussion in this chapter about *Sediment supply and transport mechanisms*). In contrast, transgressions are characterized by an excess of accommodation at the shoreline, so the entire amount of riverborn sediment is potentially trapped within backstepping fluvial and coastal systems. In this way, little terrigenous sediment is able to escape beyond the lower shoreface into the deeper shelf environment, although, as noted in the discussion of transgressive shelf-sand macroforms, tidal reworking of underlying regressive deposits may still supply sediment to the shelf environment. The difference in sediment supply between transgressive and regressive shelves is critical to understanding changes through time in the distribution of sand within coastal to shallow-marine environments.

SEQUENCES IN DEEP-WATER CLASTIC SYSTEMS

Introduction

Deep-water clastic systems have been particularly studied within the tectonic setting of divergent continental margins, where the limit between shallow- and deep-water environments is marked by the shelf edge. In such basins, the deep-water environment corresponds to the continental slope and the abyssal plain (basin-floor) settings (Fig. 2.3). Within the deep-water environment, submarine fan complexes may capture significant volumes of terrigenous sediment *via* the manifestation of gravity flows. In turn, the types and the magnitudes of gravity-flow events depend on the interplay of the same allogenic controls on sedimentation

that influence depositional processes in all other environments of the sedimentary basin (Fig. 6.27). This common thread that links depositional processes across the basin allows the types of gravity flows that deliver sediment to the deep-water environment to be studied, at least to some extent, within the predictive framework of sequence stratigraphy.

Due to their location within the basin, the deep-water systems generally form the *most remote* portion of each stratigraphic sequence relative to the coeval shoreline, being among the most difficult types of deposits to interpret in sequence stratigraphic terms (Posamentier and Allen, 1999). Besides their remote location within the sequence, the deep-water systems may also lack the *physical connection* with the correlative coastal and shallow-water facies, especially when the shoreline is far away from the shelf edge (Figs. 5.7, 5.26, and 5.57). During such stages, the distal portion of the shelf receives little sediment, and condensed sections may develop in the best case scenario. Alternatively, the general instability at the shelf edge imposed by rapid increases in water depth or by base-level fall may lead to scouring processes in the outer shelf and upper slope settings, which isolate the products of deep-water sedimentation from the rest of the sequence. The only time when the shelf edge is the *locus* of significant sediment accumulation is when the shelf is subaerially exposed following a stage of high-magnitude base-level fall, and the coastline is close to the shelf-slope boundary. This may be the case with stages of late forced regression, lowstand normal regression and early transgression (Figs. 5.27, 5.44, and 5.56), when sediment entry points, in combination with coastal processes, provide a direct supply to the slope and basin-floor settings. Under these circumstances, the deep-water deposits are closer to their correlative coastal systems, although the lateral continuity of the sequence may still be missing because of the lack of sediment stability on the upper slope.

FIGURE 6.27 Allogenic controls on deep-water gravity flows.

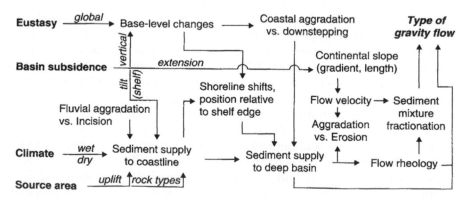

In addition to the common lack of physical connection between the deep- and shallow-water systems, the detailed analysis of their temporal and facies relationships is also hampered by the usual absence of sufficient age control data. *Time control* is the only factor that could compensate for the lack of lateral continuity within the sequence, but the amount of paleontological or radiometric information that is usually available is insufficient to establish an unequivocal correlation between unconformities on the shelf and the sediments of the deep-water environment. These practical pitfalls may be overcome by theoretical considerations, whose validity is of course as good as the correctness of the first principles involved in the construction of stratigraphic models.

Deep-water systems are also unique relative to their fluvial to shallow-water counterparts in terms of the difficulty to study *modern analogues*. Modern analogues are generally of tremendous help in understanding ancient deposits, such as for example the processes of coastal erosion that may occur in some transgressive settings (Figs. 3.20, 3.24, and 3.26). The relative inaccessibility of the deep-water environments deprives the geologist of the first-hand observation of modern processes, which also explains why deep-water systems are perhaps less understood relative to their nonmarine, coastal, and shallow-water correlatives.

Physical Processes

Four major processes contribute to the accumulation of clastic sediments in deep-water environments: (1) progradation of shelf-edge deltas onto the upper slope, following stages of high-magnitude base-level fall; (2) gravity flows; (3) contour currents; and (4) pelagic sedimentation. The pelagic sedimentation from suspension is a background process that takes place continuously irrespective of base-level changes, although cyclic variations in the input of hemipelagic material logically follow the transgressions and regressions of the shoreline. Contour currents are also generally non-diagnostic for sequence stratigraphy, reflecting mass balance processes of thermohaline circulation independent of base-level fluctuations. The timing of shelf-edge deltaic progradation, as well as the timing and the nature of gravity flows, may, however, be modeled more closely in relation to changes in base level, and are presented briefly below.

Progradation of Shelf-edge Deltas

Shelf-edge deltas may prograde onto the continental slope during late stages of base-level fall, as well as during stages of subsequent early base-level rise (Figs. 5.4, 5.27, 5.44, 6.28, and 6.29). Forced regressive shelf-edge deltas, and their correlative open-shoreline deposits, are dominated by offlapping geometries of the prograding lobes, which are truncated by the subaerial unconformity (Figs. 3.22, 4.15, and 5.4). Lowstand normal regressive shelf-edge deltas and their correlative strandplains assume coastal aggradation coeval with progradation, and therefore are characterized by the presence of deltaic and/or alluvial topsets (Figs. 3.22 and 5.4).

The two types of shelf-edge deltas are placed within the same systems tract by Posamentier *et al.* (1988), as part of the early and late portions of their 'lowstand' package (Fig. 1.7), but belong to different systems tracts in the view of Hunt and Tucker (1992). Truncated shelf-edge deltas are part of the falling-stage systems tract, and correspond to the 'slope component' of Hunt and Tucker (1992). They are coeval with the forced regressive submarine fans (the 'basin floor component' of Hunt and Tucker, 1992), with a timing that corresponds to the stratigraphic gap absorbed by the subaerial unconformity. Shelf-edge deltas with delta plain topsets are part of the overlying lowstand systems tract (early-rise normal regression), which is equivalent to the lowstand 'wedge' of Posamentier *et al.* (1988) (Figs. 1.7 and 5.4).

Shelf-edge deltas represent important sediment entry points into the deep-water environment (Figs. 6.28 and 6.29), and are responsible for the most significant accumulations of sand within the submarine fans. As illustrated in Fig. 5.63, the sandy sediment supply is expected to peak at the end of the forced regression of the shoreline, marking the boundary between forced regressive (below) and normal regressive deposits (above). This conclusion is based on the difference in sand-trapping capabilities between forced regressive and lowstand normal regressive deltas and adjacent fluvial systems.

Forced regressive shelf-edge deltas retain relatively little sand, as accommodation at the shoreline is negative. Sediment reworked as a result of fluvial incision adds to the riverborn sediment, and the bulk of this supply is fed to the submarine fans primarily *via* high-density turbidity flows. The resulting basin-floor and slope fans include the coarsest sediment fraction of the deep-water systems, with the dominant architectural element being represented by sandy frontal splays (Fig. 5.27).

In contrast, lowstand normal regressive fluvial systems and shelf-edge deltas trap sand as a result of positive accommodation on the continental shelf, which results in an abrupt decrease in the amount of sand that is delivered to the deep-water environment. The manifestation of gravity flows still continues during this stage, but involving overall finer sediment

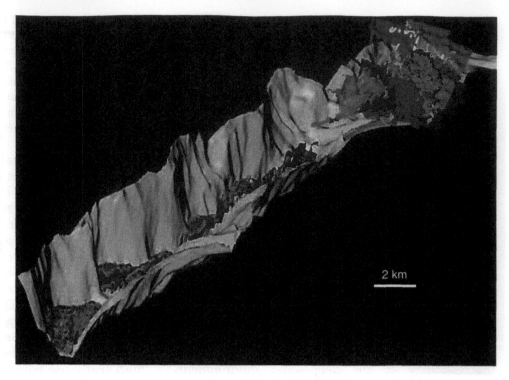

FIGURE 6.28 Upper slope channel and associated shelf edge delta (De Soto Canyon area, Gulf of Mexico; image courtesy of H.W. Posamentier). Note the moderate sinuosity meander channel pattern that can be observed almost to the deltaic clinoform toe-sets. For scale, the delta is about 2 km wide at the shelf edge (uppermost part of the channel), and the maximum depth of the channel is about 275 m.

(lower sand/mud ratio) and therefore a lighter sediment/water mixture. These lower-density turbidity currents travel farther into the basin relative to the sandier forced regressive flows, and result primarily in the accumulation of leveed channels on the basin floor

(Fig. 5.44). The leveed channels of the lowstand normal regressive submarine fan systems mimic the geomorphological characteristics of meandering fluvial systems, and terminate into smaller and more distal frontal splays (Fig. 5.44).

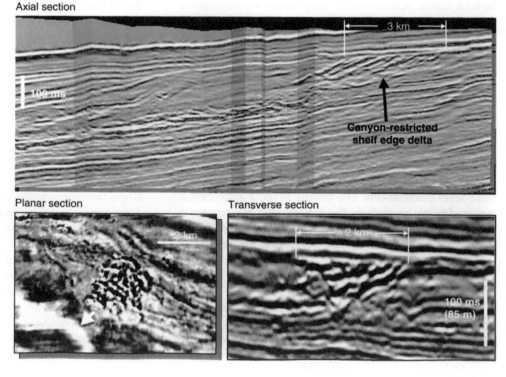

FIGURE 6.29 Canyon-restricted shelf edge delta, observed in axial, planar, and transverse views (De Soto Canyon area, Gulf of Mexico; images courtesy of H.W. Posamentier). The arrow in the planar section indicates the direction of progradation. The delta is about 2 km wide and 3 km long.

The reason why the low-density turbidity currents associated with lowstand normal regressive shelf-edge deltas travel farther into the basin relative to the high-density turbidity flows fed by forced regressive shelf-edge deltas is two fold: firstly, the motion of lower-density flows require shallower *slope gradients*, so the low-density turbidity currents are more suited to travel across the low-gradient basin floor (Fig. 5.37); secondly, the higher *proportion of mud* within low-density turbidity flows sustains the formation of levees over larger distances, which helps in keeping the flow confined across the basin floor. The higher proportion of mud within low-density turbidity flows also explains why the distal frontal splays of these systems are commonly smaller than the more proximal frontal splays associated with high-density turbidity currents.

Gravity Flows

Gravity flows occur almost throughout the full cycle of base-level changes, recording a maximum during the late stages of forced regression, and a minimum during the highstand normal regressions (Figs. 5.7, 5.26, 5.27, 5.44, 5.56, and 5.57). Having said that, the magnitude of the flows, the type of flows, and the sand/mud ratio are expected to vary from one systems tract to another in a predictable fashion that can be related, at least in part, to base-level fluctuations (Figs. 5.63, 6.27, and 6.30).

The main types of gravity flows that deliver sediment to the deep-water environment are the cohesive debris flows (mudflows) and the turbidity currents (Fig. 6.31). In addition, other types of gravity flows include grainflows (sandy non-cohesive debris flows), liquefied flows, and fluidized flows, which commonly do not exist as independent flows but are rather associated with the 'traction carpet' of turbidity currents (Stow *et al.*, 1996). The manifestation of one particular type of flow in preference over the other is mainly a function of sediment supply to the shelf edge ('staging area'), which in turn may be related to the position of the shoreline relative to the shelf edge, and the direction and rates of base-level changes at the shoreline

Systems tract	Dominant process	Dominant products		Stratigraphic surface
		Continental slope	Basin floor	
HST	Pelagic/hemipelagic sedimentation	Condensed section	Condensed section	MFS
Late TST	Mudflows	Seafloor grooves, mudflow lobes	Mudflow lobes	
Early TST	Low-density turbidity flows[2]	Entrenched channels	Leveed channels	WTFC
LST	Low-density turbidity flows[2]	Entrenched channels	Leveed channels	MRS
Late FSST	High-density turbidity flows[1]	Leveed channels	Frontal splays	CC[3]
Early FSST	Mudflows	Seafloor grooves, mudflow lobes	Mudflow lobes	WTFC
HST	Pelagic/hemipelagic sedimentation	Condensed section	Condensed section	BSFR

FIGURE 6.30 Dominant processes and products in the deep-water setting (continental slope *vs.* basin floor) during the six stages of the base-level cycle illustrated in Figs. 5.7, 5.26, 5.27, 5.44, 5.56 and 5.57. Abbreviations: HST—highstand systems tract; FSST—falling-stage systems tract; LST—lowstand systems tract; TST—transgressive systems tract; BSFR—basal surface of forced regression; CC—correlative conformity; MRS—maximum regressive surface; MFS—maximum flooding surface; WTFC—within-trend facies contact. Notes: [1]—due to a high sediment/water ratio, these flows tend to be overloaded (i.e., sediment supply > energy flux), and hence aggradational, on both the continental slope and the basin floor; [2]—due to a low sediment/water ratio, these flows are underloaded on the steep continental slope, but tend to become overloaded on the nearly flat basin floor; [3]—*sensu* Hunt and Tucker (1992). Frontal splays tend to be larger and in a more proximal position in the case of high-density turbidity flows (less mud in the sediment/water mixture, and hence shorter levees and more sand available for splay sedimentation), and smaller and in a more distal position in the case of low-density turbidity flows (the higher mud content sustains the formation of levees over larger distances, and proportionally less sand is available for the construction of levees).

Characteristics of gravity-flows		Occurrence		Controls
Mudflows (cohesive debris flows)	Volume of sediment	Low or high, depending on the magnitude of erosion	During early forced regressions and late transgressions	Erosion of the shelf edge region caused by the lowering of the wave base during base-level fall or by hydraulic instability during rapid base-level rise
	Sediment-water ratio (density of the flow)	High (plastic flows)		
	Sand-mud ratio	Low, due to the dominance of pelagic sediments and the lack of terrigenous clastic input		
	Maximum grain size	Depends on the nature of sediments eroded at or around the shelf edge		
Turbidity flows (involving riverborne sediment)	Volume of sediment Sediment-water ratio (density of the flow) Sand-mud ratio	High: during late forced regressions Low: during lowstand normal regressions and early transgressions		Proportional to supply. Inversely proportional to the amounts of fluvial and coastal accommodation: *the more sediment is trapped in aggrading fluvial and coastal systems, the less is available for the deep-water environment*
	Maximum grain size	Largest at the end of base-level fall. It gradually decreases during lowstand normal regression and subsequent transgression		Proportional to fluvial energy, and to the gradient of fluvial landscapes: *landscape gradients tend to steepen during base-level fall and to shallow during base-level rise as a result of coastal aggradation*

FIGURE 6.31 Characteristics of the main types of gravity-flow deposits. Other types of gravity flows, such as grain flows, liquefied flows and fluidized flows, which do not commonly form independent flows, may be associated with the traction carpet of turbidity currents.

and at the shelf edge (Fig. 6.32). As a matter of principle, most of the terrigenous sediment accumulation in the deep-water environment takes place during forced regressions, when accommodation on the continental shelf is negative.

The following section describes the variety of depositional elements that may be encountered within deep-water settings, and which may form in relation to the different types of gravity flows. As lithologies and, implicitly, reservoir properties, may vary greatly with the type of depositional element, their analysis and recognition prior to drilling is a critical step in the process of petroleum exploration. The imaging of depositional elements based on the analysis of high-resolution 3D seismic data is becoming a routine procedure in the workflow of sequence stratigraphic modeling, and is referred to as 'seismic geomorphology' (Posamentier, 2000, 2004a).

Depositional Elements

The basic depositional elements of the deep-water clastic systems include submarine-canyon fills, turbidity-flow channel fills, turbidity-flow levees and overbank sediment waves, turbidity-flow splay complexes, and mudflow macroforms (e.g., Galloway and Hobday, 1996; Stow et al., 1996; Posamentier, 2000, 2003, 2004a; Kolla et al., 2001; Piper and Normark, 2001; Posamentier and Walker, 2002; Posamentier and Kolla, 2003; Weimer and Slatt, 2004). Some of these depositional elements form in relation to channelized flows (e.g., confined within canyons or leveed channels); others relate to non-channelized flows (e.g., turbidity-flow splays, or mudflow lobes) or overbank sedimentation (e.g., levees and sediment waves adjacent to channelized flows). Following a comparison with fluvial systems, the 'overbank' environment of the deep sea corresponds to the seafloor in the inter-channel areas, which is subject to pelagic or hemipelagic sedimentation, but also to additional influx of sediment, particularly fine-grained, that escapes the channelized flows. These fine-grained pelitic sediments are the equivalent of fluvial floodplain facies.

The sediment that is subject to gravity-flow transport may be composed of an initially homogeneous mixture of sand and mud, with the ratio between the different grain-size fractions depending on several controls such as (1) terrigenous sediment supply;

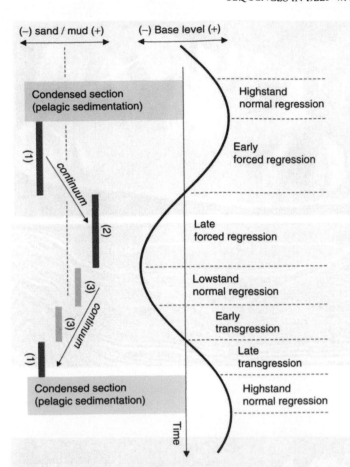

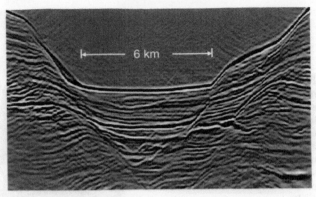

FIGURE 6.33 Submarine canyon backfilled with fine-grained sediment slumped from the flanks, during canyon abandonment (Gulf of Mexico; image courtesy of H.W. Posamentier). The muddy nature of the canyon fill is revealed by the saggy (concave up) geometry of the seismic reflections.

FIGURE 6.32 Dominant types of gravity flows that supply sediment to the deep-water environment, in relation to specific stages of shoreline shift (modified from Posamentier and Kolla, 2003). Note that there is a continuum between the end-member types of gravity flows as changes in sediment supply are gradational through time. Key: (1) cohesive debris flows (mudflows); (2) high-density turbidity currents and grainflows, forming proximal frontal splays; (3) lower-density turbidity currents, forming leveed channels and distal frontal splays.

(2) position of shoreline (sediment entry points) relative to the shelf edge; (3) instability at the shelf edge; (4) seafloor gradients; and (5) direction and rates of base-level changes. This initially homogeneous mixture is subject to fractionation during the manifestation of gravity flows, which results in an uneven distribution of the finer *vs.* coarser sediment fractions between the various depositional elements. This of course makes a difference in the relative reservoir quality of the different elements of the submarine fan complex, and implicitly in the strategy of exploration and the establishment of drilling priorities.

Submarine-canyon Fills

Submarine canyons develop on continental slopes, being associated with steeper gradients, and are generally risky targets for petroleum exploration as they tend to be mud-filled with slumped pelitic sediments from the banks. The muddy nature of the canyon fill is documented by sagging seismic reflections inside the canyon relative to the banks as a result of differential sediment compaction (Fig. 6.33). Due to the repetitive nature of gravity flows, the canyon fill may be reworked many times during its lifetime, so even where sand is present within the canyon the reservoirs tend to be heavily compartmentalized (Fig. 6.34). Submarine canyons, on the continental slope, may be associated with any of the gravity-flow types discussed above, from mudflows to turbidity currents.

Turbidity-flow Channel Fills

Channels may develop on the continental slope and on the basin floor, under variable physiographic conditions, and are generally associated with turbidity currents. Submarine channels may display features similar to the ones of fluvial systems, ranging from straight to meandering, and from incised to aggradational (Posamentier and Walker, 2002). As in the case of fluvial systems, the balance between aggradation and incision may be related to changes in the ratio between the flow energy and its sediment load. For example, low-density turbidity flows tend to form entrenched channels on continental slopes, being underloaded on steep-gradient seafloors, whereas high-density turbidity currents, which are potentially overloaded on continental slopes due to their high sediment load, may form aggrading leveed channels on steep-gradient seafloors (Figs. 5.27, 5.40, 5.44, 6.30, and 6.35). Similar to meandering rivers, submarine channels may display a high sinuosity (Fig. 5.46), and

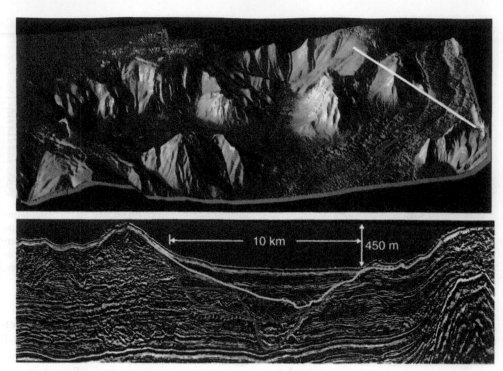

FIGURE 6.34 Modern seafloor seismic imaging (upper image) and cross-sectional view (lower image; location indicated by the white line in the upper image) of the Mississippi canyon (Gulf of Mexico; images courtesy of H.W. Posamentier). The 2D seismic line shows the complex nature of the canyon fill, which recorded multiple stages of aggradation and erosion related to the activity of gravity flows. The arrow in the upper image shows the current direction of gravity flows. For scale, the unfilled space to the top of the canyon is about 450 m high, and the canyon fill is 720 m thick.

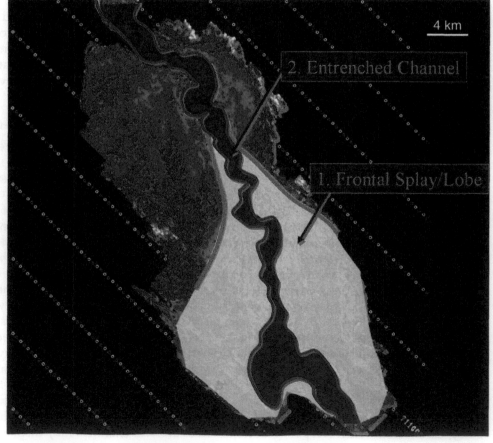

FIGURE 6.35 Interpreted seismic data showing turbidity-flow channel entrenchment into an older frontal splay (continental slope setting, De Soto Canyon area, Gulf of Mexico; image courtesy of H.W. Posamentier). The frontal splay is associated with high-density turbidity currents and is interpreted to have accumulated during late stages of forced regression. The entrenched channel is associated with low-density turbidity flows, and is interpreted to have formed during subsequent lowstand normal regression and ensuing early transgression.

be associated with levees and crevasse splay deposits. Channel fills are sandier relative to their levee deposits, as the coarser (heavier) sediment is preferentially trapped within the channel during the flow, and may form good and continuous reservoirs (Fig. 5.41).

The length of the channel across the seafloor is generally inversely proportional to the density of the turbidity current. High-density turbidites (sandier, as the ones related to late stages of forced regression) accumulate close to the base of the slope, whereas low-density

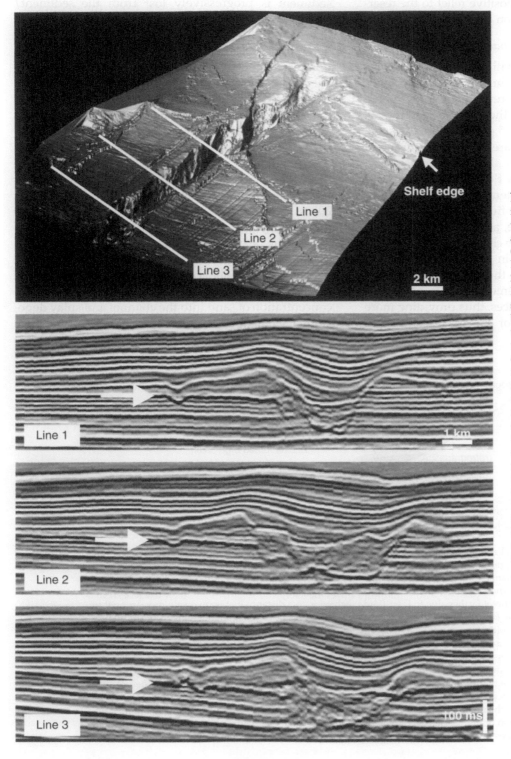

FIGURE 6.36 Three-dimensional illuminated surface (top image) and three cross-sectional seismic lines in an upper continental-slope setting (De Soto Canyon area, Gulf of Mexico; modified from Posamentier, 2004a; images courtesy of H.W. Posamentier). The 3D illuminated surface shows the structure at the base of leveed channels, at the time of the formation of grooves associated with debris flows in the early stages of base-level fall. Arrows on the seismic lines indicate the stratigraphic position of the dip-oriented grooves that are observed on the 3D illuminated surface. These data show the change in the types of gravity flows that operate on the continental slope during forced regression, from mudflows and associated grooves (early forced regression) to high-density turbidites and associated leveed channels (late forced regression). For scale, the channel width is approximately 1.8 km. The channel depth varies from 175 m at leveed channel to 275 m at shelf edge.

turbidites (muddier, as the ones related to lowstand normal regressions and early transgressions) travel farther into the basin, allowing for longer channels, usually with a higher sinuosity and higher levees, to develop. The sandy nature of channel fills, in contrast to their finer-grained levee and overbank facies, is documented by the inversion of topography that follows compaction, which confers a positive relief to the reservoir as seen on seismic reflection data (e.g., Fig. 5.47).

Turbidity-flow Levees and Overbank Sediment Waves

Levees may be associated with either slope or basin-floor channels, and may form in relation to both high- and low-density turbidity currents, commonly where the flows are overloaded and the depositional trend is aggradational (e.g., Figs. 5.40, 5.48, and 6.36). As such, high-density turbidity flows are particularly prone to forming leveed channels on the continental slope, whereas low-density turbidity flows are more likely to generate leveed channel systems on the basin floor (Fig. 6.30). Levees capture the finer-grained fraction of the original sediment mixture, as the finer and lighter sediments may escape from the channel during the flow. Levees are better developed in relation to the lower-density turbidity currents on the basin floor, as these are richer in fine-grained sediment, tend to be longer-lived and travel greater distances, thus providing more time and better conditions for the fractionation and

spillover of the fine-grained sediment fraction from channel onto overbank. Therefore, in the progression of high- to low-density turbidity currents in Fig. 6.37, the channels of the distal splay systems, which most likely form during stages of lowstand normal regression and early transgression, would be associated with the best developed levees. At the same time, as the muddy sediment is progressively lost from the sediment/water mixture with increasing travel distance, levees tend to thin down-system (Fig. 6.38). Where the levee height decreases below a critical threshold, the flow becomes unconfined, leading to a rapid dissipation of energy and the formation of frontal (terminal) splays.

Levee deposits also tend to be thicker and sandier at outer channel bends, as a function of flow momentum and the associated spillover patterns (Posamentier, 2004a; Fig. 6.39). It has been noted that levee height is commonly in a range of meters, with the outer bend levees having two or three times the relief of the inner bend levees (Piper and Normark, 1983; Posamentier, 2003). High-resolution seismic imaging of levee morphology also reveals that the inner, steeper side of levees may be marked by slump scars associated with sediment instability and mass wasting along levee walls (Fig. 5.47). The formation of levees is important from an exploration prospective, because it allows the concentration of sand into the channel and splay depositional elements.

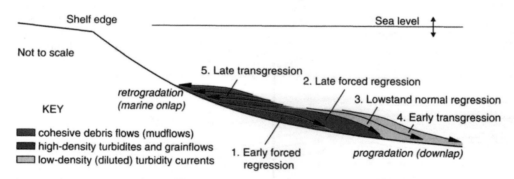

FIGURE 6.37 Idealized architecture of a submarine fan complex that may form during the base-level cycle in Fig. 6.32. The *fan progrades* during the first four time steps (early forced regression to early transgression) in response to the change in the type of gravity flow, based on the assumption that the travel distance of the flow depends on its rheological behavior: mudflows travel the shortest distance, due to their discrete shear strength; turbidites travel farther, due to their fluidal behavior, to a distance that is inversely proportional to the flow density. The *fan retrogrades* during transgression, as a result of the gradual change from fluidal to plastic behavior (turbidites to mudflows, respectively) which accompanies the decrease with time in the sand/mud ratio. According to this general scenario, submarine fans are more likely to onlap the continental slope during transgressions (the 'marine onlap' of Galloway, 1989). Note that the vertical stacking of gravity-flow products in this diagram is the artifact of the 2D style of representation. In a 3D basin, autocyclic shifts in the *locus* of sedimentation of the different lobes may result in their accumulation along different dip-oriented profiles. This is a common problem with all 2D models. However, the general principle that relates the travel distance to the rheological behavior of the flow still remains valid. This idealized model of the relative *locus* of deposition of the various deep-water facies may be altered by variations in sediment supply that could modify the types of gravity flows, as well as by uneven seafloor topography.

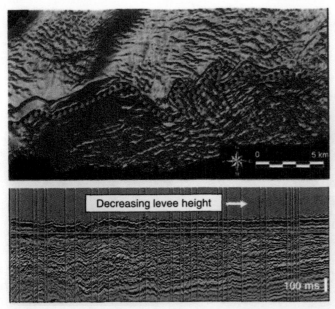

FIGURE 6.40 Seafloor reflection dip magnitude map showing sediment waves related to 'flow stripping' in the overbank area of channelized basin-floor turbidity flows (offshore Nigeria; modified from Posamentier and Kolla, 2003; image courtesy of H.W. Posamentier).

FIGURE 6.38 Seismic traverse (lower image) along the levee crest of the turbidity-flow channel shown on the seismic horizon map (upper image) (basin-floor setting, Gulf of Mexico; modified from Posamentier and Kolla, 2003; images courtesy of H.W. Posamentier). The datum in the seismic traverse (green line) marks the top of levee deposits. The levee height decreases in a basinward direction up to the point where no more fine-grained sediment is available in the turbidity flow to sustain the formation of levees. Where the levee height decreases below a critical threshold, the flow becomes unconfined and a frontal splay forms at the end of the leveed channel.

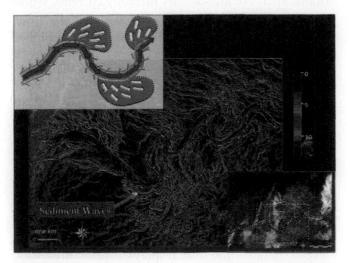

FIGURE 6.39 Reflection dip magnitude map showing sediment waves related to 'flow stripping' in the overbank area of channelized basin-floor turbidity flows (Pleistocene leveed channel, offshore eastern Borneo, Kalimantan, Indonesia; modified from Posamentier, 2004a; image courtesy of H.W. Posamentier).

Beyond levees, for several kilometers into the overbank environment, linear features defined as 'sediment waves' may form in relation to overspill and flow stripping from the main channel (Posamentier et al., 2000; Posamentier and Kolla, 2003; Posamentier, 2003; Figs. 6.39 and 6.40). These sediment waves are morphologically similar to the straight-crested dune fields described for fluvial and shallow-water systems in the context of flow regime charts, excepting that they develop over much larger scales. The spacing between the deep-water sediment waves may vary greatly, from 50 to 250 m (Posamentier, 2003), probably reflecting the variable energy (velocity) of the overbank flow. From the orientation of sediment waves in relation to the main channel, it may be inferred that they form preferentially in the extension of outer bends, with their crests oriented perpendicular to the direction of overbank flow (Fig. 6.39). Sediment waves are not necessarily associated with crevasse splays, and, being related to secondary flows that escape the main channel while the system is active, are generally composed of finer-grained sediment fractions relative to the levee deposits. The petroleum reservoir quality of deep-water depositional elements therefore decreases away from the main channel, from channel fills to levees and overbank sediment waves.

Turbidity-flow Splay Complexes

Splays may be classified into crevasse (lateral) splays (lateral relative to a channel—less important; Fig. 6.41) and frontal (terminal) splays (in front of a

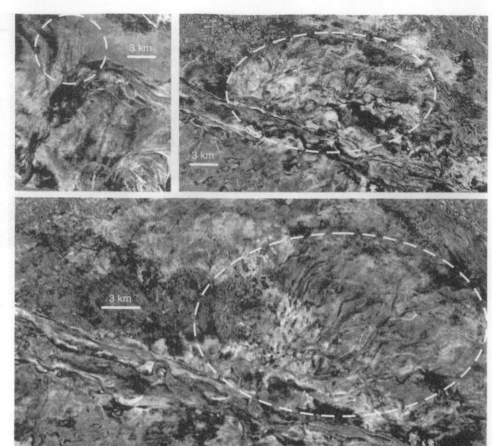

FIGURE 6.41 Seismic amplitude extraction maps showing examples of turbidity-flow crevasse splays in a deep-water basin-floor setting (Gulf of Mexico; images courtesy of H.W. Posamentier).

channel, representing the most distal depositional element of the system – more important; Figs. 5.42, 6.42, and 6.43), and generally form where the energy of a flow drops to the point that triggers the accumulation of the bulk of its sediment load. As opposed to the mudflow lobes, which freeze on deceleration and tend to be more proximal relative to the shelf edge, turbidity-current splays involve a gradual loss of sediment from the flow until the flow energy is totally exhausted, and therefore tend to be located farther away relative to the shelf edge. Within a given divergent continental margin setting, the location of frontal splays depends on the type of turbidity current, as discussed in the case of channels. The frontal splays of high-density turbidity flows may form aprons at the base of the continental slope, with shorter leveed channels on the continental slope, whereas the frontal splays of the low-density turbidity currents form basin-floor lobes, with well-developed leveed channels associated with them. The frontal splays of high-density turbidity currents tend to be larger, because sediment supply is higher and only little sediment of the original mixture is trapped within

channel-levee systems. The opposite is valid for the frontal splays of low-density turbidity currents, whose size decreases with the travel distance of the flow. At the same time, distal splays may have a higher textural maturity (cleaner sand, even though in lesser amounts and potentially finer-grained), as the finest-grained fractions of the original sediment/water mixture separate from the flow and are trapped within levees and overbank sediment waves in the process of sediment transport.

This discussion demonstrates the complexity of turbidite reservoirs, whose distribution (location within the basin) and quality (grain size and textural maturity) depend on a number of competing mechanisms that control sediment supply and the type of gravity flow. The ratio between the two main depositional elements of the turbidity flow deposits, i.e., leveed channels and frontal splays, is primarily a function of the density of the current, which in turn is a reflection of the initial sediment composition (sediment/water and sand/mud ratios). The reservoirs of the high-density currents are dominated by proximal frontal splays, which include the bulk of the original

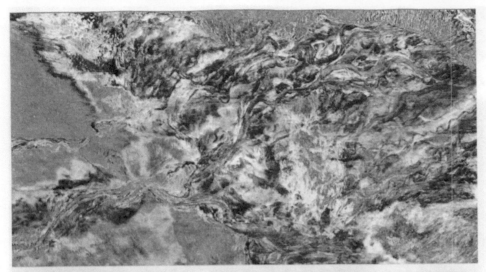

2 km

FIGURE 6.42 Seismic amplitude extraction maps showing examples of turbidity-flow frontal splays in a deep-water basin-floor setting (Gulf of Mexico; images courtesy of H.W. Posamentier).

sand-rich sediment. In contrast, the reservoirs of the low-density currents are volumetrically dominated by leveed channel fills, which use most of the sand of the original mixture (Figs. 6.30 and 6.44). The remaining part of the sand forms relatively smaller distal frontal splays, located where levees thin below a critical height, whereas most of the mud contributes to the formation of levee deposits and overbank (basin-floor) sediment waves (Posamentier and Walker, 2002; Posamentier and Kolla, 2003). Frontal splays generally include sheet-bedded sandstones, in contrast to the mudflow deposits that may take the form of sheets or lobes with chaotic internal structure, and both deposit types may achieve thicknesses in excess of 100 m (Posamentier and Walker, 2002; Posamentier and Kolla, 2003).

Mudflow (Cohesive Debris Flow) Macroforms

Mudflow macroforms, primarily in the shape of sheets and lobes, are most likely to accumulate in the deep-water environment during stages of early forced regression and late transgression, when sediment entry points into the marine basin are far from the shelf edge, and the wave or hydraulic energy regimes generate instability in the shelf edge region (Figs. 5.26, 5.57, and 6.30). During such stages of the base-level cycle, the dominance of fine-grained sediments in the distal shelf environment, which serves as the staging area for gravity flows, promotes the manifestation of mudflows over turbidity currents.

Mudflow deposits may form a significant portion of the deep-water submarine fan complexes (e.g., Fig. 6.44), and therefore they are often encountered in the process of offshore petroleum exploration. Due to their dominant fine-grained composition, mudflow deposits do not form petroleum reservoirs, and so their recognition as such, and separation from turbidity-flow depositional elements, is a critical risk-reducing factor in the pre-drilling stage of exploration. Criteria for the identification of mudflow deposits have been developed based on their geomorphological, stratigraphic, and structural features that can be observed on 3D seismic data (e.g., Posamentier and Kolla, 2003), which in turn are explained by the plastic rheological behavior of the sediment/water mixture during the manifestation of the flow. These criteria include significant erosional relief at the base of mudflow sheets or lobes (Fig. 5.33); the presence of grooves at the base of

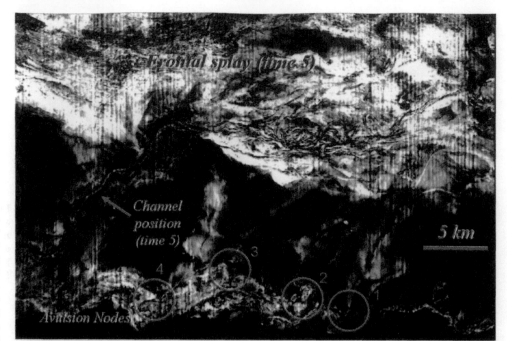

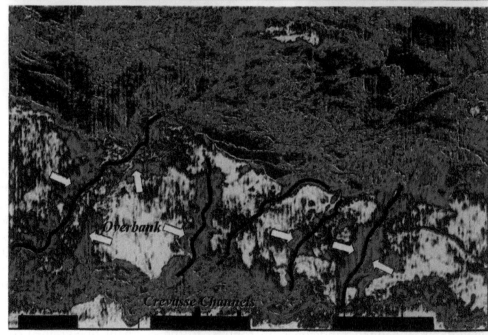

FIGURE 6.43 Seismic expression of a basin-floor frontal splay (distributary channel complex) on an amplitude extraction map (Gulf of Mexico; modified from Posamentier, 2003; images courtesy of H.W. Posamentier). Also note the position of the turbidity-flow channel during previous time steps, and the updip migration of avulsion nodes with time. This pattern of migration of avulsion nodes is attributed to allogenic controls (e.g., changes in discharge and sediment yield of gravity flows triggered by base-level shifts), as opposed to the pattern of downdip migration of avulsion nodes which is interpreted as the result of autogenic mechanisms (Posamentier, 2003).

mudflow deposits (Fig. 5.34), or preserved upslope as a result of flow motion (Fig. 6.36); internal convolute architecture of mudflow deposits (Fig. 5.35); and the presence of internal thrust faults and associated compressional ridges at the top of mudflow sheets or lobes (Fig. 5.36).

The dense nature of mudflows, in which particles are maintained in suspension during the manifestation of the flow by the cohesiveness of the sediment/water mixture (matrix strength) rather than the turbulence of the water or any other clast-supporting mechanism,

confers on the flow a non-channelized character, even though transverse cross-sectional views through mudflow lobes may resemble the morphology of a channel (Fig. 6.45). In such cases, the channel shape is an artifact of erosional relief (Fig. 5.33), and it is not representative for the style of sediment transport. A key characteristic of mudflows, which explains some of their diagnostic features, is their tendency to freeze on deceleration. This behavior, caused by the discrete shear strength of the sediment/water mixture, is responsible for the contorted seismic facies observed

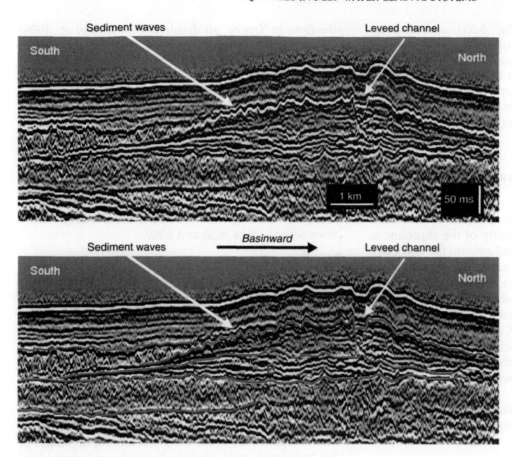

FIGURE 6.44 Uninterpreted and interpreted seismic lines showing a typical (and complete) basin-floor succession of gravity-flow deposits formed in response to a full cycle of base-level changes (modified from Posamentier and Kolla, 2003; seismic line courtesy of H.W. Posamentier). A—mudflow deposits (chaotic internal facies) interpreted to correspond to the early falling-stage systems tract; B—turbidity-flow frontal splay (well-defined parallel reflections), interpreted to represent the late falling-stage systems tract; C—leveed channel and correlative overbank facies (strong reflections associated with the sandy channel fill and weak reflections/transparent facies associated with the finer-grained overbank deposits), interpreted to include the lowstand systems tract and the early portion of the transgressive systems tract; D—mudflow deposits (chaotic internal facies), interpreted to represent the late portion of the transgressive systems tract. Note the gradual progradation of gravity-flow deposits into the basin from A to C, and the retrogradation from C to D (compare with Fig. 6.37).

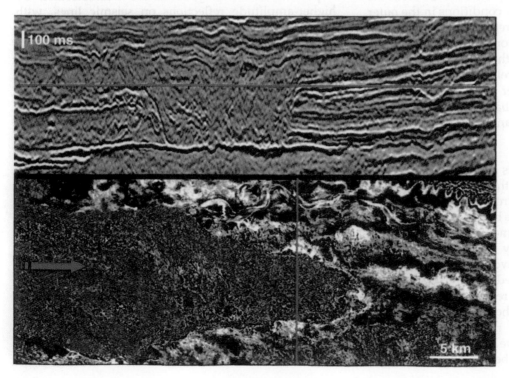

FIGURE 6.45 Transverse and planar sections (upper and lower seismic images, respectively) through a Pleistocene mudflow lobe accumulated at the toe of the continental slope (Gulf of Mexico; images courtesy of H.W. Posamentier). Note that in transverse view the lobe resembles the shape of a channel due to the substantial erosional relief created by the motion of mudflow deposits. However, the flow is not 'channelized' in a conventional sense.

within mudflow deposits (Fig. 5.35), and also for the development of thrust faults and compressional ridges (Fig. 5.36). These stratigraphic and structural features relate to the fact that the frontal part of the flow tends to freeze first, as approaching lower gradients towards the base of the continental slope, creating a barrier which the remaining part of the flow crashes into.

Cyclicity of Deep-water Systems in Relation to Shoreline Shifts

As sediment supply to the deep-water environment depends strongly on the proximity of the shoreline to the shelf edge (staging area), and also on the trajectory of the shoreline (transgressive, forced regressive or normal regressive), a predictive stratigraphic cyclicity of deep-water gravity flows and depositional elements may be established in relation to shoreline shifts (Figs. 5.37, 5.63, 6.30, 6.32, 6.37, and 6.44). This section summarizes not only the predicted changes in the type of gravity flows that are expected during consecutive stages of a base-level cycle, but also the contrasts in depositional elements and depositional trends between the slope and the basin-floor settings of the deep-water environment during each stage of shoreline shift (Fig. 6.30).

Highstand Normal Regressions

No significant gravity flows into the deep-water environment are expected during highstand normal regressions, as sediment entry points are far from the shelf edge and the bathymetric conditions in the outer shelf to upper slope settings are relatively stable (Fig. 5.7). Pelagic sedimentation is the dominant process on the distal continental shelf, and in the deep-water environment, which results in the development of condensed sections throughout the outer shelf and the deep-water (continental slope and basin-floor) portions of the basin (Figs. 5.7, 5.63, 6.30, and 6.32).

Early Forced Regressions

Even though the shoreline is still far from the shelf edge, the rapidly changing bathymetric conditions in the distal shallow-water environment coupled with the lowering of the storm wave base during base-level fall trigger instability and sediment reworking around the shelf edge, and implicitly the manifestation of gravity flows. These flows involve mainly the fine-grained sediment accumulated on the outer shelf during the highstand normal regression. As a result, the dominant gravity flows in the early stages of forced regression are represented by mudflows (Figs. 5.26, 5.63, and 6.32). These cohesive debris flows have a discrete shear strength (plastic behavior) which causes

the flow to 'freeze' on deceleration (Stow et al., 1996). This property translates into transparent to chaotic facies on seismic lines, significant erosion of the underlying substrate, basal grooves, internal thrust faults and associated compressional ridges at the top of such deposits (Figs. 5.33–5.36). Due to their plastic rheological behavior, mudflows travel shorter distances relative to the turbidity currents, and their products may be part of, but not necessarily restricted to, slope fans/aprons (Fig. 6.37). Seafloor processes are dominated by erosion on the continental slope (e.g., see grooves in Fig. 6.36) and sediment accumulation towards the toe of the slope and in the proximal basin-floor setting (Figs. 5.26 and 6.30).

Late Forced Regressions

These are the most important stages for the construction of submarine fans, when the largest amount of terrigenous sediment is delivered to the deep-water environment. The proximity of sediment entry points to the shelf edge, together with the negative accommodation on the largely subaerially exposed continental shelf (Fig. 5.27), result in large amounts of sand being transferred into the deep basin via high-density turbidity currents (Figs. 5.63, 6.32, and 6.36). Sedimentation rates in the deep-water environment during late forced regressions are the highest among all stages of the base-level cycle, and the bulk of sandy turbidites accumulate in proximal frontal splays located close to the limit between the slope and the abyssal plain (Fig. 5.42). In contrast to the underlying mudflow deposits, these sandy lobes are characterized by higher-amplitude reflections on seismic lines, with well-defined layer-cake architecture (Figs. 5.32 and 6.44), indicating (1) continuous deceleration of the flow with the shallowing of the seafloor gradient, (2) gradual deposition of the sediment load until the energy of the flow is totally exhausted, and (3) the relatively coarse nature of the sediment.

As the turbidity currents have a fluidal behavior, with no inherent shear strength, they do not freeze on deceleration. This flow property allows them to travel farther into the basin, and on lower seafloor gradients, relative to the underlying mudflow deposits (Figs. 6.37 and 6.44), although they may also be found superimposed in some locations (Fig. 5.32). As with the products of mudflows, high-density turbidites may also be part of, even though not restricted to, slope fans/aprons. High-density turbidites are generally expected to aggrade in both continental slope and basin-floor settings due to the overloaded nature of the flow, even on steep gradients where the flow energy is highest. Continental slopes are dominated by the aggradation of leveed channels (Figs. 5.39–5.42, and 6.36), whereas basin floors receive sediments predominantly within frontal splays (Figs. 5.27, 5.42, 6.30, and 6.44).

Lowstand Normal Regressions

The creation of accommodation on the continental shelf during early stages of base-level rise results in a net decrease in the amount of sand that is delivered to the deep basin, as a significant amount of riverborn sediment (and particularly its coarser fractions) is now trapped in amalgamated fluvial channel fills and aggrading coastline systems. Trapping of terrigenous sediment on the continental shelf, starting with the onset of base-level rise, triggers the change from high- to low-density turbidites across the correlative conformity *sensu* Hunt and Tucker (1992) (Fig. 5.63). While the density of turbidity flows is measured by the sediment/water ratio, the change from high- to low-density flows at the onset of base-level rise means not only a decrease in the volume of sediment that is made available to the deep-water environment, but also a decrease in the sand/mud ratio of the sediment involved in the flow. Thus, the terrigenous sediment influx that feeds the late forced regressive high-density turbidity currents includes *all* sediment fractions of the riverborne sediment, as accommodation across the continental shelf is negative. In contrast, as accommodation becomes positive on the continental shelf, the coarser riverborne sediment is trapped first within aggrading fluvial to coastal systems, leaving the *finer-grained fractions* still available to feed the lowstand normal regressive low-density turbidity flows. This explains the abrupt decrease in the sand/mud ratio across the correlative conformity indicated in Fig. 5.63.

The surplus of riverborne sediment that exceeds the amount of available accommodation on the continental shelf is therefore fed to submarine fans by low-density turbidity currents (Fig. 5.44). As these diluted flows carry an overall finer-grained sediment load relative to the previous late forced regressive currents, the formation of levees on the basin floor is sustained over larger distances, enabling the flows to travel farther into the basin across the abyssal plain (Fig. 6.37). This explains why low-density turbidity flows are commonly associated with smaller and more distal frontal splays, a feature that reflects their characteristic low sand/mud ratio. Due to the net decrease in the amount of riverborn sediment that is available to the deep-water environment, sedimentation rates in the submarine fans during lowstand normal regressions also decrease accordingly. The low-density turbidity flows of lowstand normal regressions tend to be underloaded on the continental slope, generating entrenched channels (Fig. 5.45), and overloaded on the basin floor where leveed channels form the dominant depositional element (Figs. 5.44–5.48, 6.30, and 6.44). The underloaded character of low-density flows on continental slopes relates to their low sediment/water ratio, as the amount of sediment load is insufficient to compensate for the high energy of the flow on relatively steep seafloor gradients. Such flows may only become overloaded (and hence aggradational) on nearly flat surfaces, where the energy flux drops below the threshold required to maintain the entire amount of sediment load in suspension.

Early Transgressions

In the early stage of transgression, the shoreline is still in the shelf edge region (Fig. 5.56), and therefore there is a chance that some riverborn sediment, together with the sediment reworked in the upper shoreface by wave-ravinement processes, may be transported into the deep basin by low-density turbidity currents. Such currents are, however, expected to be even more diluted relative to the previous regressive flows because the bulk of the riverborn and coastal erosion-derived sediment is now trapped in backstepping beaches, estuary-mouth complexes, and healing-phase wedges (Figs. 5.56, 5.63, and 6.32). In contrast to the case of lowstand normal regressions, the sand-trapping efficiency of the transgressive fluvial and coastal systems is much higher, being enhanced by the higher rates of base-level rise which allow accommodation to outpace sedimentation. The diluted turbidity currents of the early transgressive stage may travel farther into the basin relative to the previous lowstand normal regressive flows, due to the lower density of the sediment/water mixture (Fig. 6.37). At the same time, sedimentation rates in the deep-water environment are also expected to decrease, following the trend that started with the onset of base-level rise. This trend is driven by the increase with time in the rates of base-level rise (Fig. 3.19), which results in (1) increasingly efficient trapping of riverborn sediment within fluvial and coastal systems; (2) a corresponding decrease in the amount of sand, and sediment in general, that is available to the deep-water environment; (3) dilution of the turbidity currents, accompanied by a decline in the sand/mud ratio and a corresponding decrease in reservoir quality of turbidites; (4) decrease in sedimentation rates in the deep-water environment; and (5) increase in the travel distance of gravity flows (i.e., the 'progradational trend' noted in Figs. 5.63 and 6.37). In parallel with the continued trend of progradation of turbidites into the basin, deep-water healing-phase wedges start forming following the onset of transgression, from the toe of the slope and gradually onlapping the continental slope (Figs. 3.22 and 5.56). The accumulation of deep-water healing-phase deposits continues throughout the transgressive stage, and the characteristics of this type of 'transgressive slope aprons' are discussed further in the following section.

Similar to the lowstand normal regressive flows, the low-density turbidity currents of early transgressions may only become overloaded on nearly horizontal basin floors, where the energy of the flow drops gradually to zero. As a result, the early transgressive flows tend to be entrenched on the continental slope (Fig. 5.45), with age-equivalent leveed channels on the basin floor (Figs. 5.46–5.48, 5.56, 6.30, and 6.44).

Late Transgressions

During late transgressions the shoreline is far from the shelf edge (Fig. 5.57), which, coupled with the efficient sediment trapping within nonmarine and coastal systems, reduces dramatically the chance of any river-born sediment to reach the deep-water portion of the basin. The fine-grained sediments that accumulate on the shelf and upper slope are, however, subject to reworking due to the general hydraulic instability created by rapidly increasing water depths at the shelf edge, which explains the manifestation of mudflows in the deep-water environment (Figs. 5.57, 5.63, and 6.32). These mudflows, owing to their plastic behavior (discrete shear strength), travel shorter distances relative to the previous turbidity currents, and so they may backstep and with time onlap the continental slope, forming a 'transgressive slope apron' associated with 'marine onlap' (Galloway, 1989; Figs. 4.2 and 6.37). This landward shift with time of the point of sediment onlap onto the continental slope is referred to as a 'retrogradational trend' in Figs. 5.63 and 6.37.

The transgressive mudflow deposits may display the same characteristics on seismic lines and horizon slices as the cohesive debris flows of the early forced regression (Figs. 5.33–5.36, and 6.44). As in the case of early forced regressive mudflows, gravity-flow-related seafloor processes during late transgressions are dominated by erosion on the continental slope (e.g., see grooves in Fig. 6.36) and sediment accumulation towards the toe of the slope and in the proximal basin-floor setting (Figs. 5.57 and 6.30).

In addition to the gravity flows described above, which build submarine fans associated primarily with point-sources of sediment supply, the deep-water environment may also accumulate *healing-phase wedges* during shoreline transgression (Figs. 3.22, 5.56, and 5.57). These wedges may also include significant volumes of sediment (Fig. 3.22), but in contrast to the submarine fans, sediment sources may be considered linear and the transport of sediment is primarily by diffusion rather than channelized (see the 'overbank' location of the cross-section in Fig. 3.22 relative to the closest channelized flow; Figs. 5.56 and 5.57; more details about healing-phase wedges are provided in Chapter 5). The deep-water healing-phase wedges develop during the entire stage of shoreline transgression, by gradually

onlapping the continental slope (Figs. 5.56 and 5.57). This landward shift (backstepping) with time of the point of sediment onlap onto the continental slope also qualifies as 'marine onlap,' as in the case of late transgressive mudflow deposits (Fig. 4.2). It can be noted that two types of 'transgressive slope aprons' may be identified in deep-water settings, each associated with retrogradation and marine onlap: one type consists of late transgressive mudflow deposits, which are part of submarine fan complexes (Fig. 6.37), and a second type consists of healing-phase deposits that have a potentially better spatial development along strike (Figs. 3.22, 5.56, and 5.57). The type of transgressive slope apron that may be intercepted along dip-oriented 2D seismic lines depends on the location of the cross-sectional profiles relative to the submarine fan complexes and their feeding canyons or channels. As such, 2D transects that are placed in the 'interfluve' areas of submarine fan complexes are most likely to capture the architecture of transgressive healing-phase wedges (Fig. 3.22).

Summary

The change in the type of gravity flows that operate during a full cycle of base-level shifts (Figs. 5.37, 5.63, 6.30, 6.32, and 6.44) triggers changes in the *locus* of deposition of submarine lobes, as illustrated in Fig. 6.37. Assuming a smooth bathymetric profile of the seafloor and no significant changes in sediment supply during the base-level cycle, a progradation through time of the submarine fan lobes is generally expected from the mudflow deposits of the early forced regression to the low-density turbidites of early transgression. Following this progradational trend, a retrogradation of the submarine fan lobes is expected during transgression, accompanying the transition from low-density turbidites to mudflows (Fig. 6.37). Note that the 1D and 2D models (Figs. 5.63 and 6.37) do not necessarily reflect the reality of a 3D basin, because they ignore the autocyclic lateral shifts in the location of the various architectural elements of the deep-water depositional systems. This means that the composite profile in Fig. 5.63 may not necessarily reflect the vertical facies shifts in a single location, but it may be composed of several sections of different ages that are located in different areas of the submarine fan complex. Keeping this in mind, one has to be careful with the interpretation of the deep-water basal surface of forced regression (Fig. 4.27) which, unless it corresponds indeed to the base of the earliest gravity-flow products of forced regression, may be confused with the facies contacts at the base of younger lobes of the submarine fan complex.

The quality and distribution of the deep-water reservoirs depend primarily on sediment supply (which in turn is controlled by other first-order mechanisms, as discussed above), basin physiography, and types of gravity flows. Given a smooth bathymetric profile of the basin, *slope fans/aprons* may include more texturally immature sediments, due to the shorter transport distance, and may form as a result of mudflows or high-density turbidity currents. The products of the latter flows, in spite of the limited degree of sorting, may form potentially the best and the largest reservoirs of the deep-water systems, as they are related to the high sediment supply, with the highest sand/mud ratio, which is commonly associated with the late stages of forced regression. The dominant depositional element of this type of reservoirs is represented by frontal splays. *Basin-floor fans* are mainly related to lower-density turbidity currents, which are able to travel greater distances, and which produce reservoirs mainly dominated by leveed channels. These types of fans also have frontal splays, which may be more texturally mature (as mud is separated and trapped within levees in the process of sediment transport) but volumetrically less important relative to the leveed channels.

Deep-water clastic systems have received less attention in the past relative to their fluvial to shallow-water correlatives, partly because of the technical difficulties in exploring and drilling deeper offshore areas. Technological advances in seismic exploration and drilling techniques allowed for a change in focus in recent years, bringing turbidite reservoirs to the forefront of petroleum exploration. Offshore exploration is of course more challenging and expensive, so every effort should be made prior to drilling to generate detailed and accurate stratigraphic models. Simple models like the ones illustrated in Figs. 5.63, 6.32, and 6.37 only capture general theoretical principles, and need to be re-evaluated on a case-by-case basis, taking into account the realities of each particular basin.

The relative inaccessibility of the present day deep-water environments deprives the geologist of the first-hand observation of modern processes, which explains why deep-water systems are generally less understood relative to their fluvial, coastal, and shallow-water correlatives. The lack of easily accessible modern analogues in deep-water environments is, however, compensated by the technological advances in the fields of seismic data acquisition and processing, which allow for the high-resolution imaging of the 3D architecture and evolution through time of deep-water systems (e.g., Figs. 5.33–5.36 and 5.39–5.48). Recent work on the characterization of deep-water petroleum reservoirs and other depositional elements has been published by Posamentier and Kolla (2003) and Weimer and Slatt (2004). In the absence of easy access to modern analogues, to observe gravity flows in action in present day deep-water environments, outcrop analogues are particularly useful to study the small-scale sedimentology and physical (reservoir) characteristics of turbidites and other gravity-flow-related facies (Figs. 4.27, 6.46, and 6.47), as well as their larger-scale architecture (e.g., Wickens, 1994; Scott, 1997; Scott and Bouma, 1998; Bouma and Stone, 2000).

SEQUENCES IN CARBONATE SYSTEMS

Introduction

The application of sequence stratigraphy to carbonate depositional systems was a topic of debate in the late 1980s, particularly with respect to how a sequence framework developed essentially for clastic systems can be adapted to reflect the realities of carbonate environments (Vail, 1987; Sarg, 1988; Schlager, 1989). Following up on these early contributions, significant progress was made in the early 1990s when the fundamental principles of carbonate sequence stratigraphy, as well as the differences between the clastic and carbonate stratigraphic models, were elucidated (Coniglio and Dix, 1992; James and Kendall, 1992; Jones and Desrochers, 1992; Pratt *et al.*, 1992; Schlager, 1992; Erlich *et al.*, 1993; Hunt and Tucker, 1993; Long, 1993; Loucks and Sarg, 1993; Tucker *et al.*, 1993). The current status of carbonate sequence stratigraphy has been summarized by Schlager (2005).

'Principles' of sequence stratigraphy, and the definition of the fundamental sequence stratigraphic concepts, are independent of the type of depositional environments established within a sedimentary basin, and are discussed in this book based primarily on the processes and products of clastic environments. Nevertheless, the types of shoreline shifts, the systems tract nomenclature in relation to base-level changes, the types of stratigraphic surfaces or stratigraphic sequences, may all be applied to carbonate depositional systems as well. Notable differences, however, between the stratigraphic models of clastic and carbonate systems relate mainly to the geometry of systems tracts and the sediment budget across the basin during the various stages of the base-level cycle. Such differences stem from the all-important *sedimentation* variable, whose interplay with accommodation controls the type of shoreline shifts, the depositional trends within the basin, and implicitly the formation and architecture of systems tracts.

In contrast with basins dominated by siliciclastic environments, whose bulk of sediment is terrigenous

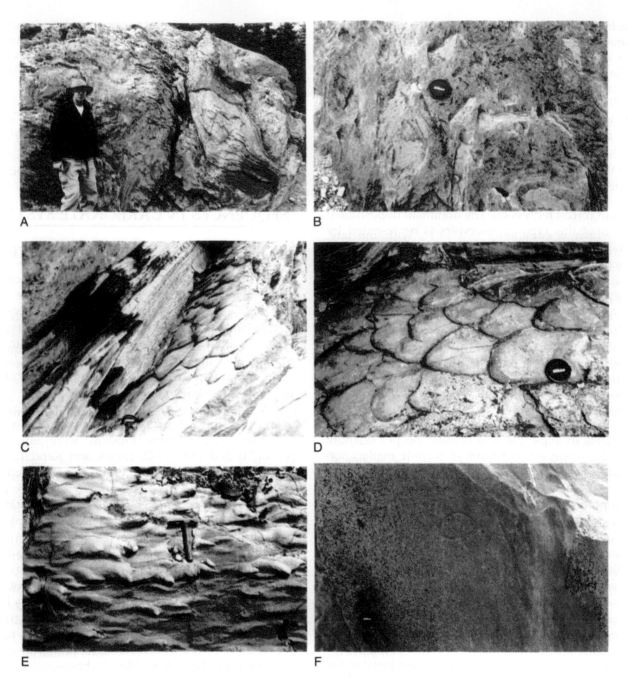

FIGURE 6.46 Sedimentological features of deep-water facies in outcrop. A—slump deposits in a continental slope setting, showing internal deformation of coherent but unlithified sediment. Slumping indicates instability at the shelf edge, generally related to periods of time of rapidly changing bathymetric conditions, such as during forced regressions and transgressions. Lithology in this example is represented by calcareous sandstones and siltstones (Devonian, Sassenach Formation, Jasper National Park, Alberta); B—rip-up clasts of pelagic material at the base of the slump structures in photograph A. The pelagic material accumulated on the continental slope prior to the slumping event (Devonian, Sassenach Formation, Jasper National Park, Alberta); C—distal frontal splay facies, showing flute marks at the base of a turbidite rhythm that consists of the divisions B to E of the Bouma sequence. The contact in the photograph separates hemipelagic sediments above (but older stratigraphically; division E) from parallel-stratified sandstone (below, but younger as the succession is overturned; division B). Proximal frontal splay facies that are likely part of the same submarine fan complex are shown in Fig. 4.27B (Precambrian, Miette Group, Jasper National Park, Alberta); D—flute marks at the base of a turbidite rhythm (detail from photograph C). Note the paleoflow direction from left to right (Precambrian, Miette Group, Jasper National Park, Alberta); E—flute marks at the base of a turbidite rhythm. Note the paleoflow direction from right to left (Paleogene, accretionary prism of Barbados); F—turbidite rhythm showing a fining-upward trend (younging direction from left to right), consisting of the divisions A to C of the Bouma sequence (Paleogene, accretionary prism of Barbados).

FIGURE 6.47 Sedimentological features of deep-water turbidites in outcrop. A—convolute bedding in the division C of turbidite facies (Paleogene, accretionary prism of Barbados); B—asymmetrical (current) ripples at the top of the division C of a turbidite rhythm (Paleogene, accretionary prism of Barbados); C—carbonaceous shale within the pelitic fraction (division E) of distal splay turbidite facies (Late Permian, Collingham Formation, Ecca Pass, Karoo Basin); D—volcanic ash within the pelitic fraction (division E) of distal splay turbidite facies (Late Permian, Collingham Formation, Ecca Pass, Karoo Basin); E—distal frontal splay facies, less than 50 m in total thickness, showing low-density turbidites composed mainly of the divisions D (parallel laminated silt) and E (pelitic) of the Bouma sequence (Late Permian, Collingham Formation, Ecca Pass, Karoo Basin); F—Proximal frontal splay facies, showing a 70 cm thick high-density turbidite rhythm dominated by divisions A (massive sandstone) and B (parallel-laminated sandstone) of the Bouma sequence. Note sole marks at the base of the overlying turbidite rhythm. The total thickness of this proximal frontal splay is about 1000 m (Late Permian, Ripon Formation, Ecca Pass, Karoo Basin).

in nature and supplied by 'extra-basinal' sources, carbonate platforms and associated deep-water systems rely on 'intra-basinal' sediment that is generated primarily within the shallow-water carbonate factory. 'Pure' carbonate systems, which receive little or no riverborn or wind-born clastic input, sustain processes of aggradation based entirely on the chemical or biochemical precipitation of carbonates within the basin. The productivity of such 'carbonate factories,' which dictates the rates of sedimentation (seafloor aggradation) depends on a number of factors including climate, amount of clastic influx, surface area of the carbonate platform, water depth and illumination, nutrients, salinity and rates of base-level changes (Walker and James, 1992). Following the initial precipitation of carbonates, sediment reworking and redistribution within the basin may occur as a result of mechanical erosion by waves and various types of currents, and bioerosion. The bulk of this sediment is generated on the carbonate platform top, and part of it may be remobilized and transported to the deeper portions of the basin by gravity (density) flows and storm surges (e.g., Hine et al., 1981, 1992).

Sediment supply is therefore a key to understanding how sequence stratigraphy works in the case of carbonate depositional systems, and how carbonate models differ from the 'standard' clastic sequence frameworks. Fundamentally, changes in base level have a reciprocal effect on the availability of sediment in carbonate vs. clastic basins. As shown by studies of the sedimentation rates during the late Quaternary base-level cycles in various low- and high-latitude continental margin settings (Droxler and Schlager, 1985; Schlager, 1992), deep-water clastic deposits accumulate most rapidly during lowstands in base level, when terrigenous sediment is delivered most efficiently across the subaerially exposed continental shelf to the shelf edge ('lowstand shedding'), whereas the rates of aggradation of deep-water carbonate deposits are highest during base-level highstands, when the carbonate factory on the continental shelf is most productive ('highstand shedding'). This opposite response of carbonate and clastic systems to base-level changes is a consequence of the intra- vs. extra-basinal origin of the sediment, respectively. In addition to this first-order contrast between carbonate and clastic systems, the response of carbonate platforms to changes in base level also depends on their geometry and relation to the basin margins. Carbonate ramps, for example, are more comparable to the geometry of siliciclastic continental shelves, whereas carbonate shelves and banks are fundamentally different from clastic shelves, being characterized by flat tops, steep slopes, and often high relief (Fig. 6.48; Burchette and Wright, 1992; James and Kendall, 1992). As such, it has

been realized that the sequence stratigraphy of carbonate shelves and banks differs from that of carbonate ramps, and that the opposite response between carbonate shelves/banks and clastic shelves, with respect to sediment supply to the deep-water basin, is not fully realized in the case of carbonate ramps (Burchette and Wright, 1992; James and Kendall, 1992; MacNeil and Jones, in press). This section of the book emphasizes on carbonate shelves, which typify the fundamental differences between carbonate and clastic systems. The key aspects of the carbonate sequence stratigraphic model, for a shelf-type platform (Fig. 6.48), are presented below.

The Carbonate Sequence Stratigraphic Model

With sediment supplied by extra-basinal sources, siliciclastic systems may aggrade to sea level from any depth, providing that sufficient sediment input is available. This basic principle explains all the geometric features of systems tracts presented in Figs. 5.7, 5.26, 5.27, 5.44, 5.56, and 5.57. In contrast, carbonate shelves are in antiphase with this clastic model, as the amount of carbonate sediment, intra-basinal in nature, is proportional to the productivity of the shallow-water carbonate factory on the platform top: lowering of the base level, followed by the subaerial exposure of the platform top, generally shuts down the carbonate factory, whereas a rising base level generates accommodation for the development of the carbonate platform.

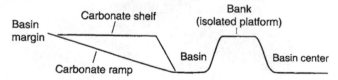

FIGURE 6.48 Types of carbonate platforms, based on geometry, slope gradients, and the relation to the basin margin (modified from James and Kendall, 1992). The major types of carbonate platforms include carbonate shelves, carbonate ramps, and isolated platforms (banks). Carbonate shelves have different geometries from continental siliciclastic shelves, being characterized by a relatively flat top, steep slopes, and often high relief. The margin of these shelf-type platforms may be rimmed by reefs or some form of barrier complex, or unrimmed. Carbonate ramps are more comparable to the geometry of siliciclastic continental shelves. In contrast with carbonate shelves and ramps, isolated platforms (banks) are disconnected from the mainland. It is being increasingly realized that the sequence stratigraphy of carbonate systems varies with the type of carbonate platform. Owing to their geometry and relation to the basin margins, carbonate ramps show the closest affinity to the sequence stratigraphy of clastic shelves. In contrast, carbonate shelves and banks are fundamentally different from clastic shelves, particularly with respect to the patterns of sediment supply to the basin during various stages of the base-level cycle. This section of the book emphasizes on shelf-type platforms, which typify the fundamental differences between carbonate and clastic systems.

In addition to this, another limiting factor for the production of carbonate sediment is the fact that carbonate platforms may only rebound from maximum water depths that correspond to the limit of the photic zone, as the rates of carbonate production at aphotic depths are negligible (Schlager, 1992).

It can be noted that, in contrast to clastic systems, where the rates of aggradation are a function of sediment supply coupled with local energy flux, irrespective of water depth, carbonate systems are much more sensitive to water depth and environmental conditions in general. 'Highstand shedding' of sediment from the shelf into the deep-water portion of the basin, therefore, is only possible where carbonate platforms are within the photic zone, allowing platform carbonates to be actively produced, and where sedimentation rates exceed the rates of generation of accommodation. These conditions are best fulfilled during times of highstand normal regression, when a significant portion of the carbonate shelf is submerged, and assuming that water depth does not exceed the photic limit. It may be inferred that not all highstand systems tracts are conducive to carbonate platform growth and highstand shedding of carbonate sediment into the deep-water environment (MacNeil and Jones, in press). Indeed, any rises in base level during previous transgressive stages, at rates that exceed the growth potential of the carbonate platform, may terminate the growth of the platform and the production of carbonate sediment. Such stages of rapid flooding and drowning of the carbonate platform result in the formation of 'drowning unconformities,' which are unique to carbonate environments and mark a fundamental switch in the style of sedimentation and stratal stacking patterns, from carbonate to clastic systems (Schlager, 1989, 1992).

Drowning Unconformities

Within the framework of carbonate sequence stratigraphy, drowning unconformities represent arguably the most important departure from the repertoire of stratigraphic surfaces that characterizes clastic successions. Because of their major significance, and their commonly strong signature on seismic lines, drowning unconformities are often referred to as 'sequence boundaries' in mixed carbonate/siliciclastic successions (Schlager, 1992). Whether the choice of drowning unconformities as sequence boundaries is appropriate or not, is a matter of choice and possibly a topic of debate, as explained below. What is really important is to recognize drowning unconformities as such, and to avoid possible confusions with other sequence stratigraphic surfaces that may have an equally prominent signature on seismic data. For example, it has been noted that the geometry of drowning unconformities resembles somewhat the physical attributes of subaerial unconformities, as both are potentially associated with high-amplitude reflections with an irregular relief across the continental shelf, although the two surfaces are fundamentally different and form during opposite stages of the base-level cycle (Schlager, 1989, 1992). According to Schlager (1992), the misinterpretation of drowning unconformities as subaerial unconformities may explain, in some cases, erroneous reconstructions of the history of base-level changes in some basins, and the discrepancy between the results obtained from sequence stratigraphy relative to other independent techniques. Criteria for the identification of drowning unconformities are reviewed below.

Highstand Systems Tracts

The basic stages of evolution of a carbonate shelf, each corresponding to the formation of a systems tract, are presented in Fig. 6.49. As a general principle, stages of highstand normal regression are most favorable to the development of carbonate systems, both on the continental shelf and within the deep-water setting, for two reasons. Firstly, the large-scale flooding of the platform that is common during highstand stages, as following transgressions, provides a significant surface area for carbonate production. Secondly, base-level rises during highstand stages, generating accommodation for platform growth, but with relatively low rates, allowing the carbonate platform to keep up with the rate of creation of accommodation. This ensures that no drowning occurs, and, as the rates of base-level rise decrease with time during the highstand stage, the volume of carbonate sediment that exceeds the amount of available accommodation is shed to the deep-water environment, generating significant accumulations of clastic carbonates on the slope and on the basin floor ('highstand shedding'). Therefore, under highstand conditions, *production* outpaces *accumulation* on the platform top, and the excess of carbonate sediment is transferred to the deeper-water environment ('basin') mainly by storm surges and gravity flows (e.g., Neumann and Land, 1975). These deep-water clastic carbonates are generally preserved providing that accumulation takes place above the calcium carbonate compensation depth. The formation of such a highstand systems tract composed of shallow- and deep-water carbonate systems may be considered as the first stage in the evolution of a carbonate shelf (Fig. 6.49). Note that accommodation is measured to the base level, which is below the sea level due to the energy of waves and currents, and not to the sea level (see Chapter 3 for more details). This explains why highstand shedding takes place while a shallow-water environment is still maintained on the platform top (i.e., the water column

FIGURE 6.49 Generalized life cycle of a carbonate shallow- to deep-water system in a sequence stratigraphic framework, from the initial growth of the carbonate platform to its burial by siliciclastic systems (compiled information from James and Kendal, 1992; Jones and Desrochers, 1992; and Schlager, 1992). Distinct stages of this cycle may include: initial platform growth (1), karstification (2), regeneration and rimmed-shelf development (3), renewed platform growth (4), drowning (5) and burial (6). These stages do not necessarily occur in this full succession. For example, stages 1–4 may repeat with time without a drowning unconformity being formed (i.e., the model of James and Kendall, 1992). Following several such cycles, an increase in the rates of base-level rise may lead to the drowning of the carbonate platform, when the carbonate factory is shut down and a 'drowning unconformity' develops across the basin (stage 5). Note that, as the carbonate platform backsteps gradually in the process of drowning during rapid transgression, the drowning unconformity is diachronous, younging landward. Following the formation of a drowning unconformity, the water on the shelf is too deep to revitalize the carbonate factory during subsequent highstand, and therefore the drowning unconformity is commonly downlapped by prograding highstand clastic systems (stage 6). Abbreviations: HST—highstand systems tract; FSST—falling-stage systems tract; LST—lowstand systems tract; TST—transgressive systems tracts; SL—sea level.

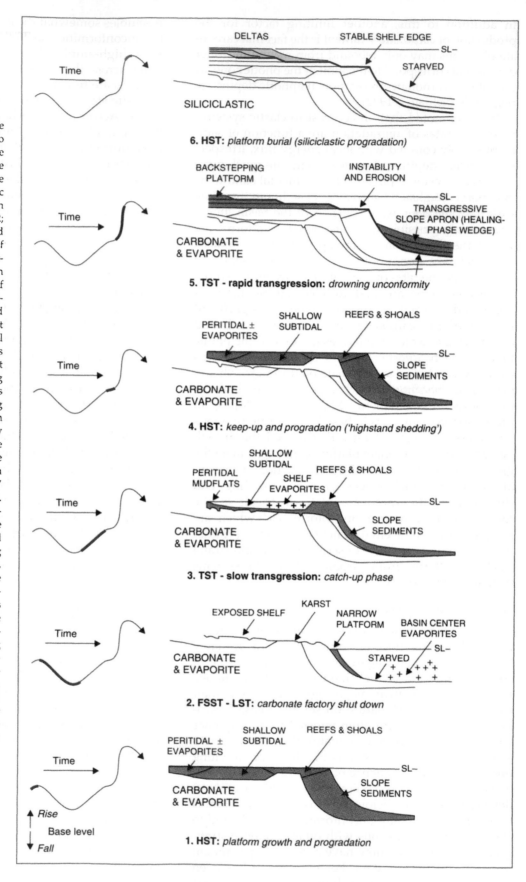

between the sea level and the base level). Under this highstand regime, the amount of accommodation created on the platform top by base-level rise is consumed entirely by sedimentation, which means that available accommodation is zero, even though water depth is positive, and that the base level and the seafloor are superimposed (see Fig. 3.8 to visualize the difference between available accommodation and water depth).

As suggested in Fig. 6.49, carbonate shelves may sustain the formation of highstand systems tracts during different stages of their evolution. A 'pure' carbonate succession that records several cycles of base-level changes commonly starts with a highstand systems tract, which marks the initiation of the platform, includes as many internal highstand systems tracts as the number of cycles recorded, and terminates with a final highstand systems tract that marks the switch from carbonate to siliciclastic sedimentation. It can be concluded that three types of highstand systems tracts may be distinguished in the context of carbonate sequence stratigraphy: an 'initial' highstand systems tract, which leads to the early development of the carbonate platform (stage 1 in Fig. 6.49); 'internal' highstand systems tracts, which succeed relatively slow transgressions that are survived by the carbonate platform (e.g., stage 4 in Fig. 6.49); and a 'final' highstand systems tract, which follows the drowning of the carbonate platform and initiates the burial of the carbonate succession by prograding siliciclastics (stage 6 in Fig. 6.49). The latter type of highstand systems tract marks the return to clastic systems on the continental shelf (Fig. 5.7), and accumulates on top of the drowning unconformity (for an example, see the case study of the Wilmington Platform: Fig. 5–9 in Schlager, 1992).

The 'initial' and 'internal' highstand systems tracts of a carbonate succession display the characteristic features of carbonate shelves, as described above. These include the development of carbonate facies to base level on the platform top (shallow-water setting), and the accumulation of thick deposits of clastic limestones in the deep-water environment as a result of 'highstand shedding.' During such stages, carbonate platforms 'keep up' with the rise in base level, reflecting a balance between accommodation and carbonate productivity, and the surplus of carbonate sediment leads to the progradation of the shelf edge (Jones and Desrochers, 1992). In contrast, the 'final' highstand systems tract consists almost entirely of a 'highstand prism' on the continental shelf, with a correlative condensed section of pelagic sediments in the starved deeper portion of the basin (Fig. 5.7). This drastic change in sediment budget across the continental margin reflects the difference in the patterns of sediment dispersal between clastic and carbonate depositional environments.

Falling-stage—Lowstand Systems Tracts

Following stages of highstand normal regression, when most accommodation across the carbonate platform is consumed and as a result water depths are very shallow, any fall in base level, even of relatively low magnitude, tends to lead to rapid forced regression and the subaerial exposure of the platform top. Subaerial exposure of the platform top continues during subsequent lowstand normal regressions, which is why the falling-stage to lowstand interval may be studied as one stage with distinct consequences for the evolution of the carbonate shelf (stage 2 in Fig. 6.49). This principle does not necessarily apply to carbonate ramps, which show closer affinity to the stratigraphic architecture of clastic shelves (e.g., MacNeil and Jones, in press).

The fundamental implication of base-level fall within the context of a carbonate shelf is that the carbonate factory is shut down following its subaerial exposure. Consequently, the carbonate platform is subject to karstification, as fluvial systems advance across the continental shelf and adjust to lower elevations of the shoreline. Fluvial incision, coupled with the dissolution of carbonates, leads to the development of an array of karst structures which may be preserved in the rock record in the process of burial during subsequent stages of base-level rise. The karst topography at the top of the exposed carbonate platform describes the relief associated with the subaerial unconformity within carbonate successions. These unconformities serve as depositional sequence boundaries, and may separate highstand carbonates below from transgressive carbonates above (Fig. 6.49). It should be noted, however, that processes of karstification are climate-dependent and that under arid climatic conditions karst may not develop but calcrete profiles, with less topographic relief, may form instead.

If the forced regressive shoreline falls below the elevation of the shelf top, which is likely considering the shallow depths of the highstand platforms, the much steeper slope may only support the development of a relatively narrow belt of carbonate deposits (Fig. 6.49). Hence, only a small amount of carbonate sediment is expected to be shed to the deep-water environment during the falling-stage to lowstand intervals. Sediment starvation in the deep-water environment may, however, promote the precipitation of other chemical deposits on the seafloor, notably of basin-center evaporites in the case of restricted basins (James and Kendall, 1992; Fig. 6.49).

Transgressive Systems Tracts

In addition to forced regressions, transgressions represent another switch that may, under particular circumstances, shut down the carbonate factory. In general, transgressions pose a threat to carbonate platforms because the rates of base-level rise are higher than the rates of aggradation at the shoreline, which commonly leads to a deepening of the water in most areas of the platform. If water deepens more than the photic limit, the platform is drowned and the carbonate factory is shut down. If the platform remains within the photic zone in spite of the deepening of the water, the carbonate factory 'survives' the transgression, and the production of carbonate sediment continues and eventually catches up with the newly created accommodation as the rising base level decelerates and transgression gives way to highstand normal regression. It can be noted that two transgressive scenarios may be envisaged, with contrasting consequences for the evolution of carbonate platforms: slow transgressions, associated with *internal cycles* of carbonate successions, which do not interrupt the production of carbonates (e.g., stage 3 in Fig. 6.49); and rapid transgressions, associated with *terminal cycles* of carbonate successions, which lead to the drowning of carbonate platforms and the change from carbonate to clastic systems (e.g., stage 5 in Fig. 6.49). It is important to note that, within the context of carbonate sequence stratigraphy, the concept of 'drowning' refers to a situation where transgression follows highstand without an intervening stage of base-level fall (as shown in Fig. 6.49). This is in contrast with the concept of 'flooding', as used within the context of clastic sequence stratigraphy, where the inferred deepening of the water may occur following a stage of base-level fall (see Chapter 4 for more details on the concept of 'flooding surface').

Slow transgressions create an excess of accommodation across the carbonate shelf, which results in the formation of shallow-water subtidal depozones between the shoreline and the rimmed shelf edge. These depozones, or lagoons, are commonly of low energy, being protected from the open sea by distal-shelf barrier reefs (Fig. 6.49). The formation of barrier reefs in the distal region of the continental shelf during transgression may be controlled by a combination of factors, including: pre-existing karstic topography, as areas closer to the shelf edge are less exposed to dissolution during previous stages of forced regression, hence maintaining higher elevations; the distal location relative to the source areas of clastic sediment; and the proximity to the active lowstand carbonate platform. While the shelf is flooded during slow transgressions, the relatively low rates of base-level rise may allow the distal-shelf reefs to grow to base level, keeping up with the newly created accommodation (i.e., no water deepening in the distal-shelf reef region during transgression). At the same time, the rest of the carbonate platform is submerged, but with water depths within the limits of

the photic zone. This allows the carbonate factory to survive transgression, and the production of carbonates to continue until it eventually catches up with the rising base level during the subsequent highstand stage. Although a transfer of carbonate sediment from the shelf to the deep-water environment may occur during slow transgressions, such sediment supply to the slope and basin-floor settings is far less than the 'highstand shedding' due to the availability of accommodation on the shelf top, which traps most of the carbonate sediment.

Rapid transgressions, associated with high rates of base-level rise, result in the drowning of the carbonate platform (i.e., water depth exceeding the photic limit), which shuts down the carbonate factory. Where rapid transgressions follow stages of active platform growth across the continental shelf (Fig. 6.49), the transgressive platforms display characteristic backstepping geometries, becoming progressively narrower in the process of drowning. The case study of the Miocene Platform in the Pearl River Mouth Basin, South China Sea, provides an example of such a backstepping carbonate platform (Erlich *et al.*, 1990; Schlager, 1992; Fig. 5–10 of Schlager, 1992). The cessation of carbonate productivity during rapid transgressions results in the formation of drowning unconformities. As the carbonate factory is shut down on the platform top, also disabling the delivery of new carbonate sediment to the deep-water environment, drowning unconformities have a basin-wide development, extending across the shelf and within the deep-water setting (Fig. 6.49).

Drowning represents the final stage in the evolution of a carbonate platform, prior to the return to a clastics-dominated environment. Once the platform is drowned below the photic limit, filling of the available accommodation during subsequent highstand normal regression may only be achieved by means of siliciclastic progradation. Sedimentary processes during drowning already resemble clastic patterns of sediment dispersal. This is particularly evident in the distal shelf to deep-water settings, as the lack of carbonate production coupled with hydraulic instability at the shelf edge caused by rapid base-level rise result in the erosion of the shelf edge region and the formation of a healing-phase wedge that onlaps the continental slope, just as in the case of 'pure' clastic systems (e.g., compare Fig. 6.49 with Figs. 5.56 and 5.57). Healing-phase wedges consist of fine-grained sediment with a transparent facies on seismic lines, which accumulates in gently dipping layers, with an angle of repose that is lower relative to the seaward flank of the carbonate platform. As observed in the case of the Wilmington Platform (Meyer, 1989; Schlager, 1989), the drowning unconformity is onlapped by the healing-phase deposits, which are interpreted as being formed during the early phases of transgression. The formation of healing-phase wedges is most likely in the case of unrimmed carbonate shelves, but it may be inhibited where shelf edges are reefal, blocking the sediment transfer into the basin, or where the starved shelf seafloor is indurated by intense marine cementation during drowning, preventing the erosion of the shelf edge and thus reducing the amount of sediment that can be delivered to the basin (e.g., Sarg, 1988). On the shelf, the formation of the drowning unconformity continues during the backstepping of the carbonate platform, gradually expanding shoreward (Fig. 6.49). It is therefore important to note that drowning unconformities are potentially diachronous, younging towards the basin margins, being formed during a period of time that may span the entire duration of the transgressive stage.

In summary, criteria for recognizing drowning unconformities that form during rapid transgressions, at the end of the carbonate platform life cycle, include: high-amplitude reflections on seismic lines associated with a significant contrast of acoustic impedance between carbonate facies below and clastic facies above (see case studies in Schlager, 1992); the pattern of carbonate platform backstepping on the continental shelf, which indicates drowning as opposed to subaerial exposure (stage 5 in Fig. 6.49); onlapping by a transgressive slope apron (healing-phase wedge) in the deep-water setting (stage 5 in Fig. 6.49); and downlapping by highstand deltas in the continental shelf setting (stage 6 in Fig. 6.49). This discussion reveals that the drowning unconformity, which is unique to carbonate systems, may have the significance of a maximum regressive surface in the deep-water setting, where it is onlapped by the transgressive slope apron, and of a (younger) maximum flooding surface on the continental shelf ('downlap surface' on seismic lines). The fact that the drowning unconformity is downlapped by highstand deltas provides an unequivocal criterion for separating this surface from the subaerial unconformity. The latter is not downlapped by deltaic systems, as lowstand deltas prograde beyond the seaward termination of the subaerial unconformity, but it is rather onlapped by lowstand and/or transgressive fluvial systems, or reworked by transgressive ravinement surfaces (see Chapters 4 and 5 for more details). Stages 5 and 6 in Fig. 6.49 capture the most significant stratigraphic features of the drowning unconformity, showing its position at the contact between backstepping platform carbonates below and prograding clastic deltas above, on the continental shelf, and at the base of the transgressive slope apron in the deep-water environment. These diagrams are based on the case studies of the Miocene Platform

in the Pearl River Mouth Basin, South China Sea (drowning unconformity as a high-amplitude reflection at the top of a backstepping platform), and of the Late Jurassic–Early Cretaceous Wilmington Platform (drowning unconformity as a high-amplitude reflection at the base of a shelf delta and at the base of a slope apron (Meyer, 1989; Schlager, 1989, 1992; Erlich et al., 1990) (seismic lines in Schlager, 1992).

Discussion: Sequence Boundaries in Carbonate Successions

The drowning unconformity was identified as a 'type 3' sequence boundary by Schlager (1999) within the context of carbonate sequence stratigraphy, in contrast to the 'type 1' and the 'type 2' sequence boundaries used in the case of clastic systems (Vail et al., 1984). The fundamental differences between types 1, 2, and 3 sequence boundaries are summarized in Fig. 6.50. According to Vail et al. (1984), a type 1 sequence boundary forms during a stage of rapid eustatic sea-level fall, resulting in a relative sea-level fall both at the shelf edge and at the shoreline, whereas a type 2 sequence boundary forms when the rate of eustatic sea-level fall is less than the rate of subsidence at the shelf edge (relative sea-level rise at the shelf edge), but greater than the rate of subsidence at the shoreline (relative sea-level fall at the shoreline), resulting in the formation of a subaerial unconformity that is characterized by minor erosion and a limited lateral extent across the continental shelf (Fig. 5.1). The introduction of types 1 and 2 sequence boundaries in sequence stratigraphy was meant to make the distinction between 'major' and 'minor' subaerial

Sequence boundaries	Depositional system	Relative sea-level changes		Stratigraphic surfaces
		Shoreline	Shelf edge	
Type 1	Clastic and carbonate	Fall	Fall	Subaerial unconformities and their correlative conformities
Type 2		Fall	Rise	
Type 3	Carbonate	Rise	Rise	Drowning unconformities

FIGURE 6.50 Definition of types 1, 2, and 3 sequence boundaries according to Vail et al. (1984) and Schlager (1999). Both types 1 and 2 sequence boundaries include unconformable and conformable portions (subaerial unconformity and its correlative conformity; Vail et al., 1984; Galloway, 1989). In contrast, the type 3 sequence boundary (drowning unconformity) may be a maximum regressive surface in the deep-water setting ('basin') and a maximum flooding surface at the top of the carbonate platform (Fig. 6.49). The concept of type 3 sequence boundary is therefore fundamentally different from the types 1 and 2 depositional sequence boundaries. The type 1 vs. type 2 terminology has been abandoned in recent years, in favor of a single depositional sequence boundary. In this context, the type 3 terminology becomes redundant, and the 'type 3 sequence boundary' should be referred to as the 'drowning unconformity' (see text for more details).

unconformities (significant erosion and areal extent *vs.* minor erosion and limited areal extent), respectively (see Chapter 5 for more details). It should be noted that both types 1 and 2 sequence boundaries involve the formation of *subaerial unconformities* (Vail *et al.*, 1984, reiterated subsequently by Galloway, 1989; Fig. 5.1), in contrast to the concept of type 3 sequence boundary of Schlager (1999) that refers to a *drowning unconformity* that forms during rapid relative sea-level rise across the entire carbonate platform following a stage of highstand (Fig. 6.50). Therefore, even though the type 2 sequence boundary of Vail *et al.* (1984) assumes a relative sea-level rise *at the shelf edge*, one must not confuse between the types 2 and 3 sequence boundaries, as they are fundamentally different concepts. The separation of a distinct 'type 3' sequence boundary by Schlager (1999) was therefore fully warranted at a conceptual level. Nevertheless, as the 'type 1' *vs.* 'type 2' terminology has been abandoned in recent years (see Chapter 5 for a further discussion on this topic), the usage of the 'type 3 sequence boundary' terminology has become redundant as well, and one should use the term of 'drowning unconformity' instead.

The question still remains whether drowning unconformities, as opposed to subaerial unconformities or other types of stratigraphic surfaces, are an appropriate choice for sequence boundaries in carbonate successions, as proposed by Schlager (1989, 1992, 1999). To some extent, the applicability of this approach depends on the scale of observation and the nature of the stratigraphic succession under analysis. Owing to their mode of formation, and their position at the contact between carbonate facies below and clastic facies above, multiple drowning unconformities may only be found in mixed clastic—carbonate successions (Fig. 6.51). In such cases, drowning unconformities relate to major cycles of changing sedimentation regimes, and bound 'sequences' consisting of a couplet of clastic and overlying carbonate stratigraphic units (Fig. 6.51). At smaller scales, however, drowning unconformities may not be used to describe the internal cyclicity of 'pure' carbonate successions, because no episodes of drowning are recorded during such depositional intervals. For example, the repetition of stages 1–4 in Fig. 6.49 generates stratigraphic cyclicity, as described by the carbonate sequence stratigraphic model of James and Kendall (1992), but no drowning unconformities are accounted for as sequence boundaries as the production of carbonates may be uninterrupted for several cycles of base-level changes. In such cases, the mapping of drowning unconformities as sequence boundaries may underestimate the true number of sequences that are present within the succession under analysis, as the products of several cycles of base-level

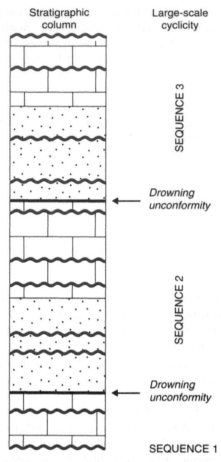

FIGURE 6.51 Hypothetical stratigraphic column of a mixed carbonate/siliciclastic succession in which drowning unconformities are used as sequence boundaries, following the method proposed by Schlager (1989, 1992). Wavy lines indicate subaerial unconformities (depositional sequence boundaries), which may occur within both carbonate and siliciclastic stratigraphic units. Note that each individual carbonate or siliciclastic succession may include several depositional sequences. In this example, sequences bounded by drowning unconformities reflect a large-scale cyclicity of changing sedimentation regimes, from clastic to carbonate, but the smaller-scale cycles that describe the internal architecture of carbonate and siliciclastic deposits do not have corresponding 'sequences' in this approach. Drowning unconformities may have the significance of shallow-water maximum flooding surfaces and deep-water maximum regressive surfaces associated with rapid transgressions. Other maximum flooding and maximum regressive surfaces associated with slower transgressions may, however, be present in this succession (not shown), within depositional sequences. See text for details.

changes (i.e., depositional sequences bounded by subaerial unconformities) may be amalgamated into one drowning unconformity-bounded 'sequence' (Fig. 6.51). Such a drowning unconformity-bounded 'sequence' would include strata that are genetically *unrelated*, which violates the definition of a 'sequence' (Fig. 1.9).

Case studies of mixed carbonate—siliciclastic successions have been documented for a wide range of temporal scales, from 10^1–10^2 Ma (e.g., Long and Norford, 1997) to 10^0 Ma cycles of changing sedimentation regimes (e.g., Vecsei and Duringer, 2003).

The *caveat* of the generalization that drowning unconformities are always placed at the contact between carbonate facies below and clastic facies above is that this is typical of carbonate platforms attached to the mainland, where clastic sediment supply is available following the stage of drowning. Isolated carbonate platforms ('banks'; Fig. 6.48), however, which are detached from the mainland and lack a source of clastic sediment supply, may resume carbonate production following drowning, once the seafloor reaches again the photic zone, without an intervening stage of clastic sedimentation. In such cases, drowning unconformities may occur *within* carbonate successions (i.e., carbonate facies below and above), and are typically marked by hardgrounds that form by processes of marine cementation during stages of sediment starvation when the carbonate factory is shut down. Even in the case of isolated banks, however, one must make the distinction between subaerial unconformities (base-level fall following highstand) and drowning unconformities (rapid base-level rise following highstand; Fig. 6.49). Similar to the discussion of carbonate platforms attached to the mainland, the mapping of drowning unconformities as 'sequence boundaries' within a succession of carbonate bank facies may result in the amalgamation of several depositional sequences into one drowning unconformity-bounded 'sequence,' and therefore the interpreter may miss to recognize several cycles of base-level changes.

Another pitfall of drowning unconformities is their potential for being highly diachronous. As discussed above, the sequence stratigraphic significance of drowning unconformities may vary from maximum regressive surfaces, in the deep-water setting, to maximum flooding surfaces on the continental shelf. The period of time required for the formation of a drowning unconformity may span the entire stage of shoreline transgression, during which interval the surface gradually expands (and youngs) in a shoreward direction. Thus, the landward termination of the drowning unconformity may be significantly younger than its deep-water portion, and age-equivalent to the maximum flooding surface that tops the deep-water healing-phase wedge. The lack of chronostratigraphic significance diminishes the value of drowning unconformities in a sequence stratigraphic framework, even though they may be mapped with relative ease on seismic lines as high-amplitude (but time-transgressive) reflections. The time-transgressive character of drowning

unconformities, and their formation within the marine environment during stages of abrupt water deepening, makes them equivalent to the within-trend flooding surfaces discussed in the case of clastic systems in Chapter 4. Drowning unconformities may therefore be regarded as a special type of flooding surface, applicable to carbonate systems, which form as the seafloor drowns to water depths in excess of the photic limit. It can be noted that not all flooding surfaces in carbonate environments qualify as drowning unconformities, but only those associated with rapid transgressions. On the continental shelf, such flooding surfaces become maximum flooding surfaces where no other transgressive deposits accumulate on top of the backstepping carbonate platforms (Fig. 6.49). As seen on seismic data (Schlager, 1992), this is commonly the case as the carbonate productivity decreases dramatically in the process of drowning, during rapid transgression.

Besides the limitations outlined above, the shallow- and deep-water portions of drowning unconformities (maximum flooding and maximum regressive surfaces, respectively) are already employed as sequence boundaries by two different sequence stratigraphic models. As such, using drowning unconformities as sequence boundaries in shallow-water successions is similar to the genetic sequence stratigraphic approach, with the exception that not all maximum flooding surfaces are drowning unconformities, but only the ones associated with rapid transgressions. Similarly, using drowning unconformities as sequence boundaries in deep-water successions resembles the T–R sequence stratigraphic approach, with the exception, again, that not all deep-water maximum regressive surfaces are drowning unconformities, but only those which mark the onset of rapid transgressions.

It may be concluded that all three sequence stratigraphic models described above in this chapter provide the means for a more detailed sequence stratigraphic analysis of carbonate successions, as subaerial unconformities (depositional sequence boundaries), maximum flooding surfaces (genetic stratigraphic sequence boundaries) and maximum regressive surfaces (T–R sequence boundaries) may all occur more frequently than drowning unconformities in the carbonate rock record. Notwithstanding the limitations imposed by using drowning unconformities as sequence boundaries, their identification in the carbonate or mixed carbonate-siliciclastic rock record still remains of fundamental importance for the reconstruction of the major stages in the evolution of the basin, and for the understanding of the sediment composition and dispersal patterns that characterize various stratigraphic intervals.

7

Time Attributes of Stratigraphic Surfaces

INTRODUCTION

A central and yet controversial topic in sequence stratigraphy is the assessment of stratigraphic surfaces in a chronostratigraphic framework. Are the bounding surfaces of sequences and systems tracts time lines, i.e., generated at the same time everywhere within the basin? The answer to this question is of paramount importance for stratigraphic correlation, and although this problem has been debated for some time (Miall, 1991, 1994; Catuneanu et al., 1998b; Posamentier and Allen, 1999), agreement is yet to be reached. Part of the problem derives from the way concepts are defined, often with contradictory meanings, as explained in this chapter.

The assumptions made in the early days of seismic and sequence stratigraphy, in the 1970s and 1980s, emphasized heavily the role of eustatic changes in sea level in the generation of the preserved stratigraphic record, and implicitly on the formation of systems tract and sequence boundaries. As eustatic changes in sea level are global in nature, it was implied that sequence stratigraphic surfaces that correspond to the four events of the base-level cycle (Fig. 4.7) are isochronous, being formed everywhere at the same time. Subaerial unconformities, assumed to have been formed during global stages of sea-level fall, were also assigned global synchronicity. These early ideas led to the construction of the global cycle chart (Vail et al., 1977; Haq et al., 1987, 1988), whose fundamental premise was that sequence stratigraphic conformities are world-wide correlatable time lines, while sequence-bounding unconformities correspond to stratigraphic hiatuses of global significance. Thus, each set of sequence stratigraphic surfaces found in any basin around the world was invariably traced back and correlated to the 'global master curve,' assuming that each stratigraphic cycle is globally synchronous. As a result, sequence

cycles were seen as geochronologic units that provide the means to subdivide the sedimentary rock record into genetic chronostratigraphic intervals (Vail et al., 1977, 1991).

The eustasy-driven stratigraphic approach provided an easy, but oversimplified way of looking at the temporal significance of stratigraphic surfaces. In reality, no surface can possibly be an absolute time line, and the degree of diachroneity may vary greatly with the mode of formation and the dependence on parameters characterized by variable rates along dip and strike, such as tectonism and sedimentation. While a continuum is expected within the diachroneity range, the qualifiers used to describe the degree of diachroneity in this book are 'low' vs. 'high' (Fig. 4.9): a 'low diachroneity' refers to a difference in age that falls below the resolution of the current dating techniques, which therefore is undetectable by biostratigraphic, magnetostratigraphic, or radiometric means; a 'high diachroneity' refers to a more significant difference in age between the different portions of a stratigraphic surface, which can be emphasized using the available dating techniques.

This chapter analyzes the degrees of diachroneity that various stratigraphic surfaces may be associated with, by using simple numerical models. The temporal significance of stratigraphic surfaces is arguably one of the most difficult topics to deal with, and to quantify, in sequence stratigraphy. Firstly, addressing this topic requires computational simulations, which, although increasingly employed in stratigraphic analyses (e.g., Jordan and Flemings, 1991; Johnson and Beaumont, 1995; Catuneanu et al., 1998b; Harbaugh et al., 1999), are still not fully embraced by the majority of geologists as part of their workflow routine. Secondly, the methods of defining stratigraphic surfaces are to some extent loose, from the generic nature of the reference curve relative to which sequence stratigraphic surfaces are defined (e.g., Fig. 4.7), to the ambiguous and sometimes conflicting criteria that are used to pinpoint surfaces in the rock record. There are historical reasons why the meaning of the reference curve for the definition of sequence stratigraphic surfaces is not fully constrained, and is perhaps even misunderstood, and these reasons are fully explored in the following section of this chapter. Conflicting criteria that persist in the definition of some surfaces also have historical roots, stemming from deeply entrenched 'taboos' in geological thinking. Many of these deeply rooted principles in geology have been re-evaluated in recent years (e.g., the concept of uniformitarianism, and its implications for establishing a hierarchy of stratigraphic cycles—more discussions in Chapter 8; the position of sequence boundaries in fully fluvial successions relative to coarse-grained facies—see discussions in Chapter 6; the relationship between shoreline shifts, sediment grading and water-depth changes—discussed in this chapter; the 'time-barrier' vs. time-transgressive character of sequence-bounding unconformities—also discussed in this chapter, etc.), as afforded by advances in data acquisition, laboratory techniques, or numerical modeling.

As discussed above, considerable confusion with respect to where some of the conformable sequence stratigraphic surfaces are placed in the rock record is sourced by alternative approaches to their definition, such as, for example, the placement of a marine maximum flooding surface at the top of a 'fining-upward' succession or at the top of a 'deepening-upward' succession. While such alternative definitions overlap to some extent, and are traditionally considered as equivalent, a quantitative analysis of the interplay of the controls on grain size and water-depth changes reveals that different definition criteria may be satisfied by discrete surfaces that are spatially offset, as opposed to pointing to one single surface as intended. This chapter addresses all these aspects involved in the definition of sequence stratigraphic surfaces, which have a direct impact on their temporal attributes, from the 'master' reference curve to the criteria employed for mapping them in the rock record.

REFERENCE CURVE FOR THE DEFINITION OF STRATIGRAPHIC SURFACES

The timing of all sequence stratigraphic surfaces and systems tracts is defined relative to *one curve* that describes a full cycle of sea-level, relative sea-level, or base-level changes, depending on the model that is being employed (e.g., Figs. 4.7 and 6.1). For example, the original correlative conformity of Posamentier *et al.*

(1988) was considered to form during early sea-level fall (Fig. 6.1), which was later revised to the onset of sea-level fall (Posamentier *et al.*, 1992b) or the onset of relative sea-level fall (Posamentier and Allen, 1999). The correlative conformity of Hunt and Tucker (1992) is taken at the end of relative sea-level fall (Fig. 6.1), and so on. Irrespective of the conceptual approach, each model shows *one curve* relative to which all surfaces and systems tracts are defined.

This theoretical curve is generally presented as a generic sinusoid, with an unspecified position within the basin. The generic nature of this reference curve originates from the early seismic and sequence stratigraphic models of the late 1970s to late 1980s, which were based on the assumption that eustasy is the main driving force behind sequence formation at all levels of stratigraphic cyclicity (Figs. 3.2 and 6.1). Since eustasy is global in nature, there was no need to specify where that reference curve is positioned within the basin under analysis. Subsequent realization in the 1990s, based on much earlier work, that tectonism is as important as eustasy in controlling stratigraphic cyclicity, led to the replacement of the eustatic curve with other reference curves, of relative sea-level (eustasy plus tectonism) or base-level (relative sea level plus energy of the depositional environment)

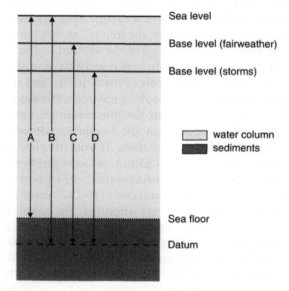

FIGURE 7.1 Concepts of water depth, sea level, relative sea level, and base level. For the significance of the datum, refer to Chapter 3, as well as Figs. 3.12 and 3.15. Changes in distances A, B, and C/D reflect water-depth (bathymetric) changes, relative sea-level changes, and base-level changes, respectively. Note that the position of the base level is a function of environmental energy, which marks the difference between the concepts of relative sea-level and base-level changes (see also Fig. 3.15). Sea-level changes are independent of datum, seafloor, and sedimentation, and are measured relative to the center of Earth.

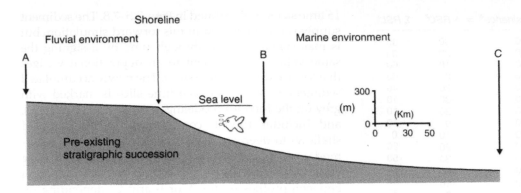

FIGURE 7.2 Dip-oriented cross-section through a hypothetical extensional basin. Locations A, B, and C are characterized by different subsidence rates, as illustrated in Fig. 7.3.

changes (Figs. 3.15 and 7.1). The shortcoming of these conceptual advances was that the new reference curves were still regarded as generic, with an unspecified position within the basin, despite the fact that subsidence is invariably differential along both dip and strike.

Figure 7.2 presents a dip-oriented cross-section through a hypothetical basin in an extensional setting, e.g., a divergent continental margin. In such settings, subsidence rates vary along dip, increasing towards the basin (Pitman, 1978; Angevine, 1989; Jordan and Flemings, 1991). The reference locations A, B, and C are therefore characterized by different subsidence rates, as illustrated in Fig. 7.3. Changing subsidence rates implies that any point within the basin is characterized by its own curve of relative sea-level/base-level fluctuations (Fig. 7.4), so no one curve is representative for the entire basin. Due to differences in subsidence rates, the three curves of relative sea-level changes in Fig. 7.4 are *offset relative to one another*, not only in terms of magnitudes but also in terms of timing of the high and low peaks along the curves. The amount of temporal offset would be even higher for basins undergoing a more pronounced differential subsidence. Under these circumstances, which one of these curves should be chosen as a reference to define the timing of systems tracts and bounding surfaces (Fig. 4.7)?

None of the 'static' curves (related to specific locations within the basin) in Fig. 7.4 is the perfect candidate for *the* reference curve we need. The curve of relative sea-level changes at location A, which may approximate the *basin margin*, is not suitable because it does not describe changes in accommodation in the shoreline region, which have a direct impact on the direction of shoreline shift. The same applies to the curve that characterizes location C, which may approximate the *basin center*, because this location is again far from the shoreline region so it does not have a direct control on the direction of shoreline shift. Location B may offer the closest approximation for the reference curve, as being closer to an average shoreline position, but it is

not perfect because with time, the shoreline may move closer or farther away relative to this location.

Two out of the four main events of a full base-level cycle refer specifically to changes in the direction of shoreline shifts, from regression to transgression and *vice versa* (Figs. 1.7 and 4.7). Even the other two events, i.e., the onset and end of base-level fall, are also taken, irrespective of the sequence model of choice, to signify a change in the type of shoreline shift, from normal to forced regression and *vice versa*, respectively (Fig. 3.19). Hence, all four main events of the reference curve of base-level changes are *linked to the shoreline*, implying changes in the direction and/or the type of shoreline shift. What is commonly overlooked is the fact that, as we move away from the shoreline, changes in base level recorded in other locations may differ significantly from the reference curve that describes changes in the direction and/or the type of shoreline shift. For example, a forced regression (base-level fall at the shoreline) may well be coeval with a base-level rise offshore, due to variations in subsidence rates (the case envisaged by Vail *et al.*, 1984, for the formation of 'type 2' sequence boundaries), and so on (Figs. 7.2–7.4). This means that the curves of base-level changes that characterize discrete locations within the basin are offset relative to one another, as explained above (Fig. 7.4), which requires us to specify where exactly along each dip-oriented section the reference curve of base-level changes is taken.

The shoreline is dynamic, as it continuously changes its position within the basin as a function of the local balance between accommodation and sedimentation. The reference curve of base-level shifts should therefore describe the *changes in accommodation at the shoreline*, wherever the shoreline is within the basin at each time step. This means that the actual shifts in base level along the reference curve, in terms of magnitude and timing, can be quantified by interpolating between the 'static' A, B, and C curves, according to the location of the shoreline at every time step.

	Time*	Δ Eustasy#	- Δ Subsidence#	= Δ RSL#	Σ RSL+
Location A in Figure 7.2	0 to 1	30	0	30	30
	1 to 2	20	0	20	50
	2 to 3	10	0	10	60
	3 to 4	0	0	0	60
	4 to 5	−20	0	−20	40
	5 to 6	−30	0	−30	10
	6 to 7	−20	0	−20	−10
	7 to 8	0	0	0	−10
	8 to 9	10	0	10	0
	9 to 10	20	0	20	20
	10 to 11	30	0	30	50
	11 to 12	20	0	20	70
	12 to 13	10	0	10	80
	13 to 14	0	0	0	80
	14 to 15	−5	0	−5	75
Location B in Figure 7.2	0 to 1	30	−5	35	35
	1 to 2	20	−5	25	60
	2 to 3	10	−5	15	75
	3 to 4	0	−5	5	80
	4 to 5	−20	−5	−15	65
	5 to 6	−30	−5	−25	40
	6 to 7	−20	−5	−15	25
	7 to 8	0	−5	5	30
	8 to 9	10	−5	15	45
	9 to 10	20	−5	25	70
	10 to 11	30	−5	35	105
	11 to 12	20	−5	25	130
	12 to 13	10	−5	15	145
	13 to 14	0	−5	5	150
	14 to 15	−5	−5	0	150
Location C in Figure 7.2	0 to 1	30	−10	40	40
	1 to 2	20	−10	30	70
	2 to 3	10	−10	20	90
	3 to 4	0	−10	10	100
	4 to 5	−20	−10	−10	90
	5 to 6	−30	−10	−20	70
	6 to 7	−20	−10	−10	60
	7 to 8	0	−10	10	70
	8 to 9	10	−10	20	90
	9 to 10	20	−10	30	120
	10 to 11	30	−10	40	160
	11 to 12	20	−10	30	190
	12 to 13	10	−10	20	210
	13 to 14	0	−10	10	220
	14 to 15	−5	−10	5	225

FIGURE 7.3 Changes in sea level, subsidence, and relative sea level along the profile in Fig. 7.2, during a period of time of 1.5 Ma (modified from data provided by H. W. Posamentier). Incremental changes in sea level, subsidence, relative sea level, and cumulative relative sea level are shown for time steps of 100.000 years. The curve of sea-level changes is the same for the three reference locations in Fig. 7.2. Subsidence rates increase towards the basin, and are considered to be stable during the 1.5 Ma time interval: 0 m/10⁵ years for location A, 5 m/10⁵ years for location B, and 10 m/10⁵ years for location C. Eustasy combined with subsidence allows for the calculation of the relative sea-level change (Δ RSL) for each time step. The cumulative relative sea level (Σ RSL) is calculated in the last column of the table. Key: * (x 10⁵ years), # m/10⁵ years, + m.

A forward modeling simulation, based on the data shown in Figs. 7.2–7.4 helps to illustrate these points. Starting from the initial reference profile in Fig. 7.2, and based on the sea level and subsidence data in Fig. 7.3, the stratigraphic architecture is gradually built for the

15 time steps, as illustrated in Figs. 7.5–7.8. The sediment supply is not quantified in this forward simulation, but is maintained constant through time by assigning the same volume of sediment to all depositional wedges that form at different time steps. The newly accumulated sedimentary wedge of each time slice is marked with grey on the forward modeling diagrams (Figs. 7.5–7.8), and includes fluvial, coastal, and undifferentiated shallow- to deep-water facies. The amount of accommodation created or destroyed at the shoreline during each time slice was calculated by interpolation between profiles A and B, or B and C, depending on the location of the shoreline during that particular time interval. This amount of available accommodation is critical to establish how much aggradation or erosion takes place at the shoreline during each time step, which also has a direct influence on the rates of progradation or retrogradation. The types of shoreline shifts are marked on the cross-sections, and the differential subsidence was taken into account by gradually tilting the profile from one time step to another accordingly. The final stratigraphic architecture and facies relationships are shown in Fig. 7.9.

Figure 7.10 illustrates the curve of relative sea-level changes at the shoreline, which was obtained by interpolation between the curves that characterize locations A, B, and C (Figs. 7.2–7.4), according to the shoreline position at every time step. This is the reference curve that sequence stratigraphic models are centered around, which describes changes in accommodation *at the shoreline*. The interplay between sedimentation and *this* curve of base-level changes controls the transgressive and regressive shifts of the shoreline, as marked in the diagram, and implicitly the timing of all systems tracts and bounding surfaces. This reference curve is 'dynamic,' as it follows the shoreline in its shifts back and forth along the dip-oriented cross-section, as opposed to the 'static' curves of specific locations (e.g., A, B, and C in Fig. 7.2).

SHORELINE SHIFTS, GRADING, AND BATHYMETRY

Controls on Sediment Grading and Water-depth Changes

In addition to understanding the meaning of the reference curve of relative sea-level changes (Fig. 7.10), it is also important to observe the bathymetric trends in different locations within the basin during the transgressive and regressive shifts of the shoreline. As discussed in more detail in Chapter 3, water depth at any location within the marine basin depends on the interplay

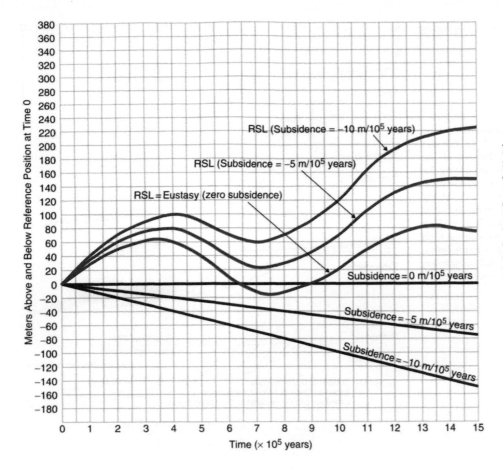

FIGURE 7.4 Subsidence, eustatic, and relative sea-level curves plotted based on the data provided in Fig. 7.3, for the 1.5 Ma time interval (modified from data provided by H. W. Posamentier). Note that for location A (Figs. 7.2 and 7.3), where subsidence is zero, the sea-level curve coincides with the relative sea-level curve. For locations B and C (Figs. 7.2 and 7.3), the relative sea-level curves account for the combined effects of eustasy and subsidence.

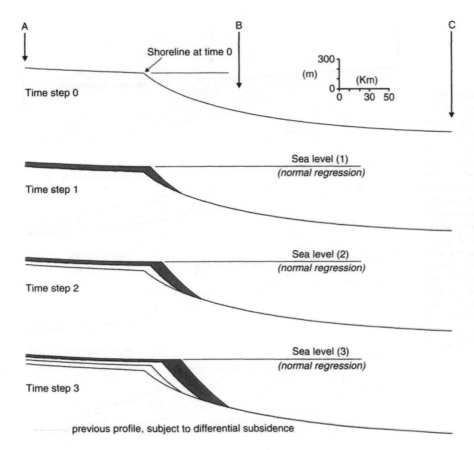

FIGURE 7.5 Forward modeling simulation (time steps 1–3) based on the data provided in Figs. 7.2–7.4. See text for additional explanations. Note that the amount of accommodation that is being created at the shoreline decreases with time, which triggers an increase in the progradation rates. This amount of accommodation is calculated by interpolation between the relative sea-level curves at locations A and B (Fig. 7.4), according to the position of the shoreline during each time slice. Each cross-section is tilted relative to the previous one according to the rates of differential subsidence. All facies prograde in relation to the basinward shift of the shoreline.

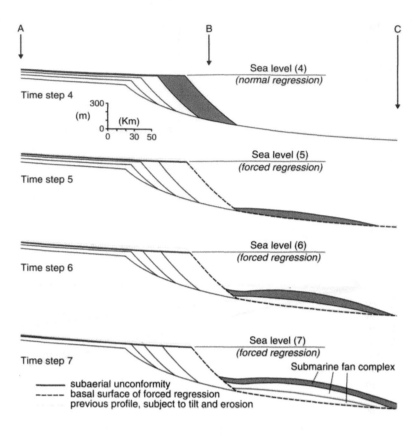

FIGURE 7.6 Forward modeling simulation (time steps 4–7) based on the data provided in Figs. 7.2–7.4. See text for additional explanations. Note that the top of the normal regressive deposits is subject to erosion during the forced regression of the shoreline. As a result, the subaerial unconformity forms. The amount of erosion at the shoreline is calculated by interpolation between the relative sea-level curves at locations A and B (Fig. 7.4), according to the position of the shoreline during each time slice. Each cross-section is tilted relative to the previous one according to the rates of differential subsidence. No offlapping shallow-marine facies are represented, for simplicity.

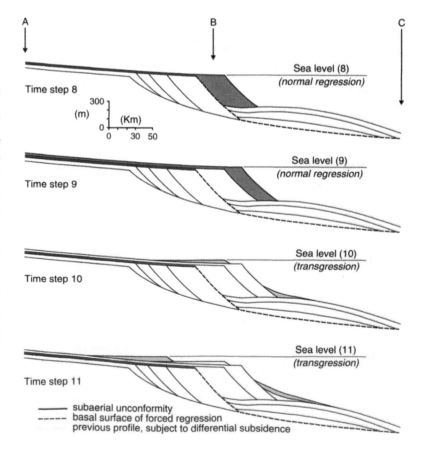

FIGURE 7.7 Forward modeling simulation (time steps 8–11) based on the data provided in Figs. 7.2–7.4. See text for additional explanations. Note that the change from normal regression (time steps 8 and 9) to transgression (time steps 10 and 11) depends on the interplay of sedimentation and accommodation at the shoreline. Since sediment supply is not quantified in this exercise, the transgression was arbitrarily selected to start with time step 10, because of the accelerated rates of base-level rise noted in Fig. 7.4. The amount of available accommodation at the shoreline is calculated by interpolation between the relative sea-level curves at locations A, B, and C (Fig. 7.4), according to the position of the shoreline during each time slice. Each cross-section is tilted relative to the previous one according to the rates of differential subsidence. The rates of progradation and retrogradation change with time, and are linked to the rates of creation of accommodation at the shoreline: the lower the accommodation at the shoreline, the higher the progradation rates; the higher the accommodation at the shoreline, the higher the retrogradation rates.

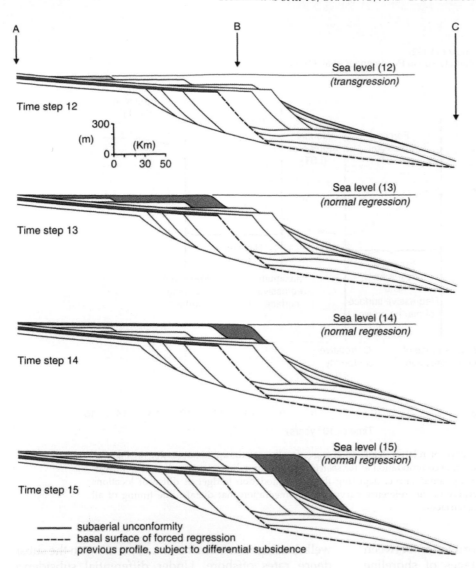

FIGURE 7.8 Forward modeling simulation (time steps 12–15) based on the data provided in Figs. 7.2–7.4. See text for additional explanations. Note that the change from transgression (time step 12) to normal regression (time steps 13 to 15) depends on the interplay of sedimentation and accommodation at the shoreline. Since sediment supply is not quantified in this exercise, the normal regression was arbitrarily selected to start with time step 13, because of the lower rates of base-level rise noted in Fig. 7.4. The amount of available accommodation at the shoreline is calculated by interpolation between the relative sea-level curves at locations A, B, and C (Fig. 7.4), according to the position of the shoreline during each time slice. Each cross-section is tilted relative to the previous one according to the rates of differential subsidence. The rates of progradation and retrogradation change with time, and are linked to the rates of creation of accommodation at the shoreline: the lower the accommodation at the shoreline, the higher the progradation rates; the higher the accommodation at the shoreline, the higher the retrogradation rates.

of subsidence, sea-level change, and sedimentation. Subsidence and sea-level rise contribute towards water deepening, whereas sea-level fall and sedimentation promote water shallowing. In contrast, sediment grading trends that develop during the same transgressive and regressive shifts of the shoreline are controlled by a different set of parameters, which may be independent of water depth, namely sediment supply and depositional energy. This section examines the intricate relationship between shoreline shifts, grading, and

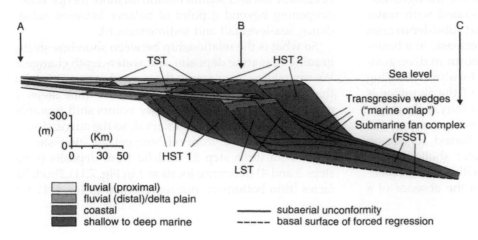

FIGURE 7.9 Stratigraphic architecture of the succession modeled in Figs. 7.5–7.8 (modified from data provided by H. W. Posamentier). Abbreviations: LST—lowstand systems tract; TST—transgressive systems tract; HST—highstand systems tract; FSST—falling-stage systems tract.

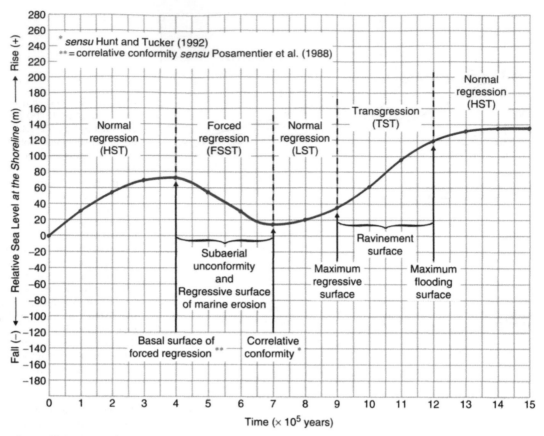

FIGURE 7.10 Reference curve of relative sea-level changes for the stratigraphic succession in Fig. 7.9. This curve indicates changes in accommodation *at the shoreline*, as the shoreline transgresses and regresses with time. It is therefore a 'dynamic' curve, depicting the accommodation budget at different locations through time. It is the interplay of *this* reference curve and sedimentation that dictates the timing of all systems tracts and bounding surfaces.

bathymetry, and the extent of the correlation between the latter two variables during stages of shoreline transgression or regression.

Figure 7.11 shows two curves of water-depth changes in two specified locations, for the 15 time steps of the forward model run. Water depth is calculated as the vertical distance between the sea level and the seafloor at each time step. Note that the transgression of the shoreline can be safely associated with water deepening in the basin, assuming that subsidence rates increase, and sedimentation rates decrease, in a basin-ward direction. This is usually the norm in divergent-type settings, as considered in this forward modeling simulation. The normal regression of the shoreline is coeval with water shallowing in the vicinity of the shore-line, and with water deepening offshore, based on the same assumptions as above. The forced regression of the shoreline corresponds to water shallowing in Fig. 7.11, but it may also be coeval with water deepening beyond the submarine fan or in the absence of a

well-developed submarine fan, depending on the subsidence rates offshore. Under differential subsidence conditions, it is conceivable that low subsidence rates may be outpaced by eustatic fall at the shoreline (base-level fall, and forced regression of the shoreline, resulting in water shallowing in the vicinity of the shoreline), whereas higher subsidence rates may outpace the sum of eustatic fall and sedimentation offshore (hence water deepening beyond a point of balance between subsidence, sea-level fall and sedimentation).

So what is the relationship between shoreline shifts, grading of marine deposits, and water-depth changes? We can take as an example the normal regressive delta that progrades into the basin between time steps 1 and 4 (Fig. 7.11). Sediment entry points shift towards the basin during this time interval, so the succession is coarsening upward, from offshore pelagic (time step 1) to prodelta (time step 2) to delta front deposits (time steps 3 and 4) (reference location 1 in Fig. 7.11). Prodelta facies (thin bottomset, not represented in Fig. 7.11 for

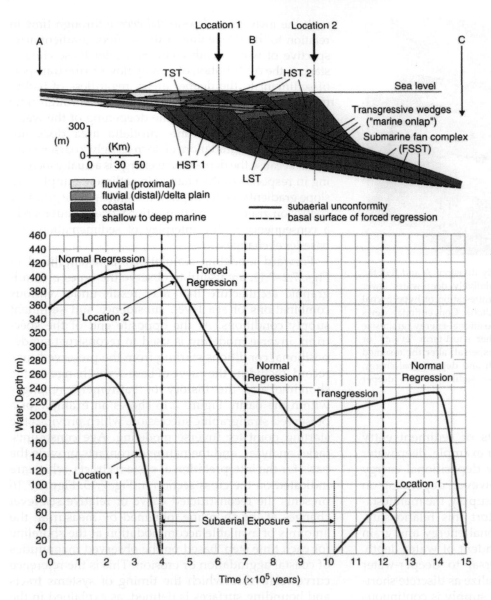

FIGURE 7.11 Water-depth changes through time, at locations 1 and 2 indicated on the cross-section. See text for explanations. Abbreviations: LST—lowstand systems tract; TST—transgressive systems tract; HST—highstand systems tract; FSST—falling-stage systems tract.

simplicity) are characterized by the presence of silt-size riverborn sediments, which gradually prograde over the finer pelagic basinal facies. In turn, the silty prodelta sediments are prograded by the sand-dominated delta front facies, which generates the overall coarsening-upward trend. At the same time, the bathymetric profile shows a deepening of the water during time steps 1 and 2, followed by shallowing as soon as the point of observation (location 1 in Fig. 7.11) becomes part of the subtidal (delta front) environment. The deepening of the water during the deposition of shelf and overlying prodelta facies (time steps 1 and 2) is due to the fact that sedimentation rates are too low to keep up with the rates of creation of accommodation. For time steps 3

and 4, the shallowing of the water is driven by the much higher sedimentation rates in the delta front environment, which outpace the rates of base-level rise. This example shows that the water-depth/grain-size relationship is not necessarily linear, as commonly assumed, meaning that an increase in water depth does not have to be accompanied by a decrease in size of the sediment delivered to that particular location.

The generally assumed linear relationship between water depth, depositional energy, and grain size is modified by other parameters such as slope (gradient of seafloor) and the associated gravity-driven subaqueous flows (Figs. 6.15, 7.12, and 7.13), and, most importantly, by changes in sediment supply associated

FIGURE 7.12 Sandy turbidites (mainly divisions A and B of the Bouma sequence) accumulated in a relatively deep-water lower delta front—prodelta setting, *via* the manifestation of hyperpycnal flows (river-dominated Ferron delta, Utah). Conventional views account for a linear water-depth/depositional-energy/grain-size relationship. Gravity-driven flows (either short-term 'events' or long-term diffusion-style of sediment dispersal) alter the assumed linear relationship between water depth and depositional energy, explaining the progradation of coarser sediments into deeper (or deepening) water.

with shoreline shifts (i.e., shifts of sediment entry points into the marine basin). For example, deep water is generally correlated with low depositional energy and fine-grained sediments. However, depending on seafloor gradients and sediment supply, the manifestation of gravity flows may distort this relationship, leading to increases in depositional energy and grain size in a manner that is independent of water depth. Gravity-driven sediment dispersal to deeper-water environments may either materialize as discrete short-term events, or, where sediment supply is continuous and the prograding clinoforms are sufficiently steep, as long-term diffusion. The latter mechanism of sediment transport is commonly invoked to explain the formation of healing-phase wedges (see Chapters 5 and 6 for details), and may simply mean an increase in the frequency of small-magnitude gravity flows to the point that resembles continuous sediment transfer from staging areas (e.g., upper delta front) to depositional areas (e.g., lower delta front to prodelta).

Figure 7.14 provides a simple model of deposition on a subsiding seafloor that is subject to *differential subsidence*, in which deltaic progradation generates a coarsening-upward succession in both shallowing and coeval deepening waters, in areas proximal and distal relative to the shoreline, respectively. In this example, the reverse grading is directly associated with the progradation of sediment entry points (regressive shoreline shift) and

with the increase in *depositional energy* through time in relation to the steepening of the seafloor gradient, irrespective of water-depth changes. Under these circumstances, the contribution of gravity flows to the transport of sediment to the deeper areas, regardless of bathymetric trends, is increasingly important through time (Figs. 6.15, 7.12 and 7.13). The deepening of the water in the lower delta front—prodelta area does not prevent the progradation of increasingly coarser sediment, because the depositional energy is actually increasing in response to the change in sediment supply and slope gradients. Water depth is therefore not the primary control on grain size and depositional energy, but merely a consequence of the interplay of sedimentation and base-level changes.

Forward modeling simulations, as exemplified in Figs. 7.2–7.11, are useful to observe the types of stratigraphic architecture that may result under various combinations of sea level, subsidence and sediment supply conditions. At the opposite end of the spectrum, inverse modeling is used to reconstruct syndepositional conditions starting from the preserved rock record, including the curve of base-level changes at the shoreline and the history of bathymetric changes within the basin. Figure 7.15 shows an example of a preserved stratigraphic succession, which provides the starting point for inverse modeling. Age constraints, facies analyses, and the nature of contacts provide the basis for the interpretation of systems tracts, which are indicated on the cross-section in Fig. 7.15. Figure 7.16 presents the reconstructed curve of relative sea-level changes at the shoreline, obtained by measuring the amounts of available accommodation at the shoreline for each time step, based on the observed magnitudes of coastal aggradation or erosion. This is the reference curve relative to which the timing of systems tracts and bounding surfaces is defined, as explained in the previous section of this chapter. Note that the inverse modeling approach taken to reconstruct this curve is very different from the forward modeling approach in Fig. 7.10. Here, no 'static' curves are available for interpolation (see locations A, B, and C in Figs. 7.2–7.4), and in fact such static curves are impossible to reconstruct based on the data provided in Fig. 7.15. Hence, the curve of relative sea-level changes *at the shoreline* is the only reference curve that can be constructed starting from the preserved rock record. Also note the relationship between shoreline shifts and bathymetric changes in Fig. 7.17. For example, the highstand systems tracts, which prograde and display coarsening-upward profiles, tend to include a deepening-upward succession at their base, overlain by a shallowing-upward package (both part of the overall coarsening-upward trend). Hence, the maximum water depth, as inferred from

FIGURE 7.13 Sandy turbidites accumulated in the lower delta front—prodelta setting of a river-dominated deltaic succession (Panther Tongue, Gentle Wash Canyon, Utah). Note the importance of hyperpycnal (gravity) flows in the transport of sediment to deeper areas; hence, no direct relationship may be established between water depth and grain size, or between water-depth changes and grading trends. Instead, grain size correlates with depositional energy, which if often independent of bathymetry. A, B—sole marks at the base of turbidite rhythms; C, D—sandy turbidites, dominated by the divisions A and B of the Bouma sequence; E, F—cyclicity of turbidite rhythms in the distal portion of the deltaic system.

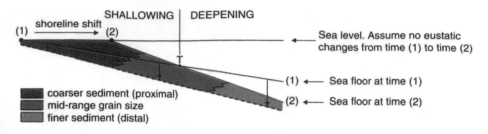

FIGURE 7.14 Progradation of a coarsening-upward succession onto a seafloor that is subject to varying bathymetric conditions. The threshold 'T' separates areas of water shallowing and deepening. This threshold is placed where sedimentation and subsidence are in a perfect balance. Landwards from T, sedimentation outpaces subsidence. Seawards from T, accommodation is created more rapidly than it is consumed by sedimentation. The succession that accumulates in the deepening water is still coarsening-upward, as the sediment entry points shift towards the sea (regression). This is the case of the Mahakam delta in Indonesia (sedimentation > subsidence nearshore, and subsidence > sedimentation offshore).

benthic foraminiferal paleobathymetry for example, often occurs *within* the highstand (normal regressive) progradational wedge (Naish and Kamp, 1997; T. Naish, pers. comm., 1998). Similar conclusions have been reached by Vecsei and Duringer (2003), who have demonstrated that the 'maximum depth interval' within the Middle Triassic marine succession of the Germanic Basin is younger than the maximum flooding surface, hence occurring within the highstand systems tract (Fig. 7.18). In this case study, the difference in age between the maximum flooding surface and the maximum water-depth interval is attributed to variations in sedimentation rates between the basin margin and the basin center, which is in agreement with previous results of numerical modeling (e.g., Catuneanu *et al.*, 1998b).

Figure 7.19 presents the time-distance (Wheeler) diagram for the cross-section in Fig. 7.15. This type of chronostratigraphic chart is useful to illustrate the shifts in the *locus* of sediment accumulation, as well as the patterns of facies migration through time. With the

inverse modeling approach (Figs. 7.15–7.19) the best we can do is to reconstruct the 'dynamic' curve of base-level changes *at the shoreline*, without being able to quantify the relative contributions of eustasy and subsidence to this curve. The forward modeling simulation (Figs. 7.2–7.11), as well as the example in Fig. 7.14, account for syndepositional differential subsidence that implies an increasingly important role for gravity flows through time, in relation to the gradual steepening of the seafloor gradient. Nevertheless, the same coarsening-upward trends in both deepening and shallowing water may also be generated without changes in seafloor gradients, with accommodation exclusively generated by sea-level rise (Fig. 7.20). This means that the accumulation of marine coarsening-upward successions under changing bathymetric conditions, including deepening water, does not necessarily require an increase in the participation of gravity flows, but can also be explained solely by changes in *sediment supply* as a result of shoreline

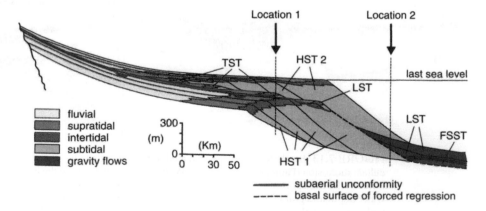

FIGURE 7.15 Stratigraphic architecture of five systems tracts (cross-section courtesy of H. W. Posamentier). The interpreted succession of systems tracts is as follows: highstand systems tract #1 (HST 1)—time steps 1–4; falling-stage systems tract (FSST) —time steps 5–7; lowstand systems tract (LST)—time steps 8–9; transgressive systems tract (TST)—time steps 10–12; and highstand systems tract #2 (HST 2)—time steps 13–15.

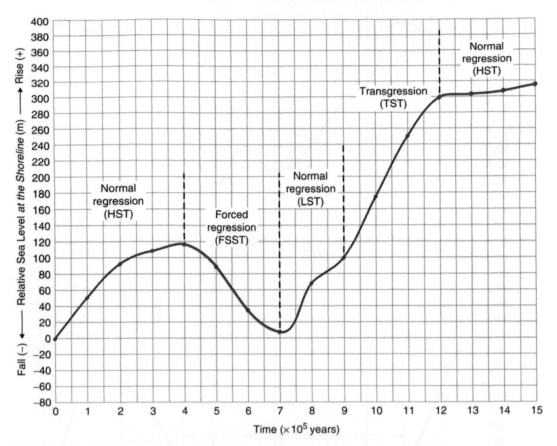

FIGURE 7.16 Reconstructed curve of relative sea-level changes at the shoreline, based on the amount of coastal aggradation or erosion at each time step. Changes in accommodation at the shoreline are measured based on the vertical distance between consecutive shoreline positions. In the case of the early highstand systems tract (HST 1 in Fig. 7.15), whose top is eroded by the subaerial unconformity, the position of the normal regressive shorelines (not preserved) is reconstructed based on the overall architecture and the inferred progradation rates (the latter are inversely proportional to the rates of coastal aggradation).

shifts (Fig. 7.20). Gravity flows, in addition to tidal currents, storm surges and hypopycnal plumes, are, of course, still important as a mechanism of sediment transport beyond the fairweather wave base, but their contribution to the distribution of sediment within the basin does not have to change through time in order to explain depositional patterns such as the ones illustrated in Fig. 7.20. In the absence of differential subsidence, the progradation of a coarsening-upward succession *in deepening water* requires that the rates of coastal aggradation (base-level rise at the shoreline, as inferred from the thickness of the delta plain topset) be higher then the rates of aggradation in the basinal to prodelta areas (as reflected by the thickness of the deltaic bottomset; Fig. 7.20). This condition is not necessary in basins affected by differential subsidence, where progradation in deepening water may occur even without coastal aggradation (Fig. 7.14).

Discussion

In conclusion, the relationship between shoreline shifts, sediment grading, and water-depth changes is much more complex than commonly inferred. Shoreline shifts exert a critical control on grading, as sediment entry points prograde and retrograde relative to discrete locations within the marine basin where changes in grain size are observed. On the other hand, the correlation between grain size and water depth, generally assumed to be linear, is altered by fluctuations in *sediment supply* and *depositional energy*. Sediment supply is to a large extent controlled by shoreline shifts, while depositional energy is affected by the various transport mechanisms that shed riverborne sediment beyond the fairweather wave base, into the deeper-water environment. Such transport mechanisms are discussed in more detail in Chapter 6 (section on *Sediment supply and transport mechanisms*), and include

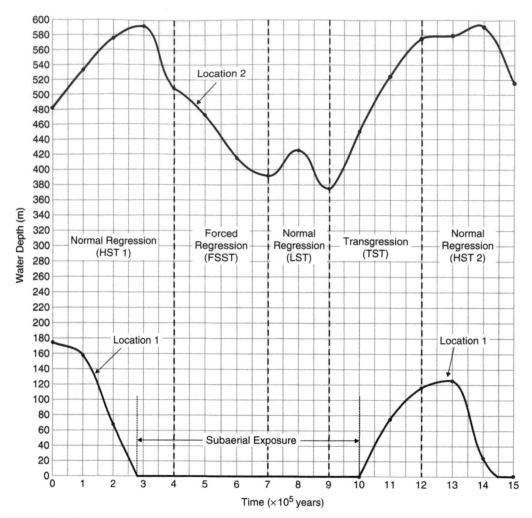

FIGURE 7.17 Water-depth changes at the reference locations 1 and 2 in Fig. 7.15. Note that in a vertical profile, a normal regressive systems tract (prograding, coarsening-upward deposits) may be associated with both deepening- and shallowing-upward trends. See text for more details. Abbreviations: LST—lowstand systems tract; TST—transgressive systems tract; HST—highstand systems tract; FSST—falling-stage systems tract.

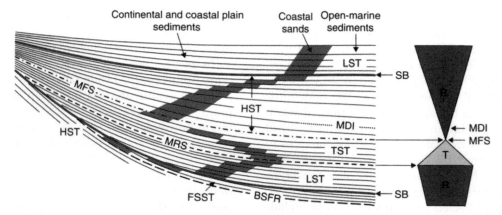

FIGURE 7.18 Sequence stratigraphic model for a low-gradient ramp in an intracratonic basin (modified from Vecsei and Duringer, 2003). Note that the maximum depth interval is younger than the maximum flooding surface, and occurs within the highstand systems tract. The difference in age between the maximum flooding surface and the maximum depth interval is attributed to differences in sedimentation rates between the basin margin and the basin center. Abbreviations: HST—highstand systems tract; FSST—falling-stage systems tract; LST—lowstand systems tract; TST—transgressive systems tract; SB—depositional sequence boundary; BSFR—basal surface of forced regression; MRS—maximum regressive surface; MFS—maximum flooding surface; MDI—maximum depth interval; R—regression and coarsening upward; T—transgression and fining upward.

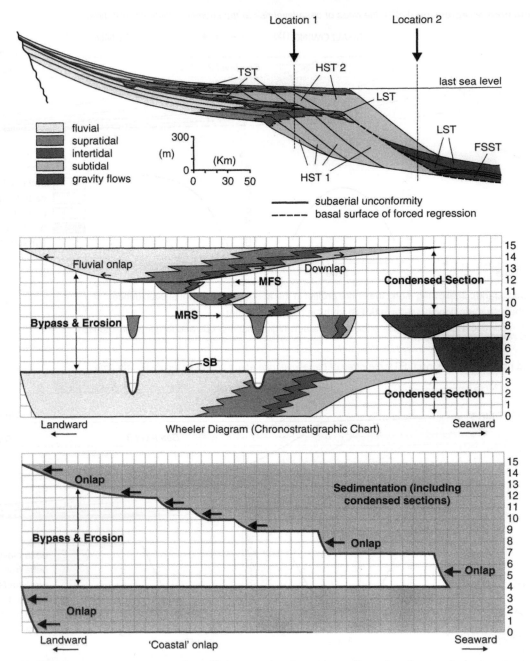

FIGURE 7.19 Wheeler diagram illustrating the time/distance relationships of the facies shown in Fig. 7.15 (courtesy of H. W. Posamentier). Note that the 'coastal' onlap curve, which was used in the early days of seismic and sequence stratigraphy for the construction of the global cycle chart, is in fact composed mainly of a combination of fluvial and marine onlap. For this reason, the assumption that 'coastal' onlap indicates stages of sea-level rise is misleading, and resulted in the representation of sea-level fall as instantaneous events.

tidal currents, storm currents, gravity flows, and hypopycnal plumes.

The progradation of a coarsening-upward succession into a deepening basin is possible during both forced and normal regressions, irrespective of the trends of destruction or creation of accommodation at the shoreline. Deepening of the water in distal deltaic

environments (lower delta front to prodelta) may be caused by (1) topsets being thicker (aggrading faster) than bottomsets, in normal regressive settings (Figs. 7.20 and 7.21), and/or (2) differential subsidence, in forced or normal regressive settings (Figs. 7.14 and 7.21). In either case, deltaic progradation in deepening water (that is, beyond the point of balance between accommodation

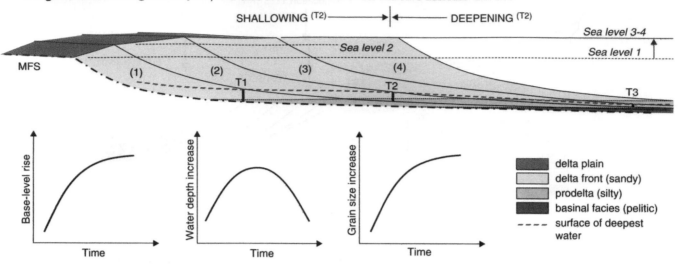

1. Highstand normal regression (HST): *the rates of base-level rise at the shoreline decrease with time*

SHALLOWING (T2) ——→|←—— DEEPENING (T2)

- Time 4 (highstand): shallowing along the entire profile, in the absence of differential subsidence
- most of the maximum flooding surface (MFS) forms in deepening water
- the surface corresponding to the peak of deepest water is highly diachronous and forms *within* the HST

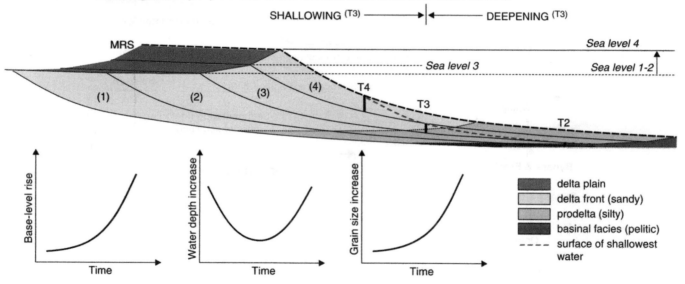

2. Lowstand normal regression (LST): *the rates of base-level rise at the shoreline increase with time*

SHALLOWING (T3) ——→|←—— DEEPENING (T3)

- Time 1 (lowstand): shallowing along the entire profile, in the absence of differential subsidence
- most of the maximum regressive surface (MRS) forms in deepening water
- the surface corresponding to the peak of shallowest water is highly diachronous and forms *within* the LST

FIGURE 7.20 Base level, water depth, and grading trends in highstand and lowstand normal regressive settings. No differential subsidence is taken into account, and the creation of accommodation is entirely attributed to sea-level rise (in contrast to Fig. 7.14). Similar illustrations of deltaic progradation, *with the lower delta front prograding into deepening water,* have been produced by Berg (1982) and Bhattacharya and Walker (1992; their Fig. 25), accounting for thinner deltaic bottomsets relative to the topsets (see also Figs. 7.9 and 7.15). The changing rates of base-level rise at the shoreline are matched by the rates of topset (delta plain) aggradation, and are compensated by inversely proportional rates of progradation. 'T' is the threshold of no water-depth change at each time step, where accommodation and sedimentation are in a perfect balance. This threshold shifts along dip through time as a function of the decreasing or increasing rates of base-level rise. The addition of differential subsidence would place the threshold T closer to the shoreline at each time step, as the area subject to water deepening would expand towards the land. Note that surfaces of deepest and shallowest water join the maximum flooding and maximum regressive surfaces in the vicinity of the shoreline, but diverge from them in a basinward direction. Also note the opposite bathymetric trends of the highstand and lowstand systems tracts: initial deepening followed by shallowing, and shallowing followed by deepening, respectively. This means that both maximum flooding and maximum regressive surfaces form in deepening water.

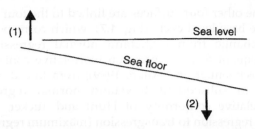

FIGURE 7.21 Schematic representation of the dominant controls on water deepening during the progradation of coarsening-upward successions. These two controls may act independently or in conjunction: (1)—coastal aggradation, at rates higher than the rates of aggradation of the distal prograding system (topset > bottomset); (2)—differential subsidence, with rates increasing basinward (subsidence > sedimentation + eustatic fall).

and sedimentation: threshold 'T' in Fig. 7.20) results in the formation of longer clinoforms through time (also inferred by Berg, 1982, and Bhattacharya and Walker, 1992, accounting for thinner deltaic bottomsets relative to topsets). *The higher the rates of aggradation at the shoreline, or the more pronounced the pattern of differential subsidence, the more evident the offset between grading and bathymetric trends* (threshold 'T' closer to the shoreline in Fig. 7.20). These principles may also apply to transgressive settings, where proper deltas develop (Figs. 5.51–5.54). In fact, the high rates of base-level rise that are commonly associated with transgressive coastlines make transgressive deltas most likely candidates for progradation into deepening water, in areas relatively close to the shoreline.

The discussion in this section was focused on tectonic settings where subsidence rates are either *constant* along dip (Fig. 7.20) or *increase* in a basinward direction (Figs. 7.2 and 7.14). In such basins, the accumulation of fining-upward successions in deepening water during transgressions is commonly the norm (with the exception of transgressive deltas, as discussed above), and departures from conventional thinking (i.e., which accounts for a linear grain-size/water-depth correlation) apply during regressions, when coarsening-upward successions may prograde into deepening water. Such situations may be common within the context of divergent continental margins, other extensional basins, or any basin that fulfills the conditions outlined above. Clastic sedimentation in foreland basins, however, may follow a different pattern, due to the fact that subsidence rates *decrease* in a basinward direction (Fig. 2.63). Under these circumstances, the progradation of coarsening-upward successions in shallowing water during regressions may become the norm, while the 'anomaly' may rather consist of the accumulation of fining-upward deposits in shallowing water during transgressions, in relation to the retrogradation of sediment entry points and the pattern of subsidence

across the basin. For example, if subsidence rates vary between 10 and 0 m/Ma along the dip of the foredeep (proximal to distal), no sea-level changes are recorded, and sedimentation rates decrease from 3 to 1 m/Ma in a basinward direction in response to the retrogradation of the shoreline, then the fining-upward transgressive deposits accumulate in deepening water proximally, but in shallowing water distally. Carbonate platforms on continental shelves may also record bathymetric 'anomalies,' such as the growth of distal-shelf barrier reefs during slow transgressions, which may keep up with the rise in base level (i.e., no change in water depth), although the shoreline is transgressive and the rest of the flooded shelf experiences a deepening of the water (Fig. 6.49).

This discussion indicates that more caution needs to be exercised when using bathymetric terms to describe observed trends of sediment grading, or observed or inferred shoreline shifts, and that interpretations of bathymetric changes should be confirmed independently, such as by using biostratigraphic or ichnological data. In this context, the equivalence between transgressive and 'deepening-upward' trends, or between regressive and 'shallowing-upward' trends, as promoted by Embry (2002, 2005), may be misleading if used as a generalization. Such equivalence is only safely valid for shallow areas in the vicinity of the shoreline (e.g., between the shoreline and the threshold 'T' in Fig. 7.20). Beyond the threshold 'T' in Fig. 7.20, the deltaic clinoforms (coarsening-upward, regressive succession) prograde into deepening water. In fact, any time a well-developed topset forms, the likelihood is that the toe of the delta front clinoforms prograde into deepening water.

Ultimately, the transgressive or regressive shoreline shifts, followed closely by overall grading trends within the marine basin (fining- and coarsening-upward, respectively), reflect the balance between accommodation and sedimentation *at the shoreline*. This balance controls the patterns of sediment supply into the basin, and implicitly grading trends, as sediment entry points retrograde or prograde with time. Elsewhere within the basin, water depth depends on the interplay of *local* accommodation and sedimentation, which may differ significantly from the conditions established in the vicinity of the shoreline. Therefore, water-depth changes may follow patterns that are independent of grading trends. However, such patterns are predictable to some extent, depending on the subsidence regimes of each sedimentary basin, and therefore they may be modeled using numerical simulations. The differentiation between grading and water-depth changes is important because the definition of some sequence stratigraphic surfaces revolves around these concepts (see numerical modeling in the following section of this chapter).

The next question is what criteria are best to employ for tracing sequence stratigraphic surfaces in the rock record as we move away from coeval paleoshorelines, and, linked to this issue, what is the temporal significance of surfaces defined on the basis of different criteria?

METHODS OF DEFINITION OF STRATIGRAPHIC SURFACES

Introduction

All seven sequence stratigraphic surfaces are linked to the reference curve of base-level changes at the shoreline (Fig. 4.7). Three of these surfaces (the subaerial unconformity, the wave/tidal transgressive ravinement surface, and the regressive surface of marine erosion) form *during* particular stages of shoreline shifts, whereas the other four (the correlative conformities of the onset and end of forced regression, the maximum regressive surface and the maximum flooding surface) are related to *changes* in the direction and/or type of shoreline shift.

A general consensus is now reached with regards to the temporal significance of sequence stratigraphic surfaces that form *during* particular stages of shoreline shift. The subaerial unconformity corresponds to a hiatus that forms and extends basinward for as long as the shoreline is subject to *forced regression*. The regressive surface of marine erosion is a highly diachronous scour that, again, forms during the *forced regression* of the shoreline. Similarly, the transgressive ravinement surfaces are highly diachronous scours that form in parallel with the *transgression* of the shoreline. The concept of diachroneity is illustrated schematically in Fig. 7.22.

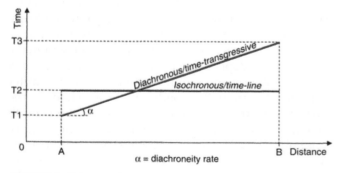

FIGURE 7.22 Isochronism and diachronism of stratigraphic surfaces. A diachronous surface is time–transgressive, i.e., 'of varying age in different areas or that cuts across time planes or biozones' (Bates and Jackson, 1987). An isochronous surface is a time line, i.e., it forms everywhere at the same time.

The other four surfaces are linked to the four *events* of the base-level cycle (Fig. 4.7), which correspond to the change from highstand normal regression to subsequent forced regression (correlative conformity of Posamentier and Allen, 1999); from forced regression to subsequent lowstand normal regression (correlative conformity of Hunt and Tucker, 1992); from regression to transgression (maximum regressive surface); and from transgression to regression (maximum flooding surface). As 'event-significant' surfaces, all four are assigned a 'low' diachroneity in Fig. 4.9. Both types of correlative conformities are defined on the basis of regional stratal stacking patterns, and are taken at the base and at the top of marine forced regressive deposits, respectively (Fig. 6.2). The timing of these correlative conformities depends primarily on the interplay of subsidence and eustasy at the shoreline. As the rates of subsidence may vary along strike, correlative conformities may be somewhat diachronous. The timing of the maximum regressive and maximum flooding surfaces depends not only on the interplay of subsidence and eustasy at the shoreline, but also on variations in sedimentation rates along the coastline. It may be inferred, therefore, that these surfaces can be even more diachronous than the correlative conformities, and so somewhat less useful for the construction of a chronostratigraphic framework, particularly along strike-oriented profiles. However, as pointed out by Posamentier and Allen (1999), 'even though the age of the bounding surfaces separating the systems tracts might be diachronous on the scale of the basin, the difference in age is commonly below biostratigraphic resolution. Even if their diachroneity can be established on a regional scale, the surfaces and systems tracts can be physically continuous over wide areas and record important phases of the evolution of the basin.' This is particularly true in basins without rapid or local changes in subsidence rates along strike, such as divergent continental margins or intracratonic basins, where systems tracts and their bounding (event-significant) surfaces may be continuous over large areas. In more tectonically-active basins, however, characterized by significant changes in subsidence patterns along strike, systems tract boundaries may be much more diachronous, and with a more limited lateral extent.

Correlative Conformities

Onset-of-fall Correlative Conformity

The correlative conformity of Posamentier and Allen (1999) (equivalent to the 'basal surface of forced regression' of Hunt and Tucker, 1992) approximates the seafloor at the onset of forced regression of the shoreline, and therefore it is generally regarded as a time line in the rock record, particularly along dip. In reality, a low diachroneity rate is recorded in relation to the rates of sediment transport in an offshore direction (Catuneanu et al., 1998b), as it takes time for the first gravity flows associated with forced regression to reach the deeper parts of the basin. The rates of offshore transport of terrigenous sediment along the depositional dip within a marine basin vary from 10^{-1}–10^0 m/s in the case of low-gradient shelf settings to 10^1–10^2 m/s in the case of turbidity flows associated with steeper gradients in continental slope settings (Reading, 1996). This low diachroneity rate is generally undetectable relative to the resolution of the current biostratigraphic or radiometric dating techniques. Along strike, however, the degree of diachroneity may be more significant, and may vary greatly depending on subsidence patterns.

The definition provided by Posamentier and Allen (1999) for their 'correlative conformity' is 'the sedimentary surface at the onset of relative sea-level (or base-level) fall.' This definition omits to specify that reference is made to changes in relative sea level *at the shoreline*. This omission relates to the inherited generic nature of the reference curve of relative sea-level (base-level) changes, as discussed earlier in this chapter, and allows for alternative interpretations of the temporal attributes of this correlative conformity, other than the ones intended by its authors. As subsidence is differential along both dip and strike, taking into account the onset of relative sea-level fall in every discrete location within the basin (rather than only at the shoreline) results in the definition of a highly diachronous surface that diverges from the earliest clinoform of forced regression referred to by Posamentier and Allen (1999) (see full modeling results in Catuneanu et al., 1998b— their Figs. 10–14). This diachronous surface may not extend across the entire basin, but only in areas where subsidence rates are within the range of variation of the rates of sea-level change (Catuneanu et al., 1998b). Consequently, depocenters that record high rates of subsidence, outpacing the rates of sea-level change at all times, experience continuous relative sea-level rise, thus not allowing for the formation of surfaces defined at the onset of *local* stages of relative fall. Such theoretical surfaces do not have a signature in the rock record that can be defined on the basis of stratal stacking patterns (i.e., they do not form sequence or systems tract boundaries), and hence they have little value for sequence stratigraphy. These are not the correlative conformities referred to by Posamentier and Allen (1999), which are *independent* of the offshore variations in the rates of relative sea-level change, thus extending across the entire basin at the base of all forced regressive marine deposits. This discussion shows how important

it is to specify where the reference curve of relative sea-level changes is taken, i.e., *at the shoreline*, in order to avoid any possible confusion. Once again, the shoreline trajectory (transgressive, normal regressive or forced regressive) represents the fundamental switch that controls sediment supply to the marine basin, and the timing of all systems tracts and bounding surfaces, in a manner that is independent of the offshore variability in subsidence rates.

End-of-fall Correlative Conformity

The correlative conformity of Hunt and Tucker (1992) is also defined on the basis of stratal stacking patterns, separating offlapping forced regressive lobes from the overlying aggradational lowstand normal regressive deposits (Haq, 1991: 'a change from rapidly prograding parasequences to aggradational parasequences'). This definition implies a diachronous correlative conformity, younger basinward, with a diachroneity rate that matches the rate of offshore sediment transport. As this diachroneity rate is low, as explained above for the basal surface of forced regression (correlative conformity of Posamentier and Allen, 1999), this surface is also often approximated with a time line in the rock record (Embry, 1995: 'The subaerial unconformity is developed and migrates seaward during base level fall and reaches its maximum extent at the end of the fall. ..., the depositional surface in the marine realm at this time of change from base level fall to base level rise is the correlative conformity,' portrayed as a time line in his fig. 1). The quasi-time line significance of such surfaces is of course only valid along depositional dip sections, as varying subsidence rates along strike may offset the transition between base-level fall and base-level rise along the shoreline.

As in the case of the correlative conformity of Posamentier and Allen (1999), the existing definitions for the correlative conformity of Hunt and Tucker (1992) fail to specify that this surface marks the end of base-level fall *at the shoreline*, along each dip-oriented profile. As subsidence rates vary throughout the basin, connecting the dots that signify the end of base-level fall in each discrete location would generate a highly diachronous surface that diverges from the sequence boundary defined by Hunt and Tucker (1992) on the basis of stratal stacking patterns (see full modeling results in Catuneanu et al., 1998b—their Figs. 10–14). If we consider the situation envisaged by Vail et al. (1984) for 'type 2' sequences, with the base level falling at the shoreline but rising at the shelf edge, the correlative conformity that accounts for a base-level fall-to-rise transition in each discrete location would only develop within a limited area on the continental shelf, beyond which subsidence rates become too high to allow for any falls in base level. This highly diachronous end-of-fall surface cannot be used as a systems tract or sequence boundary, and it does not have a sedimento-logical or stratigraphic signature in the rock record. In contrast, Hunt and Tucker's (1992) correlative conformity, which represents the youngest clinoform of forced regression (Fig. 6.2), is *independent* of the offshore variations in subsidence rates, and its timing is controlled solely by the balance between subsidence and sea-level change *at the shoreline*. As such, along dip-oriented transects, the offshore portion of the correl-ative conformity may form under rising or falling base-level conditions, depending on the subsidence patterns within the basin. This discussion reiterates the impor-tance of the shoreline for the transfer of terrigenous sediment into the marine basin, and as a reference for the overall stratigraphic architecture of the basin fill, including the timing of all sequence stratigraphic surfaces, independent of the offshore variations in subsi-dence rates.

Maximum Regressive and Maximum Flooding Surfaces

Definition

The main controversy with respect to the methods of definition, and the impact of these definitions on the temporal attributes of the sequence stratigraphic surfaces in question, rests with the marine portions of the maximum regressive and maximum flooding surfaces. They are currently defined on the basis of (1) overall grading and stratal stacking patterns; or (2) bathymetric (water-depth) changes. Although these two approaches are often considered equivalent, being used interchange-ably (e.g., Embry, 2002, 2005), they allow for different temporal significances for each surface (Catuneanu et al., 1998b). As discussed earlier in this chapter, grading trends in marine deposits and water-depth changes do not necessarily correlate, as, for example, coarsening-upward successions may prograde into both shallow-ing and deepening water (Figs. 7.14 and 7.20).

The marine portion of the maximum regressive surface is part of the transgressive–regressive (T–R) sequence boundary, and only a systems tract bound-ary in the view of the depositional and genetic strati-graphic models (Fig. 4.6). It may be defined either: (1) on the basis of stratal stacking patterns, as a conformable surface that separates regressive strata (progradational, coarsening-upward trend) below from transgressive strata (retrogradational, fining-upward trend) above; or: (2) on the basis of bathymetric (water-depth) changes, as a conformable surface record-ing the start of a deepening episode, i.e., formed when the water depth reaches the shallowest peak (Embry, 2002, 2005).

Although these two definitions are considered equivalent, they allow for different temporal signifi-cances for the maximum regressive surface. The former relates to the shoreline shifts and the associated changes in stacking patterns, which bring the maxi-mum regressive surface to a quasi-time line in a depo-sitional dip section, independent of the offshore variations in subsidence and sedimentation rates, as there is only one point in time where the shoreline is at its most basinward position. A low diachroneity rate is, however, expected in relation to the rates of sediment transport along dip, as in the case of the correlative conformities discussed above. The sediment supplied to the basin during shoreline regression generates a coarsening-upward marine succession related to the basinward facies shift, overlain by finer (and fining-upward) transgressive strata. This provides a litho-logical criterion to pinpoint the maximum regressive surface in outcrops or subsurface logs (e.g., Figs. 4.32 and 4.37).

The second definition implies a potentially highly diachronous maximum regressive surface, as water-depth changes depend on varying sedimentation and subsidence rates across the basin. In this light, it is recognized that this type of 'maximum regressive surface' is younger in areas with higher sedimentation and lower subsidence rates, where the transition from shallowing to deepening occurs later, although this diachroneity is considered 'low' (Embry, 2002, 2005). The diachroneity rate of this type of surface, defined on water-depth changes, is investigated below in this section by means of numerical modeling. Besides the actual degree of diachroneity, which, on its own, is a critical issue, the maximum regressive surface defined at the top of a 'shallowing-upward' succession may suffer from another significant limitation, which is the fact that it may not develop across the entire marine portion of the sedimentary basin. This is evident in the actively subsiding portions of sedimentary basins where subsidence rates outpace the sum of sea-level change and sedimentation; in such regions, water deepening may be continuous during several cycles of base-level changes at the shoreline, and hence the development of surfaces defined on the basis of bathy-metric changes is *restricted* to the shallow-water portions of the basin where subsidence rates are lower and sedi-mentation rates are higher (Catuneanu et al., 1998b). This aspect concerning the lateral extent of surfaces defined on the basis of water-depth changes is also explored in the numerical models presented below.

Similarly, the marine portion of the maximum flood-ing surface may also be defined in two alternative ways: (1) on the basis of stratal stacking patterns, mark-ing the change from fining-upward, retrogradational (transgressive) strata below to coarsening-upward, progradational (regressive) strata above ('downlap surface' of Galloway, 1989); or: (2) on the basis of bathymetric (water-depth) changes, being formed when the water reaches the deepest peak (i.e., at the top of a deepening-upward succession; Embry, 2002).

Again, these two methods define surfaces which are not necessarily superimposed. In the former approach, the maximum flooding surface corresponds to the moment in time when the shoreline is at its landward-most position along each depositional dip section (Fig. 5.5). In other words, the timing of the maximum flooding surface depends on the change in the patterns of sediment supply associated with the shift in shoreline trajectories, from transgressive to highstand normal regressive, irrespective of the offshore variations in subsidence or water depth. Even so, a low diachroneity is recorded along dip in relation to the rates of offshore sediment transport. In addition to this, a more significant diachroneity may exist along the depositional strike, as variations in subsidence and sedimentation rates may cause temporally offset transitions from transgression to regression along the shoreline (Gill and Cobban, 1973; Martinsen and Helland-Hansen, 1995). Where offshore sedimentation rates are very low around the time of maximum shoreline transgression, determining the maximum flooding surface within a condensed section may be very difficult; in such cases, the more readily recognizable base of the overlying terrigenous progradational wedge (limit between the condensed section and the overlying progradational shoreface facies in Fig. 5.5) may be approximated as the downlap surface. This 'maximum flooding surface' (in reality, a facies contact within the highstand systems tract) is, however, highly diachronous, with the rates of highstand shoreline regression, which can be emphasized using volcanic ash layers as time markers (Ito and O'Hara, 1994). The real maximum flooding surface, which corresponds to the peak of finest sediment at the top of the retrograding succession, has a much lower diachroneity and lies at the heart of the condensed section. Most importantly, maximum flooding (and maximum regressive) surfaces defined on the basis of stratal stacking patterns have a *basin-wide extent*, as they reflect major changes in sediment supply and depositional trends that are triggered by shifts in shoreline trajectories, in a manner that is independent of the offshore variability in water depth.

The second method of definition implies a potentially highly diachronous maximum flooding surface along both dip and strike, as the timing of the peak of deepest water depends on the variations in sedimentation and subsidence rates across the basin. As noted by Naish and Kamp (1997), T. Naish (pers. comm., 1998), Catuneanu et al. (1998b) and Vecsei and Duringer (2003), the maximum water depth often occurs *within*

the highstand (normal regressive) progradational wedge (Fig. 7.18). Thus, the boundary between prograding and retrograding geometries ('downlap surface') corresponds to a physical surface, recognizable on the basis of stratal stacking patterns (e.g., Figs. 4.39 and 4.40); in contrast, the surface that marks the peak of deepest water may be undeterminable lithologically, and may only be identified by using benthic foraminiferal paleobathymetry. The latter 'maximum flooding surface,' taken at the top of a deepening-upward succession, is younger in areas of lower sedimentation and higher subsidence rates, where the transition from deepening to shallowing occurs later, although this diachroneity is considered 'low' (Embry, 2002). The diachroneity rate that may characterize this type of surface, defined on water-depth changes, is quantified below by means of numerical modeling. Besides the actual degree of diachroneity, what hampers the applicability of maximum flooding surfaces defined at the top of 'deepening-upward' successions most is their *spatial restriction* to shallower areas within the marine basin, where cycles of water deepening and shallowing accompany the transgressive—regressive shifts of the shoreline. Outside of these areas, the actively subsiding portions of the basin may record continuous water deepening during several base-level cycles at the shoreline, as a result of subsidence rates outpacing the sum of sea-level change and sedimentation (Catuneanu et al., 1998b).

End-member boundary conditions can be applied to surfaces formed as a result of the complex interplay between eustasy, subsidence, and sedimentation, such as the maximum regressive and maximum flooding surfaces defined on the basis of water-depth changes. The temporal significance of these surfaces will be compared with the timing of surfaces defined on the basis of stratal stacking patterns.

Two-dimensional Model

To illustrate the effect that subsidence and sedimentation rates have on the timing of maximum regressive and maximum flooding surfaces defined on the basis of bathymetric changes, a simple two-dimensional geometrical basin model applied to a marine shelf setting is constructed, which is referred to in the following as Profile A. The model considers eustasy as the highest-frequency variable, to facilitate comparison with the depositional sequence model of Posamentier et al. (1988), but similar results may be obtained by taking subsidence as the higher-frequency parameter instead. The input values used for the variable rates of change are obtained from the literature (Pitman, 1978; Pitman and Golovchenko, 1983; Angevine, 1989; Galloway, 1989; Jordan and Flemings, 1991; Macdonald, 1991; Frostick and Steel, 1993).

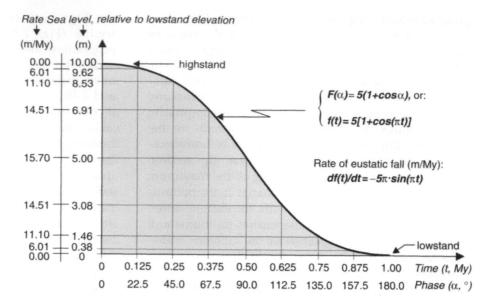

FIGURE 7.23 Curve of eustatic changes considered in the numerical model (modified from Catuneanu *et al.*, 1998b). Eustasy is assumed to vary sinusoidally with an amplitude of 10 m and period of 2 Ma. Only half the eustatic cycle (180° phase, or 1 Ma) from highstand to lowstand is shown. The rate of eustatic change at each 0.125 Ma time step, given by the first derivative of the sine curve, is shown on the left.

The assumptions of the model are as follows:

1. Eustasy varies sinusoidally with an amplitude of 10 m and a period of 2 Ma (Fig. 7.23). For the sake of brevity, only half of the eustatic cycle is shown, from highstand to lowstand. The rate of eustatic fall increases from zero at highstand (0 Ma) to a maximum of 15.7 m/Ma at the inflexion point (0.5 Ma), and then decreases to zero at lowstand (1 Ma).

2. The modeled portion of the basin is 200 km across, and the tectonic subsidence rate is constant at any particular point, but increases basinward from 20 m/Ma at the proximal end of the profile to 40 m/Ma at the distal end. This is similar to the simple divergent margin models of Pitman (1978), Angevine (1989) and Jordan and Flemings (1991).

3. The sedimentation rates change along dip in a linear manner, decreasing from 15 m/Ma at the proximal end of the basin profile to 5 m/Ma at the distal end. This reflects the tendency of coarser-grained sediments to be trapped closer to the shoreline. Since the sedimentation rate at any point along the basin profile is a function of its distance from the shoreline, it must therefore vary through time as the shoreline transgresses and regresses. Incorporation of this facies shift into the model would necessitate recalculating the sedimentation rate at each point along the profile at each model time step. However, at the scale of the modeled basin profile (200 km) this is insignificant (less than 150 m for one time step; Catuneanu *et al.*, 1998b), so the sedimentation rate is approximated to be constant at any given point through time, for simplicity.

The interplay of tectonic subsidence and sedimentation along the basin profile gives the rate and direction of motion of the seafloor relative to the center of Earth. Since the lateral facies shifts are negligible at the scale of the model, the rate of vertical seafloor motion remains constant at any particular point of the profile through the course of the eustatic half-cycle. This rate does vary spatially, however, reflecting the differen-tial subsidence and sedimentation rates along the dip-oriented basin profile.

The model advances in increments of 0.125 Ma. For each incremental time step, the addition of the rate of sea-level change (Fig. 7.23) to the rates of subsidence and sedimentation at each point along the profile, allows calculation of the rates of water-depth changes across the basin. The result is a graphic output (Fig. 7.24) that shows which portions of Profile A undergo water shallowing and which ones undergo water deepening. The boundary between these two zones marks the point at which water depth is stationary, and this point shifts with time along dip in response to the changing balance between accommodation and sedimentation.

Model Results

The model starts at eustatic highstand, where the rate of eustatic change is zero. The rate of water-depth change at this time is thus equal to the rate of vertical motion of the seafloor, and is positive along the entire length of the profile (Fig. 7.24). The water is therefore deepening throughout Profile A.

The successive incremental time steps of the model through a 1 Ma eustatic half-cycle from highstand to

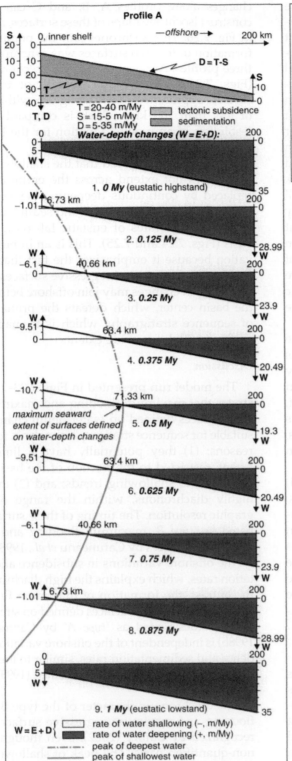

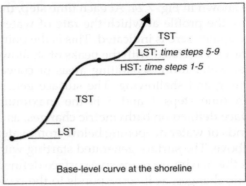

FIGURE 7.24 Timing and *locus* of formation of surfaces that mark the peaks of shallowest and deepest water, along basin Profile A, as a function of the interplay between eustasy (E), subsidence (T) and sedimentation (S) (modified from Catuneanu *et al.*, 1998b). The input values for subsidence and sedimentation are provided on the top cross section. The interplay of subsidence and sedimentation defines the rates of vertical shifts of the seafloor relative to the center of Earth (D). The rate of water-depth change (W) is shown across the basin profile for nine incremental time steps through a eustatic half cycle from highstand to lowstand. Maximum flooding and maximum regressive surfaces defined on bathymetric changes form where the rate of water-depth change equals zero at each time step. In this example, highstand normal regression (decelerating base-level rise) is succeeded by lowstand normal regression (accelerating base-level rise) without an intervening stage of forced regression. Abbreviations: LST—lowstand systems tract; TST—transgressive systems tract; HST—highstand systems tract.

lowstand are shown in Fig. 7.24. At each time step, the distance along the profile at which the rate of water-depth change equals zero is indicated. This is the point at which surfaces associated with the peaks of shallowest or deepest water form, separating areas of coeval water deepening and shallowing. The surface generated between time steps 1 and 5 is the maximum flooding surface defined on bathymetric changes, as it separates trends of water deepening below from water shallowing above. The surface generated starting with time step 5 is the maximum regressive surface defined on bathymetric changes, as it corresponds to the peak of shallowest water. As the rate of eustatic fall increases from time step 1 to 5 (Fig. 7.23), a progressively larger value of net subsidence is required to balance it and maintain the stationary water-depth condition that ends the deepening-upward trend. This point of balance shifts basinward through time, and therefore the maximum flooding surface defined on bathymetric changes is younger offshore than it is towards the basin margin.

Time step 5 (0.5 Ma) is the inflexion point on the falling limb of the sinusoidal eustatic curve and represents the maximum rate of eustatic fall. This is balanced by subsidence and sedimentation at a distance of 71.3 km along the profile (Fig. 7.24). Basinward of this point, shallowing is not possible anymore, so the water is continuously deepening. Under these circumstances, no more surfaces defined on water-depth changes can form beyond this point.

During time steps 5 to 9 the rate of eustatic fall decreases to zero. As the rate of eustatic fall decreases, it is balanced by a progressively lower value of vertical seafloor shift, and the point of balance between shallowing- and deepening-water conditions moves towards the basin margin. The surface that marks the peak of shallowest water is therefore older offshore than it is towards the basin margin (Fig. 7.24). At time step 9 (eustatic lowstand), continued subsidence results in a water deepening trend across the entire profile.

Strike Variability

To further illustrate the diachroneity of surfaces defined on water-depth changes, two more basin profiles (B and C) are added to the model. These represent dip-sections across the same basin at 50 and 100 km along strike from Profile A. Profiles B and C are assigned slightly different values of subsidence and sedimentation rates, to reflect the type of strike variability that is commonly found in the real world. All three models are run through the same eustatic half-cycle.

The graphic incremental time steps for Profiles B and C are shown in Fig. 7.25. The time/distance data for the formation of the two surfaces defined on water-depth

changes along profiles A, B, and C can be used to construct isochrone maps of these surfaces, as illustrated in Fig. 7.26. The isochrones join the points where the formation of the two surfaces was synchronous on the three profiles. As seen on the cross sections and maps (Figs. 7.24–7.26), the diachroneity of the modeled surfaces approaches 0.5 Ma along dip, within a distance of less than 200 km, which is compatible with the resolution of ammonite zonation for the Jurassic and the Cretaceous (Obradovich, 1993).

It should also be noted that the shallowing-upward trend does not extend across the entire basin but is replaced by continuous deepening beyond the point where the sum of subsidence and sedimentation starts to outpace the rates of eustatic fall on a permanent basis (Figs. 7.24 and 7.25). This is an important observation because it emphasizes the fact that maximum flooding and maximum regressive surfaces defined on water-depth changes may join offshore before reaching the basin center, which defeats the primary purpose of sequence stratigraphy which is to map bounding surfaces with basin-wide extent.

Discussion

The model run presented in Figs. 7.23–7.26 demonstrates that maximum flooding and maximum regressive surfaces defined on water-depth changes are not suitable for sequence stratigraphic analysis, for two main reasons: (1) they potentially have a limited lateral extent, restricted to the portion of the basin that may be subject to shallowing trends; and (2) they may be highly diachronous, within the range of biostratigraphic resolution. The timing of these surfaces (designated as 'type B' maximum flooding and maximum regressive surfaces by Catuneanu et al., 1998b) depends on the offshore variations in subsidence and sedimentation rates, which explains the high diachroneity rates. In contrast, the formation of maximum flooding and maximum regressive surfaces defined on stratal stacking patterns (designated as 'type A' by Catuneanu et al., 1998b) is independent of the offshore variations in subsidence and sedimentation rates, similar to the correlative conformities of Posamentier and Allen (1999) and Hunt and Tucker (1992).

The diachronous character of the type B maximum flooding and maximum regressive surfaces has been recognized by Embry (2002), even though only on a non-quantified basis. The peak of shallowest water is correctly envisaged to be younger in areas of higher sedimentation rates, i.e., adjacent to the shoreline, but the degree of diachroneity is thought to be limited to the duration of the lowstand normal regression: 'the change from a shallowing-upward trend to a deepening-upward one [i.e., the 'maximum regressive surface' of

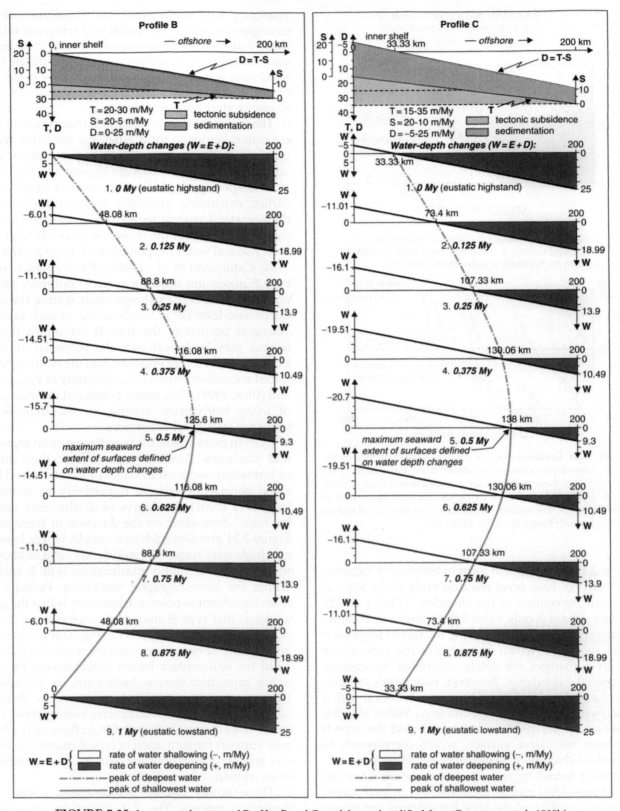

FIGURE 7.25 Incremental output of Profiles B and C model runs (modified from Catuneanu *et al.*, 1998b). These profiles are assigned different rates of subsidence and sedimentation relative to Profile A, but the same eustatic curve shown in Fig. 7.23 was used. See Fig. 7.24 for the key to abbreviations.

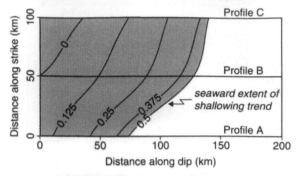

1. Timing of the surface that marks the peak of deepest water (i.e., maximum flooding surface defined on the basis of water-depth changes)

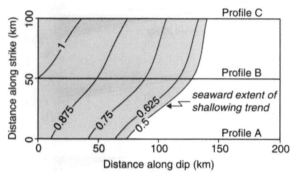

2. Timing of the surface that marks the peak of shallowest water (i.e., maximum regressive surface defined on the basis of water-depth changes)

FIGURE 7.26 Isochrone maps of the two surfaces defined on the basis of water-depth changes (modified from Catuneanu *et al.*, 1998b). Note that these surfaces have a limited extent in an offshore direction, and join along the line that marks the maximum extent of shallowing trends. The isolines on the map show the *locus* of surface formation for each time step of the model run.

Embry, 2002] ... begins to form in basinward localities at the start of base level rise and ends at the start of landward movement of the shoreline' (Embry, 2002). This assessment is only valid if the water was shallowing across the entire basin during the forced regression of the shoreline, which is generally the case where base-level changes are solely controlled by eustasy. Differential subsidence, however, may result in base-level rise and water deepening offshore, coeval with a fall in base level at the shoreline (e.g., Vail *et al*, 1984; Catuneanu *et al.*, 1998b). This means that the type B maximum regressive surface, which corresponds to the peak of shallowest water, may actually start forming during forced regression, and therefore *before* the start of base-level rise at the shoreline; in such cases, the earliest portion of the type B maximum regressive surface becomes coeval with the subaerial unconformity, and is older than the correlative conformity of Hunt and Tucker (1992). This surface continues to form,

younging landwards, until the shoreline starts its transgressive shift. The spatial and temporal relationships that can be demonstrated between the subaerial unconformity, its correlative conformity, and the type B maximum regressive surface indicates the inadequacy of using bathymetric criteria for the definition of any sequence stratigraphic surface.

The type B maximum flooding surface (peak of deepest water) begins to form when the shoreline reaches maximum transgression, starting in nearshore areas, where sedimentation rates are highest and the change from deepening- to shallowing-upward trends occurs earlier, expanding gradually basinward throughout the highstand normal regression. In basins where the forced regression of the shoreline is coeval with base-level rise and water deepening offshore (e.g., Vail *et al.*, 1984; Catuneanu *et al.*, 1998b), the formation of the type B maximum flooding surface continues *after* the end of highstand normal regression, during the subsequent base-level fall at the shoreline. In such cases, the youngest portion of the type B maximum flooding surface may be coeval with the subaerial unconformity, and is therefore younger than the basal surface of forced regression (correlative conformity of Posamentier and Allen, 1999). This, again, points out the inadequacy of using bathymetric criteria for the definition of sequence stratigraphic surfaces.

Even in cases where the type B maximum regressive and maximum flooding surfaces have their interval of formation restricted to stages of lowstand and highstand normal regressions, respectively, as postulated by Embry (2002), the degree of diachroneity can still be 'high,' depending on the duration of these stages. Figure 7.24 provides such an example, where lowstand and highstand stages last sufficiently long to allow for the formation of highly diachronous type B surfaces, within the biostratigraphic resolution. Perhaps even more significant to point out, however, is that the generalization that type B maximum regressive and maximum flooding surfaces form during stages of lowstand and highstand normal regressions, respectively, is only valid for sedimentary basins characterized by subsidence rates that *increase* basinward, as modeled by Catuneanu *et al.* (1998b). In contrast, as discussed earlier in this chapter, sedimentary basins whose subsidence rates *decrease* basinward (e.g., flexural forelands) may support the formation of type B maximum regressive and maximum flooding surfaces during transgressions instead.

These points can be demonstrated easily using numerical models. The data input for the forward simulations in Figs. 7.24 and 7.25 account for a situation where subsidence rates are generally greater than the rates of eustatic fall. This is a case of continuous relative sea-level rise, where the transgressive and (normal) regressive shifts of the shoreline are mainly controlled by sedimentation and varying rates of sea-level change. Under such circumstances, highstand normal regressions (decelerating base-level rise) are immediately succeeded by lowstand normal regressions (accelerating base-level rise) without intervening stages of forced regression (e.g., Fig. 7.24). Profile D (Fig. 7.27) depicts the same basin profile considered in Figs. 7.24 and 7.25, but this time affected by lower subsidence rates that are within the range of eustatic fluctuations. The results of the model run show that the relative sea level falls and rises through time at the shoreline, whereas offshore, beyond the point of maximum extent of relative sea-level fall (i.e., where subsidence starts outpacing the highest rate of eustatic fall), the basin is subject to continuous relative sea-level rise. This is the situation envisaged by Vail *et al.* (1984) for the formation of type 2 sequence boundaries: eustatic fall outpacing subsidence at the shoreline (forced regression) coeval with relative rise at the shelf edge (subsidence outpacing eustatic fall). The interplay of relative sea-level changes and sedimentation determines the timing of formation of surfaces marking the peaks of shallowest and deepest water (Fig. 7.27). Note that these surfaces (1) have limited lateral extent in a basinward direction; (2) meet at the point where water is beyond the influence of shallowing conditions (subsidence > eustatic fall + sedimentation); and (3) form not only during normal regressions of the shoreline, but also during forced regression, in parallel with the formation of the subaerial unconformity. The calculated diachroneity rate of these surfaces approaches a quarter of the period of the highest-frequency variable (eustasy in this case, with a period of 2 Ma; Catuneanu *et al.*, 1998b), which is in agreement with the quarter-cycle phase shift modeled by Angevine (1989), Christie-Blick (1991) and Jordan and Flemings (1991).

Figure 7.28 presents a Wheeler diagram that shows the timing of types A and B maximum flooding and maximum regressive surfaces, and also the changing balance between relative sea-level rise and fall along dip. Note that the influence of relative fall diminishes in a basinward direction in parallel to the increase in subsidence rates. Similarly, shallowing-water conditions are also less prevalent in a basinward direction, and they eventually stop manifesting altogether where subsidence exceeds the sum of sedimentation and maximum sea-level fall.

It is now clear that any type of surface whose timing depends on the offshore variations in subsidence and sedimentation rates is bound to have a potentially high diachroneity and limited lateral extent. Both these two attributes make such surfaces inadequate for sequence stratigraphy and regional correlations; what is needed, instead, are surfaces with basin wide extent and low diachroneity rates. These 'ideal' surfaces are represented by the type A maximum flooding and maximum regressive surfaces, the basal surface of forced regression (the correlative conformity of Posamentier and Allen, 1999), and the correlative conformity *sensu* Hunt and Tucker (1992). These four surfaces correspond to the four events of the base-level cycle at the shoreline (Figs. 1.7 and 4.7), and are independent of the offshore variations in subsidence and sedimentation rates (Fig. 7.28). They are defined based on stratal stacking patterns (separating forced regressive *vs.* normal regressive *vs.* transgressive stratal architectures), with a timing controlled by the changes in the direction and/or type of shoreline shift. Such changes in shoreline trajectories control the patterns of sediment supply to the marine basin, and hence the stratal geometries and depositional trends referred to above.

The curves that indicate changes in water depth and relative sea level along dip follow each other closely (Fig. 7.28), displaying parallel trends offset with the value of the sedimentation rates. These time-transgressive curves meet the shoreline at the four key points of the base-level cycle (Figs. 1.7 and 4.7), where they join the four quasi-isochronous sequence stratigraphic surfaces defined on stratal stacking patterns (type A in Fig. 7.28). As the correlative conformity (*sensu* Hunt and Tucker, 1992) marks the end of base-level fall *at the shoreline* but forms during base-level rise offshore, the (type A) maximum regressive surface too marks the end of shallowing near the shoreline but forms during water deepening (in extensional basins; potentially shallowing in forelands) offshore. Both these surfaces top successions that prograde into the basin during particular stages of shoreline shifts (forced and lowstand normal regressions, respectively), irrespective of how subsidence and sedimentation rates may vary offshore. Following the same idea, the basal surface of forced regression is taken at the base of all deposits accumulated during the forced regression *of the shoreline*, even though some of the deeper-water 'forced regressive' sediments accumulate under rising relative sea-level conditions (Fig. 7.28). This approach is warranted because all these deposits are laterally correlative and age-equivalent, with the sediment being supplied to the basin in relation to the forced regression *of the shoreline*.

The use of type B surfaces (Fig. 7.28) would make different systems tracts to be age-equivalent, as explained in detail by Catuneanu *et al.* (1998b) (their Figs. 13 and 14). Since the type B maximum flooding and maximum regressive surfaces may join in an area that is shallower than the basin center, everything that

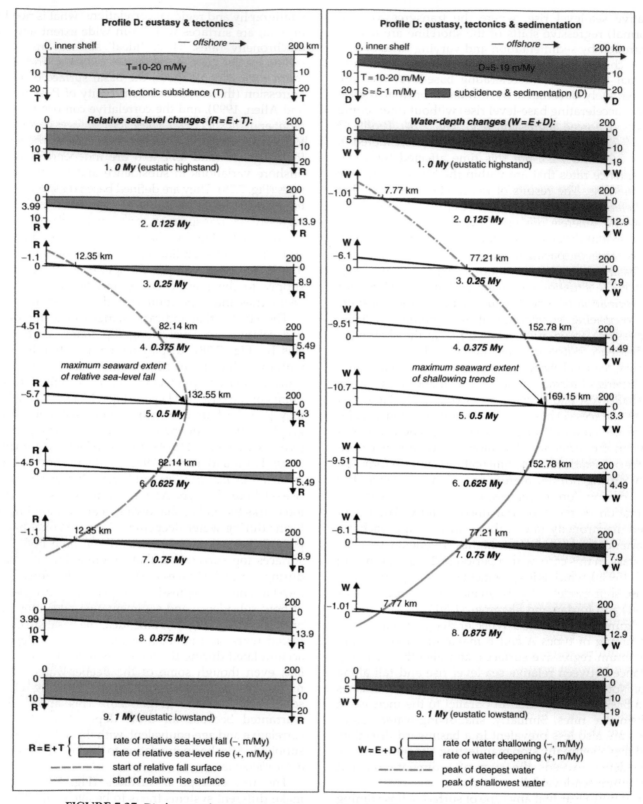

FIGURE 7.27 Diachronous formation of surfaces separating deposits accumulated under relative sea-level fall and rise conditions (left column), as well as deposits accumulated under water deepening and shallowing conditions (right column). See Fig. 7.24 for the key to abbreviations.

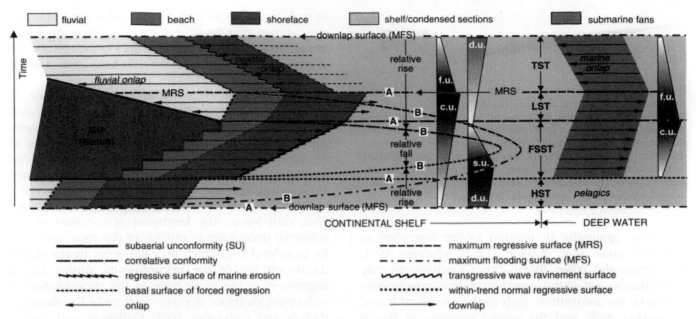

FIGURE 7.28 Temporal significance of types A and B surfaces (modified from Catuneanu *et al.*, 1998b). Type A surfaces are defined by stratal stacking patterns, and are systems tract boundaries. The timing of type A surfaces depends on base-level changes, or on the interplay between base-level changes and sedimentation, *at the shoreline*. The timing of type A surfaces is independent of the offshore variations in subsidence and sedimentation rates. The timing of type B surfaces depends on the *offshore variations* in subsidence and sedimentation rates, which confers on those surfaces high diachroneity rates (a quarter of the period of the highest-frequency variable: Catuneanu *et al.*, 1998b). Type A and B surfaces join at the shoreline, but diverge in an offshore direction. Abbreviations: HST—highstand systems tract; FSST—falling-stage systems tract; LST—lowstand systems tract; TST—transgressive systems tract; RST—regressive systems tract; c.u.—coarsening-upward; f.u.—fining-upward; d.u.—deepening-upward; s.u.—shallowing-upward. Note that grading and water-depth changes are not interchangeable concepts.

accumulates in the basin center under continuous deepening-water conditions would qualify as a transgressive systems tract, and be age-equivalent with all other systems tracts in the shallow part of the basin. Such a view would of course be misleading since the bounding surfaces that separate different stacking patterns (and associated systems tracts) are in fact quasi-time lines (with a low diachroneity that reflects the rates of sediment *transport*) along dip, independent of the offshore variations in subsidence and sedimentation rates.

An inherent and potentially higher diachroneity along strike is recorded by all types of event-significant surfaces in relation to the strike variability in subsidence and sedimentation rates *along the shoreline*. The strike diachroneity of the two correlative conformities (base and top of forced regressive deposits) is affected only by changes in subsidence rates, whereas the strike diachroneity of maximum regressive and maximum flooding surfaces is also influenced, in addition to subsidence, by fluctuations in sedimentation rates along the shoreline. As sediment supply to the shoreline may vary significantly along strike, the transitions

from regression to transgression and *vice versa* may be offset significantly from one area to another, sometimes within the range of biostratigraphic resolution (e.g., Gill and Cobban, 1973). For this reason, maximum regressive and maximum flooding surfaces tend to be less significant than the correlative conformities in a chronostratigraphic framework.

SUMMARY: TIME ATTRIBUTES OF STRATIGRAPHIC SURFACES

Subaerial Unconformity

The subaerial unconformity (Figs. 4.9 and 4.13) is generally perceived as a 'time barrier' (Winter and Brink, 1991; Embry, 2001b) based on the assumption that time lines do not cross this surface, i.e., all strata below the unconformity are older than the strata above it. Whether this is a general truth or an artifact of the lack of rigorous testing remains to be seen. The possibility of formation of diachronous unconformities which are crossed by time lines is emphasized in the

FIGURE 7.29 Formation of a subaerial unconformity in response to base-level fall at the shoreline (modified from Shanley and McCabe, 1994). Note that the effect of base-level fall on fluvial processes diminishes in a landward direction. The arrows indicate the amounts of differential fluvial incision, which decrease upstream. The position of the upstream limit of the area controlled by base-level changes at the shoreline depends on a number of variables, including the magnitude of base-level changes, the gradients of the landscape profile, and the size of the river.

literature, generally in relation to the migration of uplifted areas (e.g., Cohen, 1982; Johnson, 1991; Crampton and Allen, 1995). Other mechanisms may contribute as well to the formation of time-transgressive subaerial unconformities, such as the process of forced regression itself, and the lagged response of fluvial systems to changes in base-level at the shoreline.

It is now well established that base-level changes at the shoreline may only control fluvial processes within a limited distance upstream (Shanley and McCabe, 1994; Figs. 3.3 and 7.29). This distance varies with the landscape gradients, the size of the river and the magnitude of base-level changes, but generally ranges from tens of kilometers (e.g., 90 km in the case of the Colorado River in Texas) to more than 200 km in the case of larger rivers (e.g., 220 km for the Mississippi River; Shanley and McCabe, 1994). Beyond the landward limit of the base-level control, the river responds primarily to a combination of climatic and tectonic mechanisms (Shanley and McCabe, 1994; Blum, 1994).

In such inland areas, cycles of fluvial aggradation and degradation may be driven by changes in discharge and sediment load. These cycles may be completely out of phase relative to those driven by base-level changes (Miall, 1996).

The formation of a subaerial unconformity in response to base-level fall at the shoreline is illustrated in Fig. 7.29. As the shoreline shifts toward the basin during forced regression, expanding the areal extent of the subaerial unconformity, the landward limit of the zone of influence of base-level fall on fluvial processes shifts accordingly (i.e., assuming that the distance 'SU' in Fig. 7.29 remains constant during forced regression). This shift leaves the landward termination of the subaerial unconformity outside of the zone controlled by base-level fall, which may allow for fluvial aggradation on top of the unconformity during forced regression, at the same time as the progradation of offlapping shoreface deposits in front of the shoreline (Sylvia and Galloway, 2001; Galloway and Sylvia, 2002; Fig. 7.30). In this way, the early fluvial strata that prograde onto the subaerial unconformity may be older than the late falling-stage shoreface deposits that are truncated by the same subaerial unconformity (Fig. 7.30). This scenario, documented in the case of the Quaternary Brazos River of the Texas Gulf Coast (Sylvia and Galloway, 2001; Galloway and Sylvia, 2002), provides an example where time lines do cross a diachronous subaerial unconformity, which youngs in a basinward direction. As the subaerial unconformity keeps expanding basinward during the falling stage, the degree of diachroneity along dip matches the rates of forced regression. Along strike, the timing of the onset and end of forced regression may be modified by

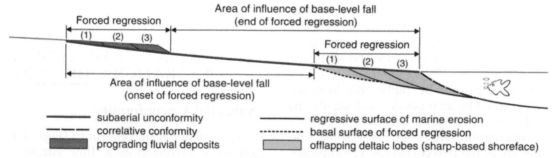

FIGURE 7.30 Fluvial and shoreface forced regressive deposits in relation to the subaerial unconformity (modified from Sylvia and Galloway, 2001). The area of influence of base-level fall is kept constant through time, and it shifts towards the basin with the rate of forced regression. The time sequence (1), (2), and (3) suggests time steps in the process of forced regression. The early fluvial strata that prograde onto the subaerial unconformity during forced regression (e.g., at time 1) are older than the late forced regressive shoreface deposits (e.g., the lobe accumulated at time 3). The latter are topped by the subaerial unconformity, which extends basinward during the forced regression. In this example, the subaerial unconformity is not a time barrier, as it is crossed by time lines.

	Stratigraphic surface	Diachroneity along dip	Diachroneity along strike
Sequence stratigraphic surface — *Event-significant*	Correlative conformity	Rate of sediment transport [1]	Rates of subsidence [2]
	Basal surface of forced regression	Rate of sediment transport [1]	Rates of subsidence [2]
	Maximum regressive surface	Rate of sediment transport [1]	Rates of subsidence and sedimentation [3]
	Maximum flooding surface	Rate of sediment transport [1]	Rates of subsidence and sedimentation [3]
Stage-significant	Subaerial unconformity	Rate of forced regression [3]	Rates of subsidence [2]
	Transgressive ravinement surfaces	Rate of shoreline transgression [3]	Rates of subsidence and sedimentation [3]
	Regressive surface of marine erosion	Rate of forced regression [3]	Rates of subsidence [2]
Within-trend facies contact	Within-trend normal regressive surface	Rate of normal regression [3]	Rates of subsidence and sedimentation [3]
	Within-trend forced regressive surface	Rate of forced regression [3]	Rates of subsidence [2]
	Within-trend flooding surface	Rate of shoreline transgression [3]	Rates of subsidence and sedimentation [3]

FIGURE 7.31 Controls on the rates of diachroneity that characterize the seven surfaces of sequence stratigraphy and the most prominent within-trend facies contacts. Degree of diachroneity: [1]—low (i.e., below the resolution of biostratigraphy or radiochronology); [2]—low to high; [3]—high (i.e., potentially within the biostratigraphic or radiometric resolution). The stratigraphic surfaces presented in this table are defined on the basis of stratal stacking patterns (for a summary of mapping criteria, see Fig. 4.9). The event-significant sequence stratigraphic surfaces (i.e., corresponding to the onset of fall, end of fall, end of regression and end of transgression events at the shoreline; Fig. 4.7) are more diachronous along strike, especially where their timing depends on variations in sedimentation and/or subsidence rates along the shoreline. These surfaces are quasi-time lines along dip, with a timing controlled by changes in shoreline trajectories and the associated patterns of sediment supply to the marine basin, irrespective of the offshore variations in subsidence rates or water depth; their diachroneity is limited to the rates of sediment transport (commonly in a range of 10^{-1}–10^2 m/s). In contrast, the stage-significant sequence stratigraphic surfaces (i.e., formed during specific stages of shoreline shift) tend to be more diachronous along dip, where their rates of diachroneity match the rates of shoreline shift. Along strike, the timing of these surfaces is controlled by variations in sedimentation and/or subsidence rates along the shoreline. Similar to the stage-significant sequence stratigraphic surfaces, the diachroneity of within-trend facies contacts is linked to the rates of shoreline shift, along dip, and to the variability recorded by the mechanisms controlling shoreline shifts (i.e., sedimentation and/or subsidence rates), along strike. Note that differential subsidence along the shoreline affects the strike diachroneity of *all* stratigraphic surfaces, while sedimentation adds only to the strike diachroneity of surfaces defined relative to the transgressive–regressive curve.

differential subsidence, which defines the degree of strike diachroneity of the subaerial unconformity (Fig. 7.31). It can be noted that subaerial unconformities may be diachronous along both dip and strike, even though their timing is independent of sedimentation.

Another mechanism that may lead to the formation of time-transgressive subaerial unconformities has been discussed by Posamentier and Allen (1999), in relation to the lagged response of fluvial systems to

downstream controls. The end of base-level fall at the shoreline may be followed closely by a cessation of fluvial incision at the river mouth, but 'because the 'message' to incise propagates upstream over a period of time [i.e., the time between base-level changes at the river mouth and the arrival of the incising knickpoint upstream; Figs. 3.31 and 3.32], upstream reaches of rivers can continue to experience incision for a time even after cessation of relative sea-level fall. ...This lagged response adds another component to the potential diachroneity of unconformable sequence boundaries.' (Posamentier and Allen, 1999).

Correlative Conformity

The correlative conformity (*sensu* Hunt and Tucker, 1992; Figs. 4.9 and 4.18) marks the top of all marine deposits that accumulate during the forced regression of the shoreline, including those which are part of the deep-water submarine fans. This sequence stratigraphic surface is defined on the basis of stratal stacking patterns (Haq, 1991: 'a change from rapidly prograding parasequences to aggradational parasequences'; i.e., the youngest clinoform associated with offlap), with a timing that corresponds to the end of base-level fall *at the shoreline* (Fig. 4.7). The correlative conformity is usually approximated with a time line (seafloor at the end of forced regression: Embry, 1995), although technically it is diachronous, younging basinward, with the rate of offshore sediment transport (Catuneanu *et al.*, 1998b). This diachroneity rate is very low, ranging from 10^{-1}–10^0 m/s in shelf-type settings to 10^1–10^2 m/s in steeper ramp settings, which is below the resolution of the current dating techniques.

It is important to note that the timing of the correlative conformity depends only on the base-level changes *at the shoreline*, and is independent of the base-level fluctuations that take place *within* the marine portion of the basin. A surface mapped at the temporal boundary between relative fall and subsequent relative rise in every discrete location within the marine portion of the basin (Fig. 7.27; type B correlative conformity in Fig. 7.28) is highly diachronous and independent of stratal stacking patterns (i.e., not a systems tract boundary).

The quasi-time line significance of the correlative conformity, as depicted above, is valid only along depositional dip transects. Along strike, variations in subsidence rates may offset the transition between base-level fall and base-level rise along the shoreline (Catuneanu *et al.*, 1998b; Fig. 7.31). The degree of such diachroneity varies with the type of sedimentary basin and the associated subsidence mechanisms. It is considered to be 'low' in the case of 'passive' margins or intracratonic basins (Posamentier and Allen, 1999), but it may become more significant in more tectonically-active basins.

Basal Surface of Forced Regression

The basal surface of forced regression (i.e., the correlative conformity of Posamentier and Allen, 1999; Figs. 4.9 and 4.25) marks the base of all marine deposits

that accumulate during the forced regression of the shoreline. It is defined on the basis of stratal stacking patterns (i.e., the oldest clinoform associated with offlap), with a timing that corresponds to the onset of base-level fall *at the shoreline*. Similar to the correlative conformity, the basal surface of forced regression is often approximated with a time line (seafloor at the onset of forced regression), but in fact it is diachronous, younging basinward, with the rate of offshore sediment transport (Catuneanu *et al.*, 1998b). This diachroneity rate is very low, ranging from 10^{-1}–10^0 m/s in shelf-type settings to 10^1–10^2 m/s in steeper ramp settings, which is below the resolution of the current dating techniques.

As in the case of the correlative conformity, the timing of the basal surface of forced regression is controlled by changes in base level *at the shoreline*, and is independent of the basinward fluctuations in base level which are induced by differential rates of subsidence (Fig. 7.28). A surface mapped at the temporal boundary between relative rise and subsequent relative fall in every discrete location within the marine portion of the basin (Fig. 7.27; type B basal surface of forced regression in Fig. 7.28) is highly diachronous and independent of stratal stacking patterns (i.e., not a systems tract boundary).

The quasi-time line significance of the basal surface of forced regression, as depicted above, is valid only along depositional dip transects. Along strike, variations in subsidence rates may offset the transition between base-level fall and base-level rise along the shoreline (Catuneanu *et al.*, 1998b; Fig. 7.31). The degree of such diachroneity varies with the type of sedimentary basin and the associated subsidence mechanisms. It is considered to be 'low' in the case of 'passive' margins or intracratonic basins (Posamentier and Allen, 1999), but it may become more significant in more tectonically-active basins.

Regressive Surface of Marine Erosion

The regressive surface of marine erosion (Figs. 4.9 and 4.29) is a highly diachronous unconformity, which expands basinward during the base-level fall *at the shoreline*, with the rate of forced regression (Fig. 7.28). The emphasis on the shoreline is necessary because forced regressions may occur at the same time as stages of base-level rise offshore (e.g., Vail *et al.*, 1984; Catuneanu *et al.*, 1998b; Fig. 7.28). This surface may become a systems tract boundary where it reworks the basal surface of forced regression (Figs. 4.20, 4.23, and 4.24). The regressive surface of marine erosion should not, however, be used indiscriminately as a systems tract boundary, because its basinward (younger) portion may be separated from the basal surface of forced regression (the true systems tract boundary) by falling-stage shelf deposits (Figs. 4.23 and 4.24). In this case, the regressive surface of marine erosion develops *within* the falling-stage systems tract.

The degree of diachroneity of the regressive surface of marine erosion encompasses the entire stage of forced regression. As such, its landward (stratigraphically lowest) portion is placed at the base of the falling-stage systems tract, whereas its basinward termination underlies the oldest shoreface deposits of the lowstand systems tract (Figs. 4.23 and 5.65).

Unlike the four event-significant sequence stratigraphic surfaces (i.e., the two correlative conformities discussed above, the maximum regressive surface and the maximum flooding surface; Fig. 4.7), which tend to be more diachronous along strike than along dip, the regressive surface of marine erosion records the opposite trends. It is expected to be highly diachronous along dip, reflecting the rates of shoreline regression, but closer to a 'time barrier' along strike. Some degree of strike diachroneity may still be recorded, however, in relation to the change in subsidence rates that may offset the timing of base-level fall along the shoreline (Fig. 7.31).

Maximum Regressive Surface

The maximum regressive surface (Figs. 4.9 and 4.32) is often defined interchangeably on the basis of either *observed stratal stacking patterns* (the limit between progradational and overlying retrogradational strata) or *inferred bathymetric changes* (the surface that forms when the water depth reaches the shallowest peak) (Embry, 2002, 2005). These definitions are not equivalent, as demonstrated by the modeling of a typical shelf setting (Catuneanu et al., 1998b).

Maximum regressive surfaces defined on stratal stacking patterns (type A in Fig. 7.28) are mapped at the top of coarsening-upward (progradational) marine successions. The coarsening-upward trend is controlled by the *shoreline shift*, i.e., the seaward migration of the sediment entry points, and is independent of the offshore variations in water depth. Regression is associated with water shallowing in the vicinity of the shoreline, where more accommodation is consumed than is created, but coeval water deepening may occur offshore (Figs. 7.11, 7.14, 7.17, and 7.20). Maximum regressive surfaces defined on stratal stacking patterns are systems tract or sequence boundaries. Their timing depends on the interplay between the rates of sedimentation and base-level rise *at the shoreline*, and it is not affected by the offshore variations in sedimentation and subsidence rates (Catuneanu et al., 1998b).

These surfaces are close to time lines in a depositional-dip section, as there is only one point in time when the shoreline changes from regressive to transgressive along that particular transect. A low diachroneity rate may be recorded, however, in relation to the rates of sediment transport, as in the case of the two correlative conformities. Along strike, type A maximum regressive surfaces may be more diachronous, as variations in subsidence and sedimentation rates may offset the timing of the end-of-regression events along the shoreline, sometimes even within the range of biostratigraphic resolution (Gill and Cobban, 1973; Fig. 7.31).

Maximum regressive surfaces defined on bathymetric changes (type B in Fig. 7.28) are much more diachronous, with a timing that depends on the offshore variations in sedimentation and subsidence rates. As indicated by the modeling of a continental shelf setting, surfaces marking the shallowest peak form *within* regressive successions, crossing the systems tract boundaries (Fig. 7.28). Consequently, the marine sediments overlying a type B maximum regressive surface (peak of shallowest water) are commonly coarser than the underlying ones (as they are all part of a coarsening-upward trend), although the former accumulate in a deepening-water setting (Fig. 7.28). The surface marking the shallowest water often occurs *within* the lowstand systems tract, possibly extending into the underlying falling-stage systems tract as well (Fig. 7.28). This surface is lithologically undeterminable, as part of a regressive coarsening-upward trend, and can only be identified by independent studies of benthic foraminiferal or trace fossil paleobathymetry (e.g., Pekar and Kominz, 2001). Type B maximum regressive surfaces are therefore independent of stratal stacking patterns, and do not form systems tract boundaries. Type B surfaces merge with the type A surfaces at the shoreline (Fig. 7.28), and so they are described by the same degree of diachroneity along strike. As the types A and B maximum regressive surfaces diverge in an offshore direction (Fig. 7.28), the identification of the latter requires estimates of water depths for benthic foraminiferal biofacies in a manner that is independent of sediment grading and stratal stacking patterns.

Maximum Flooding Surface

Maximum flooding surfaces (Figs. 4.9 and 4.39) are also often defined interchangeably on the basis of either *observed stratal stacking patterns* (top of retrogradational strata) or *inferred bathymetric changes* (peak of deepest water) (Embry, 2002). These two approaches allow for different temporal attributes, i.e., they define different surfaces that are temporally offset (types A and B in Fig. 7.28).

Maximum flooding surfaces defined on stratal stacking patterns (type A in Fig. 7.28) are systems tract or sequence boundaries. Their timing depends on the interplay between the rates of sedimentation and base-level rise *at the shoreline*, and it is not affected by offshore variations in sedimentation and subsidence rates (Catuneanu *et al.*, 1998b). These surfaces are close to time lines in a depositional-dip section, as there is only one point in time when the shoreline changes from transgressive to regressive along that particular transect. A low diachroneity rate may, however, be recorded in relation to the rates of sediment transport, as in the case of all other event-significant sequence stratigraphic surfaces. Along strike, type A maximum flooding surfaces may be more diachronous, as variations in subsidence and sedimentation rates may offset the timing of the end-of-transgression events along the shoreline, possibly within the range of biostratigraphic resolution (Gill and Cobban, 1973; Fig. 7.31).

Maximum flooding surfaces defined on bathymetric changes (type B in Fig. 7.28) are much more diachronous, with a timing that depends on the offshore variations in sedimentation and subsidence rates. In the context of the modeled continental shelf setting discussed in this chapter, these surfaces, which mark the peak of deepest water, form *within* regressive (coarsening-upward) successions, and may cross systems tract boundaries (Fig. 7.28). As noted by Naish and Kamp (1997), and also by Tim Naish (pers. comm., 1998), the surface indicating the maximum water depth (type B maximum flooding surface, identified on the basis of fossil assemblages) often occurs *within* the highstand systems tract. Depending on circumstances (see more comprehensive discussions above), the formation of the deepest-water surface may extend into the overlying falling-stage systems tract as well. This surface is lithologically undeterminable, and can only be identified by independent studies of benthic foraminiferal or trace fossil paleobathymetry (e.g., Pekar and Kominz, 2001). Type B maximum flooding surfaces are therefore independent of stratal stacking patterns, and do not form systems tract boundaries. Types A and B surfaces merge at the shoreline (Fig. 7.28), and so they are described by the same degree of diachroneity along strike. As the types A and B maximum flooding surfaces diverge in an offshore direction (Fig. 7.28), the identification of the latter requires estimates of water depths for benthic foraminiferal biofacies in a manner that is independent of sediment grading and stratal stacking patterns.

Transgressive Ravinement Surfaces

The transgressive ravinement surfaces (Figs. 4.9, 4.49, 4.52 and 4.54) are highly diachronous unconformities, which young in a landward direction with the rate of shoreline transgression (Figs. 7.28 and 7.31). Any of the two types of transgressive ravinement surfaces (wave- or tide-generated) may become a systems tract boundary where it reworks the nonmarine portion of the maximum regressive surface, or even a sequence boundary where it reworks the subaerial unconformity as well (Embry, 1995; Helland-Hansen and Martinsen, 1996; Fig. 4.53). Where underlying estuarine facies are preserved, transgressive ravinement surfaces can be traced *within* the transgressive systems tract (Figs. 4.52, 5.4–5.6).

The degree of diachroneity of transgressive ravinement surfaces encompasses the entire stage of shoreline transgression. As such, transgressive ravinement surfaces make the physical connection between maximum regressive and maximum flooding surfaces across the transgressive systems tract (Fig. 5.4). This is somewhat similar to the situation described for the regressive surface of marine erosion, which also connects two event-significant sequence stratigraphic surfaces (the two correlative conformities) across the falling-stage systems tract (Figs. 4.23 and 5.65).

As in the case of the regressive surface of marine erosion, transgressive ravinement surfaces may be more diachronous along dip than along strike, although the dependence of transgressions on sedimentation rates makes this trend less evident here (Fig. 7.31). This general trend is in contrast with the time attributes of the four event-significant surfaces, which are almost invariably more diachronous along strike than along dip (Fig. 7.31).

Within-trend Facies Contacts

All within-trend facies contacts are commonly characterized by high diachroneity, which correlates with the rates of shoreline shift, along dip, and with the variability recorded by the mechanisms controlling shoreline shifts, along strike. Each stage of shoreline shift is associated with its own within-trend facies contacts. As discussed in more detail in Chapter 5, the most prominent within-trend facies contacts include the within-trend normal regressive surface (formed during lowstand or highstand normal regressions), the within-trend forced regressive surface (formed during forced regressions), and the within-trend flooding surface (formed during transgressions). As a general trend, the diachroneity of each within-trend facies contact matches the timing of the 'stage-significant' sequence stratigraphic surface that forms during the same stage of shoreline shift (Fig. 7.31). The exception to this rule is the within-trend normal regressive surface, as there is no sequence stratigraphic surface that forms *during* normal regressions. However, the strike diachroneity of this facies contact matches the strike diachroneity of maximum regressive and maximum flooding surfaces, whose timing is linked to the end and onset of lowstand and highstand normal regressive stages, respectively. All facies contacts discussed in this section develop *within* systems tracts, and therefore they do not serve as systems tract or sequence boundaries.

The within-trend normal regressive surface (Figs. 4.9, 4.55, and 4.56) is highly diachronous, along both dip and strike. Along dip, the degree of diachroneity of this facies contact matches the rates of shoreline (lowstand or highstand) normal regression. This is the lowest-rate (slowest) type of shoreline shift, and therefore the rates of diachroneity imposed by such shoreline shifts are highest. Along strike, the dependence of normal regressions on sedimentation rates makes the within-trend normal regressive surface highly diachronous, just as for all other stratigraphic surfaces whose timing depends on variations in sedimentation rates along strike (Fig. 7.31).

The within-trend forced regressive surface (Figs. 4.9 and 4.57) is also highly diachronous along dip, with the rate of shoreline forced regression. Along strike, however, the timing of this facies contact depends only on variations in subsidence rates, and is independent of fluctuations in sedimentation rates (Fig. 7.31). This is because forced regressions themselves are defined relative to the base-level curve (as opposed to the transgressive–regressive curve), and are driven by negative accommodation at the shoreline, irrespective of the sedimentation rates on the seafloor.

The within-trend flooding surface (Figs. 4.9 and 4.60) tends to be highly diachronous both along dip and strike, due to the dependence of transgressions on sedimentation rates. Along dip, the degree of diachroneity of this facies contact matches the rates of shoreline transgression. Along strike, the timing of within-trend flooding surfaces depends on variations in subsidence and sedimentation rates along the shoreline (Fig. 7.31). This is similar to the degree of diachroneity of transgressive ravinement surfaces, which is also controlled by the same variables (Fig. 7.31). Under particular circumstances, where within-trend flooding surfaces mark episodes of abrupt relative sea-level rise (e.g., stages of rapid subsidence in tectonically-active basins), their degree of diachroneity may be low (e.g., Fig. 5.64). In such cases, the degree of diachroneity of other surfaces whose timing depends on the rate of shoreline transgression (i.e., transgressive ravinement surfaces) is also low.

Conclusions

It can be concluded that the seven surfaces of sequence stratigraphy (Fig. 4.7) may be grouped into two main categories with respect to their temporal attributes; one group includes the four 'event-significant' surfaces (corresponding to the onset of fall, end of fall, end of regression and end of transgression events at the shoreline), while the second group includes 'stage-significant' surfaces that form during specific stages of shoreline shift (Fig. 7.31). Surfaces belonging to the two groups are fundamentally different in terms of their temporal attributes, and particularly with respect to the contrast in their degrees of diachroneity along dip-oriented *vs*. strike-oriented transects.

The event-significant surfaces are near-time lines along dip, where their formation is controlled by specific events at the shoreline that change the pattern of sediment supply to the marine basin, in a manner that is independent of the offshore variations in subsidence or water depth. All four event-significant surfaces young basinward, and are characterized by the same degree of diachroneity which reflects the rates of offshore sediment transport (commonly in a range of 10^{-1}–10^2 m/s). Along strike, these surfaces are more diachronous as the timing of each associated 'shoreline event' may be offset by variations in the rates of sedimentation and/or subsidence along the shoreline. It can be noted that the strike variation in sedimentation rates affects only the timing of maximum regressive and maximum flooding surfaces, which are therefore more diachronous along strike than the two correlative conformities (Fig. 7.31).

The stage-significant surfaces display opposite trends relative to the event-significant ones, being potentially more diachronous along dip than along strike. Along dip, the degree of diachroneity of the stage-significant surfaces is high, and reflects the rates of shoreline shift. Along strike, these surfaces are still time-transgressive, due to variations in sedimentation and/or subsidence rates along the shoreline that may offset the timing of events that mark the onset and the end of each 'shoreline stage.' The strike diachroneity of subaerial unconformities and regressive surfaces of marine erosion depends only on the variations in subsidence rates along the shoreline, which may offset the timing of the onset-of-fall and end-of-fall events, and therefore the duration and timing of stages of forced regression. In addition to differential subsidence, the strike diachroneity of the transgressive ravinement surfaces is also affected by variations in sedimentation rates along the shoreline, which offset furthermore the timing of transgressive stages from one area to another. In fact, as illustrated in Fig. 7.31, the transgressive ravinement surfaces are the most diachronous of all sequence stratigraphic surfaces, with a degree of diachroneity that is potentially high along both dip and strike.

Similar to the 'stage-significant' sequence stratigraphic surfaces, the three within-trend facies contacts shown in Fig. 7.31 are also commonly characterized by high diachroneity rates, which correlate with the rates of shoreline shift, along dip, and with the variability recorded by the mechanisms controlling shoreline shifts (i.e., sedimentation and/or subsidence), along strike. Note that eustasy, which may also exert a control on shoreline shifts, is excluded from this discussion because its rates do not vary across the basin. Hence, eustasy does not contribute towards the generation of diachronous surfaces along strike. Among the three within-trend facies contacts shown in Fig. 7.31, the within-trend forced regressive surface is the only one whose strike diachroneity is independent of sedimentation rates, being solely controlled by differential subsidence along the shoreline. This is because forced regressions are driven by base-level fall at the shoreline, independent of sedimentation rates within the marine environment. It can be noted that, whether sedimentation is a factor or not, differential subsidence is always involved as a control on the strike diachroneity of any stratigraphic surface (Fig. 7.31). The reason for this is that the timing of *all* four events of the reference curve of base-level changes at the shoreline is in part dependent on subsidence, whereas sedimentation only affects the timing of surfaces defined relative to the transgressive-regressive curve (Fig. 7.31).

This discussion indicates that the timing of all stratigraphic surfaces is linked to the evolution of the *shoreline*, whose trajectories and changes thereof depend on the interplay of global (sea level) and local (subsidence, sedimentation) controls. Along dip-oriented transects, the degree of diachroneity of stratigraphic surfaces is a reflection of their 'event-significant' or 'stage-significant' nature; the former are associated with events marking a change in shoreline trajectory and sediment supply to the marine basin, being near-time lines along dip, whereas the latter correspond to actual stages of shoreline shift, amounting to a total diachroneity that measures the duration of these stages. Along strike-oriented transects, the role of local controls on the degree of diachroneity of all surfaces becomes more evident, as fluctuations in sedimentation and/or subsidence rates along the shoreline offset the timing of the four events of the reference curve of base-level changes that is valid for each individual dip-oriented transect. Events affected only by strike variations in subsidence rates (onset and end of forced regressions) are less offset temporally than the events that depend on fluctuations in sedimentation rates as well (onset and end of transgressions), which is why surfaces whose timing is influenced by sedimentation at the shoreline are generally more diachronous that surfaces whose timing is independent of sedimentation (Fig. 7.31).

8

Hierarchy of Sequences and Sequence Boundaries

INTRODUCTION

A sequence hierarchy assigns different orders to stratigraphic sequences and bounding surfaces based on their relative importance. The need for a hierarchy becomes apparent when one considers that there are numerous sequence boundaries in the rock record, often of different origins and relevant to a wide range of temporal and spatial scales, which need to be rationalized in terms of their relative nesting patterns. Within a hierarchical system, the most important sequence is recognized as of 'first-order' and may be subdivided into two or more 'second-order' sequences. In turn, a second-order sequence may be subdivided into two or more 'third-order' sequences, and so on (Fig. 8.1). The more important sequences are designated as 'high-order' (at the top of the hierarchy pyramid, i.e., of high rank), and generally have a low-frequency occurrence in the stratigraphic record. The less important sequences are of 'lower order' (i.e., lower rank, towards the base of the hierarchy pyramid) and are more frequent in the rock record (Figs. 8.1 and 8.2).

The 'high-' vs. 'low-order' terminology makes reference to the position of sequences within the pyramid (the higher the placement of a sequence within the pyramid, the higher the rank/order), and not to the numerical ordering ('first,' 'second,' etc.) (e.g., Catuneanu et al., 1997a; Holbrook, 2001; Fig. 8.1). One should note, however, that this is a 'grey area' in sequence stratigraphic nomenclature, as both approaches (e.g., 'high-order' referring to either large-scale or small-scale sequences) may be encountered in the literature. More importantly, however, is the fact that large sequences ('high-order,' according to the terminology adopted in this book) commonly consist of several smaller sequences; therefore, the big-picture stratigraphic architecture, which describes overall depositional

trends, is generally complicated, at more detailed scales of observation, by shorter-term changes in depositional trends that generate a framework of nested sequence stratigraphic surfaces of lower orders of cyclicity (Fig. 8.2). The lowest order/rank of cyclicity (e.g., 'third-order' in Fig. 8.2) describes the *actual changes* in depositional trends that can be observed in the rock record based on facies juxtaposition, affording a straight forward application of the facies-related criteria in Fig. 4.9 for the identification of sequence stratigraphic surfaces. The higher orders of cyclicity reflect *overall* depositional trends, at increasingly larger scales

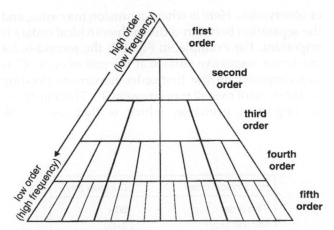

FIGURE 8.1 Diagrammatic representation of the concept of hierarchy. This pyramid approach assumes that the events leading to the formation of the most important sequences and bounding surfaces (first-order = highest within the pyramid ranking) occurred less frequently in the geological record relative to the events leading to the formation of lower-order sequence boundaries. The 'high-' vs. 'low-order' terminology makes reference to the position of sequences within the pyramid (the higher the placement of a sequence within the pyramid, the higher the rank/order), and not to the numerical ordering ('first,' 'second,' etc.).

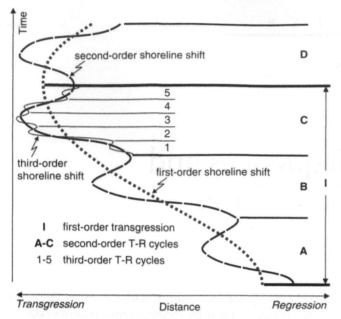

FIGURE 8.2 Superimposed patterns of shoreline shifts at different orders of cyclicity. The lowest order/rank of cyclicity ('third-order' in this example) reflects the true shift of the shoreline. The higher orders of cyclicity reflect overall trends, at increasingly larger scales of observation. Note that the second-order maximum regressive surface at the end of cycle 'C' is superimposed on the first-order maximum flooding surface (end of overall transgression 'I'). This illustrates an important principle, which is that lower-rank (higher-frequency) stratigraphic surfaces superimposed on higher-rank surfaces do not change the stratigraphic significance of the latter within the big-picture framework, as discussed in Chapter 6.

of observation. Here is where confusion may arise, and the separation between different hierarchical orders is important. For example, in Fig. 8.2, the second-order maximum regressive surface at the end of cycle 'C' is superimposed on the first-order maximum flooding surface (end of overall transgression 'I'). This illustrates an important principle, which is that lower-rank

Hierarchical order	Duration (My)
First order	50 +
Second order	3–50
Third order	0.5–3
Fourth order	0.08–0.5
Fifth order	0.03–0.08
Sixth order	0.01–0.03

FIGURE 8.3 Hierarchy system based on the duration of stratigraphic cycles (modified from Vail et al., 1991).

(higher-frequency) stratigraphic surfaces superimposed on higher-rank surfaces do not change the stratigraphic significance of the latter within the big-picture framework, as discussed in Chapter 6. Hence, once the objective of a sequence stratigraphic study is established (e.g., building a 'second-order' sequence stratigraphic framework of a sedimentary basin fill), the interpreter needs to wear the appropriate 'glasses' for the chosen resolution of stratigraphic modeling.

The critical element in developing a system of sequence hierarchy is the *set of criteria* that should be used to differentiate between the relative importance of sequences and bounding surfaces. Two different approaches are currently in use, based on the study of the Phanerozoic record: (1) a system based on boundary frequency (*cycle duration*), and (2) a system based on the magnitude of base-level changes which resulted in boundary formation (*independent of cycle duration*). The former system has historical priority, having being proposed at the dawn of seismic and sequence stratigraphy (Vail *et al.*, 1977; Fig. 3.2). This time-based hierarchy emphasizes eustasy as the main driving force behind stratigraphic cyclicity, which in turn is controlled by a combination of plate tectonic and orbital mechanisms (Fig. 3.2). As eustasy is global in nature, the philosophy behind this hierarchy system led to the construction of global cycle charts (Vail *et al.*, 1977), whose validity is currently under intense scrutiny (Miall, 1992, 1997).

It is noteworthy that the two hierarchical systems mentioned above are not mutually exclusive as the system based on boundary frequency also accounts for the concept of relative magnitude of base-level changes that resulted in the formation of different orders of sequences and sequence boundaries. It is thus implied that base-level changes associated with first-order cycles of supercontinent assembly and breakup, for example, are of much greater magnitude than the base-level changes generated by orbital forcing (Fig. 3.2). Consequently, the physical attributes and the facies shifts associated with sequence boundaries of different hierarchical orders are also expected to differ, even within the context of a hierarchy centered on the duration of stratigraphic cycles, as originally developed by Vail *et al.* (1977) and subsequently refined by Vail *et al.* (1991) (Figs. 3.2 and 8.3). In this case, the real issue becomes the documentation of how realistic it may be to consider that orderly patterns in a time framework can be established in the geological record. As discussed by Carter *et al.* (1991), a hierarchy of sequences does exist in the rock record; however, the distinctiveness of these sequences in terms of duration or periodicity is only approximate, at best.

The temporal and spatial scales of stratigraphic sequences do not define mutually exclusive ranges, and their internal nesting does not necessarily follow repetitive or ordered patterns. In spite of the lack of unequivocal temporal or spatial scales that could be associated with particular hierarchical orders, Carter et al. (1991) note that '… some orders of sequence do indeed embrace lower orders, e.g., the major first-order thermo-tectonic cycle that incorporates the complete sedimentary history of the Canterbury Basin …, which includes examples of second, third, fourth, and probably fifth-order sequences.'

One other important aspect to keep in mind is that the two alternative views on how the concept of stratigraphic hierarchy should be approached are essentially derived from Phanerozoic case studies. In spite of the limitations imposed by data availability and quality, the application of sequence stratigraphy to the Precambrian offers not only challenges, but also unique opportunities because its time window into the Earth's geological history is vastly wider relative to the duration of the Phanerozoic (Fig. 8.4). This greater time span allows one to observe Earth's processes at a broader scale, and thus gain an improved understanding of issues such as the mechanisms governing stratigraphic cyclicity and the variability thereof. The study of the Precambrian therefore provides better insights into some key issues of sequence stratigraphy, particularly the concept of hierarchy, for which the time span of the Phanerozoic is simply too short to afford any meaningful generalizations. Such new insights have been explored recently by Catuneanu et al. (2005) and Eriksson et al. (2005a, b).

	Precambrian	Phanerozoic
Time span	c. 88% of Earth's history	c. 12% of Earth's history
Facies preservation	Relatively poor (due to post-depositional tectonics, diagenesis, metamorphism)	Relatively good
Time control	Relatively poor (based on marker beds and lower resolution radiochronology)	Relatively good (marker beds, biostratigraphy, magnetostratigraphy, radiochronology)
Basin-forming mechanisms	Competing plume tectonics and plate tectonics, more erratic regime	Plate tectonics, more stable regime

FIGURE 8.4 Main contrasts between Precambrian and Phanerozoic, in terms of aspects relevant to sequence stratigraphy (from Catuneanu et al., 2005).

HIERARCHY SYSTEM BASED ON CYCLE DURATION (BOUNDARY FREQUENCY)

The hierarchy system based on cycle duration (Vail et al., 1977; Mitchum and Van Wagoner, 1991; Vail et al., 1991; Figs. 3.2 and 8.3) considers eustasy as the main driver behind sequence generation at any order of stratigraphic cyclicity. In turn, each order of cyclicity is assigned a dominant mechanism that controls eustatic changes over well-defined time scales (Fig. 3.2). Because eustasy is a global phenomenon, even though triggered potentially by localized to regional tectonism for cycles above or even within those in the Milankovitch band, sequences of different hierarchical orders are envisaged to have world-wide synchroneity. In other words, one single 'global-cycle chart' (Haq et al., 1987, 1988) would be representative to describe the stratigraphic cyclicity observed in the rock record of any basin around the world. These global cycles are regarded as 'geochronologic units defined by a single criterion – the global change in the relative position of sea level through time' (Vail et al., 1977). Sedimentary basins are, however, dominated by, and formed as a result of tectonic processes that generally operate on regional to continental scales, so stratigraphic cycles around the world are unlikely to be synchronous (see Miall, 2000, for a more detailed discussion of the global-cycle chart). In addition to the controversy brought about by the global-cycle chart, the application of the hierarchy system based on cycle duration poses two challenges to the practicing geologist: (1) from a practical perspective, time control is always required to designate and justify hierarchical orders; and (2) from a theoretical perspective, one must accept that the law of uniformitarianism applies undisputedly to the controls of stratigraphic cyclicity throughout the Earth's history.

The necessary time control to ensure, for example, that a 'third-order' sequence indeed falls within the 1–10 Ma duration bracket proposed by Vail et al. (1977) (Fig. 3.2), is often difficult to acquire even for Phanerozoic successions, and it becomes more and more unrealistic with increasing stratigraphic age. The reality is that in many cases we do not have the age data to know how much time is incorporated within a sequence, even within relatively young and well-explored sedimentary basins. In spite of this practical limitation, the hierarchy system based on cycle duration, which was originally developed based on Phanerozoic case studies (Vail et al., 1977), was eventually extrapolated to the Precambrian as well (Krapez, 1996, 1997). Krapez (1996) provides average durations for sequence orders as follows: fourth = 90–400 ka, third = 1–11 Ma, second = 22–45 Ma, and first = approximately 364 Ma.

Each of these orders of stratigraphic cyclicity is genetically related to particular tectonic (and to a much lesser extent climatic) controls whose periodicity is assumed to be more or less constant during geological time. For example, the 364 Ma duration of first-order cycles is calculated based on the assumption that nine *equal-period* global tectonic (Wilson) cycles of supercontinent assembly and breakup took place during the 3500–224 Ma interval (Krapez, 1993, 1996). The need for a hierarchy system based on cycle duration is based on the argument that 'There are no physical criteria with which to judge the rank of a sequence boundary. Therefore, sequence rank is assessed from interpretations of the origin of the strata contained between the key surfaces, and of the period of the processes that formed these strata' (Krapez, 1997).

More important than the current practical limitations related to the availability of time control, or the lack thereof, which may be resolved in the future as the resolution of dating techniques improves, is the fundamental question of whether or not the nature and periodicity of tectonic mechanisms controlling stratigraphic cyclicity were indeed constant throughout the Earth's history, as assumed by the proponents of time-based hierarchy systems. The Phanerozoic time window into the geological past is simply too small to provide an unequivocal answer to this question, and therefore the study of the Precambrian most likely holds the key to this debate. The hierarchy systems based on cycle duration are fundamentally built on the assumption that the controls on cyclicity at specific hierarchical orders are predictable, repetitive, and unchanged during the evolution of Earth. This implies that the controls on stratigraphic cyclicity are governed by the law of uniformitarianism throughout the Earth's history, allowing equal periodicity for stratigraphic cycles of the same hierarchical order, irrespective of age. However, recent work on Precambrian geology (Catuneanu and Eriksson, 1999; Eriksson *et al.*, 2004, 2005a, b) points to different conclusions, emphasizing that the tectonic mechanisms controlling the formation and evolution of sedimentary basins, for the greater part of geological time, were far more diverse and erratic in terms of origins and activity rates than originally inferred from the study of the Phanerozoic record (Fig. 8.4). Similar conclusions have been reached by studies of Milankovitch processes, which demonstrated that the periods of precession and obliquity have changed significantly with time due to continued evolution of the Earth-Moon system (Lambeck, 1980; Walker and Zahnle, 1986; Algeo and Wilkinson, 1988; Berger and Loutre, 1994). This means that *time* is largely irrelevant for designing a universally applicable hierarchy system. Instead, alternative criteria need to be identified for a more flexible conceptual framework that can be used irrespective of basin type and stratigraphic age.

The hierarchy system based on cycle duration is also problematic in the sense that the periodicity proposed for cycles above the Milankovitch band is highly speculative, and generally unsupported by empirical data. Statistical surveys suggest that there is no evidence for a time-based hierarchy in the rock record (e.g., Algeo and Wilkinson, 1988; Carter *et al.*, 1991; Drummond and Wilkinson, 1996), which is in fact apparent from the contradictions that exist between the supporters of this approach. For example, a second-order cycle is associated with a duration of 10–100 Ma by Vail *et al.* (1977), 3–50 Ma by Mitchum and Van Wagoner (1991), and 22–45 Ma by Krapez (1996); a third-order cycle has a periodicity of 1–10 Ma in the view of Vail *et al.* (1977), 0.5–3 Ma in the hierarchy of Mitchum and Van Wagoner (1991), and 1–11 Ma according to Krapez (1996); and so on. This problem is sourced from the fact that competing sequence-forming mechanisms (e.g., regional tectonism, global sea-level change, and orbital forcing), each operating over different time scales, may interplay to generate the preserved stratigraphic record. Hence, the periodicities that may describe the stratigraphic cyclicity of a particular sedimentary basin fill may be unique to that basin, rather than being consistent with a universal template of time-based hierarchy. Even for the better-documented cycles within the Milankovitch band, there is increasing evidence that non-Milankovitch processes, such as intraplate stress fluctuations, may operate within the same temporal range, thus distorting and obscuring the stratigraphic response to Milankovitch processes (Peper and Cloetingh, 1995). The direct relationship between orders of cyclicity, periodicities and triggering mechanisms is therefore problematic at best, being largely based on concepts of causation that have been shown to be unrealistic. It may be concluded that sequences of different hierarchical orders should not be expected to nest internally in a predictable and ordered pattern, but rather display a random character in terms of duration and spatial scales.

HIERARCHY SYSTEM BASED ON THE MAGNITUDE OF BASE-LEVEL CHANGES

A hierarchy system based on the magnitude of base-level changes that resulted in boundary formation provides a classification in which the order of a sequence depends on the physical attributes of its bounding surfaces, and is *independent of cycle duration*

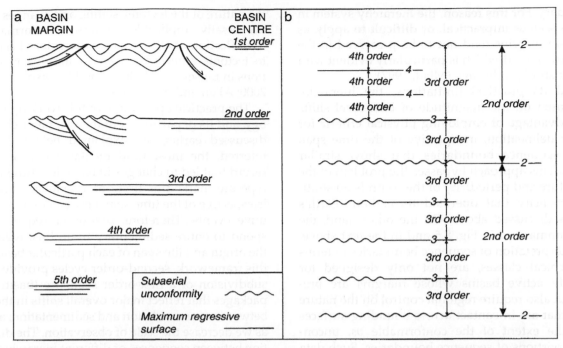

FIGURE 8.5 Hierarchy system based on the magnitude of base-level changes that resulted in the formation of bounding surfaces (modified from Embry, 1993, 1995). (a) Schematic depiction of the five orders of sequence boundaries determined from boundary characteristics which reflect base-level changes. (b) Principles of determining the order of a sequence: a sequence cannot contain within it a sequence boundary of equal or greater magnitude than its lowest magnitude boundary; the order of a sequence is equal to the order of its lowest magnitude boundary.

(Embry, 1995; Fig. 8.5). Six attributes have been chosen to establish the boundary classification: the areal extent over which the sequence boundary can be recognized; the areal extent of the unconformable portion of the boundary; the degree of deformation that strata underlying the unconformable portion of the boundary underwent during the boundary generation; the magnitude of the deepening of the sea and the flooding of the basin margin as represented by the nature and extent of the transgressive strata overlying the boundary; the degree of change of the sedimentary regime across the boundary; and the degree of change of the tectonic setting of the basin and surrounding areas across the boundary. Each of these attributes has to be assessed for each boundary, and then those boundaries with similar attributes (i.e., inferred to have been generated by similar base-level changes) are assigned to the same class of boundary. The various established classes are ordered in the hierarchy on the basis of the relative amounts of base-level shift that are inferred to be associated with each class. The class with the attributes that suggest the highest amount of base-level shift is placed at the top of the hierarchy pyramid (high rank), while the class with the inferred least amount of base-level shift is placed at the base

(low rank). Five different orders of sequence boundaries have been defined on the basis of these criteria (Embry, 1995; Fig. 8.5). With the establishment of a hierarchy system as described above, the recognition of an orderly succession of sequences is based on the principle that a sequence cannot contain within it a sequence boundary that has an equal or greater magnitude than the magnitude of its lowest magnitude boundary (Embry, 1995). This means, for example, that a second-order sequence cannot contain a first-order boundary within it, but can include third- and lower-order boundaries.

Two potential pitfalls with this classification scheme have been discussed by Miall (1997, p. 330–331). One is that it implies tectonic control in sequence generation. Sequences generated by glacio-eustasy, such as the Late Paleozoic cyclothems of North America and those of Late Cenozoic age on modern continental margins, would be first-order sequences in this classification on the basis of their areal distribution, but lower-order on the basis of the nature of their bounding surfaces. The second problem is that this classification requires good preservation of the basin margin in order to properly assess the areal extent of the unconformable portion of the boundary or the degree of deformation across the boundary. For this reason, the hierarchy system in Fig. 8.5 is seen as impractical, or difficult to apply, as many basin margins tend not to be preserved in the rock record, a trend which is particularly evident with increasing stratigraphic age (Miall, 1997).

Beyond its practical limitations, the hierarchy system based on the magnitude of base-level shifts has the advantage of employing physical criteria for boundary delineation, irrespective of the time span between sequence boundaries that show similar attributes. This approach bypasses the problem of the erratic nature and periodicity of the controls on stratigraphic cyclicity that operated throughout Earth's history, as discussed above. On the other hand, the criteria summarized in Fig. 8.5 and in the text above, for the interpretation of sequence boundaries in terms of hierarchical classes, are not only designed for tectonically active basins whose margins are preserved, but also require regional control on the nature of facies that are in contact across bounding surfaces and on the extent of the conformable vs. unconformable portions of sequence boundaries. Such data are commonly not available in the early stages of exploration of a sedimentary basin, and even when available, may not provide unequivocal solutions in the case of tectonically 'passive' basins, where stratigraphic cyclicity is primarily controlled by sea-level changes.

DISCUSSION

Because the sequence hierarchy systems that are currently in use present conceptual and/or practical limitations, the practitioner of sequence stratigraphy still faces the dilemma of how to deal with the variety of sequences that are more or less important relative to each other. No universally applicable hierarchy system that could work for all case studies, basin types, and stratigraphic ages, has been devised yet. As argued by Catuneanu (2003), the easiest solution to this problem is to deal with the issue of hierarchy on a case-by-case basis, assigning hierarchical orders to sequences and bounding surfaces based on their relative importance within each *individual basin*. This approach requires the 'first-order' partitioning of the stratigraphic record into successions that are the product of sedimentation in discrete sedimentary basins. In this context, the most important sequence boundaries in the stratigraphic record, designated as 'first-order,' are genetically related to shifts in the tectonic setting that led to changes in the type of sedimentary basins. At first-order level of stratigraphic cyclicity, the emphasis on

the nature of the tectonic setting and changes thereof is universally applicable, because the formation and classification of all types of sedimentary basins is tied to tectonic criteria (see sedimentary-basin classifications in Einsele, 1992; Busby and Ingersoll, 1995; Miall, 2000; Allen and Allen, 2005).

The practical application of this working methodology, and its departure from the two hierarchy systems discussed earlier, are explained below. As already inferred, the most fundamental events in the rock record that led to changes in tectonic setting and basin type are marked as first-order sequence boundaries, irrespective of the time span between two such consecutive events. Therefore, first-order sequences correspond to entire sedimentary basin fills, regardless of the origin and life span of each particular basin. Within this framework, second-order cycles provide the basic subdivision of a first-order sequence (basin fill) into packages that reflect major overall shifts in the balance between accommodation and sedimentation; and so on, as we decrease the scale of observation. The discrimination between sequences of different hierarchical orders is based on the relative magnitudes of facies shifts across bounding surfaces and/or the magnitudes of the transgressive-regressive cycles recorded by interior seaways. These criteria are not linked to any particular sequence-forming mechanism (e.g., tectonism), and the preservation of the basin margin is not a requirement for separating between sequences of different magnitudes or hierarchical orders. In this approach, the relative importance of sequences and events leading to the generation of sequence boundaries is assessed primarily on the basis of *facies observations*, and no interpretation of the allogenic controls is necessary to separate between sequences of different rank. As such, lower-order cycles that describe the internal architecture of any larger sequence may be controlled by any sequence-forming mechanism, from tectonism to glacio-eustasy. This flexible approach may be applied to any sedimentary basin, regardless of stratigraphic age, availability of time control, or degree of preservation of the basin margins. Sequences delineated within a particular sedimentary basin are not expected to correlate to other first- and lower-order sequences of other sedimentary basins, which most likely have different timing and duration. This method may prove to be more realistic considering the fact that each basin is unique in terms of formation, evolution, and history of base-level changes.

It can be noted that the methodology proposed herein diverges from both hierarchy systems discussed in the previous sections of this chapter. In contrast to the hierarchy system based on cycle duration, this approach is independent of the duration of stratigraphic cycles, and implicitly of the recurrence interval of sequence boundaries. This also implies that no inference is made regarding the allogenic controls that may operate at any particular order of cyclicity (e.g., in contrast with Fig. 3.2). Compared to Embry's (1995) method of sequence delineation, this approach is simpler in the sense that no boundary attributes that make reference to the dominance of tectonic controls (e.g., 'the degree of deformation that strata underlying the unconformable portion of the boundary underwent during the boundary generation') are taken into account. Instead, facies observations devoid of allogenic-control interpretations are preferred to assess the relative importance of sequences and bounding surfaces. Reference to 'the areal extent of the unconformable portion of the boundary' (Embry, 1995), which requires a good preservation of the basin margins, is also avoided. Besides the convenience of providing a simpler checklist of boundary attributes that one needs to deal with, the real advantage of the proposed method is that it can be applied to any sedimentary basin, irrespective of origin, dominant controls on depositional trends, age, and degree of preservation. Nevertheless, both the method proposed herein and Embry's (1995) approach imply that the magnitude of base-level changes increases with the hierarchical rank of the studied bounding surface, and that temporal and spatial scales are irrelevant in the process of sequence ranking within the hierarchy pyramid.

These conclusions are supported by statistical surveys of the duration and thickness of stratigraphic sequences, which demonstrated there is no evidence for a hierarchy in the rock record that can be linked to the periodicity of recurrence of the same-order bounding surfaces (Algeo and Wilkinson, 1988; Carter et al., 1991; Drummond and Wilkinson, 1996). As pointed out by Drummond and Wilkinson (1996), 'discrimination of stratigraphic hierarchies and their designation as nth-order cycles [based on cycle duration] may constitute little more than the arbitrary subdivision of an uninterrupted stratigraphic continuum.' The link between hierarchical orders and temporal durations is artificial and meaningless to a large extent, as multiple sequence-generating mechanisms that do not readily fall into simple temporal classifications may interact and contribute to the architecture of sequences in the rock record (Miall, 1997). The combination of such independent controls often results in sequences whose durations and thicknesses have log-normal distributions that lack significant modes (Drummond and Wilkinson, 1996). Similar conclusions are reflected by the work of Algeo and Wilkinson (1988), as well as that of Peper and Cloetingh (1995), who demonstrated that calculated periodicities of stratigraphic cycles may

have random distributions relative to any given sequence-forming mechanism.

An added bonus of this case-by-case basin approach is the simplification of terminology because modifiers such as 'first-order,' 'second-order,' etc., have straightforward meanings, reflecting relative importance independent of time connotations. In contrast, the time-based hierarchy systems are permeated by unnecessary and sometimes conflicting jargon. For example, the 'supersequence' of Krapez (1996) is referred to as a 'sequence' by Vail et al. (1977); the 'sequence' of Krapez (1996) corresponds to the 'mesothem' of Ramsbottom (1979), or to the 'megacyclothem' of Heckel (1986); the 'paracycle' of Krapez (1996) is equivalent to the 'major cycle' of Heckel (1986), etc. Further, the term 'paracycle' (and its corresponding 'parasequence') is particularly confusing, because parasequences (as bounded by 'flooding surfaces') may not even technically be a type of sequence, depending on what the flooding surface actually is (see discussions in Chapters 4 and 6). Beyond the terminology issue, which may be trivial to some extent, the real problem consists of the fact that each of these terms (e.g., megasequence, supersequence, parasequence, etc.) is associated with a specific time connotation (cycle duration), which requires a time control that is unavailable in most Precambrian, and many Phanerozoic, sedimentary basins.

The Western Canada Sedimentary Basin provides an example of how the proposed hierarchy system works in the case of a Phanerozoic succession, when basin-forming mechanisms have been generally more stable in terms of processes and rates (Fig. 8.4). This basin evolved as a divergent continental margin during the Paleozoic—Middle Mesozoic interval, and as a retroarc foreland system from the Middle Mesozoic to the Early Tertiary (Ricketts, 1989). The change in tectonic setting (type of basin) occurred about 180 Ma ago, with the onset of subduction and compression along the western plate boundary of North America. Hence, the 180 Ma event marks the formation of a first-order sequence boundary ('forebulge unconformity,' and its correlative conformity; Catuneanu, 2004a) that divides the fill of the Western Canada Sedimentary Basin into two first-order sequences. In turn, the first-order foreland sequence (Middle Mesozoic—Early Tertiary) may be subdivided into several second-order sequences that reflect major cycles of basin-scale transgressions and regressions of the Western Interior seaway. Each of these second-order sequences is punctuated by a set of shorter-term 'third-order' transgressive-regressive cycles, and so on. The origin of these stratigraphic cycles has been attributed to either eustatic (e.g., Plint, 1991) or tectonic controls (e.g., Catuneanu et al., 2000), and, even in the case of the latter, no deformation has been recorded across sequence boundaries due to the nature of the dominant flexural tectonics. It is also important to note that the syn-sedimentary basin margin of the Western Canada foreland system is not preserved to date, even though this is a relatively young sedimentary basin, due to a combination of orogenic front progradation (approximately 165 km during the Late Cretaceous–Paleocene stage of basin evolution; Price, 1994), which cannibalized the syn-orogenic basin margin, and post-orogenic uplift and erosion that removed at least 3 km of stratigraphic section along the basin margin during the post-Paleocene isostatic rebound (Issler et al., 1999; Khidir and Catuneanu, 2005). Under these circumstances, the application of Embry's (1995) criteria to assess the boundary rank is difficult, as facing both limitations discussed by Miall (1997) and presented earlier in this chapter. Also, the orders of stratigraphic cycles are not resolved on the basis of their temporal durations, which may vary greatly between sequences of the same rank, but rather reflect associated changes in the magnitude of facies shifts that can be observed in the field (e.g., Catuneanu et al., 1997a).

An increasing number of case studies of Precambrian sequence stratigraphy have become available in recent years, affording a better understanding of the complexity of the controlling factors on sedimentation and stratigraphic cyclicity at scales larger than originally made possible by the study of the Phanerozoic record. More meaningful generalizations can now be formulated as a result of this research (see Eriksson et al., 2005a, b for pertinent syntheses). The stratigraphy and tectonic evolution of the Kaapvaal craton of South Africa during the Late Archean – Early Proterozoic interval provide a relevant example for this discussion. This approximately 1 Ga record of Kaapvaal evolution was marked by a combination of plate tectonic and plume tectonic regimes, whose relative importance determined the type of tectonic setting and sedimentary basin established at any given time. The shifting balance between these two allogenic controls on accommodation resulted in a succession of discrete basins, starting with the Witwatersrand (accommodation provided by subduction-related tectonic loading), followed by the Ventersdorp (accommodation generated by thermal uplift-induced extensional subsidence), and lastly by the Transvaal (accommodation created by extensional and subsequent thermal subsidence). The end of the Transvaal cycle was marked by a relatively short-lived plume tectonics event, which led to the emplacement of the Bushveld igneous complex. The sedimentary fill of these three basins,

each of which is genetically related to different tectonic settings, represents an unconformity-bounded first-order depositional sequence.

The temporal duration of the Kaapvaal first-order cycles varied greatly with the type of tectonic setting, from approximately 5 Ma in the case of the plume-related Ventersdorp thermal cycle, to >600 Ma in the case of the extensional Transvaal Basin. This first-order cyclicity was independent of the Wilson-type cycles of supercontinent assembly and breakup, being rather a reflection of the interplay between plate tectonics (e.g., extension or subduction-related tectonic loading) and plume tectonics. It is noteworthy that the first-order cycles controlled by plate tectonics lasted about two orders of magnitude longer (approximately 10^2 Ma) relative to the plume tectonics cycles (approximately 10^0 Ma). Each of the Kaapvaal first-order cycles are subdivided into second-order cycles, whose temporal duration also varies greatly, from approximately 1 Ma in the case of the Ventersdorp plume tectonics-controlled basin to approximately 100 Ma in the case of the plate tectonics-controlled basins.

The case study of the Kaapvaal craton suggests that time, and implicitly the frequency of occurrence of same-order sequence boundaries in the rock record, are irrelevant to the hierarchy of stratigraphic cycles. This is a consequence of the fact that processes controlling the formation and evolution of sedimentary basins in the geological past were far more erratic than originally inferred from the study of the Phanerozoic record (Fig. 8.4). The extrapolation of principles developed from Phanerozoic case studies (e.g., Vail et al., 1977; Fig. 3.2) to the entire geological record based on the law of uniformitarianism is therefore inadequate for providing a unified approach to the concept of sequence hierarchy. This conclusion reinforces the idea that the classification of sequences and bounding surfaces should be approached on a case-by-case basis, starting from the premise that each sedimentary basin fill (i.e., the product of sedimentation within a particular tectonic setting/type of basin) corresponds to a first-order stratigraphic sequence. In turn, first-order basin fill successions are subdivided into second- and lower-order sequences as a function of the shifts in the balance between accommodation and sedimentation at various scales of observation, irrespective of time spans and the nature of the allogenic mechanisms that controlled the internal architecture of the basin fills. Such sequences, which reflect at least in part the influence of sub-global controls, are not expected to correlate to other first- and lower-order sequences of other basins, which most likely have different timing and duration.

9

Discussion and Conclusions

FUNDAMENTAL PRINCIPLES

Scope and Applications

Sequence stratigraphy studies the change in depositional trends in response to the interplay of accommodation and sedimentation, from the scale of individual depositional systems to entire sedimentary basin fills. As accommodation is controlled by allogenic mechanisms that operate at basinal to global scales, the change in depositional trends is commonly synchronized amongst all environments established within a basin, thus providing the basis for the definition of systems tracts and the development of regional models of facies predictability.

All standard sequence stratigraphic models (Fig. 1.7) account for the presence of an interior seaway within the basin under analysis and are centered on the direction and types of shoreline shifts, which control the timing of all systems tracts and sequence stratigraphic surfaces. Under these circumstances, the presence of a paleoshoreline within the basin under analysis justifies a systems-tract nomenclature that makes specific reference to transgressions and regressions. In overfilled basins, however, dominated by nonmarine sedimentation, the definition of systems tracts is based on changes in fluvial accommodation ('low' vs. 'high'), as inferred from the shifting balance between channel and overbank fluvial architectural elements (see discussions in Chapters 5 and 6). A preliminary assessment of the 'big-picture' makeup of the basin fill is therefore important before deciding what sequence stratigraphic approach is most appropriate for that particular basin. The nature of processes and associated facies that dominate a sedimentary basin depends on both sediment supply and the allogenic mechanisms controlling accommodation. The latter vary with stratigraphic age, in terms of origins and rates, and the contrasts between Precambrian and Phanerozoic are discussed in more detail in Chapter 8, as well as in a subsequent section of this chapter.

The applications of sequence stratigraphy vary with the scale of observation, from resolving details of petroleum reservoir compartmentalization and connectivity within the confines of individual depositional systems, to building basin-scale stratigraphic frameworks and reconstructing the controls that governed the evolution of sedimentary basins in the geological past. Sequence stratigraphy is now routinely applied for *reservoir studies* in stages of exploration and field development (e.g., Mutti, 1992; Ainsworth *et al.*, 1999; Lang *et al.*, 2001; Ainsworth, 2005); *modern analog studies*, in order to improve our understanding of depositional processes and facies relationships (e.g., Lang *et al.*, 2005); *basin-scale studies* of regional stratal stacking patterns and facies relationships (e.g., Long and Norford, 1997; Gibson-Poole *et al.*, 2002); *quantitative stratigraphic modeling*, including computer simulations of stratal development (e.g., Flemings and Jordan, 1989; Johnson and Beaumont, 1995; Ainsworth *et al.*, 1999, 2000; Harbaugh *et al.*, 1999; Changsong *et al.*, 2001); and for gaining insights into the stratigraphic architecture of sedimentary basins placed in different *tectonic settings* (e.g., intracratonic basins: Jackson *et al.*, 1990; Lindsay *et al.*, 1993; Vecsei and Duringer, 2003; foreland systems: Devlin *et al.*, 1993; Posamentier and Allen, 1993; Hart and Plint, 1993; Plint *et al.*, 1993; Catuneanu *et al.*, 1997b, 1999, 2000; Donaldson *et al.*, 1998, 1999; Giles *et al.*, 1999; Miller and Eriksson, 2000; divergent continental margins: Posamentier *et al.*, 1988; Simpson and Eriksson, 1990; Boyd *et al.*, 1993; Donovan, 1993; rifts and other active extensional basins: Embry, 1993; 1995; Gawthorpe *et al.*, 1994; Davies and Gibling, 2003; pull-apart basins: Ryang and Chough, 1997; etc.).

Seismic stratigraphy, which is the precursor of modern sequence stratigraphy, was specially designed to facilitate the exploration for hydrocarbons: 'Seismic Stratigraphy—Applications to Hydrocarbon Exploration' (Payton, 1977). The methods of seismic and sequence stratigraphy are now increasingly popular, and routinely employed as part of the exploration strategies for other natural commodities as well, including coal and mineral resources. What used to be an exclusive asset of the petroleum industry, has been proven to have value for all types of exploration, and the mining industries too are now making use of the benefits of the sequence stratigraphic process-based (genetic) approach. As the resolution of sequence stratigraphic modeling has increased in recent years, in parallel with technological advances in the fields of subsurface data acquisition and processing, sequence stratigraphy has also become increasingly involved in the process of production optimization, following the stages of exploration. The application of sequence stratigraphic work has therefore expanded significantly to encompass all stages of economic basin analysis, from exploration in frontier areas to production in 'mature' basins. The exploration facet of sequence stratigraphy enables predictions of the distribution of coal, placers, and petroleum source rocks, seals, and reservoir facies within the basin. In the production stage, sequence stratigraphy is used to decipher the high-resolution internal architecture of 'pay-zones,' providing insights into the fluid migration pathways within petroleum reservoirs (e.g., Ainsworth, 2005; Pyrcz *et al.*, 2005) or into the geometry and stacking patterns of coal seams (e.g., Banerjee *et al.*, 1996; Bohacs and Suter, 1997) and mineral placers (e.g., Catuneanu and Biddulph, 2001).

In addition to the traditional outcrop and subsurface methods of stratigraphic analysis, numerical simulations of facies development play an increasingly important role in constructing and testing sequence stratigraphic models in both siliciclastic (e.g., Ainsworth, 2005; Pyrcz *et al.*, 2005) and carbonate (e.g., Schlager, 2005) successions. Quantitative modeling is now routinely involved in sequence stratigraphic research, with applications ranging from simulations of regional-scale stratigraphic architecture and basin development to detailed 'pay-zone' studies. The skills required for a complete sequence stratigraphic study have therefore diversified tremendously in recent years, and a team effort combining a wide range of specialties is the preferred approach to this type of work.

'Integration' is an important keyword in sequence stratigraphy, and as suggested throughout the book, insights from outcrop, core, well-log and seismic data should, ideally, be combined for comprehensive and reliable studies. Each type of data contributes with particular insights to the final interpretation (Fig. 2.71). The lack of data is a limiting factor, and hampers the resolution and reliability of the sequence stratigraphic model. For example, information from scattered outcrops should be integrated into a coherent model by using the continuous subsurface imaging provided by seismic data, wherever possible. On the other hand, the use of seismic data without calibration with core or well logs can lead to false interpretations (e.g., the interpretation of depositional systems in Fig. 2.43 would have been impossible without the mutual calibration with well-log data). Similarly, the lack of calibration of well logs with rock data (core or nearby outcrops), and their correlation outside of the context provided by seismic imaging, can also lead to erroneous interpretations (e.g., see the equivocal well-log signatures in Figs. 2.31–2.34 and 2.36). The integration of all these data sets is therefore the key to the most effective and reliable application of the sequence stratigraphic method. A more detailed discussion of the workflow of sequence stratigraphic analysis is presented in Chapter 2.

The successful application of the sequence stratigraphic method requires a three-dimensional modeling of a sedimentary succession by integrating stratigraphic observations in section view (e.g., stratal terminations and stacking patterns) with geomorphological features that can be observed in plan view (e.g., Figs. 2.48, 2.57, and 2.67). Such modeling is made possible by 3D seismic surveys, which afford the preliminary assessment of the 'big-picture' stratigraphic framework of the basin under analysis (see Chapter 2 for further details on methodology and practical workflow). Following the initial 'big-picture' analysis, smaller-scale areas of interest can be defined and zoomed in on for more detailed studies. The increased resolution of the modern methods of subsurface data acquisition and processing affords not only a control of the geometry of discrete depositional elements, but also insights into process sedimentology (e.g., Figs. 2.58 and 2.59). In fact, owing to its genetic approach, sequence stratigraphy is inseparable from process sedimentology (Fig. 1.2). For example, the application of facies-related criteria required for the identification of sequence stratigraphic surfaces (Fig. 4.9) is impossible without a thorough understanding of the processes involved in the formation of sedimentary facies and of the conformable or unconformable contacts that separate them. In addition, the application of the sequence stratigraphic method also requires integration of other disciplines including classical stratigraphy, geophysics, geomorphology, isotope geochemistry and basin analysis (Fig. 1.1).

The Importance of Shoreline Shifts

The shoreline, with its transgressive and regressive shifts, represents the central element around which all standard sequence stratigraphic concepts have been defined. The reference curve of base-level changes (Figs. 1.7 and 4.7) describes changes in accommodation *at the shoreline*, wherever the shoreline happens to be in the basin at any given time (Figs. 7.10 and 7.16; see discussion in Chapter 7). The interplay of eustasy, subsidence and sedimentation at the shoreline controls the timing of the four main events of the reference base-level curve: the onset of forced regression (onset of base-level fall at the shoreline), the onset of lowstand normal regression (onset of base-level rise at the shoreline), the end of regression, and the end of transgression (Fig. 4.7). Each of these four events marks a change in the stratigraphic architecture of the basin fill, as each type of shoreline shift generates specific stratal stacking patterns and facies relationships. Consequently, each of the four events corresponds to the formation of a quasi-isochronous sequence stratigraphic surface in the rock record, along dip-oriented transects, that separates packages of strata (systems tracts) with distinct stratigraphic signatures. In addition to the four *event-significant* contacts, three more sequence stratigraphic surfaces form *during* specific stages of shoreline shift, namely forced regressions and transgressions (Fig. 4.7). Full details regarding the diagnostic features of all seven sequence stratigraphic surfaces and their temporal attributes are provided in Chapters 4 and 7, respectively.

It is apparent that the shoreline shifts (normal regressions *vs.* forced regressions *vs.* transgressions; see Chapter 3 for definitions) represent the main driving force behind the sequence stratigraphic framework of a basin fill, by controlling sediment supply to the marine portion of the basin, overall grading and stacking patterns, and the timing of all surfaces and systems tracts. This is why systems tract boundaries are quasi-isochronous along dip, as there is only one moment in time when the shoreline changes its direction and/or type of shift along each particular dip-oriented transect. A low diachroneity that reflects the rates of sediment *transport* (and not the sedimentation rates) is still recorded (see Chapter 7 for details). Fluctuations in the *sedimentation* rates away from the shoreline are only important in controlling the thickness of systems tracts, without influencing the timing of their boundaries. Systems tract boundaries have a potentially much higher diachroneity along strike, because the timing of the four main events of the base-level cycle (Fig. 4.7) may be offset by fluctuations in the subsidence and sedimentation rates along the shoreline.

Offshore, subsidence and sedimentation rates may vary both along dip and strike, modifying the bathymetric conditions (e.g., the timing of changes from water deepening to shallowing, and *vice versa*) at any location

within the marine basin. Similarly, variations in subsidence rates modify the timing of relative sea-level changes from one area to another within the marine basin. These variations do not interfere with the timing of systems tract boundaries. For example, the two correlative conformities mark changes in relative sea level *at the shoreline*, but may form during rising relative sea level offshore (Fig. 7.28). Similarly, the maximum regressive and maximum flooding surfaces mark changes in bathymetric trends near the shoreline, but may form under deepening-water conditions offshore (Fig. 7.28). Surfaces of shallowest and deepest water are lithologically undeterminable, and form *within* the regressive systems tract (Fig. 7.28). Such surfaces may be mapped based on techniques such as trace fossil and benthic foraminiferal paleobathymetry, and meet the systems tract boundaries at the shoreline. It is therefore important to separate stratal stacking patterns from paleobathymetry, and to describe the lithological changes in terms of *observed* grading trends rather than *inferred* water-depth changes (see Chapter 7 for a full discussion).

Theory *vs.* Reality in Sequence Stratigraphy

Sequence stratigraphic models idealize reality in the sense that they provide simplified two- or three-dimensional representations of how the architecture of facies and stratigraphic surfaces is expected to be in the field. The central theme of all standard models is that the predictable stacking pattern of systems tracts and stratigraphic surfaces is mainly controlled by the interplay of base-level changes and sedimentation *at the shoreline*. This interplay controls the direction and/or the type of shoreline shifts, as well as the timing of all systems tract and sequence boundaries. Under this assumption, the subaerial unconformity is the time equivalent of the falling-stage systems tract, the maximum flooding surface has a predictable position above the subaerial unconformity, and so on (Figs. 4.6, 4.7 and 5.4–5.6). Although these expected relationships are valid in most cases, possible deviations from the model predictions should be carefully evaluated. For example, the influence of base-level changes at the shoreline on fluvial processes only extends for a limited distance upstream (Fig. 7.29). The extent of the base-level control depends on the balance between the magnitude of base-level changes, climatic influences, and source area tectonism (Shanley and McCabe, 1994). There are instances when the role of climate is so dominant that processes of fluvial aggradation and incision are primarily controlled by changes in the balance between river discharge and sediment load, with a timing that is offset relative to the base-level fluctuations at the shoreline (Blum, 1994). The resulting fluvial sequences and subaerial unconformities will therefore not fit the position and timing predicted by standard sequence models. There are also cases where a subaerial unconformity forms during transgression, in relation to processes of coastal erosion (Leckie, 1994; Fig. 3.20).

It is interesting to note that such exceptions from the predictions of the standard sequence stratigraphic models affect mostly the nonmarine portions of sedimentary basins, while the depositional trends in marine strata tend to be more consistent with the predicted sequence stratigraphic frameworks. For this reason, a good control on the changes in depositional trends and the timing thereof is desirable on both sides of the paleoshoreline. The correlation between age-equivalent marine and nonmarine strata provides important clues to assess the origin and timing of sequences and sequence boundaries that develop in fluvial successions. For example, subaerial unconformities formed during transgressions (Fig. 3.20) may be identified as such by their temporal correlation with retrograding shallow-marine facies that onlap the transgressive wave-ravinement surface. A similar analysis is needed to fully understand the nature of the subaerial unconformities that form in relation to interglacial climatic stages, during times of ice melting (increased river discharge) and base-level rise (Blum, 1994).

One other common problem in the real world is the possible lack of preservation of systems tracts, or of portions thereof. In this case, stratigraphic surfaces that are normally expected to be separated by strata may be superimposed. Examples include transgressive ravinement surfaces that rework subaerial unconformities, regressive surfaces of marine erosion that rework the basal surface of forced regression, maximum flooding surfaces that rework maximum regressive surfaces, and subaerial unconformities that rework the underlying maximum flooding surfaces. In such situations, the observed surface should be labeled using *the name of the younger surface*, as the latter overprints the attributes of the original contact.

The Importance of the Tectonic Setting

The diversity of sequence models that are currently in use (Figs. 1.6 and 1.7) may in part be attributed to the fact that their proponents draw their own research experience from different types of sedimentary basins. Hence, each model is designed to fit the field observations from a particular tectonic setting. For example, the models of Posamentier *et al.* (1988) and Galloway (1989) describe divergent continental margins; Van Wagoner and Bertram (1995), as well as Plint and Nummedal (2000) refer to foreland basin deposits; whereas Embry

(1995) proposed a transgressive–regressive (T–R) sequence model based on the study of the Sverdrup rift basin. Each of these tectonic settings is unique in terms of tectonics, subsidence rates, sediment flux, physiography, and topographic gradients within the basin and along the basin margins, and as a result differences in stratal architecture and the development and preservation of particular depositional systems are expected as well. As pointed out by Diessel *et al.* (2000) and Davies and Gibling (2003), the subsidence history of any sedimentary basin controls the distribution of accommodation in time and space, and as a result, the stratal architecture of each basin fill reflects the unique regional and temporal variations in subsidence rates that characterize different types of tectonic settings.

A summary of the basic contrasts between low- and high-gradient settings in terms of the resulting stratigraphic architecture of the basin fill is presented in Fig. 9.1. Notably, low-gradient ('shelf-type') settings, characterized by a relatively flat topography at the shoreline, have a much better potential for accumulating fluvial lowstand deposits over much of the extent of the subaerial unconformity, and also a much better potential for the accumulation and preservation of estuarine facies. In contrast, high-gradient ('ramp-type') settings, with a steep topography at the shoreline, are unlikely to preserve either fluvial lowstand or estuarine deposits. Topography is not, of course, the only control on the accumulation and preservation of lowstand fluvial and transgressive estuarine deposits, as favorable accommodation conditions must be met as well. A common theme emerges, however, which is that the accumulation and preservation of lowstand fluvial and transgressive estuarine deposits are favored by similar sets of conditions, which means that the presence of estuarine facies in the rock record is likely to indicate the presence of underlying fluvial lowstand deposits as well. The lack of estuarine and underlying fluvial lowstand deposits in ramp settings may explain why the T–R sequence model works so well in rift and other fault-bounded basins, where the transgressive ravinement surfaces commonly rework the subaerial unconformities. This may not necessarily be the case in shelf-type settings such as continental shelves, filled foredeeps, or intracratonic basins, where thick fluvial lowstand and estuarine deposits are often preserved.

The variability imposed by the existing range of tectonic settings to the stratigraphic model indicates, once again, that no single sequence stratigraphic template ('model') will fit the entire range of case studies and geological circumstances. The interpreter needs to have the ability to adapt to local conditions, and use the set of fundamental core concepts as a starting point for building a unique model for a particular basin.

Basin types / Shoreline shifts	Low-gradient ('shelf') settings (continental shelves, filled foreland basins, intracratonic basins)	High-gradient ('ramp') settings (continental slopes, underfilled forelands, rift and strike-slip basins)
Transgressions	Estuaries are likely to form. The preservation of estuarine facies is a function of the rates of base-level rise and wind/wave energy.	Estuaries are unlikely to form, due to the steep topography, higher fluvial energy, wave erosion, and slope instability.
Normal regressions	Deltas have diagnostic topsets, as a result of aggradation in the delta plains.	
Normal regressions	Fluvial aggradation extends over a relatively large distance upstream (inconspicuous onlap onto the subaerial unconformity at lowstand).	Fluvial aggradation is restricted to a relatively small area adjacent to the shoreline (pronounced onlap). Fluvial strata have low preservation potential.
Forced regressions	Deltas have diagnostic offlapping geometries (delta plain erosion or bypass).	
Forced regressions	The regressive surface of marine erosion forms in the lower shoreface in wave-dominated settings.	No erosion in the lower shoreface, as the sea floor is already steeper then the shoreface equilibrium profile.

FIGURE 9.1 Contrast between the distinctive features of low- and high-gradient settings, in terms of processes and products of transgressions, normal regressions, and forced regressions (modified from Catuneanu, 2002). Many of these differences are due to contrasts in basin-forming mechanisms, subsidence patterns and related basin physiography, which in turn vary with the tectonic setting. Therefore, the sequence stratigraphic model needs to be adapted as a function of tectonic setting, which is an area where more work and documentation are required.

As pointed out by Posamentier and Allen (1999), 'what does not change from setting to setting is the set of first principles upon which sequence stratigraphy is based. It is this set of first principles that constitutes the backbone of the sequence-stratigraphic approach.' These 'first principles' are not unique to sequence stratigraphy, but are rather cornerstone natural laws that govern the broad field of sedimentary geology (Fig. 2.1).

Uses and Abuses in Sequence Stratigraphy

The greatest danger in sequence stratigraphy is dogma. An interpreter may easily fall into the trap of trying to fit observations into rigid templates provided by various standard models. Such attempts may stem from convenience, lack of adequate knowledge, or the desire to demonstrate the universal applicability of a particular model. A fresh look that acknowledges the uniqueness of a case study is more valuable than an interpretation that gives too much credit to entrenched ideas. In this context, one can also say that *data are as important as an open mind*, because observing data is of little use if they are forced to fit into inadequate interpretation templates. In fact, there is no right or wrong between the various approaches summarized in Fig. 1.7. Some approaches may be more adequate than others, depending on geological circumstances and available data sets, so the choice of sequence stratigraphic model should be decided on a case-by-case basis. 'Flexibility' is therefore another important keyword in sequence stratigraphy. Such quality requires not only willingness to adapt to new circumstances and to accept new ideas, but also a sound understanding of the *processes* that led to the formation of the preserved

rock record. As argued above in this chapter, process sedimentology is inseparable from sequence stratigraphy, and attempts to apply sequence stratigraphy as a stand-alone method, independent of insights afforded by process sedimentology, can only limit the reliability and the depth of the sequence stratigraphic interpretation. A *process-based* approach to sequence stratigraphy, rather than a dogmatic application of entrenched ideas that stem from *model-driven* or conventional thinking, is therefore recommended.

PRECAMBRIAN VS. PHANEROZOIC SEQUENCE STRATIGRAPHY

All 'standard' ideas and concepts of sequence stratigraphy have been developed from the study of the Phanerozoic record, which accounts for only about 12% of Earth's history. As demonstrated in recent publications (e.g., Eriksson et al., 2005a, b), the relatively narrow window into the geological past that is offered by the Phanerozoic record is insufficient to allow for meaningful generalizations, and a thorough understanding, of the mechanisms controlling accommodation and stratigraphic cyclicity. This has profound implications on the selection of criteria that should be used to classify stratigraphic sequences, and helps to resolve existing debates generated from the study of the Phanerozoic record.

The nature of basin-forming mechanisms has changed during Earth's evolution, from competing plume and plate tectonics in the Precambrian to a more stable plate-tectonic regime in the Phanerozoic (Fig. 8.4; Eriksson and Catuneanu, 2004b; Eriksson *et al.*, 2005a, b). Basins related to plume-controlled first-order cycles (i.e., plume tectonics) are prone to a dominantly nonmarine sedimentation regime because the net amount of thermal uplift generally exceeds the amount of subsidence created *via* extension above the ascending plume. As plume tectonics was much more prevalent in the Precambrian relative to the Phanerozoic, the low- and high-accommodation systems tracts seem to be more commonly applicable with increasing stratigraphic age. In contrast, basins related to plate tectonic activity are dominated by subsidence, and so they are prone to be transgressed by interior seaways. Classical sequence stratigraphy may thus be applied to such settings, where falling-stage, lowstand, transgressive, and highstand systems tracts may be recognized in relation to particular stages of shoreline shifts. Even such subsidence-dominated basins, however, may reach an overfilled state under high sediment-supply conditions, in which case the recognition of fully fluvial systems tracts (low *vs.* high accommodation) becomes the only option for the sequence-stratigraphic approach. Case studies of such overfilled plate-tectonic-related basins have been documented for both Precambrian and Phanerozoic successions (e.g., Boyd *et al.*, 1999; Zaitlin *et al.*, 2000, 2002; Wadsworth *et al.*, 2002, 2003; Leckie and Boyd, 2003; Eriksson and Catuneanu, 2004a; Ramaekers and Catuneanu, 2004).

Arguably the most important contribution of Precambrian research to sequence stratigraphy is the better understanding of the mechanisms controlling stratigraphic cyclicity in the rock record, and hence of the criteria that should be employed in a system of sequence stratigraphic hierarchy. There is increasing evidence that the tectonic regimes which controlled the formation and evolution of sedimentary basins in the more distant geological past were much more erratic in terms of origin and rates than formerly inferred solely from the study of the Phanerozoic record (Fig. 8.4; Eriksson *et al.*, 2005a, b). In this context, time is largely irrelevant as a parameter in the classification of stratigraphic sequences, and it is rather the stratigraphic record of changes in the tectonic setting that provides the key criteria for the basic subdivision of the rock record into basin-fill successions separated by first-order sequence boundaries. These first-order basin-fill successions are in turn subdivided into second- and lower-order sequences that result from shifts in the balance between accommodation and sedimentation at various scales of observation, irrespective of the time span between two same-order consecutive events. Sequences identified in any particular basin are not expected to correlate to other first- and lower-order sequences of other basins, which are likely characterized by different timing and duration. More details on the concept of sequence stratigraphic hierarchy are provided in Chapter 8.

MOVING FORWARD TOWARD STANDARDIZING SEQUENCE STRATIGRAPHY

As discussed throughout the book, at least three different approaches to sequence stratigraphic analysis are currently promoted by different 'schools' (Figs. 1.6 and 1.7). The inherent confusions caused by this variety of opinions have a negative impact on the 'consumer' (i.e., the practitioner who applies this method to the analysis of the rock record), on the communication of ideas and results between practitioners embracing alternative approaches to stratigraphic analysis, and also on the previous attempts to standardize sequence stratigraphic concepts in international stratigraphic codes (see Chapter 1 for more details). Despite this lack of cohesiveness in the field of sequence stratigraphy, common ground is bound to exist since all stratigraphers, regardless of their background and preferences, are essentially describing the same rocks, only using a different style for their conceptual packaging into sequences and systems tracts. Finding this common ground is the key for making real progress towards standardizing the fundamental concepts of sequence stratigraphy. This also requires scientific objectivity to prevail over dogmas and egos.

Standardizing sequence stratigraphy is in fact within reach, and can be achieved by looking back at the basic principles that represent the foundation of this relatively young and developing discipline. Fundamentally, sequence stratigraphy analyzes the sedimentary response to base-level changes, and the depositional trends that emerge from the interplay of accommodation and sedimentation. Base-level changes (accommodation) and sedimentation are therefore the 'structural pillars' of the sequence stratigraphic architecture. Hence, all four main events of the base-level cycle need to be accounted for, in a balanced and fair approach, in order to extract the essence of what sequence stratigraphy is all about (Fig. 4.7).

In search for a standardized approach to sequence stratigraphy, the following basic principles need to be considered:

1. Sequence stratigraphic surfaces are surfaces that can serve, at least in part, as systems tract or sequence boundaries. The set of seven sequence stratigraphic surfaces are defined relative to the four main events of a reference base-level cycle (Fig. 4.7). These surfaces

are well established, even though their assigned degree of usefulness and/or importance may vary with the model.

2. As a function of subsidence patterns, the magnitude and timing of base-level changes may vary within a sedimentary basin, from one area to another. The reference curve relative to which sequence stratigraphic surfaces and systems tracts are defined (Figs. 4.6 and 4.7) describes changes in base level *at the shoreline*, wherever the shoreline happens to be within the basin at any given time (see Chapter 7 for details).

3. The four main events of the reference base-level cycle mark changes in the direction and/or type of shoreline shift (i.e., forced regressions, normal regressions, transgressions; Figs. 3.19 and 4.7). These changes control the formation and timing of all sequence stratigraphic surfaces and systems tracts. The exception is represented by overfilled basins, dominated by nonmarine sedimentation, where the definition of systems tracts is based on changes in the ratio between channel and overbank fluvial architectural elements.

4. Recognition of sequence stratigraphic surfaces in the rock record is data-dependant. For example, subaerial unconformities represented by paleosols may be impossible to identify on well logs, in the absence of core. This does not mean that 'subtle' subaerial unconformities need to be discarded from the list of sequence stratigraphic surfaces. It simply means that the available data may be insufficient to pinpoint the position of that surface in particular sections of the rock record, and that additional data are required. Similarly, correlative conformities in shallow-water successions, as well as maximum regressive surfaces in deep-water successions, may be impossible to identify in outcrop, core or well logs, in the absence of seismic data.

5. Inherent difficulties in recognizing any of the sequence stratigraphic surfaces, depending on case studies and the available data sets, do not negate their existence or validity. In most cases, this is just a reflection of the lack of sufficient data. Integration of outcrop, core, well-log, and seismic data affords the most effective application of the sequence stratigraphic method.

6. Different genetic types of deposits (i.e., forced regressive, normal regressive, transgressive) need to be separated as distinct systems tracts, data permitting, as this is the key to the predictive aspect of sequence stratigraphy. Each such genetic wedge (systems tract) is characterized by different sediment dispersal patterns, as well as distribution and type of economic deposits (Figs. 5.7, 5.14, 5.26, 5.27, 5.44, 5.56, and 5.57).

7. Sequence stratigraphic surfaces that form independently of sedimentation (i.e., onset-of-fall and end-of-fall 'correlative conformities') are closer to time lines than surfaces that mark the end of regression and onset of transgression (i.e., maximum regressive and maximum flooding surfaces, respectively; Figs. 3.19 and 4.7) (see discussion in Chapter 7).

8. The highest-frequency (lowest-order/rank) cycles in the rock record reflect the *true changes* in depositional trends. All higher-order/rank cycles represent *overall trends*, which approximate the true facies shifts at different scales of observation (Fig. 8.2). Lower-rank stratigraphic surfaces superimposed on higher-rank surfaces do not change the stratigraphic significance of the latter within the bigger-picture framework. A sequence stratigraphic framework constructed at a particular hierarchical level should consistently include sequence stratigraphic surfaces of equal rank.

9. Where two or more sequence stratigraphic surfaces are superimposed, always use the name of the youngest surface.

The lack of formal inclusion of sequence stratigraphic concepts in the current international stratigraphic codes may be attributed largely to trivial differences in terminology and the style of conceptual packaging of the (same) rock record into sequences and systems tracts (Figs. 1.6 and 1.7). The choice of how we name the packages of strata between specific sequence stratigraphic surfaces varies with the model (Fig. 1.7), which is why the systems tract nomenclature becomes less important than the correct identification of the type of shoreline shift that is associated with that particular package of strata. Even the selection of what surface (or set of surfaces) should serve as the 'sequence boundary' (Figs. 6.1 and 6.2) becomes subjective and trivial to some extent, as the correct interpretation of sequence stratigraphic surfaces and of the origin of strata that separate them is far more important for the success of the sequence stratigraphic method. Irrespective of the model of choice, the 'pulse' of sequence stratigraphy is fundamentally represented by *shoreline shifts*, whose type and timing control the formation of all genetic packages of strata (systems tracts) and bounding surfaces. Beyond nomenclatural preferences, each stage of shoreline shift (normal regression, forced regression, transgression) corresponds to the formation of a systems tract with unique characteristics in terms of the nature of processes and products across a sedimentary basin (see Chapter 5 for a full discussion of all systems tracts). These fundamental principles are common among all models, and allow for a unified sequence stratigraphic approach.

Beyond all arguments and disagreements, the original definition of a 'sequence' by Mitchum (1977) still fits all different approaches to sequence stratigraphic analysis, and it is therefore recommended to be kept 'generic,' without specific connotations with regards to the nature of the unconformities (subaerial or marine) and their correlative conformities ('basal surface of forced regression,' 'correlative conformity,' 'maximum regressive surface' or 'maximum flooding surface'). This approach is similar to the formal definition of allostratigraphic discontinuity-bounded units (generic, with the definition of 'discontinuities' left at the discretion of the practicing geologist), and provides a first step towards formalizing sequence stratigraphic concepts. Once the concept of a 'sequence' is agreed upon, more discussions can follow as to what sequence stratigraphic surfaces are most appropriate candidates for sequence boundaries. The discussion in Chapter 6 leads to the conclusion that, for the nonmarine portion of a basin, the subaerial unconformity represents the best choice for a sequence boundary, as being associated with the most significant hiatus in a stratigraphic succession and separating strata that are genetically related. For the marine portion of the basin, the correlative conformity *sensu* Hunt and Tucker (1992) is the only sequence stratigraphic surface that represents the true correlative of the basinward termination of the subaerial unconformity, both temporally and spatially, and it is therefore recommended here as the logical counterpart of the subaerial unconformity for the definition of a throughgoing sequence boundary across an entire sedimentary basin. This choice has been made by the proponents of the depositional sequence models III and IV (Figs. 1.6 and 1.7). Once again, however, the selection of a sequence boundary is arbitrary to some extent, and less important than the correct identification of all sequence stratigraphic surfaces that are present within a study area, and of the genetic nature of intervening strata.

CONCLUDING REMARKS

Sequence stratigraphy is a modern approach to analyzing the sedimentary rock record within a time framework (Fig. 1.8). Arriving at what is known today as sequence stratigraphy took more than a century of conceptual developments, as many of its 'first principles' have been defined within the context of the broad field of sedimentary geology, long before the terms seismic or sequence stratigraphy have been coined and incorporated as such into the literature. Among the fundamental concepts of sequence stratigraphy,

the importance of *base-level changes* on sedimentation was recognized since the nineteenth century (Gilbert, 1895; Barrell, 1917; Wanless and Shepard, 1936; Wheeler and Murray, 1957), while the concept of '*sequence*,' as an unconformity-bounded unit, was in circulation for several decades (Wheeler, 1959; Sloss, 1962, 1963). Building on this foundation, the refinements brought about by sequence stratigraphy include the recognition and interpretation of *systems tracts* (Brown and Fisher, 1977; Fig. 1.9), and the integration of *seismic data* with rock data for a more comprehensive and continuous imaging of the rock record (Payton, 1977; Posamentier and Allen, 1999). These refinements represent significant advances in the understanding of the sedimentary rock record, and triggered what has been described as the third and most recent paradigm in sedimentary geology (Miall, 1995; see Chapter 1 for further details).

Consequently, sequence stratigraphy is neither the first nor the last step in the evolution of methods of analysis of sedimentary basin fills. The complexity and accuracy of geological models produced to resolve academic or economic issues improved through time in response to corresponding advances in concepts and technology. Classical geology remains the foundation of everything we know today, by providing the means to understanding the 'first principles' of sedimentary geology (Fig. 2.1). Cornerstone advances that led or contributed to the development of modern sequence stratigraphy include the concepts of base level (nineteenth century), unconformity-bounded sequence (1950s and 1960s), flow regimes (1950s and 1960s), plate tectonics (1960s), basin analysis (1970s), seismic stratigraphy (1970s), and the notion of systems tracts (1970s and 1980s). From here, modern sequence stratigraphy bloomed in the 1980s, and its principles are still being refined today.

Stimulated by technological advances in the fields of three-dimensional seismic data acquisition and processing, seismic geomorphology also developed in parallel with sequence stratigraphy, starting with the 1990s. As defined by Posamentier (2000, 2004a), seismic geomorphology deals with the imaging of depositional systems, and elements thereof, using three-dimensional seismic data. While seismic geomorphology can be performed independently of a base-level-controlled framework of stratigraphic architecture, it can also be successfully integrated into sequence stratigraphy for producing comprehensive models that combine insights from section view (classical seismic stratigraphy: reflection terminations, stratigraphic discontinuities, reflection geometries and inferred depositional systems) and map view (seismic geomorphology: imaging of geological features, and particularly depositional systems and depositional elements). Such three-dimensional control of the sedimentary basin fill is important at any stage, from exploration in frontier areas to the development of production fields, because it enhances the accuracy of facies predictability.

What will follow next in the development of present-day sequence stratigraphy is difficult to predict. Current efforts are concentrated on reducing the error margin of stratigraphic models and interpretations, during both exploration and production stages, as well as reducing the costs of exploration and production. As with the introduction of seismic geomorphology, technological advances will dictate the next cornerstone that can be achieved. For example, virtual core simulations using electric logs (micro resistivity images; Fig. 2.30) represent a relatively new technique that is currently being developed, which permits the reconstruction of sedimentary structures (insights into process sedimentology/stratigraphy) at mm scale by providing continuous virtual coring of boreholes, thus eliminating the costs of mechanical coring. Such new techniques, and data sets, will continue to be integrated into sequence stratigraphy in order to advance our knowledge and understanding of the evolution and architecture of sedimentary basin fills.

Lastly, the real world is far more complex than we can ever model, so one needs to keep an open mind when trying to find patterns that match the predictions of any sequence stratigraphic model. Sequence stratigraphic principles offer theoretical guidelines of how the facies and time relationships are expected to be under specific circumstances such as subsidence patterns, sediment supply, topographic gradients, etc., but these circumstances may change significantly with the type of sedimentary basin, as each tectonic setting is unique in terms of subsidence mechanisms, sediment supply and dispersal patterns, physiography, etc. This is one of the main sources for the conflicting ideas between the various models currently in use, as their proponents draw their conclusions from case studies derived from different tectonic settings. The study of similarities and differences between the sequence stratigraphic architectures of basins formed in different tectonic settings will help identify a broader platform of theoretical principles that should place all current ideas into a more general context. Such syntheses are still being formulated, and the incorporation of the variability imposed by changes in the tectonic setting to the sequence stratigraphic model represents a logical next step in the evolution of sequence stratigraphy.

References

Abbott, S. T. (1998). Transgressive systems tracts and onlap shellbeds from Mid-Pleistocene sequences, Wanganui Basin, New Zealand. *Journal of Sedimentary Research*, **68**, pp. 253–268.

Ainsworth, R. B. (1991). Sedimentology and high resolution sequence stratigraphy of the Bearpaw-Horseshoe Canyon transition (Upper Cretaceous), Drumheller, Alberta, Canada. M.Sc. Thesis, McMaster University, Hamilton, Ontario, p. 213.

Ainsworth, R. B. (1992). *Sedimentology and sequence stratigraphy of the Upper Cretaceous, Bearpaw–Horseshoe Canyon transition, Drumheller, Alberta*. American Association of Petroleum Geologists Annual Convention, Calgary, Field Trip Guidebook 7, p. 118.

Ainsworth, R. B. (1994). Marginal marine sedimentology and high resolution sequence analysis; Bearpaw–Horseshoe Canyon transition, Drumheller, Alberta. *Bulletin of Canadian Petroleum Geology*, Vol. **42**, no. 1, pp. 26–54.

Ainsworth, R. B. (2005). Sequence stratigraphic-based analysis of reservoir connectivity: influence of depositional architecture – a case study from a marginal marine depositional setting. *Petroleum Geoscience*, Vol. **11**, pp. 257–276.

Ainsworth, R. B., and Pattison, S. A. J. (1994). Where have all the lowstands gone? Evidence for attached lowstand systems tracts in the Western Interior of North America. *Geology*, Vol. **22**, pp. 415–418.

Ainsworth, R. B., and Walker, R. G. (1994). Control of estuarine valley-fill deposition by fluctuations of relative sea-level, Cretaceous Bearpaw-Horseshoe Canyon transition, Drumheller, Alberta, Canada. *In Incised-valley systems: Origin and Sedimentary Sequences* (R. G. Dalrymple, R. Boyd, and B. A. Zaitlin, Eds.), pp. 159–174. SEPM (Society for Sedimentary Geology) Special Publication No. **51**.

Ainsworth, R. B., Sanlung, M., and Duivenvoorden, S. T. C. (1999). Correlation techniques, perforation strategies, and recovery factors: and integrated 3D reservoir modeling study, Sirikit Field, Thailand. *American Association of Petroleum Geologists Bulletin*, Vol. **83**, no. 10, pp. 1535–1551.

Ainsworth, R. B., Bosscher, H., and Newall, M. J. (2000). Forward stratigraphic modelling of forced regressions: evidence for the genesis of attached and detached lowstand systems. *In Sedimentary Responses to Forced Regressions* (D. Hunt and R. L. Gawthorpe, Eds.), pp. 163–176. Geological Society, London, Special Publication **172**.

Aitken, J. F., and Flint, S. S. (1994). High-frequency sequences and the nature of incised-valley fills in fluvial systems of the Breathitt Group (Pennsylvanian), Appalachian foreland basin, eastern Kentucky. *In Incised Valley Systems: Origin and Sedimentary Sequences* (R. W. Dalrymple, R. Boyd, and B. A. Zaitlin, Eds.), pp. 353–368. SEPM Special Publication **51**.

Aitken, J. F., and Flint, S. S. (1996). Variable expressions of interfluvial sequence boundaries in the Breathitt Group (Pennsylvanian), eastern Kentucky, USA. *In High Resolution Sequence Stratigraphy: Innovations and Applications* (J. A. Howell and J. F. Aitken, Eds.), pp. 193–206. Geological Society of London, Special Publication **104**.

Algeo, T. J., and Wilkinson, B. H. (1988). Periodicity of mesoscale Phanerozoic sedimentary cycles and the role of Milankovitch orbital modulation. *Journal of Geology*, Vol. **96**, pp. 313–322.

Allen, D. R. (1975). Identification of sediments – their depositional environment and degree of compaction – from well logs. *In Compaction of Coarse-Grained Sediments I* (G. V. Chilingarian and K. H. Wolf, Eds.), pp. 349–402. Elsevier, New York.

Allen, G. P. (1991). Sedimentary processes and facies in the Gironde estuary: a Recent model for macrotidal estuarine systems. *In Clastic Tidal Sedimentology* (D. G. Smith, G. E. Reinson, B. A. Zaitlin and R. A. Rahmani, Eds.), pp. 29–40. Canadian Society of Petroleum Geologists, Memoir **16**.

Allen, G. P., and Posamentier, H. W. (1993). Sequence stratigraphy and facies model of an incised valley fill: the Gironde estuary, France. *Journal of Sedimentary Petrology*, Vol. **63**, no. 3, pp. 378–391.

Allen, G. P., and Posamentier, H. W. (1994). Transgressive facies and sequence architecture in mixed tide- and wave-dominated incised valleys: example from the Gironde Estuary, France. *In Incised Valley Systems: Origin and Sedimentary Sequences* (R. W. Dalrymple, R. Boyd and B. A. Zaitlin, Eds.). SEPM Special Publication **51**, pp. 225–240.

Allen, G. P., Lang, S., Musakti, O., and Chirinos, A. (1996). Application of sequence stratigraphy to continental successions: implications for Mesozoic cratonic interior basins of eastern Australia. Geological Society of Australia, Mesozoic Geology of the Eastern Australia Plate, Conference, Brisbane, September 1996, pp. 22–27.

Allen, P. A., and Allen, J. R. (2005). *Basin Analysis: Principles and Applications*. Second edition, Blackwell Science. pp. 549.

Anderson, J. B., Wolfteich, C., Wright, R., and Cole, M. L. (1982). Determination of depositional environments of sand using vertical grain size progressions. *Gulf Coast Association of Geological Societies, Transactions*, Vol. **32**, pp. 565–577.

Angevine, C. L. (1989). Relationship of eustatic oscillations to regressions and transgressions on passive continental margins. *In Origin and evolution of sedimentary basins and their energy and mineral resources* (R. A. Price, Ed.), pp. 29–35. Geophysical Monograph **48**, AGU, Washington, D. C.

Arndorff, L. (1993). Lateral relations of deltaic palaeosols from the Lower Jurassic Ronne Formation on the Island of Bornholm, Denmark. *Palaeogeography, Palaeoclimatology, Palaeoecology*, Vol. **100**, pp. 235–250.

Arnott, R. W. C. (1995). The parasequence definition – are transgressive deposits inadequately addressed? *Journal of Sedimentary Research*, Vol. **B65**, pp. 1–6.

Arnott, R. W. C., Zaitlin, B. A., and Potocki, D. J. (2002). Stratigraphic response to sedimentation in a net-accommodation-limited setting, Lower Cretaceous Basal Quartz, south-central Alberta. *Bulletin of Canadian Petroleum Geology*, Vol. **50**, no. 1, pp. 92–104.

Arnott, R. W. C., Dumas, S., and Southard, J. B. (2004). Hummocky cross-stratification and other shallow-marine structures: > 25 years and still the debate continues. Geological Association of Canada—Mineralogical Association of Canada Joint Annual Meeting, St. Catharines, Abstract on CD-ROM.

Arua, I. (1989). Clavate borings in a Maastrichtian woodground in southeastern Nigeria. *Palaeogeography, Palaeoclimatology, Palaeoecology*, Vol. **69**, pp. 321–326.

Banerjee, I., Kalkreuth, W., and Davies, E. H. (1996). Coal seam splits and transgressive-regressive coal couplets: A key to stratigraphy of high-frequency sequences. *Geology*, Vol. **24**, no. 11, pp. 1001–1004.

Bard, B., Hamelin, R. G., and Fairbanks, R. (1990). U-Th obtained by mass spectrometry in corals from Barbados: sea level during the past 130.000 years. *Nature*, Vol. **346**, pp. 456–458.

Barrell, J. (1917). Rhythms and the measurements of geological time. *Geological Society of America Bulletin*, Vol. **28**, pp. 745–904.

Bates, R. L., and Jackson, J. A. (Eds.), (1987). *Glossary of Geology* 3rd Ed. American Geological Institute, Alexandria, Virginia, p. 788.

Belknap, D. F., and Kraft, J. C. (1981). Preservation potential of transgressive coastal lithosomes on the U. S. Atlantic shelf. *Marine Geology*, Vol. **42**, pp. 429–442.

Berg, O. R. (1982). Seismic detection and evaluation of delta and turbidite sequences: their application to exploration for the subtle trap. *American Association of Petroleum Geologists Bulletin*, Vol. **66**, pp. 1271–1288.

Berger, A. L., and Loutre, M. F. (1994). Astronomical forcing through geological time. *In Orbital forcing and cyclic sequences* (P. L. de Boer and D. G. Smith, Eds.), pp. 15–24. International Association of Sedimentologists Special Publication **19**.

Beukes, N. J., and Cairncross, B. (1991). A lithostratigraphic-sedimentological reference profile for the Late Archaean Mozaan Group, Pongola Sequence: application to sequence stratigraphy and correlation with the Witwatersrand Supergroup. *South African Journal of Geology*, Vol. **94**, pp. 44–69.

Bhattacharya, J. P., and Walker, R. G. (1991). Allostratigraphic subdivision of the Upper Cretaceous Dunvegan, Shaftesbury and Kaskapau formations in the north-western Alberta subsurface. *Bulletin of Canadian Petroleum Geology*, Vol. **39**, pp. 145–164.

Bhattacharya, J. P., and Walker, R. G. (1992). Deltas. *In Facies Models: Response to Sea Level Change* (R. G. Walker and N. P. James, Eds.), pp. 157–178. Geological Association of Canada, GeoText 1.

Blum, M. D. (1990). Climatic and eustatic controls on Gulf coastal plain fluvial sedimentation: an example from the Late Quaternary of the Colorado River, Texas. *In Sequence Stratigraphy as an Exploration Tool, Concepts and Practices in the Gulf Coast* (J. M. Armentrout and B. F. Perkins, Eds.), pp. 71–83. SEPM (Society of Economic Palaeontologists and Mineralogists) Gulf Coast Section, Eleventh Annual Research Conference, Program with Abstracts.

Blum, M. D. (1991). Systematic controls of genesis and architecture of alluvial sequences: a Late Quaternary example. *In NUNA Conference on High-Resolution Sequence Stratigraphy* (D. A. Leckie, H. W. Posamentier and R. W. Lovell Eds.), pp. 7–8. Program with Abstracts.

Blum, M. D. (1994). Genesis and architecture of incised valley fill sequences: a Late Quaternary example from the Colorado River, Gulf Coastal Plain of Texas. *In Siliciclastic sequence stratigraphy: recent developments and applications* (P. Weimer and H. W. Posamentier Eds.), pp. 259–283. American Association of Petroleum Geologists Memoir 58.

Blum, M. D. (2001). Importance of falling stage fluvial deposition: Quaternary examples from the Texas Gulf Coastal Plain and Western Europe. Seventh International Conference on Fluvial Sedimentology, Lincoln, August 6–10, Program and Abstracts, p. 61.

Blum, M. D., and Price, D. M. (1998). Quaternary alluvial plain construction in response to glacio-eustatic and climatic controls, Texas Gulf coastal plain. *In Relative Role of Eustasy, Climate, and Tectonism in Continental Rocks* (K. W. Shanley and P. J. McCabe Eds.), pp. 31–48. SEPM (Society for Sedimentary Geology) Special Publication **59**.

Blum, M. D., and Tornqvist, T. E. (2000). Fluvial responses to climate and sea-level change: a review and look forward. *Sedimentology*, Vol. **47** (suppl. 1), pp. 2–48.

Blum, M. D., and Valastro, S. Jr., (1989). Response of the Pedernales River of central Texas to late Holocene climatic changes. *Association of American Geographers, Annals*, Vol. **79**, pp. 435–456.

Bohacs, K., and Suter, J. (1997). Sequence stratigraphic distribution of coaly rocks: Fundamental controls and paralic examples. *American Association of Petroleum Geologists Bulletin*, Vol. **81**, no. 10, pp. 1612–1639.

Bouma, A. H., and Stone, C. G. (Eds.) (2000). *Fine-Grained Turbidite Systems*. American Association of Petroleum Geologists Memoir **72**, p. 352.

Bowen, D. W., Weimer, P. W., and Scott, A. J. (1993). The relative success of sequence stratigraphic concepts in exploration: examples from incised valley fill and turbidite systems reservoirs. *In Siliciclastic Sequence Stratigraphy – Recent Developments and Applications* (P. Weimer, and H. W. Posamentier, Eds.), pp. 15–42. American Association of Petroleum Geologists, Memoir **58**.

Bown, T. M., and Kraus, M. J. (1981). Lower Eocene alluvial paleosols (Willwood Formation, northwest Wyoming, USA) and their significance for paleoecology, paleoclimatology, and basin analysis. *Palaeogeography, Palaeoclimatology, Palaeoecology*, Vol. **34**, pp. 1–30.

Boyd, R., Williamson, P., and Haq, B. U. (1993). Seismic stratigraphy and passive-margin evolution of the southern Exmouth Plateau. *In Sequence Stratigraphy and Facies Associations* (H. W. Posamentier, C. P. Summerhayes, B. U. Haq, and G. P. Allen, Eds.), pp. 581–603. International Association of Sedimentologists, Special Publication **18**.

Boyd, R., Diessel, C. F. K., Wadsworth, J., Chalmers, G., Little, M., Leckie, D., and Zaitlin, B. (1999). Development of a nonmarine sequence stratigraphic model. American Association of Petroleum Geologists Annual Meeting, San Antonio, Texas, USA. Official Program, p. A15.

Boyd, R., Diessel, C. F. K., Wadsworth, J., Leckie, D., and Zaitlin, B. A. (2000). Organization of non marine stratigraphy. *In Advances in the study of the Sydney Basin* (R. Boyd, C. F. K. Diessel, and S. Francis, Eds.), pp. 1–14. Proceedings of the 34th Newcastle Symposium, University of Newcastle, Callaghan, New South Wales, Australia.

Breyer, J. A. (1995). Sedimentary facies in an incised valley in the Pennsylvanian of Beaver County, Oklahoma. *Journal of Sedimentary Research*, Vol. B65, 338–347.

Bridge, J. S., and Leeder, M. R. (1979). A simulation model of alluvial stratigraphy. *Sedimentology*, Vol. **26**, pp. 599–623.

Bromley, R. G. (1975). Trace fossils at omission surfaces. *In The Study of Trace Fossils* (R. W. Frey, Ed.), pp. 399–428. Springer-Verlag, New York.

Bromley, R. G., and Ekdale, A. A. (1984). Composite ichnofabrics and tiering of burrows. *Geological Magazine*, Vol. **123**, pp. 59–65.

Bromley, R. G., Pemberton, S. G., and Rahmani, R. A. (1984). A Cretaceous woodground: the *Teredolites* ichnofacies. *Journal of Paleontology*, Vol. **58**, pp. 488–498.

Brown, A. R. (1991). Interpretation of 3-Dimensional Seismic Data. *American Association of Petroleum Geologists Memoir*, **42**, p. 341.

Brown, L. F. Jr., and Fisher, W. L. (1977). Seismic stratigraphic interpretation of depositional systems: examples from Brazilian rift and pull apart basins. *In Seismic Stratigraphy-Applications to Hydrocarbon Exploration* (C. E. Payton, Ed.), pp. 213–248. American Association of Petroleum Geologists Memoir **26**.

Brown, L. F. Jr., Benson, J. M., Brink, G. J., Doherty, S., Jollands, A., Jungslager, E. H. A., Keenan, J. H. G., Muntingh, A., and van Wyk, N. J. S., (1995). Sequence stratigraphy in offshore South African diveregent basins: an atlas on exploration for Cretaceous lowstand traps by Soekor (Pty) Ltd. American Association of Petroleum Geologists, *Studies in Geology #41*, Tulsa, Oklahoma, p. 184.

Bruun, P. (1962). Sea-level rise as a cause of shore erosion. *American Society of Civil Engineers Proceedings, Journal of the Waterways and Harbors Division*, Vol. **88**, pp. 117–130.

Buatois, L. A., Mangano, M. G., Alissa, A., and Carr, T. R. (2002). Sequence stratigraphic and sedimentologic significance of biogenic structures from a late Paleozoic marginal-to open-marine reservoir, Morrow Sandstone, subsurface of Southwest Kansas, USA. *Sedimentary Geology*, Vol. **152**, no. 1–2, pp. 99–132.

Burchette, T. P., and Wright, V. P. (1992). Carbonate ramp depositional systems. *Sedimentary Geology*, Vol. **79**, pp. 3–57.

Burns, B. A., Heller, P. L., Marzo, M., and Paola, C. (1997). Fluvial response in a sequence stratigraphic framework: example from the Montserrat fan delta, Spain. *Journal of Sedimentary Research*, Vol. **67**, No. 2, pp. 311–321.

Busby, C. J., and Ingersoll, R. V. (Eds.) (1995). *Tectonics of Sedimentary Basins*. Blackwell Science, p. 579.

Butcher, S. W. (1990). The nickpoint concept and its implications regarding onlap to the stratigraphic record. *In Quantitative dynamic stratigraphy* (T. A. Cross, Ed.), pp. 375–385. Prentice-Hall, Englewood Cliffs.

Cant, D. (1992). Subsurface facies analysis. *In Facies Models: Response to Sea Level Change* (R. G. Walker, and N. P. James, Eds.), pp. 27–45. Geological Association of Canada, GeoText 1.

Cant, D. (2004). The real significance of sequence stratigraphy to subsurface geological work. *Reservoir, Canadian Society of Petroleum Geologists*, Vol. **31**, Issue 3, p. 14.

Cant, D. J. (1991). Geometric modelling of facies migration: theoretical development of facies successions and local unconformities. *Basin Research*, Vol. **3**, pp. 51–62.

Carter, R. M., Abbott, S. T., Fulthorpe, C. S., Haywick, D. W., and Henderson, R. A. (1991). Application of global sea-level and sequence-stratigraphic models in southern hemisphere Neogene strata. *In Sedimentation, Tectonics and Eustasy: Sea-Level Changes at Active Margins* (D. I. M. Macdonald, Ed.), pp. 41–65. International Association of Sedimentologists Special Publication **12**.

Cathro, D. L., Austin, J. A., and Moss, G. D. (2003). Progradation along a deeply submerged Oligocene-Miocene heterozoan carbonate shelf: how sensitive are clinoforms to sea level variations? *American Association of Petroleum Geologists Bulletin*, Vol. **87**, no. 10, pp. 1547–1574.

Catuneanu, O. (1996). Reciprocal architecture of Bearpaw and post-Bearpaw sequences, Late Cretaceous–Early Tertiary, Western Canada Basin. Ph.D. Thesis, University of Toronto, 301 pp.

Catuneanu, O. (2001). Flexural partitioning of the Late Archaean Witwatersrand foreland system, South Africa. *Sedimentary Geology*, Vol. **141–142**, pp. 95–112.

Catuneanu, O. (2002). Sequence stratigraphy of clastic systems: concepts, merits, and pitfalls. *Journal of African Earth Sciences*, Vol. **35/1**, pp. 1–43.

Catuneanu, O. (2003). Sequence Stratigraphy of Clastic Systems. *Geological Association of Canada, Short Course Notes*, Vol. **16**, p. 248.

Catuneanu, O. (2004a). Retroarc foreland systems–evolution through time. *Journal of African Earth Sciences*, Vol. **38/3**, pp. 225–242.

Catuneanu, O. (2004b). Basement control on flexural profiles and the distribution of foreland facies: the Dwyka Group of the Karoo Basin, South Africa. *Geology*, Vol. **32**, no. 6, pp. 517–520.

Catuneanu, O., and Biddulph, M. N. (2001). Sequence stratigraphy of the Vaal Reef facies associations in the Witwatersrand foredeep, South Africa. *Sedimentary Geology*, Vol. **141–142**, pp. 113–130.

Catuneanu, O., and Bowker, D. (2001). Sequence stratigraphy of the Koonap and Middleton fluvial formations in the Karoo foredeep, South Africa. *Journal of African Earth Sciences*, Vol. **33**, pp. 579–595.

Catuneanu, O., and Elango, H. N. (2001). Tectonic control on fluvial styles: the Balfour Formation of the Karoo Basin, South Africa. *Sedimentary Geology*, Vol. **140**, pp. 291–313.

Catuneanu, O., and Eriksson, P. G. (1999). The sequence stratigraphic concept and the Precambrian rock record: an example from the 2.7–2.1 Ga Transvaal Supergroup, Kaapvaal craton. *Precambrian Research*, Vol. **97**, pp. 215–251.

Catuneanu, O., and Eriksson, P. G. (2002). Sequence stratigraphy of the Precambrian Rooihoogte-Timeball Hill rift succession, Transvaal Basin, South Africa. *Sedimentary Geology*, Vol. **147**, pp. 71–88.

Catuneanu, O., and Sweet, A. R. (1999). Maastrichtian-Paleocene foreland basin stratigraphies, Western Canada: A reciprocal sequence architecture. *Canadian Journal of Earth Sciences*, Vol. **36**, pp. 685–703.

Catuneanu, O., and Sweet, A. R. (2005). Fluvial Sequence Stratigraphy and Sedimentology of the Uppermost Cretaceous to Paleocene, Alberta Foredeep. American Association of Petroleum Geologists Annual Convention, Calgary, 23–25 June 2005, Field Trip Guidebook, p. 68.

Catuneanu, O., Sweet, A. R., and Miall, A. D. (1997a). Reciprocal architecture of Bearpaw T-R sequences, uppermost Cretaceous, Western Canada Sedimentary Basin. *Bulletin of Canadian Petroleum Geology*, Vol. **45(1)**, pp. 75–94.

Catuneanu, O., Beaumont, C., and Waschbusch, P. (1997b). Interplay of static loads and subduction dynamics in foreland basins: Reciprocal stratigraphies and the "missing" peripheral bulge. *Geology*, Vol. **25**, no. 12, pp. 1087–1090.

Catuneanu, O., Hancox, P. J., and Rubidge, B. S. (1998a). Reciprocal flexural behaviour and contrasting stratigraphies: a new basin development model for the Karoo retroarc foreland system, South Africa. *Basin Research*, Vol. **10**, pp. 417–439.

Catuneanu, O., Willis, A. J., and Miall, A. D. (1998b). Temporal significance of sequence boundaries. *Sedimentary Geology*, Vol. **121**, pp. 157–178.

Catuneanu, O., Sweet, A. R., and Miall, A. D. (1999). Concept and styles of reciprocal stratigraphies: Western Canada foreland system. *Terra Nova*, Vol. **11**, pp. 1–8.

Catuneanu, O., Sweet, A. R., and Miall, A. D. (2000). Reciprocal stratigraphy of the Campanian-Paleocene Western Interior of North America. *Sedimentary Geology*, Vol. **134**, pp. 235–255.

Catuneanu, O., Hancox, P. J., Cairncross, B., and Rubidge, B. S. (2002). Foredeep submarine fans and forebulge deltas: orogenic off-loading in the underfilled Karoo Basin. *Journal of African Earth Sciences*, Vol. **33**, pp. 489–502.

Catuneanu, O., Heredia, E., Ortega, V., Robles, J., Toledo, C., Boll, L. P., and Lopez, F. (2003a). Seismic stratigraphy of the Neogene sequences in the Gulf of Mexico. *American Association of Petroleum Geologists Annual Convention, Salt Lake City, Official Program*, Vol. **12**, pp. A27–28.

Catuneanu, O., Khidir, A., and Thanju, R. (2003b). External controls on fluvial facies: the Scollard sequence, Western Canada foredeep. *American Association of Petroleum Geologists Annual Convention, Salt Lake City, Official Program*, Vol. 12, p. A27.

Catuneanu, O., Embry, A. F., and Eriksson, P. G. (2004). Concepts of Sequence Stratigraphy. *In The Precambrian Earth: Tempos and Events* (P. G. Eriksson, W. Altermann, D. Nelson, W. Mueller and O. Catuneanu, Eds.), pp. 685–705. Developments in Precambrian Geology 12, Elsevier Science Ltd., Amsterdam.

Catuneanu, O., Khalifa, M. A., and Wanas, H. A., in press. Sequence stratigraphy of the Lower Cenomanian Bahariya Formation, Bahariya Oasis, Western Desert, Egypt. *Sedimentary Geology*.

Catuneanu, O., Martins-Neto, M., and Eriksson, P. G. (2005). Precambrian sequence stratigraphy. *Sedimentary Geology*, Vol. **176**, Issues 1–2, pp. 67–95.

Caudill, M. R., Driese, S. G., and Mora, C. I. (1997). Physical compaction of vertic palaeosols: implications for burial diagenesis and palaeo-precipitation estimates. *Sedimentology*, Vol. **44**, pp. 673–685.

Changsong, L., Eriksson, K., Sitain, L., Yongxian, W., Jangye, R., and Jingyan, L. (2001). Sequence architecture, depositional systems, and computer simulation of lacustrine basin fills in Erlian Basin, northeast China. *American Association of Petroleum Geologists Bulletin*, Vol. **85**, pp. 2017–2043.

Chaplin, J. R. (1996). Ichnology of transgressive-regressive surfaces in mixed carbonate-siliciclastic sequences, Early Permian Chase Group, Oklahoma. *In Paleozoic sequence stratigraphy; views from the North American Craton* (B. J. Witzke, G. A. Ludvigson, and J. Day, Eds.), pp. 399–418. Geological Society of America, Special Paper 306.

Christie-Blick, N. (1991). Onlap, offlap, and the origin of unconformity-bounded depositional sequences. *Marine Geology*, Vol. **97**, pp. 35–56.

Christie-Blick, N., and Driscoll, N. W. (1995). Sequence stratigraphy. *Annual Review of Earth and Planetary Sciences*, Vol. **23**, pp. 451–478.

Christie-Blick, N., Grotzinger, J. P., and Von der Borch, C. C. (1988). Sequence stratigraphy in Proterozoic successions. *Geology*, Vol. **16**, pp. 100–104.

Christie-Blick, N., Mountain G. S., and Miller, K. G. (1990). Seismic stratigraphic record of sea-level change. *In Sea-level change* (R. Revelle, Ed.), pp. 116–140. National Research Council, Studies in Geophysics. National Academy Press, Washington.

Cloetingh, S. (1988). Intraplate stress: a new element in basin analysis. *In New perspectives in basin analysis* (K. Kleinspehn, and C. Paola, Eds.), pp. 205–230. Springer-Verlag, New York.

Cloetingh, S., McQueen, H., and Lambeck, K. (1985). On a tectonic mechanism for regional sea level variations. *Earth and Planetary Science Letters*, Vol. **75**, pp. 157–166.

Cloetingh, S., Kooi, H., and Groenewoud, W. (1989). Intraplate stresses and sedimentary basin evolution. *In Origin and Evolution of Sedimentary Basins and their Energy and Mineral Resources* (R. A. Price, Ed.), *American Geophysical Union, Geophysical Monographs*, Vol. **48**, pp. 1–16.

Coe, A. L. (Ed.) (2003). *The Sedimentary Record of Sea-Level Change.* Cambridge University Press, New York, p. 287.

Cohen, C. R. (1982). Model for a passive to active continental margin transition: implications for hydrocarbon exploration. *American Association of Petroleum Geologists Bulletin*, Vol. **66**, pp. 708–718.

Collinson, J. D. (1969). The sedimentology of the Grindslow Shales and the Kinderscout Grit: a deltaic complex in the Namurian of northern England. *Journal of Sedimentary Petrology*, Vol. **39**, pp. 194–221.

Collinson, J. D. (1986). Deserts. *In Sedimentary environments and facies, 2nd Edn.* (H. G. Reading, Ed.), pp. 95–112. Blackwell Scientific Publications, Oxford.

Coniglio, M., and Dix, G. R. (1992). Carbonate slopes. *In Facies Models: Response to Sea Level Change* (R. G. Walker and N. P. James, Eds.), pp. 349–373. Geological Association of Canada, GeoText 1.

Coniglio, M., Myrow, P., and White, T. (2000). Stable carbon and oxygen isotope evidence of Cretaceous sea-level fluctuations recorded in septarian concretions from Pueblo, Colorado, U.S.A. *Journal of Sedimentary Research*, Vol. **70**, pp. 700–714.

Cotter, E., and Driese, S. G. (1998). Incised-valley fills and other evidence of sea-level fluctuations affecting deposition of the Catskill Formation (Upper Devonian), Appalachian foreland basin, Pennsylvania. *Journal of Sedimentary Research*, Vol. **68**, pp. 347–361.

Crampton, S. L., and Allen, P. A. (1995). Recognition of forebulge unconformities associated with early stage foreland basin development: Example from north Apline foreland basin. *American Association of Petroleum Geologists Bulletin*, Vol. **79**, pp. 1495–1514.

Crimes, T. P., Goldring, R., Homewood, P., van Stuijvenberg, J., and Winkler, W. (1981). Trace fossil assemblages of deep-sea fan deposits, Grunigel and Schlieren flysch (Cretaceous-Eocene, Switzerland). *Eclogae Geologicae Helvetiae*, Vol. **74**, pp. 953–995.

Cross, T. A. (1991). High-resolution stratigraphic correlation from the perspectives of base-level cycles and sediment accommodation. *In Unconformity Related Hydrocarbon Exploration and Accumulation in Clastic and Carbonate Settings* (J. Dolson, Ed.), pp. 28–41. Rocky Mountain Association of Geologists, Short Course Notes.

Cross, T. A., and Lessenger, M. A. (1998). Sediment volume partitioning: rationale for stratigraphic model evaluation and high-resolution stratigraphic correlation. *In Sequence Stratigraphy–Concepts and Applications* (F. M. Gradstein, K. O. Sandvik and N. J. Milton, Eds.), pp. 171–195. Norwegian Petroleum Society (NPF), Special Publication 8.

Cross, T. A., and Lessenger, M. A. (1999). Construction and application of a stratigraphic inverse model. *In Numerical Experiments in Stratigraphy: Recent Advances in Stratigraphic and Sedimentologic Computer Simulations* (J. W. Harbaugh, W. L. Watney, E. C. Rankey, R. Slingerland, R. H. Goldstein and E. K. Franseen, Eds.), pp. 69–83. SEPM (Society for Sedimentary Geology) Special Publication **62**.

Curray, J. R. (1964). Transgressions and regressions. *In Papers in Marine Geology* (R. L. Miller, Ed.), pp. 175–203. New York, Macmillan.

Dahle, K., Flesja, K., Talbot, M. R., and Dreyer, T. (1997). Correlation of fluvial deposits by the use of Sm-Nd isotope analysis and mapping of sedimentary architecture in the Escanilla Formation (Ainsa Basin, Spain) and the Statfjord Formation (Norwegian North Sea). Abstracts, Sixth International Conference on Fluvial Sedimentology, Cape Town, 1997, South Africa, p. 46.

Dalrymple, R. W. (1999). Tide-dominated deltas: do they exist or are they all estuaries? American Association of Petroleum Geologists Annual Convention, San Antonio, Official Program, pp. A29.

Dalrymple, R. W., Zaitlin, B. A., and Boyd, R. (1992). Estuarine facies models: conceptual basis and stratigraphic implications. *Journal of Sedimentary Petrology*, Vol. **62**, pp. 1130–1146.

Dalrymple, R. W., Zaitlin, B. A., and Boyd, R. (Eds.) (1994). Incised Valley Systems: Origin and Sedimentary Sequences. SEPM Special Publication **51**, p. 391.

Dam, G., and Surlyk, F. (1993). Cyclic sedimentation in a large wave- and storm-dominated anoxic lake; Kap Stewart Formation (Rhaetian-Sinemurian), Jameson Land, East Greenland. *In Sequence Stratigraphy and Facies Associations* (H. W. Posamentier, C. Summerhayes, B. Haq and G. P. Allen, Eds.), pp. 419–451. International Association of Sedimentologists Special Publication 18.

Davies, S. J., and Gibling, M. R. (2003). Architecture of coastal and alluvial deposits in an extensional basin: the Carboniferous Joggins Formation of eastern Canada. *Sedimentology*, Vol. **50**, p. 415–439.

Davis, W. M. (1908). *Practical Exercises in Physical Geography*. Ginn and Company, Boston, p. 148.

Demarest, J. M., and Kraft, J. C. (1987). Stratigraphic record of Quaternary sea levels: implications for more ancient strata. In *Sea Level Fluctuation and Coastal Evolution* (D. Nummedal, O. H. Pilkey and J. D. Howard, Eds.), pp. 223–239. SEPM Special Publication 41.

Devlin, W. J., Rudolph, K. W., Shaw, C. A., and Ehman, K. D. (1993). The effect of tectonic and eustatic cycles on accommodation and sequence-stratigraphic framework in the Upper Cretaceous foreland basin of southwestern Wyoming. In *Sequence Stratigraphy and Facies Associations* (H. W. Posamentier, C. P. Summerhayes, B. U. Haq and G. P. Allen Eds.), pp. 501–520. International Association of Sedimentologists Special Publication 18.

Dewey, C. P., and Keady, D. M. (1987). An allochthonous preserved woodground in the Upper Cretaceous Eutaw Formation in Mississippi. *Southeastern Geology*, Vol. 27, pp. 165–170.

Diessel, C., Boyd, R., and Wadsworth, J. (2000). On balanced and unbalanced accommodation/peat accumulation ratios in the Cretaceous coals from the Gates Formation, Western Canada, and their sequence-stratigraphic significance. *International Journal of Coal Geology*, Vol. 43, pp. 143–186.

Dominguez, J. M. L., and Wanless, H. R. (1991). Facies architecture of a falling sea-level strandplain, Doce River coast, Brazil. In *Shelf Sand and Sandstone Bodies: Geometry, Facies and Sequence Stratigraphy* (D. J. P. Swift, G. F. Oertel, R. W. Tillman and J. A. Thorne, Eds.), pp. 259–281. International Association of Sedimentologists Special Publication 14.

Donaldson, W. S., Plint, A. G., and Longstaffe, F. J. (1998). Basement tectonic control on distribution of the shallow marine Bad Heart Formation: Peace River Arch area, Northwest Alberta. *Bulletin of Canadian Petroleum Geology*, Vol. 46 (4), pp. 576–598.

Donaldson, W. S., Plint, A. G., and Longstaffe, F. J. (1999). Tectonic and eustatic controlon deposition and preservation of Upper Cretaceous ooidal ironstone and associated facies; Peace River Arch area, NW Alberta, Canada. *Sedimentology*, Vol. 46 (6), pp. 1159–1182.

Donovan, A. D. (1993). The use of sequence stratigraphy to gain new insights into stratigraphic relationships in the Upper Cretaceous of the US Gulf Coast. In *Sequence Stratigraphy and Facies Associations* (H. W. Posamentier, C. P. Summerhayes, B. U. Haq and G. P. Allen, Eds.), pp. 563–577. International Association of Sedimentologists, Special Publication 18.

Donovan, A. D. (2001). Free market theory and sequence stratigraphy. A. A. P. G. Hedberg Research Conference on Sequence Stratigraphic and Allostratigraphic Principles and Concepts, Dallas, August 26–29, Program and Abstracts Volume, p. 22.

Droxler, A. W., and Schlager, W. (1985). Glacial versus interglacial sedimentation rates and turbidite frequency in the Bahamas. *Geology*, Vol. 13, pp. 799–802.

Drummond, C. N., and Wilkinson, B. H. (1996). Stratal thickness frequencies and the prevalence of orderedness in stratigraphic sequences. *Journal of Geology*, Vol. 104, pp. 1–18.

Duval, B, Cramez, C., and Vail, P. R. (1998). Stratigraphic cycles and major marine source rocks. In *Mesozoic and Cenozoic Sequence Stratigraphy of European Basins* (P. C. De Graciansky, J. Hardenbol, T. Jacquin and P. R. Vail, Eds.), pp. 43–51. SEPM Special Publication 60.

Eberth, D., A., and O'Connell (1995). Note on changing paleoenvironments across the Cretaceous-Tertiary boundary (Scollard Fomation) in the Red Deer River valley of southern Alberta. *Bulletin of Canadian Petroleum Geology*, Vol. 43, no. 1, pp. 44–53.

Einsele, G. (1992). *Sedimentary Basins: Evolution, Facies, and Sediment Budget*. Springer-Verlag, Berlin Heidelberg, p. 628.

Ekdale, A. A., Bromley, R. G., and Pemberton, S. G. (1984). *Ichnology: the use of trace fossils in sedimentology and stratigraphy*. Society of Economic Paleontologists and Mineralogists, Short Course Notes Number 15, p. 317

Elliott, T. (1986). Clastic shorelines. In *Sedimentary Environments and Facies* (H. G. Reading, Ed.), pp. 143–177. Blackwell Scientific Publications, Oxford.

Embry, A. F. (1993). Transgressive-regressive (T–R) sequence analysis of the Jurassic succession of the Sverdrup Basin, Canadian Arctic Archipelago. *Canadian Journal of Earth Sciences*, Vol. 30, pp. 301–320.

Embry, A. F. (1995). Sequence boundaries and sequence hierarchies: problems and proposals. In *Sequence stratigraphy on the Northwest European Margin* (R. J. Steel, V. L. Felt, E. P. Johannessen and C. Mathieu, Eds.), pp. 1–11. Norwegian Petroleum Society (NPF), Special Publication 5.

Embry, A. F. (2001a). Sequence stratigraphy: what it is, why it works and how to use it. *Reservoir (Canadian Society of Petroleum Geologists)*, Vol. 28, Issue 8, p. 15.

Embry, A. F. (2001b). *The six surfaces of sequence stratigraphy*. A. A. P. G. Hedberg Research Conference on Sequence Stratigraphic and Allostratigraphic Principles and Concepts, Dallas, August 26–29, Program and Abstracts Volume, pp 26–27.

Embry, A. F. (2002). Transgressive-regressive (T–) sequence stratigraphy. In *Sequence Stratigraphic Models for Exploration and Production: Evolving Methodology, Emerging Models and Application Histories* (J. M. Armentrout and N. C. Rosen, Eds.), pp. 151–172. 22nd Annual Gulf Coast Section SEPM Foundation, Bob F. Perkins Research Conference, Conference Proceedings.

Embry, A. F. (2005). Parasequences in third generation sequence stratigraphy. American Association of Petroleum Geologists Annual Convention, June 19–22, Calgary, Alberta, Abstracts Volume, Vol. 14, p. A41.

Embry, A. F., and Catuneanu, O. (2001). Practical Sequence Stratigraphy: Concepts and Applications. Canadian Society of Petroleum Geologists, short course notes, p. 167.

Embry, A. F., and Catuneanu, O. (2002). Practical Sequence Stratigraphy: Concepts and Applications. Canadian Society of Petroleum Geologists, short course notes, p. 147.

Embry, A. F., and Johannessen, E. P. (1992). T–R sequence stratigraphy, facies analysis and reservoir distribution in the uppermost Triassic-Lower Jurassic succession, western Sverdrup Basin, Arctic Canada. In *Arctic Geology and Petroleum Potential* (T. O. Vorren, E. Bergsager, O. A. Dahl-Stamnes, E. Holter, B. Johansen, E. Lie and T. B. Lund, Eds.), p. 121–146. Norwegian Petroleum Society (NPF), Special Publication 2.

Emery, D., and Myers, K. J. (1996). *Sequence Stratigraphy*. Oxford, U. K., Blackwell, p. 297.

Eriksson, K. A., Krapez, B., and Fralick, P. W. (1994). Sedimentology of greenstone belts: signature of tectonic evolution. *Earth Sciences Reviews*, Vol. 37, p. 1–88.

Eriksson, P. G., and Catuneanu, O. (2004a). Third-order sequence stratigraphy in the Palaeoproterozoic Daspoort Formation (Pretoria Group, Transvaal Supergroup), Kaapvaal Craton. In *The Precambrian Earth: Tempos and Events* (P. G. Eriksson, W. Altermann, D. Nelson, W. Mueller and O. Catuneanu, Eds.), pp. 724–735. Developments in Precambrian Geology 12, Elsevier, Amsterdam.

Eriksson, P. G., and Catuneanu, O. (2004b). A commentary on Precambrian plate tectonics. In *The Precambrian Earth: Tempos and Events* (P. G. Eriksson, W. Altermann, D. Nelson, W. Mueller and O. Catuneanu, Eds.), pp. 201–213. Developments in Precambrian Geology 12, Elsevier, Amsterdam.

Eriksson, P. G., Altermann, W., Nelson, D., Mueller, W., and Catuneanu, O. (Eds.) (2004). *The Precambrian Earth: Tempos and*

Events. Developments in Precambrian Geology **12**, Elsevier, Amsterdam, p. 941.

Eriksson, P. G., Catuneanu, O., Nelson, D. R., and Popa, M. (2005a). Controls on Precambrian sea level change and sedimentary cyclicity. *Sedimentary Geology*, Vol. **176**, pp. 43–65.

Eriksson, P. G., Catuneanu, O., Els, B. G., Bumby, A. J., van Rooy, J. L., and Popa, M. (2005b). Kaapvaal craton: changing first- and second-order controls on sea level from c. 3.0 Ga to 2.0 Ga. *Sedimentary Geology*, Vol. **176**, pp. 121–148.

Erlich, R. N., Barrett, S. F., and Guo, B. J. (1990). Seismic and geologic characteristics of drowning events on carbonate platforms. *American Association of Petroleum Geologists Bulletin*, Vol. **74**, pp. 1523–1537.

Erlich, R. N., Longo, A. P., Jr., and Hyare, S. (1993). Response of carbonate platform margins to drowning: evidence of environmental collapse. *In Carbonate Sequence Stratigraphy – Recent Developments and Applications* (R. G. Loucks and J. F. Sarg Eds.), pp. 241–266. American Association of Petroleum Geologists Memoir **57**.

Eschard, R., Desaubliaux, G., Lecomte, J. C., van Buchem, F. S. P., and Tveiten, B. (1993). High resolution sequence stratigraphy and reservoir prediction of the Brent Group (Tampen Spur area) using an outcrop analogue (Mesaverde Group, Colorado). *In Subsurface Reservoir Characterization from Outcrop Observations* (R. Eschard and B. Doligez, Eds.), pp. 35–52. Seventh Institute Francais du Petrole, Exploration and Production Research Conference, Scarborough, England, April 12–17, 1992, Proceedings.

Ethridge, F. G., Germanoski, D., Schumm, S. A., and Wood, L. J. (2001). The morphologic and stratigraphic effects of base-level change: a review of experimental studies. Seventh International Conference on Fluvial Sedimentology, Lincoln, August 6–10, Program and Abstracts, p. 95.

Fastovsky, D. E., and McSweeney, K. (1987). Paleosols spanning the Cretaceous-Paleogene transition, eastern Montana and western North Dakota. *Geological Society of America Bulletin*, Vol. **99**, pp. 66–77.

Feldman, H. R., Gibling, M. R., Archer, A. W., Wightman, W. G., and Lanier, W. P. (1995). Stratigraphic architecture of the Tonganoxie paleovalley fill (Lower Virgilian) in northeastern Kansas. *American Association of Petroleum Geologists Bulletin*, Vol. **79**, no. 7, pp. 1019–1043.

Fielding, C. R., and Alexander, J. (2001). Fossil trees in ancient fluvial channel deposits: evidence of seasonal and longer-term climatic variability. *Palaeogeography, Palaeoclimatology, Palaeoecology*, Vol. **170**, pp. 59–80.

Fielding, C. R., Sliwa, R., Holcombe, R. J., and Jones, A. T. (2001). A new palaeogeographic synthesis for the Bowen, Gunnedah and Sydney Basins of eastern Australia. *In Eastern Australasian Basins Symposium* (K. C. Hill and T. Bernecker, Eds.), pp. 269–278. Petroleum Exploration Society of Australia Special Publication, Melbourne.

Fischer, A. G., and Bottjer, D. J. (1991). Orbital forcing and sedimentary sequences (introduction to special issue). *Journal of Sedimentary Petrology*, Vol. **61**, pp. 1063–1069.

Fisher, W. L., and McGowen, J. H. (1967). Depositional systems in the Wilcox group of Texas and their relationship to occurrence of oil and gas. Gulf Coast Association Geological Society, Transactions **17**, pp. 105–125.

Flemings, P. B., and Jordan, T. E. (1989). A synthetic stratigraphic model of foreland basin development. *Journal of Geophysical Research*, Vol. **94**, pp. 3851–3866.

Frazier, D. E. (1974). Depositional episodes: their relationship to the Quaternary stratigraphic framework in the northwestern portion of the Gulf Basin. University of Texas at Austin, Bureau of Economic Geology, Geological Circular, Vol. **4(1)**, p. 28.

Frey, R. W., and Seilacher, A. (1980). Uniformity in marine invertebrate ichnology. *Lethaia*, Vol. **13**, pp. 183–207.

Frey, R. W., Howard, J. D., and Hong, J. S. (1987). Prevalent lebensspuren on a modern macrotidal flat, Inchon, Korea; ethological and environmental significance. *Palaios*, Vol. **2**, pp. 571–593.

Frey, R. W., Pemberton, S. G., and Saunders, T. D. A. (1990). Ichnofacies and bathymetry: a passive relationship. *Journal of Paleontology*, Vol. **64**, pp. 155–158.

Frostick, L. E., and Steel, R. J. (1993). *Tectonic control and signatures in sedimentary successions*. International Association of Sedimentologists Special Publication 20, Oxford, Blackwell, p. 520.

Fursich, F. R., and Mayr, H. (1981). Non-marine *Rhizocorallium* (trace fossils) from the Upper Freshwater Molasse (Upper Miocene) of southern Germany. *Neues Jahrbuch fur Geologie und Palaontologie, Monatshefte*, Vol. **6**, pp. 321–333.

Galloway, W. E. (1989). Genetic stratigraphic sequences in basin analysis, I. Architecture and genesis of flooding-surface bounded depositional units. *American Association of Petroleum Geologists Bulletin*, Vol. **73**, pp. 125–142.

Galloway, W. E. (2001). The many faces of submarine erosion: theory meets reality in selection of sequence boundaries. A. A. P. G. Hedberg Research Conference on Sequence Stratigraphic and Allostratigraphic Principles and Concepts, Dallas, August 26–29, Program and Abstracts Volume, pp. 28–29.

Galloway, W. E. (2004). Accommodation and the sequence stratigraphic paradigm. *Reservoir, Canadian Society of Petroleum Geologists*, Vol. **31**, Issue 5, pp. 9–10.

Galloway, W. E., and Hobday, D. K. (1996). *Terrigenous clastic depositional systems*. Second edition, Springer-Verlag, p. 489.

Galloway, W. E., and Sylvia, D. A. (2002). The many faces of erosion: theory meets data in sequence stratigraphic analysis. *In Sequence Stratigraphic Models for Exploration and Production: Evolving Methodology, Emerging Models and Application Histories* (J. M. Armentrout, and N. C. Rosen, Eds.), p. 8. 22nd Annual Gulf Coast Section SEPM Foundation, Bob F. Perkins Research Conference, Conference Proceedings, Program and Abstracts.

Gawthorpe, R. L., Fraser, A. J., and Collier, R. E. L. (1994). Sequence stratigraphy in active extensional basins: implications for the interpretation of ancient basin-fills. *Marine and Petroleum Geology*, Vol. **11**, p. 642–658.

Gawthorpe, R. L., Sharp, I., Underhill, J. R., and Gupta, S. (1997). Linked sequence stratigraphic and structural evolution of propagating normal faults. *Geology*, Vol. **25**, pp. 795–798.

Ghibaudo, G., Grandesso, P., Massari, F., and Uchman, A. (1996). Use of trace fossils in delineating sequence stratigraphic surfaces (Tertiary Venetian Basin, northeastern Italy). *Palaeogeography, Palaeoclimatology, Palaeoecology*, Vol. **120**, pp. 261–279.

Gibling, M. R., and Bird, D. J. (1994). Late Carboniferous cyclothems and alluvial paleovalleys in the Sydney Basin, Nova Scotia. *Geological Society of America Bulletin*, Vol. **106**, pp. 105–117.

Gibling, M. R., and Wightman, W. G. (1994). Palaeovalleys and protozoan assemblages in a Late Carboniferous cyclothem, Sydney Basin, Nova Scotia. *Sedimentology*, Vol. **41**, pp. 699–719.

Gibling, M. R., Tandon, S. K., Sinha, R. and Jain, M., (2005). Discontinuity-bounded alluvial sequences of the southern Gangetic Plains, India: aggradation and degradation in response to monsoonal strength. *Journal of Sedimentary Research*, Vol. **75**, no. 3, p. 369–385.

Gibson-Poole, C. M., Lang, S. C., Streit, J. E., Kraishan, G. M., and Hillis, R. R. (2002). Assessing a basin's potential for geological sequestration of carbon dioxide: an example from the Mesozoic of the Petrel Sub-basin, NW Australia. *In The sedimentary basins of Western Australia 3*, (M. Keep and S. J. Moss, Eds.), pp. 439–463 Proceedings, PESA, Perth, WA.

Gilbert, G. K. (1895). Sedimentary measurement of geologic time. *Journal of Geology*, Vol. **3**, pp. 121–127.

Giles, K. A., Bocko, M. Lawton, T. F. (1999). Stacked Late Devonian lowstand shorelines and their relation to tectonic subsidence at the Cordilleran hingeline, Western Utah. *Journal of Sedimentary Research*, Vol. **69**, no. 6, pp. 1181–1190.

Gill, J. R., and Cobban, W. A. (1973). Stratigraphy and Geologic History of the Montana Group and Equivalent Rocks, Montana, Wyoming, and North and South Dakota: United States Geological Survey, Professional Paper 776, p. 73.

Gingras, M. K., Pemberton, S. G., and Saunders, T. (2001). Bathymetry, sediment texture, and substrate cohesiveness; their impact on modern Glossifungites trace assemblages at Willapa Bay, Washington. *Palaeogeography, Palaeoclimatology, Palaeoecology*, Vol. **169**, p. 1–21.

Gingras, M. K., MacEachern, J. A., and Pickerill, R. K. (2004). Modern perspectives on the *Teredolites* ichnofacies: observations from Willapa Bay, Washington. *Palaios*, Vol. **19**, pp. 79–88.

Grabau, A. W. (1913). *Principles of Stratigraphy*. A. G. Seiler, New York, p. 1185

Graciansky de, P.-C., Hardenbol, J., Jacquin, T., and Vail, P. R. (Eds.) (1998). Mesozoic and Cenozoic Sequence Stratigraphy of European Basins. SEPM (Society for Sedimentary Geology), Special Publication 60, p. 786.

Gutzmer, J., and Beukes, N. J. (1998). Earliest laterites and possible evidence for terrestrial vegetation in the Early Proterozoic. *Geology*, Vol. **26**, pp. 263–266.

Hamblin, A. P. (1997). Regional distribution and dispersal of the Dinosaur Park Formation, Belly River Group, surface and subsurface of southern Alberta. *Bulletin of Canadian Petroleum Geology*, Vol. **45**, no. 3, pp. 377–399.

Hamilton, D. S., and Tadros, N. Z. (1994). Utility of coal seams as genetic stratigraphic sequence boundaries in non-marine basins: an example from the Gunnedah basin, Australia. *American Association of Petroleum Geologists Bulletin*, Vol. **78**, pp. 267–286.

Hampson, G. J., and Storms, J. E. A. (2003). Geomorphological and sequence stratigraphic variability in wave-dominated, shoreface-shelf parasequences. *Sedimentology*, Vol. **50**, pp. 667–701.

Haq, B. U. (1991). Sequence stratigraphy, sea-level change, and significance for the deep sea. *In Sedimentation, Tectonics and Eustasy* (D. I. M. Macdonald, Ed.), pp. 3–39. International Association of Sedimentologists Special Publication 12.

Haq, B. U., Hardenbol, J., and Vail, P. R. (1987). Chronology of fluctuating sea levels since the Triassic (250 million years ago to present). *Science*, Vol. **235**, pp. 1156–1166.

Haq, B. U., Hardenbol, J., and Vail, P. R. (1988). Mesozoic and Cenozoic chronostratigraphy and cycles of sea-level change. *In Sea Level Changes–An Integrated Approach* (C. K. Wilgus, B. S. Hastings, C. G. St. C. Kendall, H. W. Posamentier, C. A. Ross and J. C. Van Wagoner, Eds.), pp. 71–108. SEPM Special Publication 42.

Harbaugh, J. W., Watney, W. L., Rankey, E. C., Slingerland, R., Goldstein, R. H., and Franseen, E. K. (Eds.) (1999). *Numerical Experiments in Stratigraphy: Recent Advances in Stratigraphic and Sedimentologic Computer Simulations*. SEPM (Society for Sedimentary Geology) Special Publication **62**, p. 362.

Harms, J. C., and Fahnestock, R. K. (1965). Stratification, bed forms, and flow phenomena (with an example from the Rio Grande). *In Primary sedimentary structures and their hydrodynamic interpretation* (G. V. Middleton, Ed.), pp. 84–115. Society of Economic Paleontologists and Mineralogists, Special Publication 12.

Hart, B. S. (2000). *3-D Seismic Interpretation: A Primer for Geologists*. SEPM (Society for Sedimentary Geology) Short Course No. 48, p. 123

Hart, B. S., and Plint, A. G. (1993). Origin of an erosion surface in shoreface sandstones of the Kakwa Member (Upper Cretaceous Cardium Formation, Canada): importance for reconstruction of stratal geometry and depositional history. *In Sequence Stratigraphy and Facies Associations* (H. W. Posamentier, C. P. Summerhayes, B. U. Haq and G. P. Allen Eds.), pp. 451–467. International Association of Sedimentologists, Special Publication 18.

Hayward, B. W. (1976). Lower Miocene bathyal and submarine canyon ichnocoenoses from Northland, New Zealand. *Lethaia*, Vol. **9**, pp. 149–162.

Heckel, P. H. (1986). Sea-level curve for Pennsylvanian eustatic marine transgressive-regressive depositional cycles along midcontinent outcrop belt, North America. *Geology*, Vol. **14**, pp. 330–334.

Heckel, P. H., Gibling, M. R., and King, N. R. (1998). Stratigraphic model for glacial-eustatic Pennsylvanian cyclothems in highstand nearshore detrital regimes. *The Journal of Geology*, Vol. **106**, pp. 373–383.

Hedberg, H. D. (1976). International stratigraphic guide: *A guide to stratigraphic classification, terminology, and procedure*. International Union of Geological Sciences, Commission on Stratigraphy, International Subcommission on Stratigraphic Classification. New York, Wiley, p. 200.

Helland-Hansen, W., and Martinsen, O. J. (1996). Shoreline trajectories and sequences: description of variable depositional-dip scenarios. *Journal of Sedimentary Research*, Vol. **66**, no. 4, pp. 670–688.

Helland-Hansen W., Lomo, L., Steel, R., and Ashton, M. (1992). Advance and retreat of the Brent delta: recent contributions of the depositional model. *In Geology of the Brent Group* (A. C. Morton, R. S. Haszeldine, M. R. Giles and S. Brown, Eds.), pp. 109–127. Geological Society of London Special Publication No **61**.

Heller, P. L., Angevine, C. L., and Winslow, N. S. (1988). Two-phase stratigraphic model of foreland-basin sequences. *Geology*, Vol. **16**, pp. 501–504.

Hernandez-Molina, F. J. (1993). Dinamica sedimentaria y evolucion durante el Pleistoceno terminal – Holoceno del Margen Noroccidental del Mar de alboran: Modelo de estratigrafia secuencial de muy alta resolucion en plataformas continentales. Ph.D. Thesis, University of Granada, Spain, p. 618.

Hine, A. C., Wilber, R. J., Bane, J. M., Neumann, A. C., and Lorenson, K. R. (1981). Offbank transport of carbonate sands along leeward bank margins, northern Bahamas. *Marine Geology*, Vol. **42**, pp. 327–348.

Hine, A. C., Locker, S. D., Tedesco, L. P., Mullins, H. T., Hallock, P., Belknap, D. F., Gonzales, J. L., Neumann, A. C., and Snyder, S. W. (1992). Megabreccia shedding from modern low-relief carbonate platforms, Nicaraguan Rise. *Geological Society of America Bulletin*, Vol. **104**, pp. 928–943.

Holbrook, J. M. (1996). Complex fluvial response to low gradients at maximum regression: a genetic link between smooth sequence-boundary morphology and architecture of overlying sheet sandstone. *Journal of Sedimentary Research*, Vol. **66**, no. 4, pp. 713–722.

Holbrook, J. M. (2001). Origin, genetic interrelationships, and stratigraphy over the continuum of fluvial channel-form bounding surfaces: an illustration from middle Cretaceous strata, southeastern Colorado. *Sedimentary Geology*, Vol. **144**, pp. 179–222.

Holbrook, J. M., and Schumm, S. A. (1998). Geomorphic and sedimentary response of rivers to tectonic deformation: a brief review and critique of a tool for recognizing subtle epeirogenic deformation in modern and ancient settings. *Tectonophysics*, Vol. **305**, pp. 287–306.

Holbrook, J. M., and White, D. C. (1998). Evidence for subtle uplift from lithofacies distribution and sequence architecture: examples from Lower Cretaceous strata of northeastern New Mexico. *In Relative role of eustasy, climate, and tectonism in continental rocks*

(K. W. Shanley and P. J. McCabe, Eds.), pp. 123–132. SEPM (Society for Sedimentary Geology) Special Publication No. 59.

Holbrook, J. M., and Wright Dunbar, R. (1992). Depositional history of Lower Cretaceous strata in northeastern New Mexico: implications for regional tectonics and depositional sequences. *Geological Society of America Bulletin*, Vol. 104, pp. 802–813.

Hunt, D. (1992). Application of sequence stratigraphic concepts to the Urgonian Carbonate Platform, SE France. Ph.D. Thesis, University of Durham, p. 410.

Hunt, D., and Tucker, M. E. (1992). Stranded parasequences and the forced regressive wedge systems tract: deposition during base-level fall. *Sedimentary Geology*, Vol. 81, pp. 1–9.

Hunt, D., and Tucker, M. E. (1993). Sequence stratigraphy of carbonate shelves with an example from the mid-Cretaceous (Urgonian) of southeast France. *In Sequence Stratigraphy and Facies Associations* (H. W. Posamentier, Summerhayes, C. P., Haq, B. U., and Allen, G. P. Eds.), pp. 307–341. International Association of Sedimentologists, Special Publication 18.

Hunt, D., and Tucker, M. E. (1995). Stranded parasequences and the forced regressive wedge systems tract: deposition during base-level fall – reply. *Sedimentary Geology*, Vol. 95, pp. 147–160.

Imbrie, J. (1985). A theoretical framework for the Pleistocene ice ages. *Journal of the Geological Society (London)*, Vol. 142, pp. 417–432.

Imbrie, J., and Imbrie, K. P. (1979). *Ice ages: solving the mystery*. Enslow, Hillside, New Jersey, p. 224.

Isbell, J. L., and Cuneo, N. R. (1996). Depositional framework of Permian coal-bearing strata, southern Victoria Land, Antarctica. *Palaeogeography, Palaeoclimatology, Palaeoecology*, Vol. 125, pp. 217–238.

Issler, D. R., Willett, S. D., Beaumont, C., Donelick, R. A., and Grist, A. M. (1999). Paleotemperature history of two transects across the Western Canada Sedimentary Basin: constraints from apatite fission track analysis. *Bulletin of Canadian Petroleum Geology*, Vol. 47, no. 4, pp. 475–486.

Ito, M., and O'Hara, S. (1994). Diachronous evolution of systems tracts in a depositional sequence from the middle Pleistocene palaeo-Tokyo Bay, Japan: *Sedimentology*, Vol. 41, pp. 677–697.

Jackson, J. M., Simpson, E. L., and Eriksson, K. A. (1990). Facies and sequence stratigraphic analysis in an intracratonic, thermal-relaxation basin: the middle Proterozoic, lower Quilalar Formation, Mount Isa Orogen, Australia. *Sedimentology*, Vol. 37, pp. 1053–1078.

James, N. P., and Kendall, A. C. (1992). Introduction to carbonate and evaporite facies models. *In Facies Models: Response to Sea Level Change* (R. G. Walker and N. P. James, Eds.), pp. 265–275. Geological Association of Canada, GeoText 1.

Jervey, M. T. (1988). Quantitative geological modeling of siliciclastic rock sequences and their seismic expression. *In Sea Level Changes–An Integrated Approach* (C. K. Wilgus, B. S. Hastings, C. G. St.C. Kendall, H. W. Posamentier, C. A. Ross and J. C. Van Wagoner, Eds.), pp. 47–69. SEPM Special Publication 42.

Johnson, D. D., and Beaumont, C. (1995). Preliminary results from a planform kinematic model of orogen evolution, surface processes and the development of clastic foreland basin stratigraphy. *In Stratigraphic Evolution of Foreland Basins* (S. L. Dorobek and G. M. Ross, Eds.), pp. 3–24. SEPM (Society for Sedimentary Geology) Special Publication 52.

Johnson, M. R. (1991). Discussion on Chronostratigraphic subdivision of the Witwatersrand Basin based on a Western Transvaal composite column. *South African Journal of Geology*, Vol. 94, pp. 401–403.

Jones, B and Desrochers, A. (1992). Shallow platform carbonates. *In Facies Models: Response to Sea Level Change* (R. G. Walker and N. P. James, Eds.), pp. 277–301 Geological Association of Canada, GeoText 1.

Jordan, T. E., and P. B. Flemings (1991). Large-scale stratigraphic architecture, eustatic variation, and unsteady tectonism: a theoretical approach: *Journal of Geophysical Research*, Vol. 96 (B4), pp. 6681–6699.

Kamola, D. L., and Van Wagoner, J. C. (1995). Stratigraphy and facies architecture of parasequences with examples from the Spring Canyon Member, Blackhawk Formation, Utah. *In Sequence Stratigraphy of Foreland Basin Deposits* (J. C. Van Wagoner and G. T. Bertram, Eds.), pp. 27–54. American Association of Petroleum Geologists Memoir 64.

Karner, G. D. (1986). Effects of lithospheric in-plane stress on sedimentary basin stratigraphy. *Tectonics*, Vol. 5, pp. 573–588.

Karner, G. D., Driscoll, N. W., and Weissel, J. K. (1993). Response of the lithosphere to in-plane force variations. *Earth and Planetary Sciences Letters*, Vol. 114, pp. 397–416.

Kauffman, E. G. (1984). Paleobiogeography and evolutionary response in the Cretaceous Western Interior Seaway of North America. *In Jurassic-Cretaceous Biochronology and Paleogeography of North America* (G. E. G. Westerman, Ed.). Geological Association of Canada, Special Paper 27, pp. 273–306.

Kerr, D., Ye, L., Bahar, A., Kelkar, B. G., and Montgomery, S. (1999). Glenn Pool field, Oklahoma: a case of improved prediction from a mature reservoir. *American Association of Petroleum Geologists Bulletin*, Vol. 83, no. 1, pp. 1–18.

Ketzer, J. M., Morad, S., and Amorosi, A. (2003a). Predictive diagenetic clay-mineral distribution in siliciclastic rocks within a sequence stratigraphic framework. *In Clay Mineral Cements in Sandstones* (R. H. Worden and S. Morad, Eds.), pp. 43–61. International Association of Sedimentologists, Special Publication No. 34.

Ketzer, J. M., Holz, M., Morad, S., and Al-Aasm, I. S. (2003b). Sequence stratigraphic distribution of diagenetic alterations in coal-bearing, paralic sandstones: evidence from the Rio Bonito Formation (early Permian), southern Brazil. *Sedimentology*, Vol. 50, pp. 855–877.

Khalifa, M. A. (1983). Origin and occurrence of glauconite in the green sandstone associated with unconformity, Bahariya Oases, Western Desert, Egypt. *Journal of African Earth Sciences*, Vol. 31, no. 3/4, pp. 321–325.

Khidir, A., and Catuneanu, O. (2003). Sedimentology and diagenesis of the Scollard sandstones in the Red Deer Valley area, central Alberta. *Bulletin of Canadian Petroleum Geology*, Vol. 51, no. 1, pp. 45–69.

Khidir, A., and Catuneanu, O. (2005). Predictive diagenetic clay-mineral distribution in siliciclastic rocks within a nonmarine sequence stratigraphic framework: The Coalspur Formation, west-central Alberta. American Association of Petroleum Geologists Annual Convention, Calgary, Abstracts Volume 14, p. A73.

Kocurek, G. (1988). First-order and super bounding surfaces - bounding surfaces revisited. *In Late Paleozoic and Mesozoic Eolian deposits of the Western Interior of the United States* (G. Kocurek, Ed.). *Sedimentary Geology*, Vol. 56, pp. 193–206.

Kolla, V. (1993). Lowstand deep-water siliciclastic depositional systems: characteristics and terminology in sequence stratigraphy and sedimentology. *Bulletin of the Center for Research and Exploration-Production Elf Aquitaine*, Vol. 17, pp. 67–78.

Kolla, V., Bourges, P., Urruty, J. M., and Safa, P. (2001). Evolution of deep-water Tertiary sinuous channels offshore Angola (west Africa) and implications for reservoir architecture. *American Association of Petroleum Geologists Bulletin*, Vol. 85, no. 8, pp. 1373–1405.

Komar, P. D. (1976). *Beach Processes and Sedimentation*, Prentice-Hall, New Jersey, p. 429.

Kominz, M. A., and Pekar, S. F. (2001). Oligocene eustasy from two-dimensional sequence stratigraphic backstripping. *Geological Society of America Bulletin*, Vol. 113, pp. 291–304.

Kominz, M. A., Miller, K. G., and Browning J. V. (1998). Long-term and short-term global Cenozoic sea-level estimates. *Geology*, Vol. 26, pp. 311–314.

Koss, J. E., Ethridge, F. G., and Schumm, S. A. (1994). An experimental study of the effects of base-level change on fluvial, coastal plain, and shelf systems. *Journal of Sedimentary Research*, Vol. B64, pp. 90–98.

Kotake, N. (1989). Paleoecology of the *Zoophycos* producers. *Lethaia*, Vol. 22, pp. 327–341.

Kotake, N. (1991). Non selective surface deposit feeding by the *Zoophycos* producers. *Lethaia*, Vol. 24, pp. 379–385.

Kraft, J. C., Chrzastowski, M. J., Belknap, D. F., Toscano, M. A., and Fletcher, C. H., III (1987). The transgressive barrier-lagoon coast of Delaware: morphostratigraphy, sedimentary sequences, and responses to relative rise in sea level. *In Sea Level Fluctuation and Coastal Evolution* (D. Nummedal, O. H. Pilkey and J. D. Howard, Eds.), pp. 129–143. SEPM Special Publication 41.

Krapez, B. (1993). Sequence stratigraphy of the Archaean Supracrustal-belts of the Pilbara Block, Western Australia. *Precambrian Research*, Vol. 60, pp. 1–45.

Krapez, B. (1996). Sequence-stratigraphic concepts applied to the identification of basin-filling rhythms in Precambrian successions. *Australian Journal of Earth Sciences*, Vol. 43, pp. 355–380.

Krapez, B. (1997). Sequence-stratigraphic concepts applied to the identification of depositional basins and global tectonic cycles. *Australian Journal of Earth Sciences*, Vol. 44, pp. 1–36.

Kraus, M. J. (1999). Paleosols in clastic sedimentary rocks: their geological applications. *Earth-Science Reviews*, Vol. 47, pp. 41–70.

Krawinkel, H., and Seyfried, H. (1996). Sedimentologic, palaeoecologic, taphonomic and Ichnologic criteria for high-resolution sequence analysis; a practical guide for the identification and interpretation of discontinuities in shelf deposits. *Sedimentary Geology*, Vol. 102, no. 1–2, pp. 79–110.

Lambeck, K. (1980). *The earth's variable rotation.* Cambridge University Press, Cambridge, p. 449.

Lander, R. H., Bloch, S., Mehta, S., and Atkinson, C. D. (1991). Burial diagenesis of paleosols in the giant Yacheng gas field, People's Republic of China: bearing on illite reactivation pathways. *Journal of Sedimentary Petrology*, Vol. 61, pp. 256–268.

Lang, S. C., Grech, P., Root, R., Hill, A., and Harrison, D. (2001). The application of sequence stratigraphy to exploration and reservoir development in the Cooper-Eromanga-Bowen-Surat Basin system. *The APPEA Journal*, Vol. 41, pp. 223–250.

Lang, S. C., Payenberg, T., Reilly, M., Krapf, C., Waclawik, V., Kassan, J., and Menacherry, S. (2005). Dryland fluvial-lacustrine reservoir analogues from the Lake Eyre Basin, Australia. *American Association of Petroleum Geologists*, Abstracts Volume, Vol. 14, p. A77.

Leckie, D. A. (1994). Canterbury Plains, New Zealand - Implications for sequence stratigraphic models. *American Association of Petroleum Geologists Bulletin*, Vol. 78, p. 1240–1256.

Leckie, D. A., and Boyd, R. (2003). Towards a nonmarine sequence stratigraphic model. *American Association of Petroleum Geologists Annual Convention*, Salt Lake City, 11–14 May 2003, Official Program, Vol. 12, p. A101.

Leckie, D. A., Fox., C., and Tarnocai, C. (1989). Multiple paleosols of the late Albian Boulder Creek Formation, British Columbia, Canada. *Sedimentology*, Vol. 36, pp. 307–323.

Leckie, D. A., Wallace-Dudley, K. E., Vanbeselaere, N. A., and James, D. P. (2004). Sedimentation in a low-accommodation setting: nonmarine (Cretaceous) Mannville and marine (Jurassic) Ellis Groups, Manyberries Field, southeastern Alberta. *American Association of Petroleum Geologists Bulletin*, Vol. 88, no. 10, pp. 1391–1418.

Leeder, M. R., and Gawthorpe, R. L. (1987). Sedimentary models for extensional tilt-block/half-graben basins. *In Continental extension tectonics* (M. P. Coward, J. F. Dewey and P. L. Hancock, Eds.), pp. 139–152. Geological Society, London, Special Publication 28.

Legaretta, L., Uliana, M. A., Larotonda, C. A., and Meconi, G. R. (1993). Approaches to nonmarine sequence stratigraphy – theoretical models and examples from Argentine basins. *In Subsurface Reservoir Characterization from Outcrop Observations* (R. Eschard and B. Doliez (Eds.), pp. 125–145. Paris, Editions Technip, Collection Colloques et Seminaires, no. 51.

Leopold, L. B., and Bull, W. B. (1979). Base level, aggradation, and grade. *American Philosophical Society, Proceedings*, Vol. 123, pp. 168–202.

Leszczynski, S., and Seilacher, A. (1991). Ichnocoenoses of a turbidite sole. *Ichnos*, Vol. 1, p. 293–303.

Lindsay, J. F., Kennard, J. M., and Southgate, P. N. (1993). Application of sequence stratigraphy in an intracratonic setting, Amadeus Basin, central Australia. *In Sequence Stratigraphy and Facies Associations* (H. W. Posamentier, C. P. Summerhayes, B. U. Haq and G. P. Allen, Eds.), pp. 605–631. International Association of Sedimentologists, Special Publication 18.

Locker, S. D., Hine, A. C., Tedesco, L. P., and Shinn, E. A. (1996). Magnitude and timing of episodic sea-level rise during the last deglaciation. *Geology*, Vol. 24, pp. 827–830.

Long, D. G. F. (1993). Limits on late Ordovician eustatic sea-level change from carbonate shelf sequences: and example from Anticosti Island, Quebec. *In Sequence Stratigraphy and Facies Associations* H. W. Posamentier, C. P. Summerhayes, B. U. Haq and G. P. Allen, Eds.), pp. 487–499. International Association of Sedimentologists, Special Publication 18.

Long, D. G. F., and Norford, B. S. (1997). Following coastlines: The paleogeography of Canada from the Late Precambrian to the Early Ordovician. Ontario Petroleum Institute, Technical Volume 36, Technical Paper 14, p. 23.

Loucks, R. G., and Sarg, J. F. (eds.) Carbonate Sequence Stratigraphy – Recent Developments and Applications. *American Association of Petroleum Geologists Memoir*, 57, p.545.

Loutit, T. S., Hardenbol, J., Vail, P. R., and Baum, G. R. (1988). Condensed sections: the key to age-dating and correlation of continental margin sequences. *In Sea Level Changes–An Integrated Approach*(C. K. Wilgus, B. S. Hastings, C. G. St.C. Kendall, H. W. Posamentier, C. A. Ross and J. C. Van Wagoner, Eds.), pp. 183–213. SEPM Special Publication 42.

Lyell, C. (1868). *Principles of Geology*. J. Murray, London, 10th edition, p. 670.

Macdonald, D. I. M. (Ed.) (1991). Sedimentation, tectonics and eustasy: sea-level changes at active margins, International Association of Sedimentologists Special Publication 12, Oxford, Blackwell, p. 518.

MacEachern, J. A., Pemberton, S. G., and Raychaudhuri, I. (1991). The substrate-controlled *Glossifungites* Ichnofacies and itsapplication to the recognition of sequence stratigraphic surfaces: subsurface examples from the Cretaceous of the Western Canada Sedimentary Basin, Alberta, Canada. *In 1991 NUNA Conference on High Resolution Sequence Stratigraphy* (D. A. Leckie, H. W. Posamentier and R. W. W. Lovell, Eds.), pp. 32–36. Geological Association of Canada, Program, Proceedings and Guidebook.

MacEachern, J. A., Raychaudhuri, I., and Pemberton, S. G. (1992). Stratigraphic applications of the *Glossifungites* ichnofacies: delineating discontinuities in the rock record. *In Applications of Ichnology to Petroleum Exploration* (S. G. Pemberton, Ed.), pp. 169–198. Society of Economic Paleontologists and Mineralogists, Core Workshop Notes 17.

MacEachern, J. A., Zaitlin, B. A., and Pemberton, S. G. (1998). High resolution sequence stratigraphy of early transgressive incised shoreface and early transgressive valley/embayment deposits of the Viking Formation, Joffre Field, Alberta, Canada. *American Association of Petroleum Geologists Bulletin*, Vol. **82**, pp. 729–756.

MacEachern, J. A., Pemberton, S. G., and Zaitlin, B. A. (1999). A late lowstand tongue, Viking Formation of the Joffre Field, Alberta. *In Isolated Marine Sand Bodies: Sequence Stratigraphic Analysis and Sedimentological Interpretation* (K. Bergman, Ed.), pp. 273–296. Society of Economic Paleontologists and Mineralogists, Special Publication **64**.

Mack, G. H., James, W. C., and Monger, H. C. (1993). Classification of paleosols. *Geological Society of America Bulletin*, Vol. **105**, pp. 129–136.

MacNeil, A. J., and Jones, B., in press. Sequence stratigraphy of a Late Devonian ramp-situated reef system in the Western Canada Sedimentary Basin dynamic responses to sea-level change and regressive reef development. *Sedimentology*.

Marriott, S. B. (1999). The use of models in the interpretation of the effects of base-level change on alluvial architecture. *In Fluvial Sedimentology VI, Proceedings* (N. D. Smith and J. Rogers, Eds.), *Special Publication of the International Association of Sedimentologists*, Vol. **28**, pp. 271–281.

Marriott, S. B., and Wright, V. P. (1993). Paleosols as indicators of geomorphic stability in two Old Red Sandstones alluvial suites, South Wales. *Journal of the Geological Society of London*, Vol. **150**, pp. 1109–1120.

Martinsen, O. J., and W. Helland-Hansen (1995). Strike variability of clastic depositional systems: does it matter for sequence stratigraphic analysis? *Geology*, Vol. **23**, pp. 439–442.

Masterson, W. D., and Eggert, J. T. (1992). Kuparuk River Field, North Slope, Alaska. *In Atlas of Oil and Gas Fields* (E. A. Beaumont and N. H. Foster, Eds.), pp. 257–284. American Association of Petroleum Geologists Treatise of Petroleum Geology.

Masterson, W. D., and Paris, C. E. (1987). Depositional history and reservoir description of the Kuparuk River Formation, North Slope, Alaska. *In Alaskan North Slope Geology* (I. Tailleur and P. Weimer, Eds.), pp. 95–107. SEPM Pacific Section and Alaska Geological Society, Vol. **50**.

Matthews, R. K. (1984). *Dynamic Stratigraphy*. Englewood Cliffs, New Jersey, Prentice-Hall, p. 489.

Meyer, F. O. (1989). Siliciclastic influence on Mesozoic platform development: Baltimore Canyon Trough, western Atlantic. *In Controls on Carbonate Platform and Basin Development* (P. D. Crevello, J. L. Wilson, J. F. Sarg and J. F. Read, Eds.), pp. 211–232. SEPM Special Publication **44**.

Miall, A. D. (1986). Eustatic sea level changes interpreted from seismic stratigraphy: a critique of the methodology with particular reference to the North Sea Jurassic record. *American Association of Petroleum Geologists Bulletin*, Vol. **70**, pp. 131–137.

Miall, A. D. (1990). *Principles of Sedimentary Basin Analysis*. Second Edition, Springer, p. 668.

Miall, A. D. (1991). Stratigraphic sequences and their chronostratigraphic correlation. *Journal of Sedimentary Petrology*, Vol. **61**, pp. 497–505.

Miall, A. D. (1992). Exxon global cycle chart: an event for every occasion? *Geology*, Vol. **20**, pp. 787–790.

Miall, A. D. (1994). Sequence stratigraphy and chronostratigraphy: problems of definition and precision in correlation, and their implications for global eustasy: *Geoscience Canada*, Vol. **21**, pp. 1–26.

Miall, A. D. (1995). Whither stratigraphy? *Sedimentary Geology*, Vol. **100**, pp. 5–20.

Miall, A. D. (1996). *The Geology of Fluvial Deposits*. Springer-Verlag, p. 582.

Miall, A. D. (1997). *The Geology of Stratigraphic Sequences*. Springer-Verlag, p. 433.

Miall, A. D. (1999). Cryptic sequence boundaries in braided fluvial successions. *American Association of Petroleum Geologists Annual Meeting* (San Antonio), Official Program, p. A93.

Miall, A. D. (2000). *Principles of Sedimentary Basin Analysis*. Third Edition, Springer, p. 616.

Miall, A. D. (2002). Architecture and sequence stratigraphy of Pleistocene fluvial systems in the Malay Basin, based on seismic time-slice analysis. *American Association of Petroleum Geologists Bulletin*, Vol. **86**, no. 7, pp. 1201–1216.

Miall, A. D., and Miall, C. E. (2001). Sequence stratigraphy as a scientific enterprise: the evolution and persistence of conflicting paradigms. *Earth-Science Reviews*, Vol. **54**, pp. 321–348.

Middleton, G. V. (1973). Johannes Walther's Law of the Correlation of Facies. *Geological Society of America Bulletin*, Vol. **84**, p. 979–988.

Milankovitch, M. (1930). Mathematische Theorie der Klimaschwankungen. *In Handbuch der Klimatologie* (W. Koppen and R. Geiger, Eds.), I (A). Gebruder Borntraeger, Berlin.

Milankovitch, M. (1941). Kanon der Erdbestrahlung und seine Anwendung auf das Eiszeitenproblem. Akad Royale Serbe 133, p. 633.

Miller, D. J., and Eriksson, K. A. (1999). Linked sequence development and global climate change: the upper Mississippian record in the Appalachian basin. *Geology*, Vol. **27**, pp. 35–38.

Miller, D. J., and Eriksson, K. A. (2000). Sequence stratigraphy of Upper Mississippian strata in the central Appalachians: a record of glacioeustasy and tectonoeustasy in a foreland basin setting. *American Association of Petroleum Geologists Bulletin*, Vol. **84**, no. 2, pp. 210–233.

Miller, K. G., Wright, J. D., and Fairbanks, R. G. (1991). Unlocking the ice house: Oligocene-Miocene oxygen isotopes, eustasy, and margin erosion. *Journal of Geophysical Research*, Vol. **96**, pp. 6829–6848.

Miller, K. G., Mountain, G. S., the Leg 150 Shipboard Party, and Members of the New Jersey Coastal Plain Drilling Project (1996). Drilling and dating New Jersey Oligocene-Miocene sequences: ice volume, global sea level, and Exxon records. *Science*, Vol. **271**, pp. 1092–1094.

Miller, K. G., Mountain, G. S., Browning, J. V., Kominz, M. A., Sugarman, P. J., Christie-Blick, N., Katz, M. E., and Wright, J. D. (1998). Cenozoic global sea-level, sequences, and the New Jersey transect: results from coastal plain and slope drilling. *Reviews of Geophysics*, Vol. **36**, pp. 569–601.

Miller, K. G., Sugarman, P. J., Browning, J. V., Kominz, M. A., Hernandez, J. C., Olsson, R. K., Wright, J. D., Feigenson, M. D., and Van Sickel, W. (2003). Late Cretaceous chronology of large, rapid sea-level changes: glacioeustasy during the greenhouse world. *Geology*, Vol. **31**, pp. 585–588.

Miller, K. G., Sugarman, P. J., Browning, J. V., Kominz, M. A., Olsson, R. K., Feigenson, M. D., and Hernandez, J. C. (2004). Upper Cretaceous sequences and sea-level history, New Jersey Coastal Plain. *Geological Society of America Bulletin*, Vol. **116**, no. 3/4, pp. 368–393.

Miller, W. (1993). Trace fossil zonation in Cretaceous turbidite facies, northern California. *Ichnos*, Vol. **3**, pp. 11–28.

Mitchum, R. M., Jr. (1977). Seismic stratigraphy and global changes of sea level, part 11: glossary of terms used in seismic stratigraphy. *In Seismic Stratigraphy–Applications to Hydrocarbon Exploration* (C. E. Payton, Ed.), pp. 205–212. American Association of Petroleum Geologists Memoir **26**.

Mitchum, R. M. Jr., and Vail, P. R. (1977). Seismic stratigraphy and global changes of sea-level, part 7: stratigraphic interpretation of seismic reflection patterns in depositional sequences. *In Seismic*

Stratigraphy–Applications to Hydrocarbon Exploration (C. E. Payton, Ed.), pp. 135–144. American Association of Petroleum Geologists Memoir **26**.

Mitchum, R. M. Jr., and Van Wagoner, J. C. (1991). High-frequency sequences and their stacking patterns: sequence stratigraphic evidence of high-frequency eustatic cycles. *Sedimentary Geology*, Vol. **70**, pp. 131–160.

Mitchum, R. M. Jr., Vail, P. R., and Thompson, S., III (1977). Seismic stratigraphy and global changes of sea-level, part 2: the depositional sequence as a basic unit for stratigraphic analysis. *In Seismic Stratigraphy–Applications to Hydrocarbon Exploration* (C. E. Payton, Ed.), pp. 53–62. American Association of Petroleum Geologists Memoir **26**.

Morrison, R. B. (1978). Quaternary soil stratigraphy – concepts, methods, and problems. *In Quaternary Soils* (W. C. Mahaney Ed.), pp. 77–108. Geo Abstracts, Norwich.

Mossop, G. D., and Shetsen, I. (compilers) (1994). Geological Atlas of the Western Canada Sedimentary Basin. Canadian Society of Petroleum Geologists and Alberta Research Council, Calgary, pp. 510.

Mutti, E. (1992). *Turbidite sandstones*. AGIP Special Publications, p. 275.

Mutti, E., Seguret, M., and Sgavetti, M. (1988). Sedimentation and deformation in the Tertiary sequences of the Southern Pyrenees. American Association of Petroleum Geologists, Mediterranean Basins Conference, September 1988, Special Publication of the Institute of Geology of the University of Parma, Italy, p. 153.

NACSN: North American Commission on Stratigraphic Nomenclature (1983). North American Stratigraphic Code. *American Association of Petroleum Geologists Bulletin*, Vol. **67**, pp. 841–875.

Naish, T., and Kamp, P. J. J. (1997). Foraminiferal depth palaeoecology of Late Pliocene shelf sequences and systems tracts, Wanganui Basin, New Zealand. *Sedimentary Geology*, Vol. **110**, pp. 237–255.

Neumann, A. C., and Land, L. S. (1975). Lime mud deposition and calcareous algae in Bight of Abaco, Bahamas – Budget. *Journal of Sedimentary Research*, Vol. **45**, pp. 763–786.

Nummedal, D. (1992). The falling sea-level systems tract in ramp settings. In SEPM Theme Meeting, Fort Collins, Colorado (abstracts), p. 50.

Nummedal, D., Riley, G. W., and Templet, P. L. (1993). High-resolution sequence architecture: a chronostratigraphic model based on equilibrium profile studies. *In Sequence Stratigraphy and Facies Associations* (H. W. Posamentier, C. P. Summerhayes, B. U. Haq and G. P. Allen, Eds.), pp. 55–68. International Association of Sedimentologists Special Publication 18.

Obradovich, J. D. (1993). A Cretaceous time scale. In Evolution of the Western Interior Basin. W. G. E. Caldwell and E. G. Kauffman (eds.) Geological Association of Canada, Special Paper 39, pp. 379–396.

Olivero, D., and Gaillard, C. (1996). Paleoecology of the Jurassic Zoophycos from southeastern France. *Ichnos*, Vol. **4**, pp. 249–260.

Olsen, T., Steel, R., Høgseth, K., Skar, T., and Røe, S. L. (1995). Sequential architecture in a fluvial succession: sequence stratigraphy in the Upper Cretaceous Mesaverde Group, Price Canyon, Utah. *Journal of Sedimentary Research*, Vol. **B65**, p. 265–280.

Olsson, R. K., Miller, K. G., Browning, J. V., Wright, J. D., and Cramer, B. S. (2002). Sequence stratigraphy and sea level change across the Cretaceous-Tertiary boundary on the New Jersey passive margin. *In Catastrophic events and mass extinctions: impacts and beyond* (C. Koeberl and K. G. MacLeod, Eds.), Geological Society of America Special Paper 356, pp. 97–108.

Paola, C. (2000). Quantitative models of sedimentary basin filling. *Sedimentology*, Vol. **47** (suppl. 1), pp. 121–178.

Paola, C., Parker, G., Mohrig, D. C., and Whipple, K. X. (1999). The influence of transport fluctuations on spatially averaged topography on a sandy, braided fluvial fan. *In Numerical Experiments in Stratigraphy: Recent Advances in Stratigraphic and Sedimentologic Computer Simulations* (J. W. Harbaugh, W. L. Watney, E. C. Rankey, R. Slingerland, R. H. Goldstein and E. K. Franseen, Eds.), pp. 211–218. SEPM (Society for Sedimentary Geology) Special Publication 62.

Paola, C. J., Mullin, C. Ellis, D. C. Mohrig, J. B. Swenson, G. Parker, T. Hickson, P. L. Heller, L. Pratson, J. Syvitski, B. Sheets, and N. Strong (2001). Experimental stratigraphy. *GSA Today*, Vol. **11** (7), pp. 4–9.

Payton, C. E. (Ed.) (1977). Seismic Stratigraphy – Applications to Hydrocarbon Exploration. *American Association of Petroleum Geologists Memoir* 26, p. 516.

Pekar, S. F., and Kominz, M. A. (2001). Two-dimensional paleoslope modeling: a new method for estimating water depths for benthic foraminiferal biofacies and paleo shelf margins. *Journal of Sedimentary Research*, Vol. **71**, pp. 608–620.

Pekar, S. F., Christie-Blick, N., Kominz, M. A., and Miller, K. G. (2001). Evaluating the stratigraphic response to eustasy from Oligocene strata in New Jersey. *Geology*, Vol. **29**, p. 55–58.

Pemberton, S. G., and Frey, R. W. (1985). The Glossifungites ichnofacies: modern examples from the Georgia coast, U.S.A. *In Biogenic Structures: Their Use in Interpreting Depositional Environments* (H. A. Curran, Ed.), pp. 237–259. Society of Economic Paleontologists and Mineralogists, Special Publication No. 35.

Pemberton, S. G., and MacEachern, J. A. (1995). The sequence stratigraphic significance of trace fossils: examples from the Cretaceous foreland basin of Alberta, Canada. *In Sequence stratigraphy of foreland basin deposits – outcrop and subsurface examples from the Cretaceous of North America* (J. C. Van Wagoner and G. Bertram, Eds.), pp. 429–475. American Association of Petroleum Geologists Memoir **64**.

Pemberton, S. G., Spila, M., Pulham, A. J., Saunders, T., MacEachern, J. A., Robbins, D., and Sinclair, I. K. (2001). Ichnology and sedimentology of shallow to marginal marine systems: Ben Nevis and Avalon reservoirs, Jeanne d'Arc Basin. Geological Association of Canada, Short Course Notes Volume 15, p. 343.

Peper, T. (1994). Tectonic and eustatic control on Albian shallowing (Viking and Paddy Formations) in the Western Canada foreland basin. *Geological Society of America Bulletin*, Vol. **106**, pp. 253–264.

Peper, T., and Cloetingh, S. (1992). Lithosphere dynamics and tectono-stratigraphic evolution of the Mesozoic Betic rifted margin (southeastern Spain). *Tectonophysics*, Vol. **203**, pp 345–361.

Peper, T., and Cloetingh, S. (1995). Autocyclic perturbations of orbitally forced signals in the sedimentary record. *Geology*, Vol. **23**, p. 937–940.

Peper, T., Beekman, F., and Cloetingh, S. (1992). Consequences of thrusting and intraplate stress fluctuations for vertical motions in foreland basins and peripheral areas. *Geophysical Journal International*, Vol. **111**, pp. 104–126.

Peper, T., Van Balen, R., and Cloetingh, S. (1995). Implications of orogenic growth, intraplate stress variations, and eustatic sea-level change for foreland basin stratigraphy – inferences from numerical modeling. *In Stratigraphic Evolution of Foreland Basins* (S. L. Dorobek and G. M. Ross, Eds.), pp. 25–35. SEPM (Society for Sedimentary Geology) Special Publication No. 52.

Pettijohn, F. J. (1975). *Sedimentary rocks*. 3rd ed. Harper and Row, p. 628.

Piper, D. J. W., and Normark, W. R. (1983). Turbidite depositional patterns and flow characteristics, Navy submarine fan, California borderland. *Sedimentology*, Vol. **30**, pp. 681–694.

Piper, D. J. W., and Normark, W. R. (2001). Sandy fans – from Amazon to Hueneme and beyond. *American Association of Petroleum Geologists Bulletin*, Vol. **85**, no. 8, pp. 1407–1438.

Pitman, W. C. (1978). Relationship between sea level change and stratigraphic sequences of passive margins: Geological Society of America Bulletin, Vol. **89**, pp. 1389–1403.

Pitman, W. C., and Golovchenko, X. (1983). The effect of sea level change on the shelf edge and slope of passive margins. *In The shelf break: critical interface on continental margins* (D. J. Stanley and G. T. Morre, Eds.), pp. 41–58. SEPM Special Publication **33**.

Pitman, W. C., and Golovchenko, X. (1988). Sea-level changes and their effect on the stratigraphy of Atlantic-type margins. *In The geology of North America, vol. I-2, The Atlantic continental margin, United States* (R. E. Sheridan and J. A. Grow, Eds.), pp. 545–565. Geological Society of America, Boulder, Colorado.

Platt, N. H., and Keller, B. (1992). Distal alluvial deposits in a foreland basin setting – the lower freshwater Molasse (lower Miocene), Switzerland: sedimentology, architecture and palaeosols. *Sedimentology*, Vol. **39**, pp. 545–565.

Playfair, J. (1802). *Introduction to the Huttonian Theory of the Earth.* General Publishing, Toronto, facsimile 1956, p. 528.

Plint, A. G. (1988). Sharp-based shoreface sequences and "offshore bars" in the Cardium Formation of Alberta; their relationship to relative changes in sea level. *In Sea Level Changes – An Integrated Approach* (C. K. Wilgus, B. S. Hastings, C. G. St. C. Kendall, H. W. Posamentier, C. A. Ross and J. C. Van Wagoner, Eds.), SEPM Special Publication 42, pp. 357–370.

Plint, A. G. (1991). High-frequency relative sea-level oscillations in Upper Cretaceous shelf clastics of the Alberta foreland basin: possible evidence for a glacio-eustatic control? *In Sedimentation, Tectonics and Eustasy: Sea-level Changes at Active Margins* (D. I. M. Macdonald, Ed.), pp. 409–428. International Association of Sedimentologists, Special Publication **12**, Blackwell.

Plint, A. G. (1996). Marine and nonmarine systems tracts in fourth-order sequences in the Early-Middle Cenomanian, Dunvegan Alloformation, north-eastern British Columbia , Canada. *In High-Resolution Sequence Stratigraphy: Innovations and Applications* (J. A. Howell and J. F. Aitken, Eds.), pp. 159–191. Geological Society of London Special Publication **104**.

Plint, A. G. (2000). Sequence stratigraphy and paleogeography of a Cenomanian deltaic complex: the Dunvegan and lower Kaskapau formations in subsurface and outcrop, Alberta and British Columbia, Canada. *Bulletin of Canadian Petroleum Geology*, Vol. **48**, no. 1, pp. 43–79.

Plint, A. G., and Nummedal, D. (2000). The falling stage systems tract: recognition and importance in sequence stratigraphic analysis. *In Sedimentary Response to Forced Regression* (D. Hunt and R. L. Gawthorpe, Eds.), pp. 1–17. Geological Society of London Special Publication **172**.

Plint, A. G., Eyles, N., Eyles, C. H., and Walker, R. G. (1992). Control of sea level change. *In Facies Models: Response to Sea Level Change* (R. G. Walker and N. P. James, Eds.), pp. 15–25. Geological Association of Canada, GeoText 1.

Plint, A. G., Hart, B. S., and Donaldson, W. S. (1993). Lithospheric flexure as a control on stratal geometry and facies distribution in upper Cretaceous rocks of the Alberta foreland basin. *Basin Research*, Vol. **5**, pp. 69–77.

Plummer, C. C., and McGeary, D. (1996). *Physical Geology.* Wm. C. Brown Publishers, p. 539.

Posamentier, H. W. (2000). Seismic stratigraphy into the next millennium; a focus on 3D seismic data. American Association of Petroleum Geologists Annual Convention, New Orleans, Abstracts Volume, Vol. 9, p. A118.

Posamentier, H. W. (2001). Lowstand alluvial bypass systems: incised vs. unincised. *American Association of Petroleum Geologists Bulletin*, Vol. **85**, no. 10, pp. 1771–1793.

Posamentier, H. W. (2002). Ancient shelf ridges–A potentially significant component of the transgressive systems tract: Case study from offshore northwest Java. *American Association of Petroleum Geologists Bulletin*, Vol. **86/1**, pp. 75–106.

Posamentier, H. W. (2003). Depositional elements associated with a basin floor channel-levee system: case study from the Gulf of Mexico. *Marine and Petroleum Geology*, Vol. **20**, pp. 677–690.

Posamentier, H. W. (2004a). Seismic geomorphology: imaging elements of depositional systems from shelf to deep basin using 3D seismic data: implications for exploration and development. *In 3D Seismic Technology: Application to the Exploration of Sedimentary Basins* (R. J. Davies, J. A. Cartwright, S. A. Stewart, M. Lappin and J. R. Underhill, Eds.), pp. 11–24. Geological Society, London, Memoir **29**.

Posamentier, H. W. (2004b). 3D seismic visualization from a geological perspective: examples from shallow and deep water environments. *Canadian Society of Petroleum Geologists, Reservoir*, Vol. **31**, Issue 1, p. 8.

Posamentier, H. W., and Allen, G. P. (1993). Variability of the sequence stratigraphic model: effects of local basin factors. *Sedimentary Geology*, Vol. **86**, p. 91–109.

Posamentier, H. W., and Allen, G. P. (1999). Siliciclastic sequence stratigraphy: concepts and applications. SEPM Concepts in Sedimentology and Paleontology No. **7**, p. 210.

Posamentier, H. W., and Chamberlain, C. J. (1993). Sequence stratigraphic analysis of Viking Formation lowstand beach deposits at Joarcam Field, Alberta, Canada. *In Sequence Stratigraphy and Facies Associations* (H. W. Posamentier, C. P. Summerhayes, B. U. Haq and G. P. Allen, Eds.), pp. 469–485. International Association of Sedimentologists Special Publication **18**.

Posamentier, H. W., and James, D. P. (1993). Sequence stratigraphy – uses and abuses. *In Sequence Stratigraphy and Facies Associations* (H. W. Posamentier, C. P. Summerhayes, B. U. Haq and G. P. Allen, Eds.), pp. 3–18. International Association of Sedimentologists Special Publication **18**.

Posamentier, H. W., and Kolla, V. (2003). Seismic geomorphology and stratigraphy of depositional elements in deep-water settings. *Journal of Sedimentary Research*, Vol. **73**, no. 3, pp. 367–388.

Posamentier, H. W., and Morris, W. R. (2000). Aspects of the stratal architecture of forced regressive deposits. *In Sedimentary Responses to Forced Regressions* (D. Hunt and R. L. Gawthorpe, Eds.), pp. 19–46. Geological Society of London, Special Publication **172**.

Posamentier, H. W., and Vail, P. R. (1988). Eustatic controls on clastic deposition II–sequence and systems tract models. *In Sea Level Changes–An Integrated Approach* C. K. Wilgus, B. S. Hastings, C. G. St.C. Kendall, H. W. Posamentier, C. A. Ross and J. C. Van Wagoner, Eds.), pp. 125–154. SEPM Special Publication **42**.

Posamentier, H. W., and Walker, R. G. (2002). Turbidite facies models; integrating subsurface and outcrop. Canadian Society of Petroleum Geologists Diamond Jubilee Convention, June 3–7, Calgary, Program and Abstracts, p. 283.

Posamentier, H. W., Jervey, M. T., and Vail, P. R. (1988). Eustatic controls on clastic deposition I – conceptual framework. *In Sea Level Changes–An Integrated Approach* (C. K. Wilgus, B. S. Hastings, C. G. St. C. Kendall, H. W. Posamentier, C. A. Ross and J. C. Van Wagoner, Eds.), pp. 110–124. SEPM Special Publication **42**.

Posamentier, H. W., Allen, G. P., and James, D. P. (1992a). High resolution sequence stratigraphy – the East Coulee Delta, Alberta. *Journal of Sedimentary Petrology*, Vol. **62**, no. 2, pp. 310–317.

Posamentier, H. W., Allen, G. P., James, D. P., and Tesson, M. (1992b). Forced regressions in a sequence stratigraphic framework: concepts, examples, and exploration significance. *American Association of Petroleum Geologists Bulletin*, Vol. **76**, pp. 1687–1709.

Posamentier, H. W., Meizarwin, P. S., Wisman, T., and Plawman, T. (2000). Deep water depositional systems–ultra-deep Makassar Strait, Indonesia. *In Deep-water reservoirs of the world* (P. Weimer, R. M. Slatt, J. Coleman, N. C. Rosen, H. Nelson, A. H. Bouma, M. J. Styzen and D. T. Lawrence, Eds.), pp. 806–816. Gulf Coast Society of SEPM Foundation, 20th Annual Research Conference.

Powell, J. W. (1875). Exploration of the Colorado River of the West and its tributaries. U. S. Government Printing Office, Washington, D. C., p. 291.

Pratt, B. R., James, N. P., and Cowan, C. A. (1992). Peritidal carbonates. *In Facies Models: Response to Sea Level Change* (R. G. Walker and N. P. James, Eds.), pp. 303–322. Geological Association of Canada, GeoText 1.

Press, F., and Siever, R. (1986). *Earth*. Fourth edition, W. H. Freeman and Company, p. 656.

Press, F., Siever, R., Grotzinger, J., and Jordan, T. H. (2004). *Understanding Earth*. 4th Edn., W. H. Freeman and Company, New York, p. 567.

Price, E. D. (1999). The evidence and implication of polar ice during the Mesozoic. *Earth Science Reviews*, Vol. **48**, pp. 183–210.

Price, R. A. (1994). Cordilleran tectonics and the evolution of the Western Canada Sedimentary Basin. *In Geological Atlas of the Western Canada Sedimentary Basin* (G. D. Mossop and I. Shetsen (compilers), pp. 13–24. Canadian Society of Petroleum Geologists and Alberta Research Council.

Pyrcz, M. J., Catuneanu, O., and Deutsch, C. V. (2005). Stochastic surface-based modeling of turbidite lobes. *American Association of Petroleum Geologists Bulletin*, Vol. **89**, no. 2, pp. 177–192.

Rahmani, R. A. (1988). Estuarine tidal channel and nearshore sedimentation of a Late Cretaceous epicontinental sea, Drumheller, Alberta, Canada. *In Tide-influenced Sedimentary Environments and Facies* (P. L. de Boer, A. van Gelder and S. D. Nio, Eds.), pp. 433–471. Dordrecht, The Netherlands, D. Reidel.

Ramaekers, P., and Catuneanu, O. (2004). Development and sequences of the Athabasca Basin, Early Proterozoic, Saskatchewan and Alberta, Canada. *In The Precambrian Earth: Tempos and Events* (P. G. Eriksson, W. Altermann, D. Nelson, W. Mueller and O. Catuneanu, Eds.), pp. 705–723. Developments in Precambrian Geology 12, Elsevier Science Ltd., Amsterdam.

Ramsbottom, W. H. C. (1979). Rates of transgression and regression in the Carboniferous of NW Europe. *Journal of the Geological Society, London*, Vol. **136**, pp. 147–153.

Reading, H. G. (Ed.) (1978). *Sedimentary Environments and Facies*. 2nd ed. Blackwell Scientific Publications, p. 615.

Reading, H. G. (Ed.) (1996). *Sedimentary Environments: Processes, Facies and Stratigraphy*. Third Edition, Blackwell Science, p. 688.

Reading, H. G., and Collinson, J. D. (1996). Clastic coasts. *In Sedimentary Environments: Processes, Facies and Stratigraphy3rd Edn.*, (H. G. Reading, Ed.), pp. 154–231. Blackwell Science.

Reinson, G. E. (1992). Transgressive barrier island and estuarine systems. *In Facies Models–Response to Sea Level Change* (R. G. Walker and N. P. James, Eds.), pp. 179–194. Geological Association of Canada, GeoText 1.

Ricketts, B. D. (1989). *Western Canada Sedimentary Basin: A Case History*. Canadian Society of Petroleum Geologists, Calgary, p. 320.

Rider, M. H. (1990). Gamma Ray log shape used as a facies indicator: critical analysis of an oversimplified methodology. *In Geological Applications of Wireline Logs* (A. Hurst, M. A. Lovell and A. C. Morton, Eds.), pp. 27–37. Geological Society of London Special Publication Classics.

Rine, J. M., Tillman, R. W., Culver, S. J., and Swift, D. J. P. (1991). Generation of Late Holocene ridges on the middle continental shelf of New Jersey, U.S.A. – evidence for formation in a mid-shelf setting, based upon comparison with a nearshore ridge. *In Shelf Sand and Sandstone Bodies: Geometry, Facies and Sequence Stratigraphy* (D. J. P. Swift, G. F. Oertel, R. W. Tillman and J. A. Thorne, Eds.), pp. 395–426. International Association of Sedimentologists Special Publication **14**.

Rossetti, D. F. (1998). Facies architecture and sequential evolution of an incised-valley estuarine fill: the Cujupe Formation (Upper Cretaceous to Lower Tertiary), Sao Luis Basin, northern Brazil. *Journal of Sedimentary Research*, Vol. **68**, pp. 299–310.

Rubidge, B. S. (Ed.) (1995). Biostratigraphy of the Beaufort Group (Karoo Supergroup). Council for Geoscience, Geological Survey of South Africa, p. 46.

Rubidge, B. S., Hancox, P. J., and Catuneanu, O. (2000). Sequence analysis of the Ecca-Beaufort contact in the southern Karoo of South Africa. *South African Journal of Geology*, Vol. **103(1)**, pp. 81–96.

Ryang, W. H., and Chough, S. K. (1997). Sequential development of alluvial/lacustrine system: southeastern Eumsung Basin (Cretaceous), Korea. *Journal of Sedimentary Research*, Vol. **67**, no. 2, pp. 274–285.

Sarg, J. F. (1988). Carbonate sequence stratigraphy. *In Sea Level Changes–An Integrated Approach* (C. K. Wilgus, B. S. Hastings, C. G. St. C. Kendall, H. W. Posamentier, C. A. Ross and J. C. Van Wagoner, Eds.), pp. 155–182. SEPM Special Publication **42**.

Saucier, R. T. (1974). Quaternary geology of the lower Mississippi valley. Arkansas Archeological Survey, Research Series No. 6, p. 26.

Savrda, C. E. (1991). Teredolites, wood substrates, and sea-level dynamics. *Geology*, Vol. **19**, pp. 905–908.

Savrda, C. E. (1995). Ichnologic applications in paleoceanographic, paleoclimatic, and sea-level studies. Palaios, Vol. **10**, pp. 565–577.

Schlager, W. (1989). Drowning unconformities on carbonate platforms. *In Controls on Carbonate Platform and Basin Development* (P. D. Crevello, J. L. Wilson, J. F. Sarg and J. F. Read, Eds.), pp. 15–25. SEPM (Society of Economic Paleontologists and Mineralogists) Special Publication **44**.

Schlager, W. (1991). Depositional bias and environmental change–important factors in sequence stratigraphy. *Sedimentary Geology*, Vol. **70**, p. 109–130.

Schlager, W. (1992). Sedimentology and sequence stratigraphy of reefs and carbonate platforms. Continuing Education Course Note Series #34, *American Association of Petroleum Geologists*, p. 71.

Schlager, W. (1993). Accommodation and supply–a dual control on stratigraphic sequences. *In Basin Analysis and Dynamics of Sedimentary Basin Evolution* (S. Cloetingh, W. Sassi, F. Horvath and C. Puigdefabregas (Eds.), pp. 111–136. *Sedimentary Geology*, Vol. **86**.

Schlager, W. (1994). Reefs and carbonate platforms in sequence stratigraphy. *In High resolution sequence stratigraphy: innovations and applications* (S. D. Johnson, Ed.), pp. 143–149. Liverpool, Abstract Volume.

Schlager, W. (1999). Type 3 sequence boundaries. *In Advances in Carbonate Sequence Stratigraphy: Application to Reservoirs, Outcrops, and Models* (P. M. Harris, A. H. Saller and J. A. Simo, Eds.), pp. 35–46. SEPM (Society for Sedimentary Geology) Special Publication No. **63**.

Schlager, W. (2002). *Sedimentology and sequence stratigraphy of carbonate rocks*. Earth and Life Sciences, Vrije Universiteit, Amsterdam, pp. 146.

Schlager, W. (2005). Carbonate Sedimentology and Sequence Stratigraphy. SEPM Concepts in Sedimentology and Paleontology #8, p. 200.

Schumm, S. A. (1993). River response to baselevel change: implications for sequence stratigraphy. *Journal of Geology*, Vol. **101**, pp. 279–294.

Schwarzacher, W. (1993). *Cyclostratigraphy and the Milankovitch theory*. Developments in Sedimentology 52, Elsevier, Amsterdam, p. 225.

Scott, E. D. (1997). Tectonics and sedimentation: The evolution, tectonic influences and correlation of the Tanqua and Laingsburg subbasins, southwest Karoo Basin, South Africa. Ph.D. Thesis, Louisiana State University, p. 234.

Scott, E. D., and Bouma, A. (1998). Influence of tectonics on basin shape and deep water sedimentary fill characteristics: Tanqua and Laingsburg Subbasins, southwest Karoo Basin. *In Gondwana 10: Event stratigraphy of Gondwana (Abstracts)* (J. Almond, J., anderson, P. Booth, A. Chinsamy-Turan, D. Cole, M. J. de Wit, B. S. Rubidge, R. Smith, J. van beven Donker and B. C. Storey , Eds.) *Journal of African Earth Sciences*, Vol. **27 (1A)**, p. 174.

Seilacher, A. (1964). Biogenic sedimentary structures. *In Approaches to Paleoecology* (J. Imbrie and N. Newell Eds.), pp. 296–316. John Wiley, New York.

Seilacher, A. (1967). Bathymetry of trace fossils. *Marine Geology*, Vol. **5**, pp. 413–428.

Seilacher, A. (1978). Use of trace fossil assemblages for recognizing depositional environments. *In Trace Fossil Concepts* (P. B. Basan, Ed.), pp. 185–201. Society of Economic Paleontologists and Mineralogists, Short Course **5**.

Selley, R. C. (1978a). *Ancient Sedimentary Environments.* 2nd ed. Cornell University Press, Ithaca, New York, p. 287.

Selley, R. C. (1978b). Concepts and Methods of Subsurface Facies Analysis. American Association of Petroleum Geologists, Continuing Education Course Notes Series 9, p. 82.

Serra, O., and Abbott, H. T. (1982). The contribution of logging data to sedimentology and stratigraphy. *Society of Petroleum Engineers Journal*, Vol. **22(1)**, p. 117–131.

Shanley, K. W., and McCabe, P. J. (1991). Predicting facies architecture through sequence stratigraphy – an example from the Kaiparowits Plateau, Utah. *Geology*, Vol. **19**, p. 742–745.

Shanley, K. W., and McCabe, P. J. (1993). Alluvial architecture in a sequence stratigraphic framework: a case history from the Upper Cretaceous of southern Utah, U.S.A. *In Quantitative Modeling of Clastic Hydrocarbon Reservoirs and Outcrop Analogues* (S. Flint and I. Bryant, Eds.), pp. 21–55. International Association of Sedimentologists Special Publication **15**.

Shanley, K. W., and McCabe, P. J. (1994). Perspectives on the sequence stratigraphy of continental strata. *American Association of Petroleum Geologists Bulletin*, Vol. **78**, pp. 544–568.

Shanley, K. W., and McCabe, P. J. (eds.) (1998). Relative role of eustasy, climate, and tectonism in continental rocks. SEPM (Society for Sedimentary Geology) Special Publication No. **59**, p. 234.

Shanley, K. W., McCabe, P. J., and Hettinger, R. D. (1992). Significance of tidal influence in fluvial deposits for interpreting sequence stratigraphy. *Sedimentology*, Vol. **39**, pp. 905–930.

Shanmugam, G. (1988). Origin, recognition and importance of erosional unconformities in sedimentary basins. *In New Perspectives in Basin Analysis* (K. L. Kleinspehn and C. Paola, Eds.), pp. 83–108. Springer-Verlag, New York.

Shanmugam, G., Poffenberger, M., and Alava, J. T. (2000). Tide-dominated estuarine facies in the Hollin and Napo ("T" and "U") formations (Cretaceous), Sacha Field, Oriente Basin, Ecuador. *American Association of Petroleum Geologists Bulletin*, Vol. **84**, no. 5, pp. 652–682.

Shepheard, W. W., and Hills, L. V. (1970). Depositional environments of the Bearpaw-Horseshoe Canyon (Upper Cretaceous) transition zone, Drumheller Badlands, Alberta. *Bulletin of Canadian Petroleum Geology*, Vol. **18**, pp. 166–215.

Siggerud, E. I. H., and Steel, R. J. (1999). Architecture and trace-fossil characteristics of a 10,000–20,000 year, fluvial-to-marine sequence, SE Ebro Basin, Spain. *Journal of Sedimentary Research*, Vol. **69B**, pp. 365–383.

Simons, D. B., Richardson, E. V., and Nordin, C. F. (1965). Sedimentary structures generated by flow in alluvial channels. *In Primary sedimentary structures and their hydrodynamic interpretation* (G. V. Middleton, Ed.), pp. 34–52. Society of Economic Paleontologists and Mineralogists, Special Publication **12**.

Simpson, E., and Eriksson, K. A. (1990). Early Cambrian progradational and transgressive sedimentation patterns in Virginia: an example of the early history of a passive margin. *Journal of Sedimentary Petrology*, Vol. **60**, pp. 84–100.

Sloss, L. L. (1962). Stratigraphic models in exploration. *American Association of Petroleum Geologists Bulletin*, Vol. **46**, pp. 1050–1057.

Sloss, L. L. (1963). Sequences in the cratonic interior of North America. *Geological Society of America Bulletin*, Vol. **74**, pp. 93–114.

Sloss, L. L., Krumbein, W. C., and Dapples, E. C. (1949). Integrated facies analysis. *In Sedimentary facies in geologic history* (C. R. Longwell, Ed.), pp. 91–124. Geological Society of America Memoir **39**.

Snedden, J. W. (1984). Validity of the use of the spontaneous potential curve shape in the interpretation of sandstone depositional environments. *Gulf Coast Association of Geological Societies, Transactions*, Vol. **34**, pp. 255–263.

Snedden, J. W. (1991). Origin and sequence stratigraphic significance of large dwelling traces in the Escondido Formation (Cretaceous, Texas, USA). *Palaios*, Vol. **6**, no. 6, pp. 541–552.

Snedden, J. W., and Dalrymple, R. W. (1999). Modern shelf sand ridges: historical review of modern examples and a unified model of ridge origin and evolution. *SEPM Concepts in Sedimentology and Paleontology*, Vol. **6**, pp. 13–28.

Snedden, J. W., and Kreisa, R. D. (1995). A sequence stratigraphic model for shelf sand ridge genesis and evolution. American Association of Petroleum Geologists Annual Convention, March 5–8, 1995, Houston, Texas, Program with Abstracts, p. 90A.

Snedden, J. W., Tillman, R. W., Kreisa, R. D., Schweller, W. J., Culver, S. J., and Winn, R. D. Jr. (1994). Stratigraphy and genesis of a modern shoreface-attached sand ridge, Peahala Ridge, New Jersey. *Journal of Sedimentary Research*, Vol. **B64**, pp. 560–581.

Soil Survey Staff (1975). Soil taxonomy. U. S. Department of Agriculture Handbook 436, p. 754.

Soil Survey Staff (1998). Keys to Soil Taxonomy, 8th Edn. U. S. Department of Agriculture, Natural Resources Conservation Services, p. 327.

Soreghan, G. S., Elmore, R. D., Katz, B., Cogoini, M., Banerjee, S. (1997). Pedogenically enhanced magnetic susceptibility variations preserved in Paleozoic loessite. *Geology*, Vol. **25**, pp. 1003–1006.

Stoll, H. M., and Schrag, D. P. (1996). Evidence for glacial control of rapid sea level changes in the Early Cretaceous. *Science*, Vol. **272**, pp. 1771–1774.

Stow, D. A. V., Reading, H. G., and Collinson, J. D. (1996). Deep seas. *In Sedimentary Environments: Processes, Facies and Stratigraphy* 3rd Edn., (H. G. Reading, Ed.), pp. 395–453. Blackwell Science.

Summerfield, M. A. (1985). Plate tectonics and landscape development on the African continent. *In Tectonic Geomorphology* (M. Morisawa and J. Hack, Eds.), pp. 27–51. Allen and Unwin, Boston.

Summerfield, M. A. (1991). *Global Geomorphology: an Introduction to the Study of Landforms.* Wiley, New York, p. 537.

Suppe, J., Chou, G. T., and Hook, S. C. (1992). Rates of folding and faulting determined from growth strata. *In Thrust Tectonics* (K. R. McClay, Ed.), pp. 105–121. London, Chapman and Hall.

Suter, J. R., Berryhill, H. L. Jr., and Penland, S. (1987). Late Quaternary sea level fluctuations and depositional sequences, southwest Louisiana continental shelf. *In Sea-level changes and coastal evolution* (D. Nummedal, O. H. Pilkey and J. D. Howard, Eds.), pp. 199–222. SEPM (Society for Sedimentary Geology) Special Publication No. **41**.

Sweet, A. R., Long, D. G. F., and Catuneanu, O. (2003). Sequence boundaries in fine-grained terrestrial facies: biostratigraphic time control is key to their recognition. Geological Association of Canada–Mineralogical Association of Canada joint annual meeting, Vancouver, May 25–28, Abstracts Volume **28**, p. 165.

Sweet, A. R., Catuneanu, O., and Lerbekmo, J. F. (2005). Uncoupling the position of sequence-bounding unconformities from lithological criteria in fluvial systems. *American Association of Petroleum Geologists Annual Convention*, 19–22 June 2005, Calgary, Alberta. Abstracts Volume, Vol. **14**, p. A136.

Swift, D. J. P. (1968). Coastal erosion and transgressive stratigraphy. *Journal of Geology*, Vol. **76**, pp. 444–456.

Swift, D. J. P. (1975). Barrier-island genesis: evidence from the central Atlantic shelf, eastern U.S.A. *Sedimentary Geology*, Vol. **14**, pp. 1–43.

Swift, D. J. P. (1976). Coastal sedimentation. *In Marine Sediment Transport and Environmental Management* (D. J. Stanley and D. J. P. Swift, Eds.), pp. 255–311. John Wiley & Sons, New York.

Swift, D. J. P., and Field, M. F. (1981). Evolution of a classic ridge field, Maryland sector, North American inner shelf. *Sedimentology*, Vol. **28**, p. 461–482.

Swift, D. J. P., and Thorne, J. A. (1991). Sedimentation on continental margins I: a general model for shelf sedimentation. *In Shelf sand and sandstone bodies–geometry, facies, and sequence stratigraphy* (D. J. P. Swift, G. F. Oertel, R. W. Tillman and J. A. Thorne, Eds.), pp. 3–31. Blackwell, Oxford, International Association of Sedimentologists, Special Publication **14**.

Swift, D. J. P., Kofoed, J. W., Saulsbury, F. B., and Sears, P. C. (1972). Holocene evolution of the shelf surface, central and southern Atlantic shelf of North America. *In Shelf Sediment Transport: Process and Pattern, Stroudsburg, Pennsylvania, Dowden*, (D. J. P. Swift, D. B. Duane and O. H. Pilkey, Eds.), pp. 499–574. Hutchinson & Ross.

Sylvia, D., and Galloway, B. (2001). Response of the Brazos River (Texas USA) to latest Pleistocene climatic and eustatic change. Seventh International Conference on Fluvial Sedimentology, Lincoln, August 6–10, Program and Abstracts, p. 264.

Tandon, S. K., and Gibling, M. R. (1994). Calcrete and coal in late Carboniferous cyclothems of Nova Scotia, Canada: Climate and sea-level changes linked. *Geology*, Vol. **22**, no. 8, pp. 755–758.

Tandon, S. K., and Gibling, M. R. (1997). Calcretes at sequence boundaries in upper Carboniferous cyclothems of the Sydney Basin, Atlantic Canada. *Sedimentary Geology*, Vol. **112**, pp. 43–67.

Taylor, A. M., and Gawthorpe, R. L. (1993). Application of sequence stratigraphy and trace fossil analysis to reservoir description: examples from the Jurassic of the North Sea. *In Petroleum Geology of Northwest Europe* (J. R. Parker, Ed.), pp. 317–336. Geological Society of London.

Tibert, N. E., and Gibling, M. R. (1999). Peat accumulation on a drowned coastal braidplain: the Mullins Coal (Upper Carboniferous), Sydney Basin, Nova Scotia. *Sedimentary Geology*, Vol. **128**, pp. 23–38.

Thorne, J. A., and Swift, D. J. P. (1991). Sedimentation on continental margins, VI. A regime model for depositional sequences, they component systems tracts, and bounding surfaces. *In Shelf Sand and Sandstone Bodies* (D. J. P. Swift, G. F. Oertel, R. W. Tillman and J. A. Thorne, Eds.), pp. 189–255. International Association of Sedimentologists, Special Publication **14**.

Tucker, M. E., Calvet, F., and Hunt, D. (1993). Sequence stratigraphy of carbonate ramps: systems tracts, models and application to the Muschelkalk carbonate platforms of eastern Spain. *In Sequence Stratigraphy and Facies Associations* (H. W. Posamentier, C. P. Summerhayes, B. U. Haq and G. P. Allen, Eds.), pp. 397–415. International Association of Sedimentologists, Special Publication **18**.

Twenhofel, W. H. (1939). *Principles of Sedimentation*. New York, McGraw-Hill, p. 610.

Uchman, A., and Demircan, H. (1999). A Zoophycos group trace fossil from Miocene flysch in southern Turkey: evidence for U shaped causative burrow. *Ichnos*, Vol. **6**, pp. 251–260.

Underhill, J. R. (1991). Controls on Late Jurassic seismic sequences, Inner Moray Firth, UK North Sea: a critical test of a key segment of Exxon's original global cycle chart. *Basin Research*, Vol. **3**, pp. 79–98.

Vail, P. R. (1975). Eustatic cycles from seismic data for global stratigraphic analysis (abstract). *American Association of Petroleum Geologists Bulletin*, Vol. **59**, pp. 2198–2199.

Vail, P. R. (1987). Seismic stratigraphy interpretation procedure. *In Atlas of Seismic Stratigraphy* (A. W. Bally, Ed.), pp. 1–10. American Association of Petroleum Geologists Studies in Geology **27**.

Vail, P. R., and Wornardt, W. W. (1990). Well log-seismic stratigraphy; an integrated tool for the 90's: Gulf Coast Section, SEPM Foundation Eleventh Annual Research Conference Program and Extended Abstracts, pp. 379–388.

Vail, P. R., Mitchum, R. M. Jr., and Thompson, S., III (1977). Seismic stratigraphy and global changes of sea level, part four: global cycles of relative changes of sea level. *American Association of Petroleum Geologists Memoir* **26**, pp. 83–98.

Vail, P. R., Hardenbol, J., and Todd, R. G. (1984). Jurassic unconformities, chronostratigraphy and sea-level changes from seismic stratigraphy and biostratigraphy. *In Interregional Unconformities and Hydrocarbon Accumulation* (J. S. Schlee, Ed.), pp. 129–144. American Association of Petroleum Geologists Memoir **36**.

Vail, P. R., Audemard, F., Bowman, S. A., Eisner, P. N., and Perez-Cruz, C. (1991). The stratigraphic signatures of tectonics, eustasy and sedimentology–an overview. *In Cycles and Events in Stratigraphy*. (G. Einsele, W. Ricken and A. Seilacher, Eds.), pp. 617–659. Berlin, Springer-Verlag.

Van Wagoner, J. C. (1995). Overview of sequence stratigraphy of foreland basin deposits: terminology, summary of papers, and glossary of sequence stratigraphy. *In Sequence Stratigraphy of Foreland Basin Deposits* (J. C. Van Wagoner and G. T. Bertram, Eds.), pp. ix-xxi. American Association of Petroleum Geologists Memoir **64**.

Van Wagoner, J. C., and Bertram, G. T. (eds.) (1995). Sequence Stratigraphy of Foreland Basin Deposits, American Association of Petroleum Geologists Memoir **64**, p. 487.

Van Wagoner, J. C., Posamentier, H. W., Mitchum, R. M. Jr., Vail, P. R., Sarg, J. F., Loutit, T. S., and Hardenbol, J. (1988). An overview of sequence stratigraphy and key definitions. *In Sea Level Changes–An Integrated Approach* C. K. Wilgus, B. S. Hastings, C. G. St. C. Kendall, H. W. Posamentier, C. A. Ross and J. C. Van Wagoner, Eds.), pp. 39–45. SEPM Special Publication **42**.

Van Wagoner, J. C., Mitchum, R. M. Jr., Campion, K. M., and Rahmanian, V. D. (1990). Siliciclastic sequence stratigraphy in well logs, core, and outcrops: concepts for high-resolution correlation of time and facies. American Association of Petroleum Geologists Methods in Exploration Series 7, p. 55.

Vecsei, A., and Duringer, P. (2003). Sequence stratigraphy of Middle Triassic carbonates and terrigenous deposits (Muschelkalk and Lower Keuper) in the SW Germanic Basin: maximum flooding versus maximum depth in intracratonic basins. *Sedimentary Geology*, Vol. **160**, pp. 81–105.

Verdier, A. C., Oki, T., and Suardy, A. (1980). Geology of the Handil field (East Kalimantan, Indonesia). *In Giant oil and gas fields of the decade: 1968–1978* (M. T. Halbouty, Ed.), pp. 399–421. American Association of Petroleum Geologists Memoir **30**.

Wadsworth, J., Boyd, R., Diessel, C., Leckie, D., and Zaitlin, B. (2002). Stratigraphic style of coal and non-marine strata in a tectonically influenced intermediate accommodation setting: the Mannville Group of the Western Canadian Sedimentary Basin, south-central Alberta. *Bulletin of Canadian Petroleum Geology*, Vol. **50**, no. 4, p. 507–541.

Wadsworth, J., Boyd, R., Diessel, C., and Leckie, D. (2003). Stratigraphic style of coal and non-marine strata in a high accommodation setting: Falher Member and Gates Formation (Lower Cretaceous), western Canada. *Bulletin of Canadian Petroleum Geology*, Vol. **51**, no. 3, p. 275–303.

Waldron, J. W. F., and Rygel, M. C. (2005). Role of evaporite withdrawal in the preservation of a unique coal-bearing succession: Pennsylvanian Joggins Formation, Nova Scotia. *Geology*, Vol. **33**, no. 5, p. 337–340.

Walker, J. C. G., and Zahnle, K. J. (1986). Lunar nodal tide and distance to the moon during the Precambrian. *Nature*, Vol. **320**, p. 600–602.

Walker, R. G. (1992). Facies, facies models and modern stratigraphic concepts. *In Facies Models: Response to Sea Level Change* (R. G. Walker and N. P. James, Eds.), pp. 1–14. Geological Association of Canada, GeoText 1.

Walker, R. G., and James, N. P. (Eds.) Facies Models: Response to Sea Level Change. Geological Association of Canada, GeoText 1, p. 454.

Walker, R. G., and Plint, A. G. (1992). Wave- and storm-dominated shallow marine systems. *In Facies Models: Response to Sea Level Change* (R. G. Walker and N. P. James, Eds.), pp. 219–238. Geological Association of Canada, GeoText 1.

Wanas, H. A. (2003). An authigenesis of glauconite in association with unconformity: a case study from the Albian/Cenomanian boundary at Gabal Shabrawet, Egypt. *Geochemistry (Moscow)*, Vol. **5**, pp. 570–576.

Wanless, H. R., and Shepard, F. B. (1936). Sea level and climatic changes related to Late Paleozoic cycles. *Geological Society of America Bulletin*, Vol. **47**, pp. 1177–1206.

Webb, G. E. (1994). Paleokarst, paleosol, and rocky-shore deposits at the Mississippian-Pennsylvanian unconformity, northwestern Arkansas. *Geological Society of America Bulletin*, Vol. **106**, pp. 634–648.

Wehr, F. L. (1993). Effects of variations in subsidence and sediment supply on parasequence stacking patterns. *In Siliciclastic Sequence Stratigraphy–Recent Developments and Applications* (P. Weimer and H. W. Posamentier, Eds.), pp. 369–378. American Association of Petroleum Geologists Memoir 58.

Weimer, P., and Slatt, R. M. (2004). Petroleum systems of deepwater settings. Society of Exploration Geophysicists, Tulsa, Distinguished Instructor Series, No. 7, p. 470.

Wheeler, H. E. (1958). Time stratigraphy. *American Association of Petroleum Geologists Bulletin*, Vol. **42**, pp. 1047–1063.

Wheeler, H. E. (1959). Unconformity bounded units in stratigraphy. *American Association of Petroleum Geologists Bulletin*, Vol. **43**, pp. 1975–1977.

Wheeler, H. E. (1964). Baselevel, lithosphere surface, and time-stratigraphy. *Geological Society of America Bulletin*, Vol. **75**, pp. 599–610.

Wheeler, H. E., and Murray, H. H. (1957). Baselevel control patterns in cyclothemic sedimentation. *American Association of Petroleum Geologists Bulletin*, Vol. **41**, pp. 1985–2011.

Wickens, H de V. (1994). Basin Floor Fan Building Turbidites of the southwestern Karoo Basin, Ecca Group, South Africa. Ph.D. Thesis, University of Port Elizabeth, South Africa, p. 233.

Wilgus, C. K., Hastings, B. S., Kendall, C. G. St.C., Posamentier, H. W., Ross, C. A., and Van Wagoner, J. C. (Eds.) (1988). Sea Level Changes–An Integrated Approach. SEPM (Society of Economic Paleontologists and Mineralogists), Tulsa, Oklahoma, Special Publication 42, p. 407.

Willis, A., and Wittenberg, J. (2000). Exploration significance of healing-phase deposits in the Triassic Doig Formation, Hythe, Alberta. *Bulletin of Canadian Petroleum Geology*, Vol. **48**, no. 3, pp. 179–192.

Winter, H. de la R. (1984). Tectonostratigraphy, as applied to the analysis of South African Phanerozoic basins. *Transactions of the Geological Society of South Africa*, Vol. **87**, pp. 169–179.

Winter, H. de la R., and Brink, M. R. (1991). Chronostratigraphic subdivision of the Witwatersrand Basin based on a Western Transvaal composite column. *South African Journal of Geology*, Vol. **94**, pp. 191–203.

Wood, L. J., Ethridge, F. G., and Schumm, S. A. (1993). The effects of rate of base-level fluctuations on coastal plain, shelf and slope depositional systems: an experimental approach. *In Sequence Stratigraphy and Facies Associations* (H. W. Posamentier, C. P. Summerhayes, B. U. Haq and G. P. Allen, Eds.), pp. 43–53. International Association of Sedimentologists Special Publication **18**.

Wood, L.J, Koss, J.E and Ethridge, F. G. (1994). Simulating unconformity development and unconformable stratigraphic relationships through physical experiments. Unconformity Controls, 1994 Symposium, Rocky Mountain Association of Geologists, Proceedings, pp. 23–34.

Wright, V. P. (1994). Paleosols in shallow marine carbonate sequences. *Earth-Science Reviews*, Vol. **35**, pp. 367–395.

Wright, V. P., and Marriott, S. B. (1993). The sequence stratigraphy of fluvial depositional systems: the role of floodplain sediment storage. *Sedimentary Geology*, Vol. **86**, pp. 203–210.

Wright, V. P., and Platt, N. H. (1995). Seasonal wetland carbonate sequences and dynamic catenas; a re-appraisal of palustrine limestones. *Sedimentary Geology*, Vol. **99**, pp. 65–71.

Ye, L. (1995). Paleosols in the Upper Guantae Formation (Miocene) of the Gudong oil field and their application to the correlation of fluvial deposits. *American Association of Petroleum Geologists Bulletin*, Vol. **79**, pp. 981–988.

Ye, L., and Kerr, D. (2000). Sequence stratigraphy of the Middle Pennsylvanian Bartlesville Sandstone, northeastern Oklahoma: a case of an underfilled incised valley. *American Association of Petroleum Geologists Bulletin*, Vol. **84**, no. 8, pp. 1185–1204.

Yoshida, S., Willis, A., and Miall, A. D. (1996). Tectonic control of nested sequence architecture in the Castlegate Sandstone (Upper Cretaceous), Book Cliffs, Utah. *Journal of Sedimentary Research*, Vol. **66**, pp. 737–748.

Yoshida, S., Miall, A. D., and Willis, A. (1998). Sequence stratigraphy and marine to nonmarine facies architecture of foreland basin strata, Book Cliffs, Utah, U.S.A.: discussion. *American Association of Petroleum Geologists Bulletin*, Vol. **82**, pp. 1596–1606.

Zaitlin, B. A., Dalrymple, R. W., and Boyd, R. (1994). The stratigraphic organization of incised-valley systems associated with relative sea-level change. *In Incised-valley systems: origin and sedimentary sequences* (R. W. Dalrymple, and B. A. Zaitlin, (Eds.), pp. 45–60. SEPM Special Publication **51**.

Zaitlin, B. A., Potocki, D., Warren, M. J., Rosenthal, L., and Boyd, R. (2000). Sequence stratigraphy in low accommodation foreland basins: an example from the lower Cretaceous Basal Quartz Formation of southern Alberta. Abstracts, GeoCanada 2000 conference, Canadian Society of Petroleum Geologists, CD-ROM.

Zaitlin, B. A., Warren, M. J., Potocki, D., Rosenthal, L., and Boyd, R. (2002). Depositional styles in a low accommodation foreland setting: an example from the Basal Quartz (Lower Cretaceous), southern Alberta. *Bulletin of Canadian Petroleum Geology*, Vol. **50**, no. 1, pp. 31–72.

Author Index

Subject Index

Printed and bound by CPI Group (UK) Ltd, Croydon, CR0 4YY

03/10/2024

01040328-0018